Chrysler Engines
1922–1998

Other SAE titles of interest:

Hall-Scott
The Untold Story of a Great American Engine Maker
Francis H. Bradford and Ric A. Dias
(Order No. R-368)

Pioneers, Engineers, and Scoundrels
The Dawn of the Automobile in America
Beverly Rae Kimes
(Order No. R-358)

The SAE Story: One Hundred Years of Mobility
Robert Post
(Order No. R-360)

Chrysler Engines 1922–1998

Willem L. Weertman

SAE International™
Warrendale, Pa.

For permission and licensing requests, contact:

SAE Permissions
400 Commonwealth Drive
Warrendale, PA 15096-0001 USA
E-mail: permissions@sae.org
Tel: 724-772-4028
Fax: 724-772-4891

Library of Congress Cataloging-in-Publication Data

Weertman, Willem L.
Chrysler engines 1922–1998 / Willem L. Weertman.
p. cm.
Includes bibliographical references and index.
ISBN 978-0-7680-1642-0
1. Chrysler automobile--Motors--History. I. Title.

TL215.C55W39 2007
629.25'04--dc22 2006039361

SAE International
400 Commonwealth Drive
Warrendale, PA 15096-0001 USA
E-mail: CustomerService@sae.org
Tel: 877-606-7323 (inside USA and Canada)
724-776-4970 (outside USA)
Fax: 724-776-1615

ISBN 978-0-7680-1642-0

SAE Order No. R-365

Printed in the United States of America.

Contents

Credits

Acknowledgments

I wish to recognize all those who were involved in the design, development, and manufacture of Chrysler engines during the span of this history. The satisfaction of creation, the long hours of design and testing, the challenge of problem resolution, and the elation of success are part of the engineering process, shared by many. I feel privileged to be among the people who did the work.

I want to thank all the people who assisted me on this project, including the active Chrysler LLC employees, retirees, members of supplier companies, and others who provided information. Among these valued and knowledgeable advisors and encouragers are the following: Floyd Allen, Elton Amsden, Tom Asmus, Mark Boratto, Al Bosley, Burke Brown, Charlie Brown III, Bob Cahill, L.X. Chapin, Bill Chapman, Mary Chappell, Jack Conners, Kraig Courtney, Jim Dean, Lou DeSimone, Barry Dressel, Bob Eccles, Eric Emmer, Dean Engle, Debbie Fornal, Barbara Fronczak, Roger Gaskill, Bob Graham, Pete Hagenbuch, Bill Hancock, Mark Hanna, Tom Hoover, John Hurst, Fred Kaehler, Doug Keyes, Sean Keyes, Mike Krag, Rod Kurowski, Bob Lee, Bob Lees, Bob Lewis, Kim Lyon, Ken Mack, Ev Moeller, Al Mukomel, Bob Pauley, Art Ponder, Ed Poplawski, Marv Raguse, Bob Rarey, Larry Rathgeb, Bruce Raymond, Bill Ridenour, Gordon Rinschler, Bernard Robertson, Brandt Rosenbusch, Michael Royce, Larry Shephard, Fred Shrimpton, Troy Simonsen, Lee Sorenson, Daniel Stern, Syd Terry, Bruce Thomas, Clay Thompson, Jeff Warner, John Wehrly, Gene Weiss, Harold Welch, Bill Westervelt, Kenneth Wilson, Dick Winkles, Paul Wise, and Keith Wright. Sadly, some of these people are no longer with us but certainly continue in spirit.

Among these people, special thanks go to Barry Dressel, manager of the Walter P. Chrysler Museum, who gave impetus to the idea of my writing this book when he requested a short summary of Chrysler engines to accompany a possible Chrysler history project. Putting together the short summary, which amounted to a few pages containing one small picture, brief specifications, and a time line for each engine, kindled for me an interest in the engines of long ago. This interest was fanned by Barry's lifting the hoods of the cars on display at the museum, including the original Chrysler B-model phaeton, so I could take measurements and photos of the engines. With much information uncovered, writing a book about the history of Chrysler engines became a logical expansion of the original request. Although the original project was shelved, the brief summaries emerged to become part of the Walter P. Chrysler Museum's website and can be viewed in the Engine Chronology section of the Chrysler Corporation Historical Products.

My special thanks also go to Brandt Rosenbusch, manager of the Chrysler Historical Collection, who provided access and guidance to all the archival records, vehicles, and engines in the care of the historical collection, without which this book simply could not have been written. With few exceptions, the hundreds of images of Chrysler vehicles and engines displayed in this book came from the well-cataloged files under Brandt's management. Brandt's maintenance and guardianship of Chrysler's, now that of Chrysler LLC, heritage is to be highly commended.

Fellow Chrysler retiree Bruce Thomas's hospitality in making the vehicles of his private collection available for engine measurements and photos was much appreciated assistance, as well as his sharing of information and memories about Chrysler Engineering people and programs.

Slant Six engine enthusiast Daniel Stern's careful review of the Slant Six engine material, which provided both corrections and excellent suggestions, was a valued contribution to the information presented in the chapter on in-line six-cylinder overhead valves engines.

My thanks also go to the staff members of the National Automotive History Collection of the Detroit Public Library, who were most helpful in supplying vitally needed information from their extensive, in-depth wealth of automotive source material. In addition, the Lawrence Technological University Library staff provided valuable assistance in locating useful reference material in their excellent book and microfilm files.

Finally, I want to thank Jeff Worsinger of SAE International and his reviewers, including James Wagner, for their guidance on how to turn my original draft into the final and much improved form presented in this book.

Willem L. Weertman
Former Chief Engineer—Engine Engineering (retired)
Engineering Division, Chrysler Corporation
February 2006

Foreword

Those familiar with the history of Chrysler will recognize that the starting date of 1922 shown in the title of this book is obviously prior to the June 6, 1925 date when Chrysler Corporation was incorporated and even prior to the January 1924 date when the first production Chrysler car was shown to the public. The reason for this is that there were two serious efforts, described in the text, to build and market a Chrysler-named car that preceded these events. Had either effort been successful, the car would have emerged around the 1922 model year. Although unsuccessful, these efforts did play a key role in bringing the right people together at the right time to lay the foundation for the successful entry of Chrysler Corporation into the automotive arena.

The 1998 final date shown in the title reflects the November 17, 1998 date when Chrysler Corporation merged with Daimler Benz AG (Aktiengesellschaft) to form DaimlerChrysler AG, marking the end of Chrysler Corporation.

The main text of this book attempts to provide general information about every production engine built by Chrysler, in the form of descriptions, images, specifications, and time lines. Each engine is defined in terms of its basic structure (e.g., cylinder bore spacing, number of main bearings, valve operation) because once the production machine tools are in place, these elements usually remain constant for the life of the engine. The specifications displayed have been selected as the most pertinent to each engine during its lifetime.

A number of engines that were manufactured by other companies and were installed in Chrysler passenger cars are included, in all cases Chrysler having had a participatory role in their being qualified for use. Engines in cars that were completely engineered and built by other companies and badged as Chrysler models have not been included. Also not included are any purchased diesels or purchased gasoline engines used by Dodge trucks, or any alternative fuel engines.

On August 5, 1987, American Motors Corporation (AMC) was merged into Chrysler Corporation. Chapter 19 describes the five AMC engines that were in production at the time. Largely omitted are the numerous engines built by AMC and its predecessor companies prior to the merger.

Also, with a few notable exceptions that include the Chrysler gas turbine described in Chapter 18, there is little mention of the multitude of experimental engines designed by the Engine Design and Research Engineering departments, many of them built and tested. Among the excluded experimental engines are two all-aluminum V-8s: one coded A612 and based on the LB engine, and a smaller V-8 coded A771. Both of these engines preceded the aluminum-block Slant Six engine that went into production in 1961. Another experimental engine, a turbo-diesel version of the Slant Six engine, coded A513, was tested extensively in the early 1980s but never became a production reality. Many other engines never achieved production status, but all made a contribution in one way or another to Chrysler's base of engine knowledge.

Engineering A-program codes, already referred to in the preceding paragraphs, were started in 1935 to identify major programs. The first designator issued, A1, was for a complete car including its engine that appears to be an L-head V-8 based on the surviving record—a parts list. In 1964, after A999 was assigned, a second series was started, beginning with A100, recycling the numbers. There were enough years separating the two series that Engineering had no problem with this situation, but care is needed by anyone delving into these programs to differentiate between an earlier program and a later program, both with the identical A-program number.

Regarding orientations, the terms "left," "right," "front," and "rear" as mentioned in this book refer to directions as viewed from the driver's position.

Until the 1950s, the use of fractions of an inch was common on engineering drawings. Rather than repeat the original fractions for historical fidelity, with few exceptions such as thread sizes, the dimensions presented in the book are shown as decimals.

With these qualifications, the story of Chrysler engines can be told.

List of Engines by Type

L-Head Engines

V-8 Engines

In-Line Six-Cylinder Overhead Valve Engines

Front-Wheel-Drive Four-Cylinder Engines

Front-Wheel-Drive V-6 Engines

Viper V-10 Engine

Truck-Only Engines

Marine and Industrial Engines

World War II Military Engines

Gas Turbine Engines

American Motors Corporation Engines

List of A-Programs

Information on the sequencing of A-program numbers can be found in the Foreword of this book.

CHAPTER 1

Chrysler-Willys— The First "Chrysler" Cars and Engines

The car bearing Walter P. Chrysler's name, shown at the January 1924 Auto Show in New York City and whose phenomenal sales were to lead directly to the formation of Chrysler Corporation the following year, was not the first car to carry his name. It was actually the third, but, most importantly, it was the first to go into production.

The first car and engine that carried the name "Chrysler" was designed by Willys Corporation engineers as a new car with a new six-cylinder engine, to be built in a town a long way from Detroit—Elizabeth, New Jersey. John North Willys, president of Willys-Overland, had a plan to produce the car and engine there in a large new building that was four stories high and 1440 ft long, with a manufacturing floor area of more than two-million square feet.[1] The site of the plant was that of a former Duesenberg plant that Willys had acquired in 1918.[2] Engineering development of the car and construction of the building were well underway when Walter P. Chrysler joined Willys as executive vice president in January 1920,[3] a move by Chrysler from his retired status—not at the request of John Willys, but at the request of a committee of bankers who were owed $46 million by Willys Corporation.[2] Chrysler brought excellent credentials to the job, having turned around Buick Motor Company, a division of General Motors, from being a money loser to being highly profitable. This was followed by his promotion to vice president in charge of operations before leaving his post at General Motors over disagreements with his boss, the charismatic and micromanaging William C. (Billy) Durant. Armed with the total authority to do whatever was necessary to return Willys to profitability,[2] Chrysler immediately went to work straightening out the company operations. A few months later, in April 1920, it was reported that the new car would be named "Chrysler,"[4] one source indicating that this was a provision in Chrysler's acceptance of his position with Willys.[2]

Further bolstering the possible emergence of a new "Chrysler" car line was the July 1920 incorporation in the state of Delaware of the Chrysler Motor Company as a division of the Willys Corporation.[5] Sometime during this formative period, a large and illuminated sign proclaiming "Chrysler—The Six Cylinder Motor Car" was erected on top of the Willys Elizabeth plant.[6] In a rush to ready the plant for production, machine tools were purchased for the new car and engine. However, all was not well in the product engineering area. As road test miles and dynamometer hours accumulated on pre-production evaluation cars and engines, serious problems started to occur. The most severe problems—including frame failure, spring breakage, and crankshaft problems—indicated that the basic designs of both the car and the engine were faulty.[6,7] The engine was known to have only three main bearings.[6]

With the future of Willys and his own personal reputation hanging on this new car, there were plenty of reasons for Walter P. Chrysler to get a thorough, unbiased evaluation of the car before proceeding further. An honest appraisal came from an engineer who was already on the Willys payroll, Allen B. (Tobe) Couture, who had worked for Duesenberg Motors before it was bought by Willys and had become an astute car test engineer. During a demonstration ride in the car, when Chrysler asked for his honest opinion, Couture, in his typical direct-to-the-point and forthright manner, told him that the car was a disaster and that it should be scrapped and a new design started![7] That must have been a shock!

Rising to the challenge of adversity, Chrysler then took an action that would lead, in only a few years, to the founding of

Chrysler Corporation. The action Chrysler took was to contact three top Studebaker engineers—Fred M. Zeder, Owen R. Skelton, and Carl Breer—and request that they join him at Willys, form an engineering group separate from the Willys engineers, and take a look at the car and engine. Although it meant leaving their Detroit homes (Studebaker Automotive Division, including Engineering, was located near downtown Detroit[6]), the three men accepted Chrysler's offer. As it turned out, they had become dissatisfied with Studebaker management and were ready for such a change.[6]

In July 1920, Zeder, Skelton, and Breer (ZSB) arrived in Elizabeth with about fifteen of the best men from their former engineering department at Studebaker.[8] They immediately launched two parallel programs: an analysis and possible fixes for the current problems, and a hush-hush program of completely new designs for a replacement car and engine.[8] By mid-November 1920, their conclusion was the same as Couture's—the original designs could not be salvaged.[8]

According to the notebook entries of Newton F. Hadley, one of the Studebaker designers who made the switch to Willys at the request of Zeder, Skelton, and Breer, the six-cylinder engine designed by the Willys engineers was rather small at 184 cu. in. with a bore and stroke of 3.125 × 4.0 in., and it had a crankshaft with three main bearings, as already noted. For reference, the four-cylinder engine of the Ford Model T, in production since 1909, was only slightly smaller at 177 cu. in. The new six-cylinder replacement engine designed by the ZSB engineering team, with a substantially larger displacement of 224 cu. in., a bore and stroke of 3.25 × 4.5 in. and a crankshaft with seven main bearings, certainly would provide a substantial upgrade in performance and quality for the new car.

The emphasis shifted completely to the new designs and, after showing their designs to Chrysler and finding him enthusiastically supportive,[8] the ZSB engineering group proceeded at full throttle. Chrysler informed the bankers that the originally planned Willys-engineered "Chrysler" car and engine would not be built,[8] evidently stopping work on it. Meanwhile, Willys Corporation issued a statement on October 30, 1920,[1] repeated on December 9, 1920,[9] saying that production of the new Chrysler Six car would begin at the plant in Elizabeth, New Jersey in the spring of 1921. The December 9 statement was a bit puzzling because it called the unchanged spring 1921 production date a "fortunate delay." The replacement ZSB-engineered car and engine, of course, did not have a hope of meeting a "spring 1921" production date such a short time period away.

Although the members of the ZSB engineering team had the freedom of a completely new design, Fred Zeder tempered their freedom by requesting that they utilize as many of the available tools as possible to minimize additional tooling costs.[2] When the new engine design was completed, the Willys production people took on the responsibility of furnishing the prototype parts for engineering testing, which they did in the remarkably short time of 90 days.[8] The engineering program went so well that by the spring of 1921, the first experimental car of the new "Chrysler" design was being tested on the roads of New Jersey and Pennsylvania at the rate of 800 miles per day. A teardown at 30,000 miles showed the car to be in perfect condition.[8] The replacement car was described as being light in weight, with outstanding performance and with the capability of 70 miles per hour.[8] Unfortunately, there are no known photographs of the vehicle. The engineering drawing on Fig. 1.1 of its brake and clutch pedals shows the ZSB engineers were not bashful about displaying the Chrysler name in front of the driver.

Fortunately, the Chrysler Historical Collection does have part prints and an assembly drawing of the ZSB engine, so it can be described in some detail. The views on Figs. 1.2 and 1.3 show the front and top of the engine, respectively. Internal details appear in the cross section on Fig. 1.4 and in the longitudinal section on Fig. 1.5.

The basic configuration was that of an in-line, six-cylinder, L-head engine. The bore and stroke were 3.25 × 4.5 in., giving a displacement of 224 cu. in. The cylinder spacing (i.e., the distance from the centerline of one cylinder to the next) of the cast iron cylinder block was not uniform but was graduated. The first two cylinders, #1 and #2, were spaced 4 in. apart. The last two cylinders, #5 and #6, also were 4 in. apart. The next cylinders, #2 and #3, were 4.25 in. apart, as were cylinders #4 and #5. The distance between the two cylinders at the center of the engine, #3 and #4, was 4.5 in. The shallow skirt of the cylinder block did not extend below the crankshaft centerline. The deck height (i.e., the distance from the crankshaft centerline to the top of the block) was 13.875 in.

Identical intake and exhaust valves were located along the right side of the cylinder block, as were the intake and exhaust manifolds. A quality feature that was unusual for high-volume engines at the time was the use of seven main bearings for the machined-all-over, forged steel crankshaft—quite an upgrade from the initially planned three-main-bearing Willys engine. The diameters of the crankshaft main journals and crankpins of the new Chrysler engine both were 1.874 in. The combustion chambers of the cast iron cylinder head were open with tapered roofs. Each chamber was highest at the spark plug and tapered down to its left side, farthest away from the spark plug, as can be seen in the cross section on Fig. 1.4. The pistons were made of cast iron, which was the predominant piston material at the time. The forged steel connecting rod had a dimension of 9.75 in. from the center of the piston pin to the center of the big-end bearing, usually referred to as the connecting rod center distance.

A rotary-gear oil pump, bolted to the bottom of the cylinder block, with its vertical shaft driven by a gear at the center of the camshaft, provided full-pressure oil to the main bearings, connecting rod big-end bearings, and camshaft bearings. This lubrication system, together with the seven-main-bearing crankshaft, became a distinctive quality feature of ZSB-designed engines.

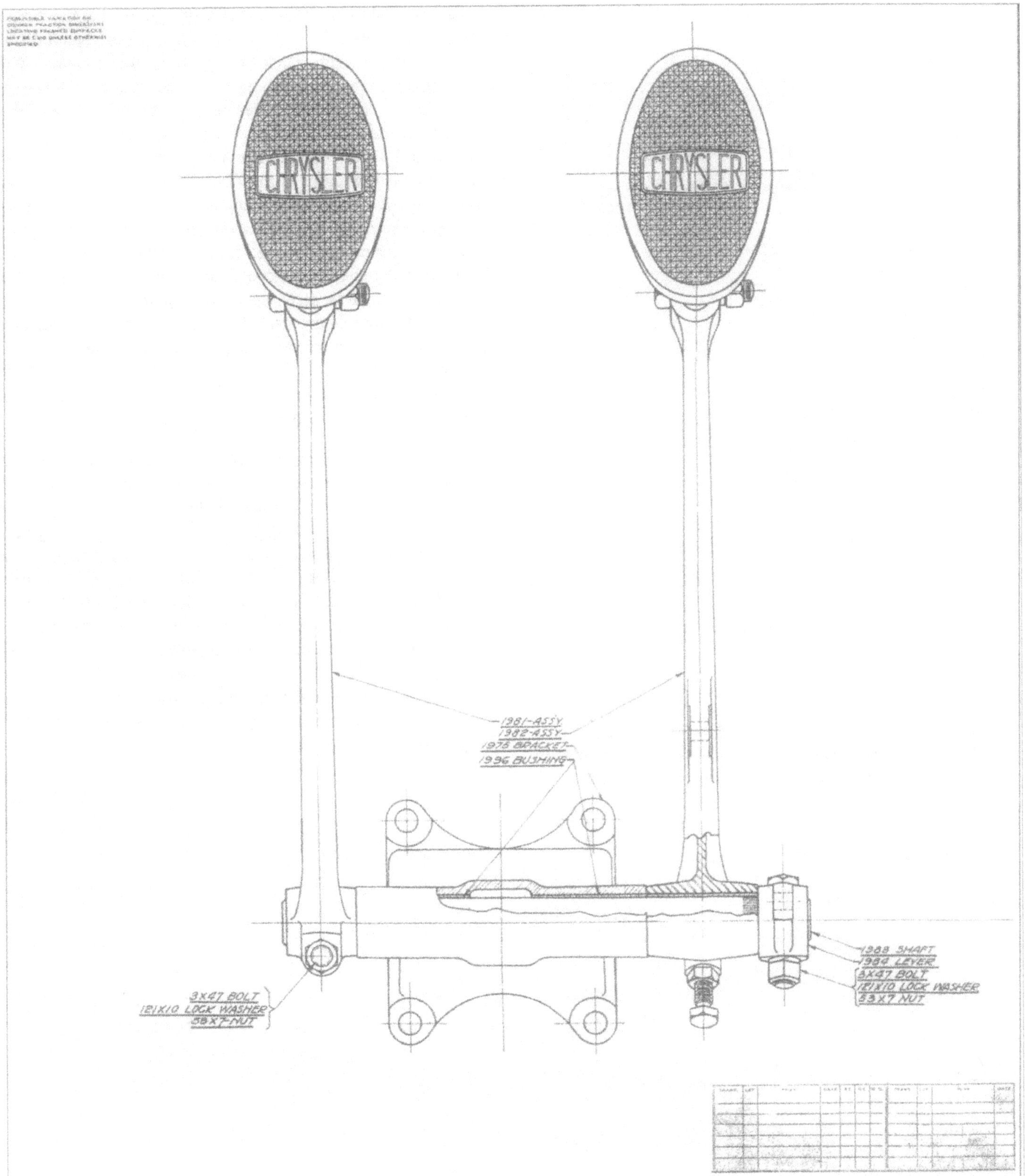

Fig. 1.1 *The Chrysler-Willys clutch (left) and brake pedal assembly drawing part #1974, dated January 14, 1921.*

Helical gears that were 1 in. wide drove the camshaft from the crankshaft. The water pump, distributor, and generator all were driven by a short accessory shaft that was gear-driven from the camshaft. This accessory shaft was mounted in a small housing (called a bracket in the parts list) that piloted into a wing of the cylinder block at its front-right corner. The generator, mounted

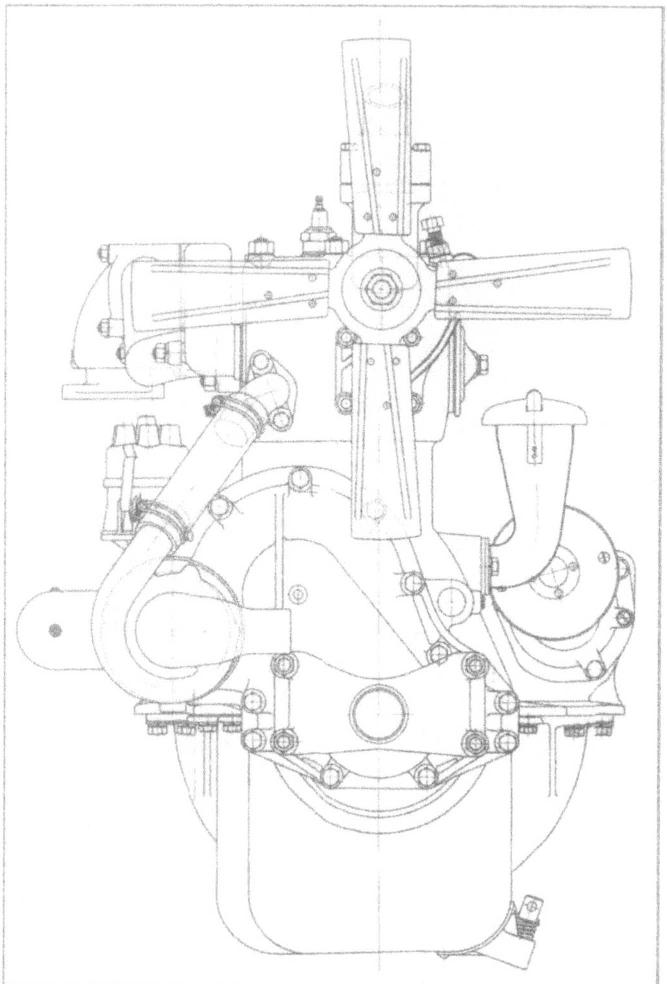

Fig. 1.2 *Front view of the ZSB-engineered 1922-intended 224-cu.-in. Chrysler-Willys L-head six-cylinder engine.*

to the rear of the accessory housing, was coupled to the rear of the accessory shaft. A gear cut into the accessory shaft drove the distributor, which was located on top of the housing. The water pump used a bracket attached to the gear case cover for its mounting. Rotational power for the water pump came from the nose of the accessory shaft via a tongue-and-groove joint. The use of a water pump for positive, forceful coolant flow was another quality feature of the ZSB-designed Chrysler-Willys engine because engines of many companies, including Maxwell and Ford, used the simpler but far less effective pumpless thermosyphon cooling. A 1.25-in.-wide flat leather belt, driven by a crowned pulley on the crankshaft, transmitted power to the fan shaft. As can be discerned on the engine assembly drawing, maintenance of the fan belt tension was accomplished by turning an adjusting screw on the fan bearing mounting bracket. Overall, the design was well executed, and if this Chrysler car and engine had gone into production, they would have been hailed as superior in many respects.

The remainder of 1920 and early 1921 apparently were used for testing. Production releases were started around May 1921. The engine was not given a unique designator, but each drawing carried a note, "Released for C." This second Chrysler-named car and engine apparently were on track to join the Willys model lineup for 1922. In an interesting disclosure in early July 1921, Walter Chrysler confided in ZSB that he was planning to purchase the Elizabeth plant and the Chrysler car from Willys Corporation and to start his own company.[8] But nothing happened. Also, despite production releases being made for the replacement Chrysler car, nothing was happening in terms of production machinery being purchased or modified to produce the ZSB design.

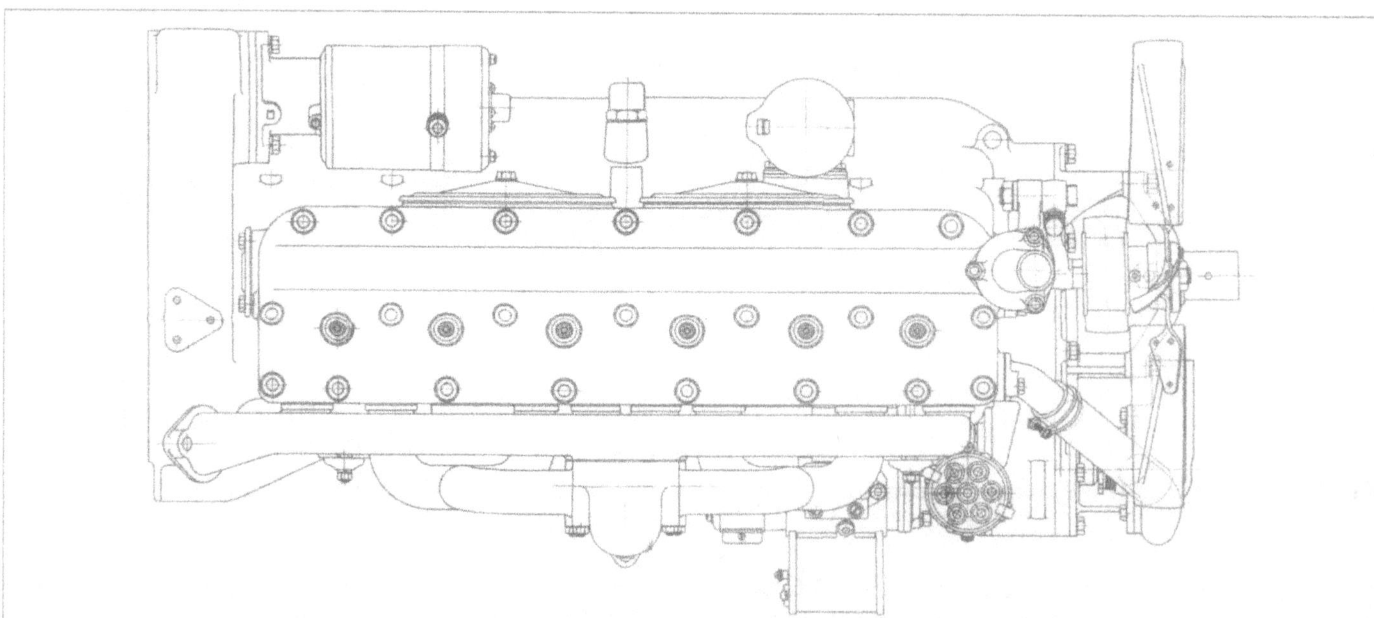

Fig. 1.3 *Top view of the ZSB-engineered 1922-intended Chrysler-Willys L-head six-cylinder engine.*

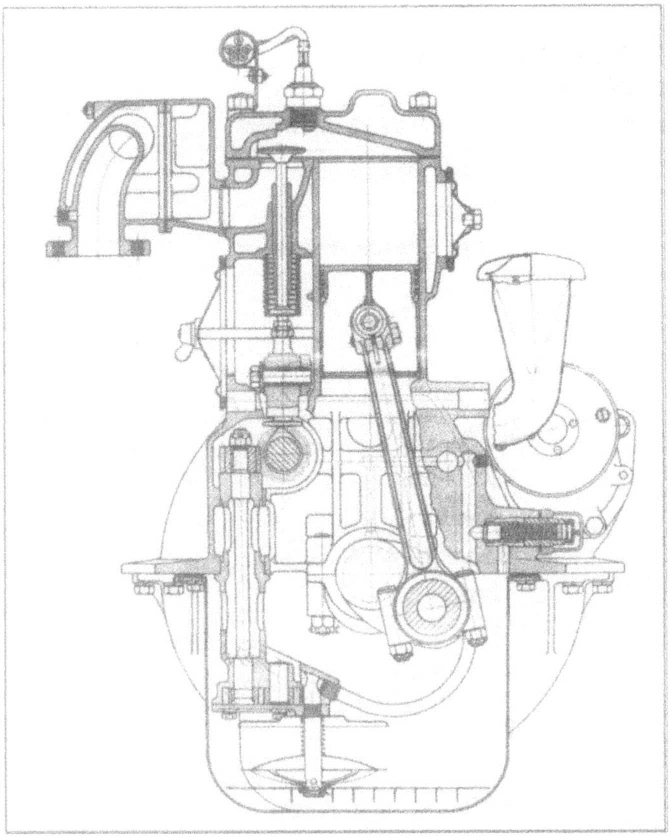

Fig. 1.4 *Cross section of the ZSB-engineered 1922-intended Chrysler-Willys L-head six-cylinder engine. Note the tapered roof, open combustion chamber, low-mounted oil pump, and cast iron piston. Motor assembly drawing line clarity computer enhanced by W.L. Weertman.*

Then, in a complete surprise to Chrysler, Zeder, Skelton, Breer, and the engineering staff, an announcement was made that on November 30, 1921, John Willys had put Willys Corporation into receivership.[8] As good as this hurried replacement, second Chrysler car and engine turned out to be, neither the car nor the engine would ever be produced.

The bankruptcy did not cause Walter Chrysler to have an employment problem because he already had a second job. Back in August 1920, the same bankers who had asked him to rescue Willys asked him to do a similar feat with Maxwell while maintaining his Willys turnaround effort. In fact, on May 12, 1921, acting as head of the Maxwell-Chalmers reorganization committee, Chrysler purchased the assets of the Maxwell and Chalmers companies at auction in Highland Park, Michigan.[10,11] Chrysler soon became chairman of the reorganized Maxwell Motor Co., a demanding job because the company had plenty of problems. Now he could walk away from Willys and spend full time with Maxwell. Chrysler's official departure from Willys came in February 1922.[12]

For Zeder, Skelton, Breer, and their engineering team, it was a different story. They were in New Jersey, and their employer, Willys Corporation, was bankrupt. They had to find work.

Table 1.1 shows specifications for the Willys-engineered Chrysler-Willys engine and the replacement ZSB-engineered Chrysler-Willys engine.

Time lines for the engines and events described in this chapter are displayed as part of Chart 2.1 located at the end of Chapter 2.

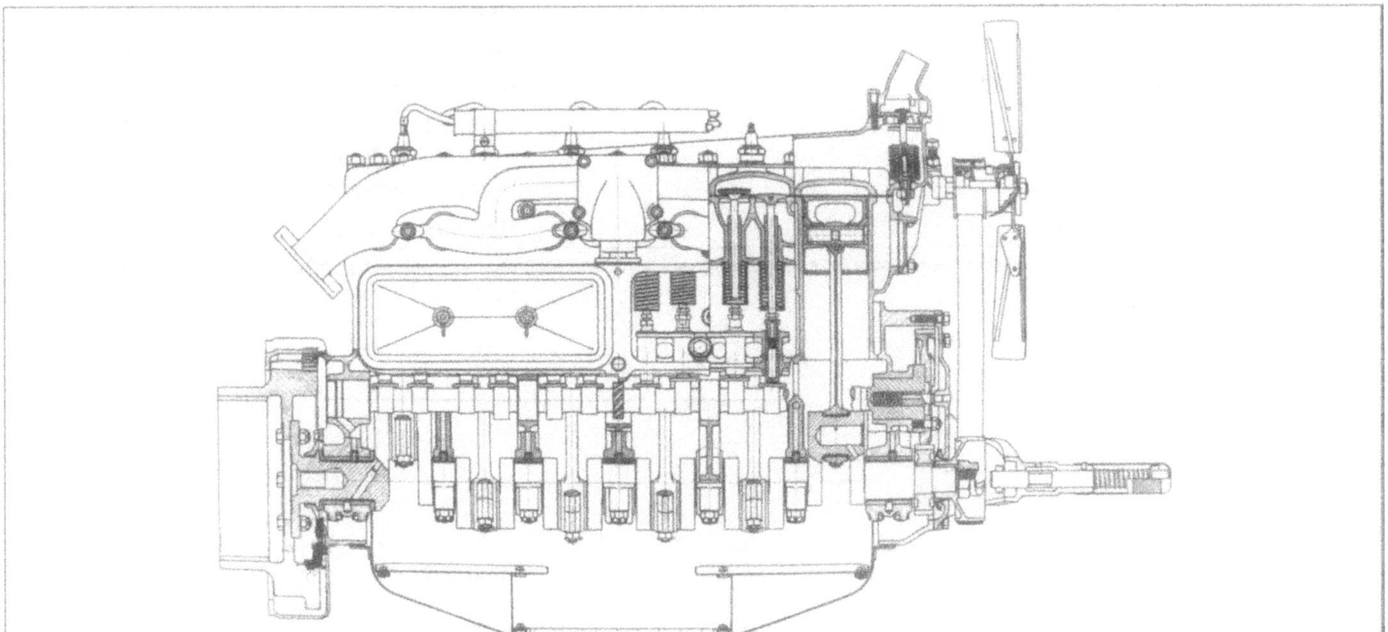

Fig. 1.5 *Longitudinal section of the ZSB-engineered 1922-intended Chrysler-Willys L-head six-cylinder engine. Note the seven-main-bearing crankshaft, the engagement device for hand cranking, and the flat fan belt. Motor assembly drawing line clarity computer enhanced by W.L. Weertman.*

TABLE 1.1
SPECIFICATIONS FOR THE CHRYSLER-WILLYS L-HEAD SIX-CYLINDER ENGINES

Year	1921 Intended	1921–1922 Intended
Car line	Chrysler-Willys	Chrysler-Willys
Model	Chrysler	Chrysler
Engineering design group	Willys	Zeder, Skelton, and Breer
Displacement (cu. in.)	184.1	224
Bore	3.125	3.25
Stroke	4.0	4.5
Compression ratio		4.5
BHP @ rpm		61 @ 2800
Torque (lb-ft) @ rpm		150 @ 800
Combustion chamber		Open
Cyl. block mat'l.—skirt type		Cast iron—shallow
Cyl. block deck height		13.875
Cyl. block length		32.81
Cyl. bore spacing		1-2, 5-6 = 4 2-3, 4-5 = 4.25 3-4 = 4.5
Ck/shaft—no. of main brgs.	3	7
Ck/shaft main journal dia.	2 @ 2.03 1 @ 2.06	1.875
Crankpin journal dia.	1.875	1.875
Piston mat'l.—no. of rings		Cast iron—3
Connecting rod centers		9.75
Valve system		L-head
Valve events*		210–230–neg. 2
Camshaft drive		Helical gear
Valves		Intake and exh. 1.625
Valve lash adjustment		Mech. tappet
Oil pump		Vert. at bottom, ctr. of block full-pressure lube
Distributor		Vert., on access hsg.
Fuel system		Updraft carburetor

* For an explanation of valve events, see the Abbreviations and Definitions section near the back of this book.

CHAPTER 2

Zeder, Skelton, and Breer's Engines A, B, C, and D

Following the November 30, 1921 bankruptcy of their employer, Willys Corporation, Zeder, Skelton, and Breer (ZSB) felt they had only one viable course of action—to stay together with their entire staff as an engineering team and look for outside work.[1] They arranged with the Willys receivers for permission to use the engineering spaces that they were already occupying at the Willys Elizabeth plant, including the engine dynamometers, for a small monthly fee.[2] By the end of the year, they had incorporated as the Zeder, Skelton, and Breer Engineering Company[3] and were ready for business.

The receivers initially tried to sell the Willys Elizabeth plant, with one prospect being the H.H. Franklin Manufacturing Company, maker of air-cooled engines.[4] Failing to achieve a satisfactory arrangement with any potential buyers, the receivers then set a date of June 9, 1922 for the auction of the plant and its contents, which included the prototype Chrysler cars, engines, designs, and blueprints created by Zeder, Skelton, and Breer.[5]

A-Model Engine

An early project for now-on-their-own Zeder, Skelton, and Breer was that of a complete car and four-cylinder engine, started in late December 1921. Designated as the A-model, engine layouts and engineering part prints had numbers preceded with an "A-." Figures 2.1 and 2.2, from ZSB Layout A-17, dated January 31, 1922, show longitudinal and cross-section views of the engine, respectively. Similar to the six-cylinder Chrysler-Willys engine on which they had been working for some time, this engine was an L-head with open tapered-roof combustion chambers. Its bore and stroke of 3.5 × 4.75 in. provided a displacement of 183 cu. in. The ZSB-favored crankshaft design with main bearings straddling each crankpin (making a total of five main bearings for the four-cylinder engine) and full-pressure oiling for all main bearings, connecting rod big-end bearings, and camshaft bearings were incorporated. Cylinder spacing of the shallow-skirt, cast iron cylinder block was 4.5 in. for the end cylinders, with the two center cylinders being 4.75 in. apart. Identical intake and exhaust valves were used. Originally, the camshaft was driven from the crankshaft by two helical-toothed gears. However, around March 1922, this design was revised to use a timing chain and sprockets. According to layouts, a design of an aluminum piston for the engine was prepared in November 1922 as a replacement for the original cast iron piston. A gear assembled to the camshaft drove the distributor, which was mounted horizontally on the gear case cover at the front of the engine. For driving the oil pump, a gear was cut into the camshaft between cylinders #3 and #4.

Although part prints, layouts, and a parts list in the archival files of the Chrysler Historical Collection give considerable technical information about the engine, and notebook entries by Newton F. Hadley indicate that at least one engine was built and run, other information regarding the identity of the client and whether the vehicle and engine ever went into production unfortunately appear to be lost.

B-Model Engine

The A-model four-cylinder engine job must have helped keep the ZSB group alive and together during this uncertain time, but Fred Zeder had far more ambitious plans for all of them. Lining up support from the wealthy Clement Studebaker, Jr., whose father was president of Studebaker for 33 years, and Rollin H. White, president of the Cleveland Tractor Co., Zeder embarked on nothing less than the formation of a new automobile company. As described in trade journals, Clement Studebaker, Jr. would be its president, Fred Zeder would provide the car, and Rollin H. White would arrange for its manufacture within the facilities of the Cleveland Tractor Co.[6] As a prelude, Zeder founded the Zeder Motor Car Company

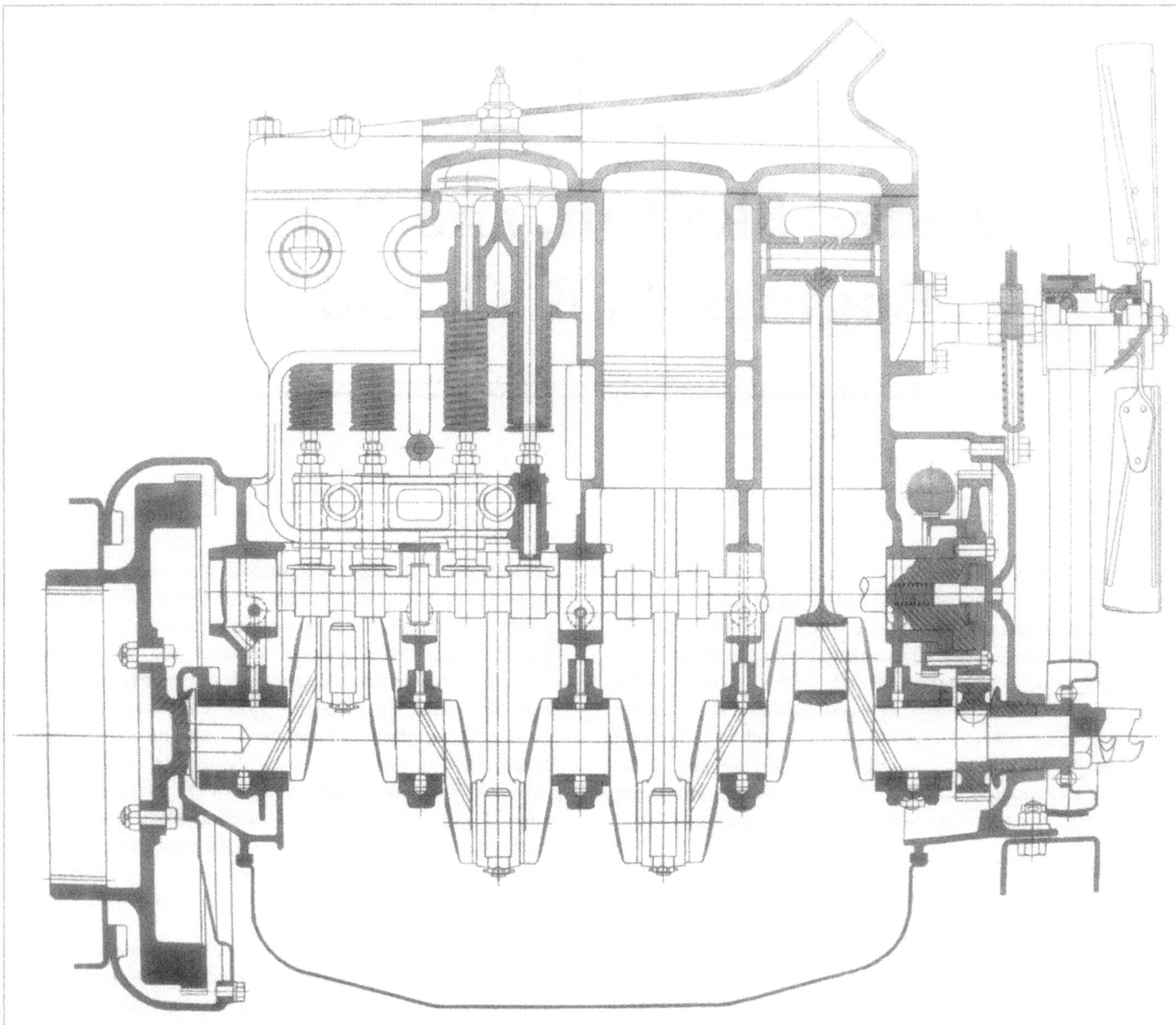

Fig. 2.1 *A longitudinal section of the ZSB-designed A-model four-cylinder engine. The diagonal drilled holes in the crankshaft deliver high-pressure oil to the connecting rod bearings. Large bolts at the bottom of the gear case cover attach the engine solidly to the front cross member of the car. From ZSB layout A-17, Motor Assembly, dated January 31, 1922. Layout line clarity computer enhanced by W.L. Weertman.*

in Ohio in March 1922.[7] Around the same time, Fred Zeder and the entire ZSB engineering group severed their connection with Willys Corporation, as Chrysler had done in February, and moved from Elizabeth, New Jersey, to a building located at 24–26 Mechanic Street in Newark, New Jersey.[7]

Regarding the product of the Zeder Motor Car Company, first reports stated that the company was considering the purchase from Willys Corporation of the rights to manufacture the replacement Chrysler car,[8,9] by now widely recognized as an excellent car.[6] The reports evidently were well enough founded for Walter P. Chrysler to state that he had withdrawn the right to use his name in connection with this enterprise and that while the receivers could sell the engineering rights in this car, the "Chrysler" name did not go with them.[10] The next reports about the Zeder car portrayed the car as an entirely new product and that at least nine months would be required to complete experimental work.[7] With the Zeder car and a possible design launch for the engine soon after the A-model, I postulate that ZSB Engineering assigned it the designator "B-model." Regardless, the concerted endeavor to create and put into production a Zeder car failed quickly, certainly before

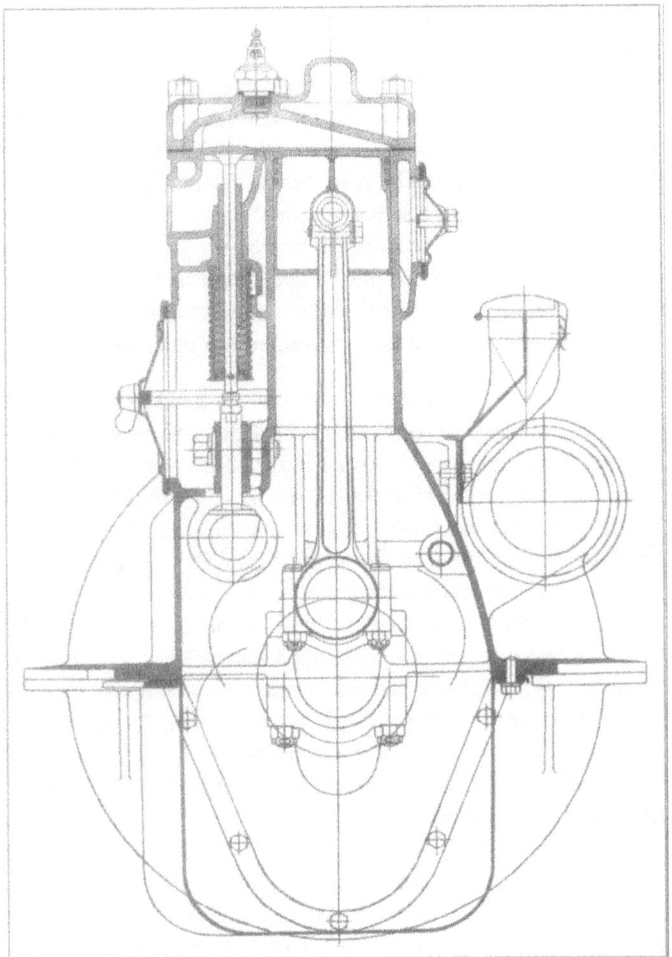

Fig. 2.2 *A cross section of the ZSB-designed A-model four-cylinder engine. From layout A-17, Motor Assembly, January 31, 1922. The open, tapered-roof combustion chamber appears similar to the chamber of the Chrysler-Willys engine (see Fig. 1.4 in Chapter 1). Layout line clarity computer enhanced by W.L. Weertman.*

any experimental cars or engines would have been built. This failure dashed the immediate hopes for ZSB Engineering to become part of a larger corporation.

Staying afloat with contract work, ZSB Engineering in May 1922 started designing a 255-cu.-in. six-cylinder "C-model" engine, further described here later. Evidence points to William C. "Billy" Durant, who needed an engine for the "Flint" car he was planning to build, as being the possible client.[1]

Then, on June 9, 1922, the much-heralded auction of the Willys Elizabeth plant took place, with two final bidders: James M. Smith, representing Walter P. Chrysler, and Joseph P. Day, representing Billy Durant. Smith's highest bid was $5,500,000, topped by Day at $5,525,000, and Smith bid no further.[11] With his winning bid, Durant got the plant that he wanted for expanded production of another car with which he was involved, called the "Star," and with the plant, the rights, drawings, and prototype cars of the Chrysler car that Willys had planned to build there. Of course, all this was another blow to ZSB Engineering because the sale of the plant to Durant quashed their final hope that Chrysler would regain the plant and their car and would resume efforts to put the car into production.

Undaunted, they returned to work on their own on a fresh design of the kind of car they knew Chrysler would want—the design quite possibly being a continuation and improvement of the Zeder car and engine. They knew that if Chrysler really liked the new design, he would find a way to put it into production and retain their services. Therefore, they worked on it through the summer, with Chrysler, according to Skelton, often visiting their design room in Newark, which was a short commuter rail ride from Chrysler's office in Manhattan.[2] Finally, Zeder, Skelton, and Breer were ready to make a presentation to Chrysler at his office in New York City, which they did on November 11, 1922.[2] They certainly hit a home run with the proposal because Chrysler immediately approved the specifications and agreed to a personal contract with ZSB Engineering for the car and engine.[2] Starting in early December 1922, the first of hundreds of design layouts with sequential numbers preceded with "BCL," signifying "B-model, Chassis Layout," were figuratively flying off the drafting tables of ZSB Engineering, as evidenced by archival microfilm records at the Chrysler Historical Collection.

The continued story of this B-model engine is presented in Chapter 3.

Contract Work for Durant and the C-Model Engine

As already noted, starting in May 1922 and looking very much like a larger edition of the Chrysler-Willys engine, ZSB Engineering designed a C-model engine having a larger bore and stroke of 3.375 × 4.75 in. and a displacement of 255 cu. in. Front and right-side views of the engine are displayed on Figs. 2.3 and 2.4, respectively. Longitudinal-section and cross-section views of the engine are shown on Figs. 2.5 and 2.6, respectively. All cylinder bores were spread farther apart, with graduated spacings of 4.25, 4.5, and 4.875 in. from end to center cylinders. A 2.625-in.-deep skirt was provided for the longer C-model cylinder block. The L-head valve arrangement and open tapered-roof combustion chambers were carryovers from the Chrysler-Willys design, as was the ZSB classic seven-main-bearing crankshaft with full-pressure oiling. The camshaft was gear-driven and had a gear cut at the center of its length for driving the oil pump, bolted to the bottom of the block. Similar to the Chrysler-Willys engine, a gear-driven accessory shaft transmitted power to the water pump at the right front of the engine and to the generator, attached to the right side of the cylinder block. On this C-model engine, the ignition distributor was vertical, mounted on the gear case cover and

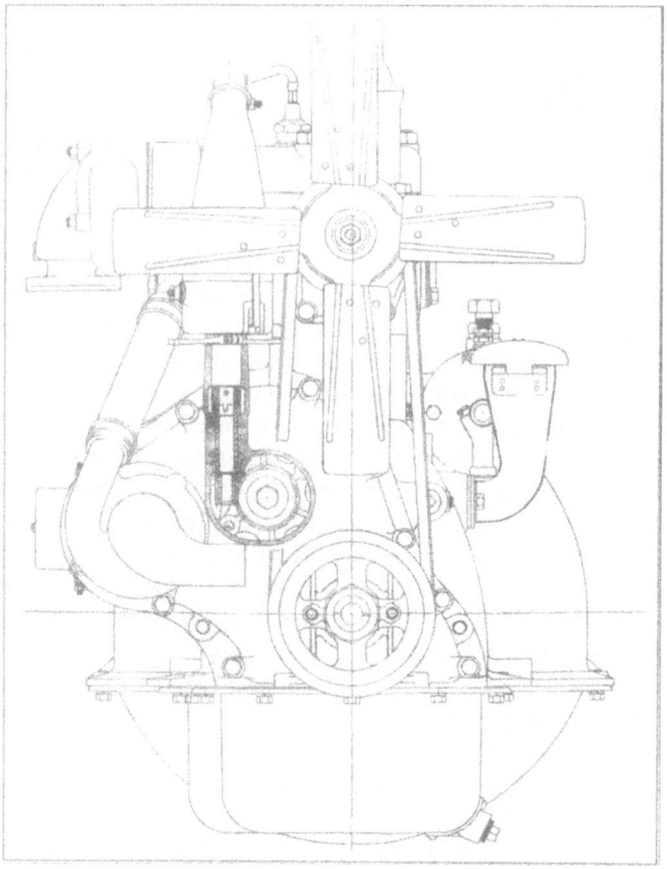

Fig. 2.3 *Front view of the 255-cu.-in. C-model six-cylinder engine. A vee belt drives the fan. The efficient-looking water pump is driven by internal gears. From Zeder, Skelton, and Breer Engineering Co. layout "Motor Assembly—Sheet 1," dated July 25, 1922.*

geared to the nose of the camshaft at the front of the engine. A newer V-belt system replaced the old flat leather belt of the Chrysler-Willys fan drive.

Evidence that at least one car and engine were built and tested comes from a listing of engine test reports at the Chrysler Historical Collection, which shows that an engine test report was written around this time by Carl Breer and Phil Kent of a Chrysler six-cylinder Model C engine,[12] and from entries in a notebook of Newton F. Hadley dated 1922.

Although the engine initially was designed at 255 cu. in as noted, around August 1922, the displacement was increased to 268.4 cu. in. by a stroke increase from 4.75 to 5.00 in., according to information recorded by designer Newton F. Hadley. This required a lengthening of the connecting rod from 10.5 to 11.0 in. and probably required the cylinder block height to be increased.

In their books,[1,13] both Carl Breer and Walter P. Chrysler state that ZSB Engineering was awarded a contract by Billy Durant to design an engine for a new car that Durant was planning and that was to be called the "Flint." In making visits to the idle, empty Willys Elizabeth plant in the spring of 1922 prior to the auction, Durant rode in one of the Chrysler-Willys prototype cars,[13] which might have given him extra incentive to buy the property. According to Chrysler, for the Flint car Durant wanted an engine with more horsepower, leading to the contract with ZSB Engineering for an engine of the same type as the engine in the Chrysler-Willys but larger.[13] The C-engine fits this description exactly; therefore, the C-engine may have been the engine that ZSB and Durant initially intended to power the Flint.

However, Continental Motors received the production contract from Durant to supply engines for the Flint, and the engine that launched the Flint in 1924 was not the ZSB C-engine. Rather, it was a Continental Model 6E L-head six-cylinder engine.[14] Using the exact same bore and stroke of 3.375 × 5.0 in. and displacement of 268.4 cu. in. as the final ZSB C-engine design, it had two important features that were characteristic of ZSB engines until that time: a seven-main-bearing crankshaft with full-pressure lubrication and an open, tapered-roof combustion chamber. Beyond these similar features, the design was totally different from the ZSB design practice with the camshaft, valves, and the intake and exhaust manifolds located on the left side of the engine rather than on the right and having a rather old-fashioned exposed rotating shaft on the right side of the engine to drive the generator. The cylinder bore spacing was different on the Continental engine. The similarities between the ZSB C-engine and the Continental engine as used in the Flint car undoubtedly came about because of Zeder, Skelton, and Breer being hired by Durant to consult with Continental Motors[1] on the new engine. The final outcome suggests a possibility that existing production tooling at Continental Motors caused the design to become heavily Continental oriented to minimize tooling costs. When the car was announced in 1923, its use of a Continental engine was declared, along with a technical description of the engine,[15] without any mention of ZSB Engineering involvement. A complete set of detail drawings of the Continental Model 6E engine, sent to ZSB Engineering by Continental Motors during the Flint engine program, is on file at the Chrysler Historical Collection.

Statements, sometimes published, that the Flint was powered by an engine designed and developed by Zeder, Skelton, and Breer—either the Chrysler-Willys engine obtained by Durant when he bought the Willys Elizabeth plant at auction, or the contracted C-engine—do not appear to be correct.

Of interest, a Flint sales brochure dated June 1924 had an entire page extolling the virtues of the seven-main-bearing-crankshaft of the engine.[16]

D-Model Engine

ZSB Engineering designed another engine for Billy Durant. In his book, Carl Breer writes of ZSB Engineering contracting with Durant for the design of a new engine for Locomobile,[1]

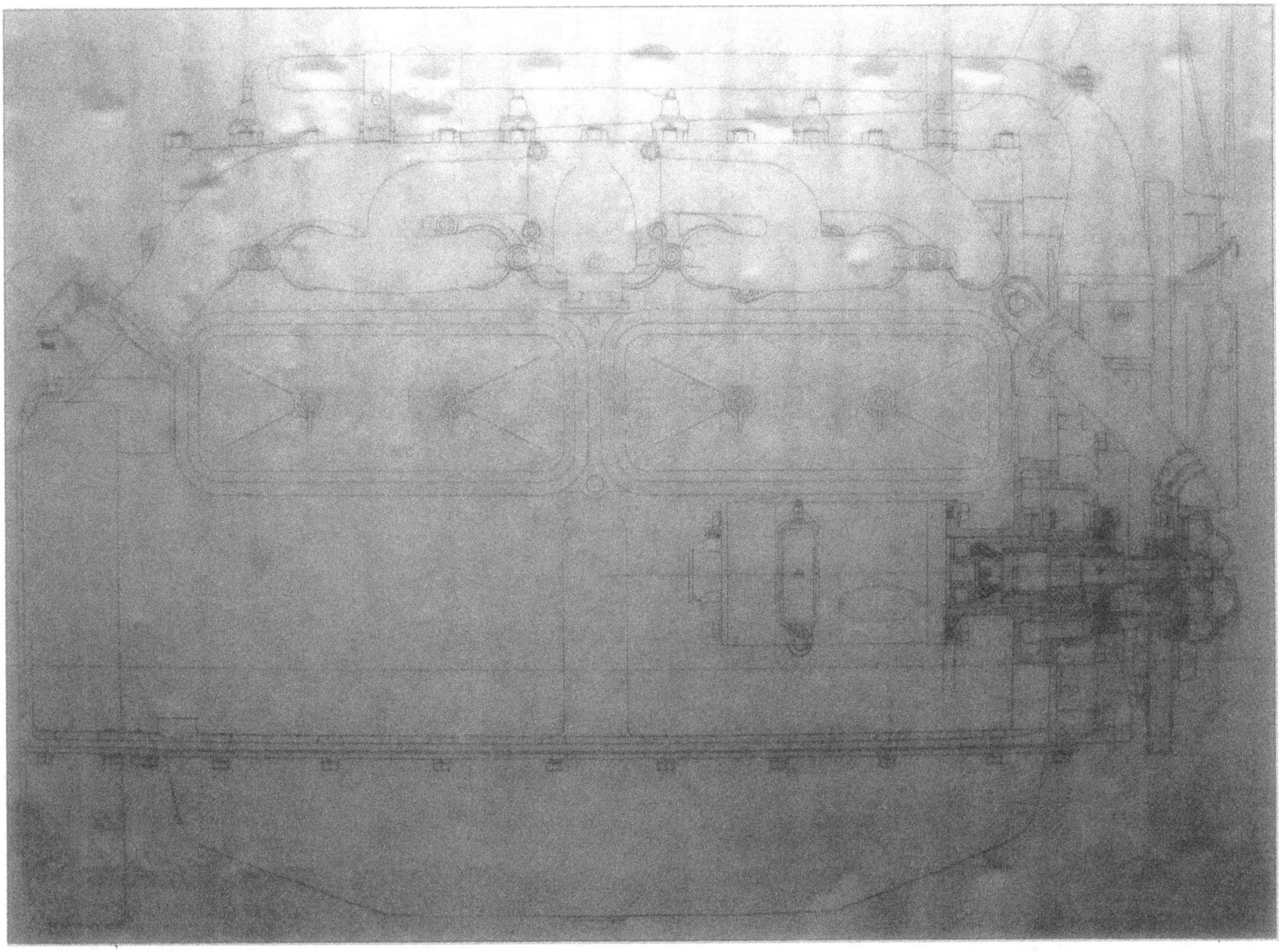

Fig. 2.4 *Right side and partial section of the C-model six-cylinder engine. From Zeder, Skelton, and Breer Engineering Co. layout "Motor Assembly—Sheet 1," dated July 25, 1922.*

a luxury car company owned by Durant Motor Corporation, whose powerplant was a huge 525-cu.-in. T-head engine of ancient technical vintage. The leadoff ZSB layout, DCL-1, "Preliminary Layout of Motor" completed by Mel Carpentier on February 28, 1923, shows a big six-cylinder, L-head engine with a bore and stroke of 3.75 × 5.25 in. displacing 348 cu. in. and its distributor mounted vertically on the chain case cover at the front of the engine, as can be seen on the view of the right side of the engine shown in Fig. 2.7. The oil pump appears to be located at the center of the engine at the bottom of the cylinder block inside the crankcase, gear-driven from the camshaft. The ZSB-favored seven main bearings and full-pressure lubrication for all major bearings were part of the design. A new feature appeared on this engine that would be repeated on all future Chrysler L-head engines—the water pump was relocated to a position high at the front of the cylinder block. Individual design studies show a number of different combustion chambers with turbulence-inducing quench areas above the piston away from the spark plug. This indicates that Zeder, Skelton, and Breer were investigating this type of combustion chamber, first developed by Sir Harry Ricardo of England, which was well known to the U.S. auto industry by 1921[17] and which will be discussed in more detail in Chapter 3.

Within a month, on March 21, 1923, a second engine assembly layout, DCL-19, was completed, evidently superseding the earlier layout. An existing microfilm copy of a blueprint of layout DCL-19 shows the engine to have an aluminum crankcase that was bolted to cast iron cylinders forming a two-piece cylinder block. The cylinder bore centers, unchanged from the prior DCL-1 layout, were spaced 4.75 in. apart except at the center of the engine, where cylinders #3 and #4 were 5.375 in. apart. The oil pump, bolted to the bottom of the cylinder block assembly, and the distributor, attached to the top of the cylinder head,

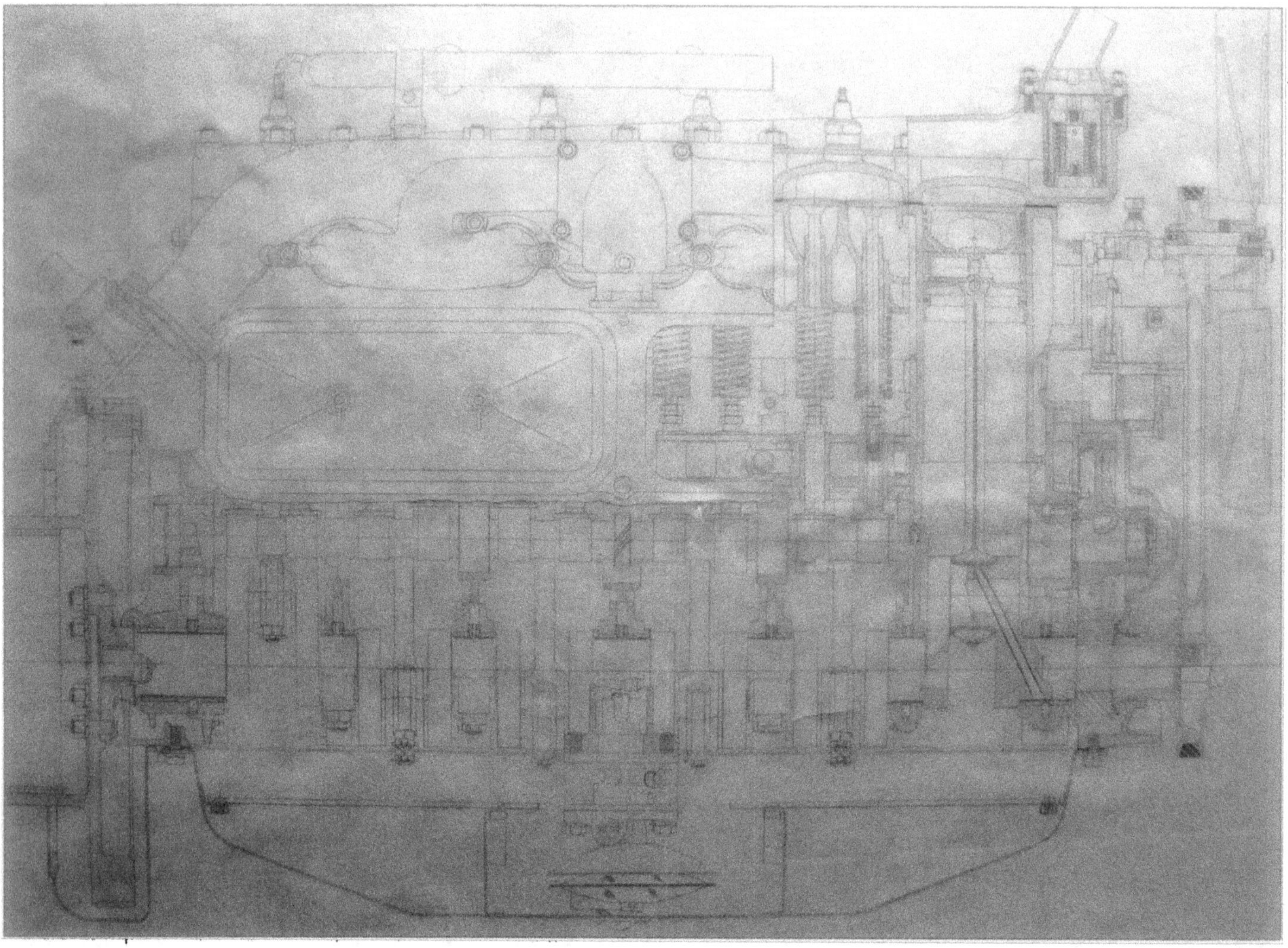

Fig. 2.5 *(Above) Longitudinal section of the C-model six-cylinder engine. Pressed-in tubes were used in the seven-main-bearing crankshaft to transport oil to the connecting rod big-end bearings. From Zeder, Skelton, and Breer Engineering. Co. layout "Motor Assembly—Sheet 1," dated July 25, 1922.*

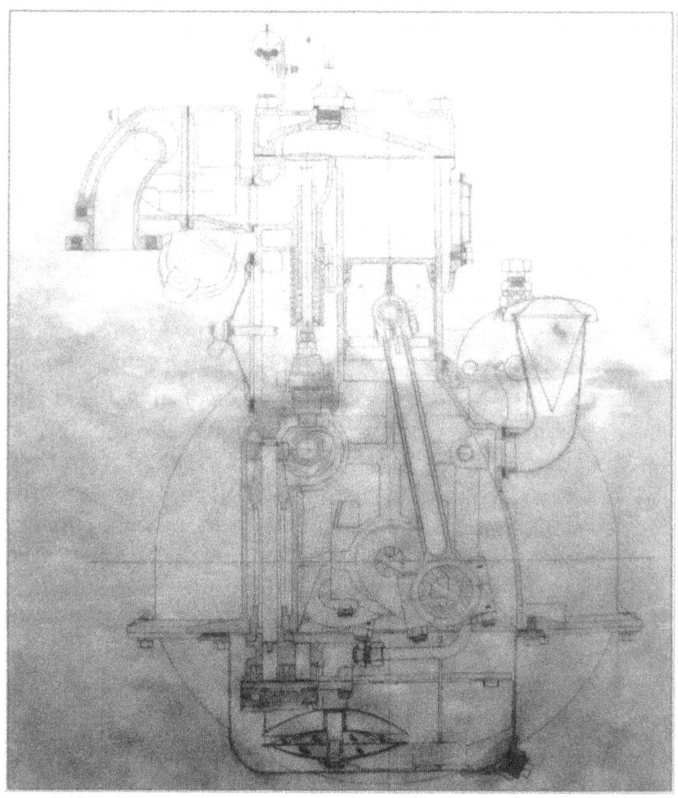

Fig. 2.6 *(Right) Cross section of the C-model six-cylinder engine. The open tapered-roof combustion chamber has been carried over from prior engines. The oil pump design is somewhat similar to the Chrysler-Willys engine oil pump (see Fig. 1.4 in Chapter 1). From Zeder, Skelton, and Breer Engineering Co. layout "Motor Assembly—Sheet 1," dated July 25, 1922.*

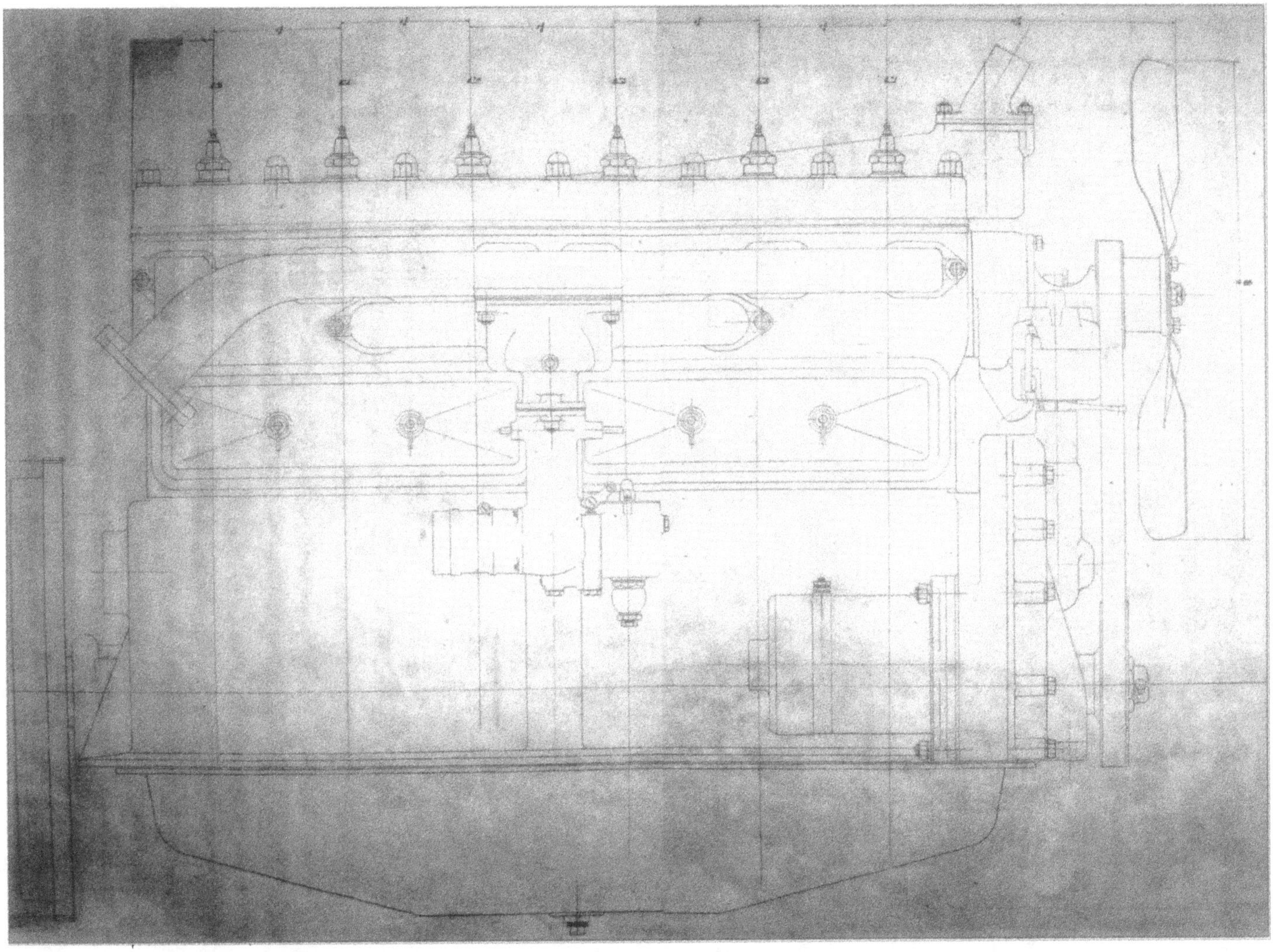

Fig. 2.7 *Right side of the initial 348-cu.-in. D-model six-cylinder engine. From Zeder, Skelton, and Breer Engineering Co. layout DCL-1, "Preliminary Layout of Motor—D-model," dated February 28, 1923.*

both were located at the right-center of the engine, driven by a vertical shaft arrangement geared to the camshaft, as shown in Fig. 2.8. The combustion chambers in the cylinder head, visible in Fig. 2.9, were provided with a turbulence-inducing quench area, showing that the ZSB experimental results were being readied for production use. Seven main bearings and full-pressure lubrication were carried over from the earlier DCL-1 design. In time sequence, shown on Chart 2.1 at the end of this chapter, this engine was being designed at the same time as the final version of ZSB's own B-engine that shared similar features, as will be described in the next chapter.

This layout of the D-engine is significant to the history of Chrysler engines because it is the earliest known existing layout that combines the spectrum of features—a turbulent combustion chamber, a seven-main-bearing crankshaft, full-pressure lubrication, a vertical driveshaft at the right-center of the engine for driving the oil pump in the crankcase and the distributor on top of the cylinder head, a V-belt-driven water pump mounted to the front-top of the cylinder block, and six exhaust ports with each exhaust valve seat surrounded by coolant—that were repeated in the all-important Chrysler B-engine being designed almost simultaneously by ZSB and the next two engines that followed.

In the lower right corner of layout DCL-19, the word "Locomobile" appears, giving a positive tie-in between the ZSB Engineering layout and the client. With the introduction of the Locomobile Model 90 in the fall of 1925 as a 1926 model, technical details of the Model 90 engine were made public.[18] A slight bore increase from 3.75 to 3.875 in. had increased the displacement of the engine to 371.5 cu. in. from the 348 cu. in. provided on layout DCL-19. The sections of the production engine displayed in the published information appear virtually identical to the sections on ZSB layout DCL-19, with one significant exception: the combustion chambers were open without any semblance of quench area. Why the Locomobile engineers chose to forego this important development with

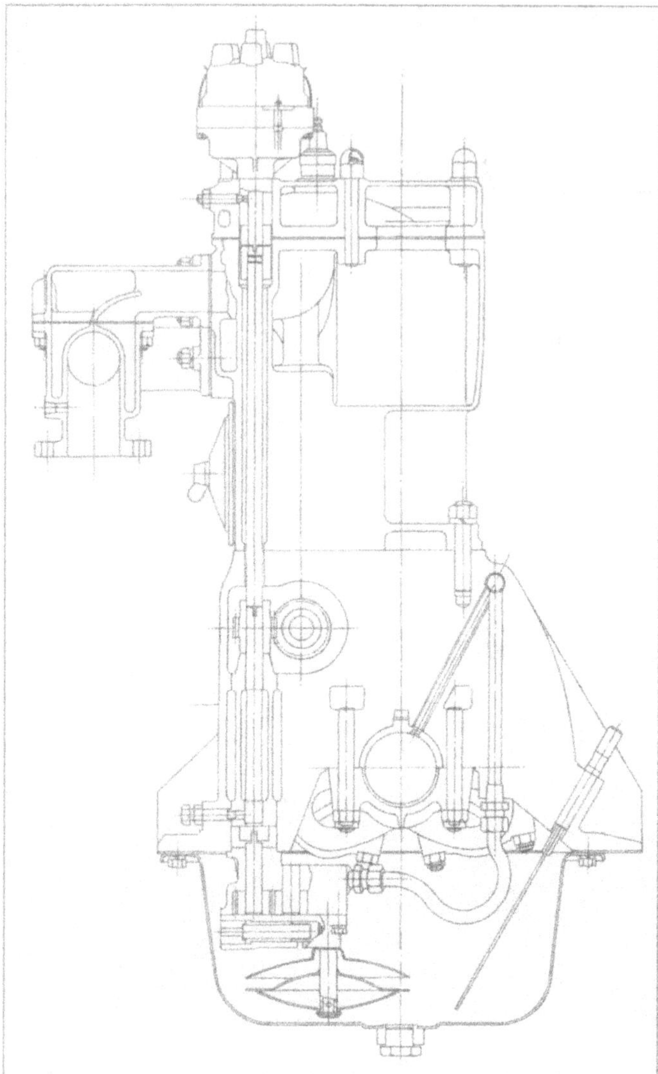

Fig. 2.8 *Cross section through the center of the final design 348-cu.-in. D-model six-cylinder engine. Line drawing made by W.L. Weertman from Zeder, Skelton, and Breer Engineering Co. layout DCL-19 dated March 21, 1923.*

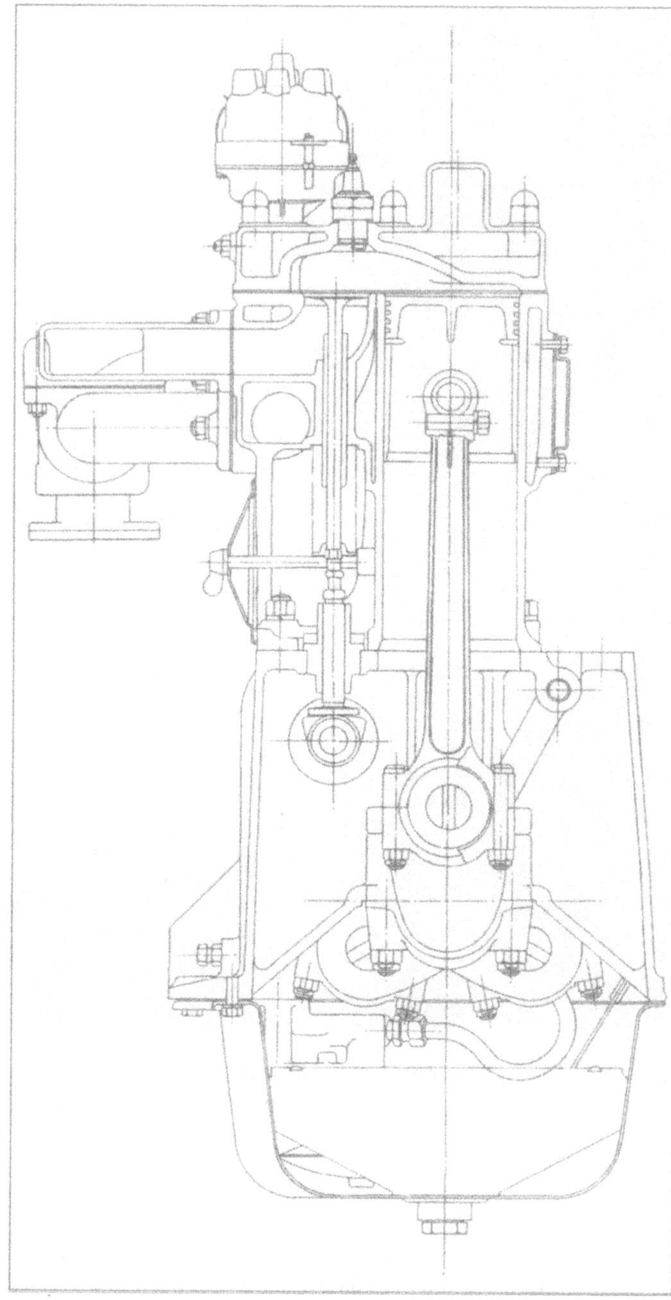

Fig. 2.9 *Cross section through cylinder #1 of the final design 348-cu.-in. D-model six-cylinder engine. Line drawing made by W.L. Weertman from Zeder, Skelton, and Breer Engineering Co. layout DCL-19 dated March 21, 1923.*

its proven power and fuel economy gains, handed to them by ZSB Engineering, remains a mystery.

Of interest, based on design layout records, this D-engine for Durant was not the only ZSB-designed engine to be given a D designator. After Zeder, Skelton, and Breer left New Jersey to join Chrysler at Maxwell in Detroit (a move discussed in Chapter 3) and after the initial Chrysler B-model car had been in production for several months, they made layouts of a D-model engine that certainly was never put into production and disappeared shortly after engineering work began on the E-model engine (also discussed in Chapter 3).

Although engines A, C, and the first D received little publicity compared to the first Chrysler B-model production engine, those engines hold a special place in the development of Chrysler engines because they kept Zeder, Skelton, and Breer Engineering Co. alive during difficult days. They also gave Zeder, Skelton, and Breer the opportunity to perfect their already impressive automotive engineering skills. Engines C and D were especially important because, as contract projects for Billy Durant, they provided ZSB with continued access to the Elizabeth plant's dynamometer facilities after the plant was sold—access they needed to work on their own engine as well as on the engines for Durant. With these skills and lessons learned, they created their three engineering masterpieces: the engines for the Chrysler B-model, Chrysler Imperial E-model, and Chrysler H-model automobiles. These engines are described in Chapter 3.

Table 2.1 shows the specifications for the ZSB engines A, C, and D. Chart 2.1 displays time lines for the events and engines described in both Chapters 1 and 2.

TABLE 2.1
SPECIFICATIONS FOR EARLY ZSB L-HEAD ENGINES

Year	1922	1922	1922
Model	A-model	C-model—1st design	C-model—Final design
No. of cylinders	Four	Six	Six
Displacement (cu. in.)	183	255	268.4
Bore	3.5	3.375	3.375
Stroke	4.75	4.75	5.0
Compression ratio		4.25:1	4.3:1
BHP @ rpm	43 @ 2400		
Torque (lb-ft) @ rpm	113 @ 1000–1600		
Combustion chamber	Open	Open	Open
Cyl. block mat'l.—skirt type	Cast iron—shallow	Cast iron—deep	Cast iron—deep
Cyl. block deck height	15.75	14.875	
Cyl. block length		33.4375	33.4375
Cyl. bore spacing	1-2, 3-4 = 4.5 2-3 = 4.75	1-2, 5-6 = 4.25 2-3, 4-5 = 4.5 3-4 = 4.875	1-2, 5-6 = 4.25 2-3, 4-5 = 4.5 3-4 = 4.875
Ck/shaft—no. of main brgs.	5	7	7
Ck/shaft main journal dia.	1.875	2.0	2.0
Crankpin journal dia.	1.875	2.0	2.0
Piston mat'l.—no. of rings	Aluminum—3	Cast iron—4	Cast iron—4
Connecting rod centers	11.5	10.5	10.5
Valve system	L-head	L-head	L-head
Valve events	216-228-1		216-228-1
Camshaft drive	Chain	Helical gear	Helical gear
Valves—head diameter	Identical intake and exh. 1.69	Intake and exh. 1.75	Intake and exh. 1.75
Valve lash adjustment	Mech. tappets	Mech. tappets	Mech. tappets
Oil pump	Gear drive from camshaft Full-pressure lube	Mtd. at right, bottom of block; gear drive from ca/s; full pressure	Mtd. at right, bottom of block; gear drive from ca/s; full pressure
Distributor		Vert. mount on access housing	Vert. mount on access housing
Fuel system	Updraft carburetor	Updraft carburetor	Updraft carburetor

(Continued on next page)

TABLE 2.1
(Continued)

Year	1923	1923
Model	D-model—1st design	D-model—Final design
No. of cylinders	Six	Six
Displacement (cu. in.)	348	348
Bore	3.75	3.75
Stroke	5.25	5.25
Compression ratio	4.3:1	4.4:1
BHP		
Torque (lb-ft)		
Combustion chamber		Turbulent
Cyl. block mat'l.—skirt type	Cast iron cylinders Al crankcase—deep	Cast iron cylinders Al crankcase—deep
Cyl. block deck height	16.75	17.00
Cyl. block length		
Cyl. bore spacing	All except 3-4 = 4.75 3-4 = 5.375	All except 3-4 = 4.75 3-4 = 5.375
Ck/shaft—no. of main brgs.	7	7
Ck/shaft main journal dia.		
Crankpin journal dia.		
Piston mat'l.—no. of rings		
Connecting rod centers	11.5	11.5
Valve system	L-head	L-head
Valve events		
Camshaft drive	Chain	Chain
Valves—head diameter		
Valve lash adjustment	Mech. tappets	Mech. tappets
Oil pump	Mtd. at right, bottom of block; gear drive from ca/s; full pressure	Mtd. at right, bottom of block; gear drive from ca/s; full pressure
Distributor	Mtd. on chain case cover; gear drive from camshaft	Mtd. on top of cyl. head; vert. shaft from camshaft
Fuel system	Updraft carburetor	Updraft carburetor

Chart 2.1
Time Lines—Early Engines and Formation of Chrysler

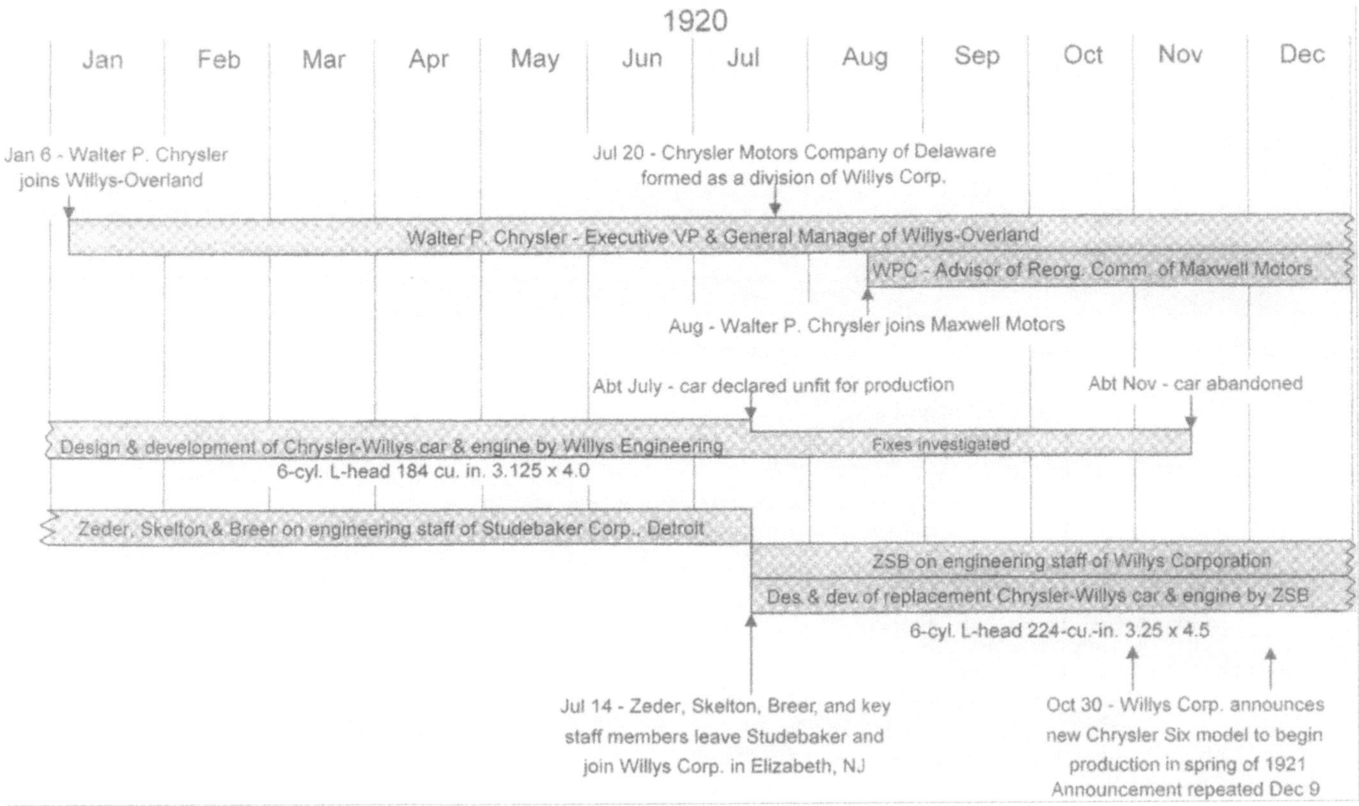

Chart 2.1
(Continued)

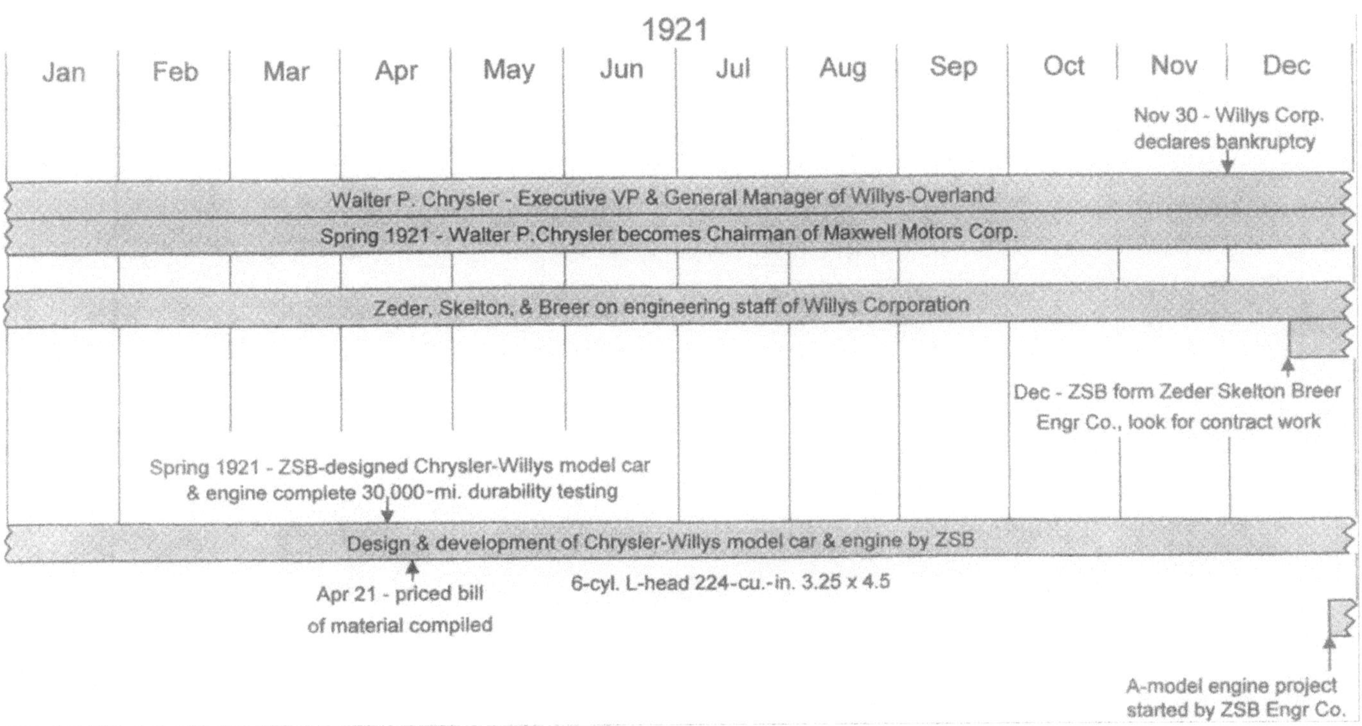

Chart 2.1
(Continued)

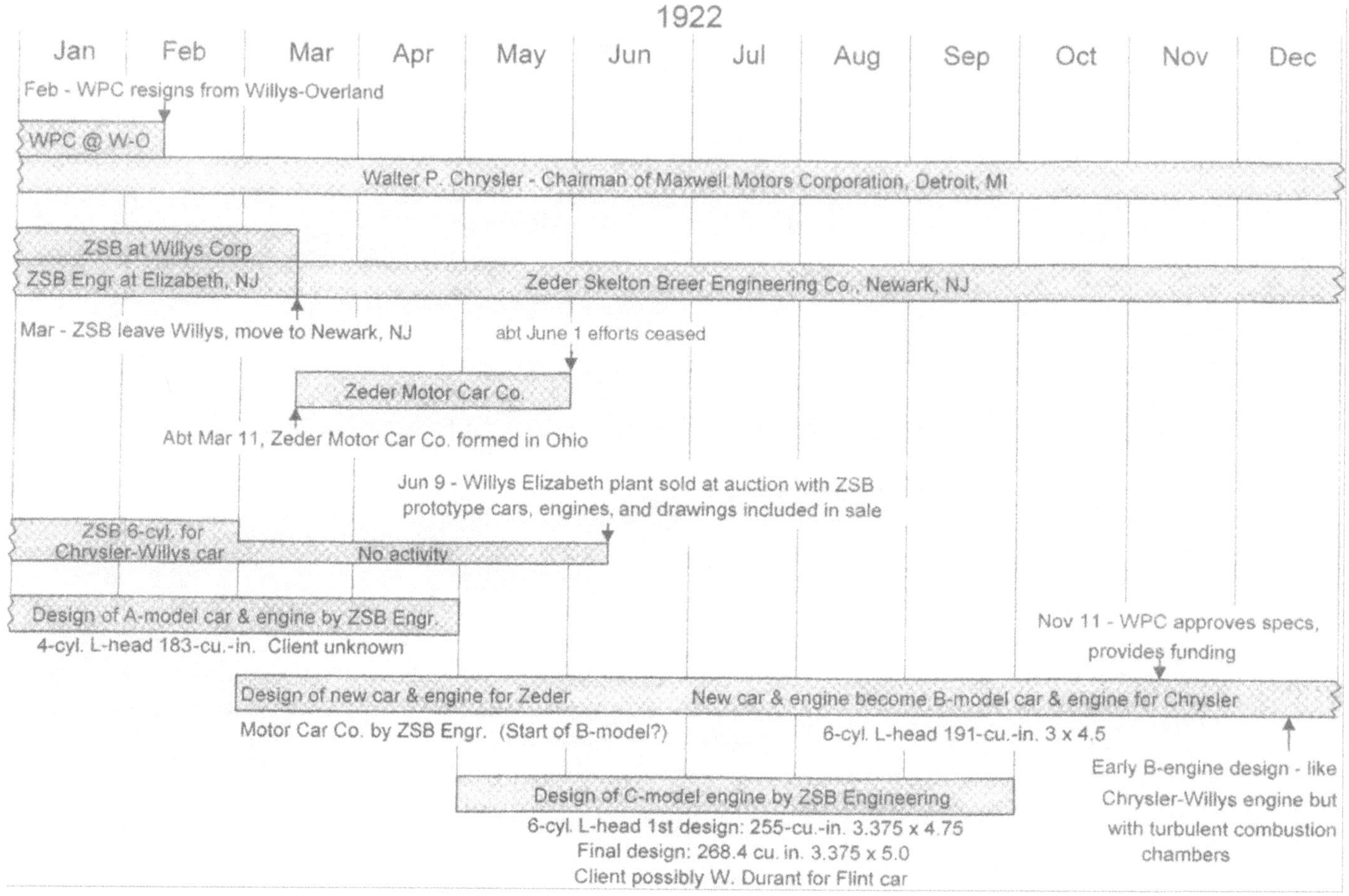

Chart 2.1
(Continued)

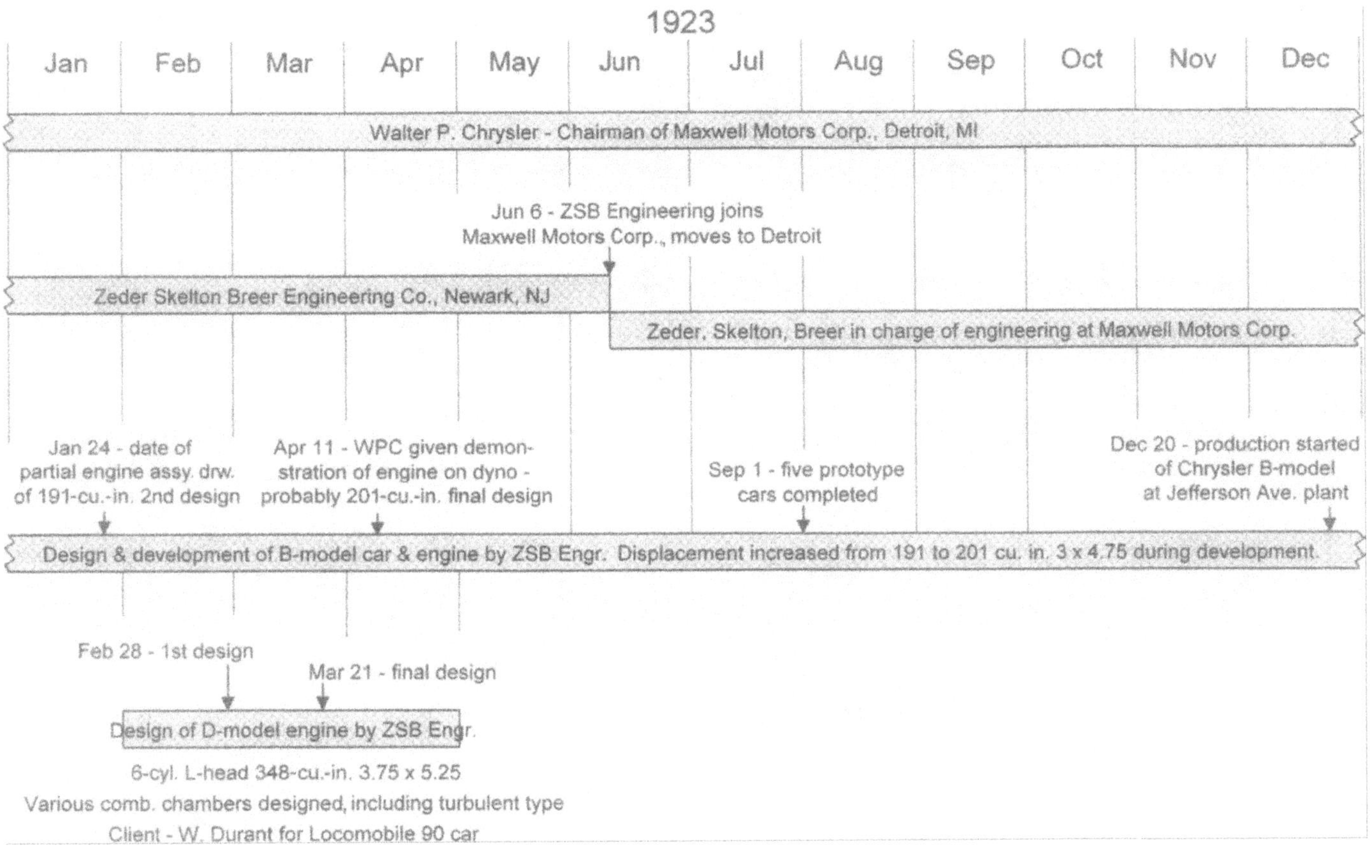

Chart 2.1
(Continued)

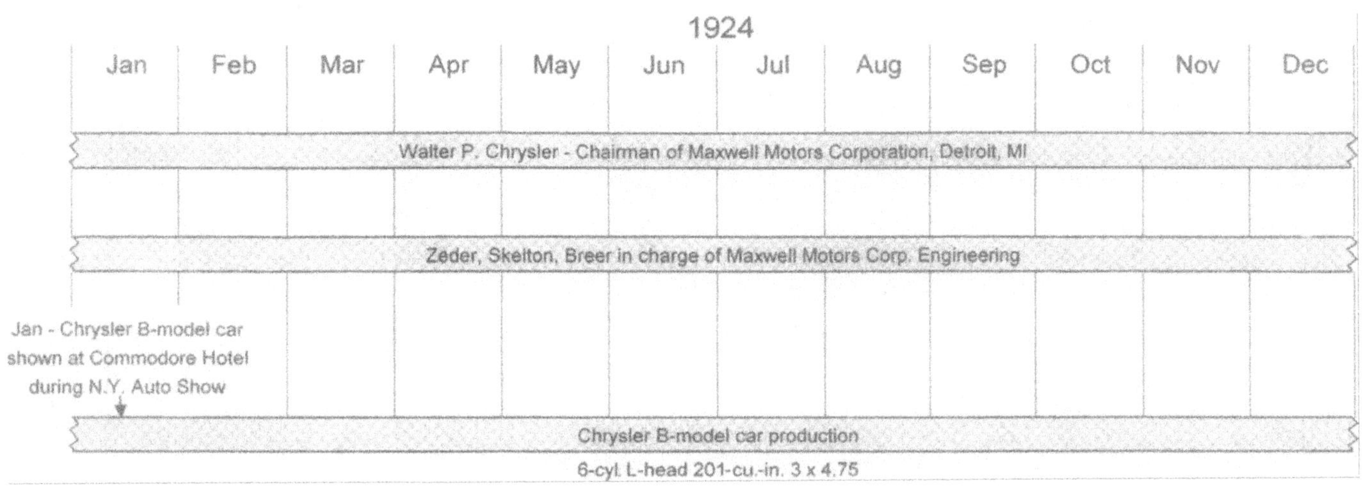

Chart 2.1 *(Cont.)*

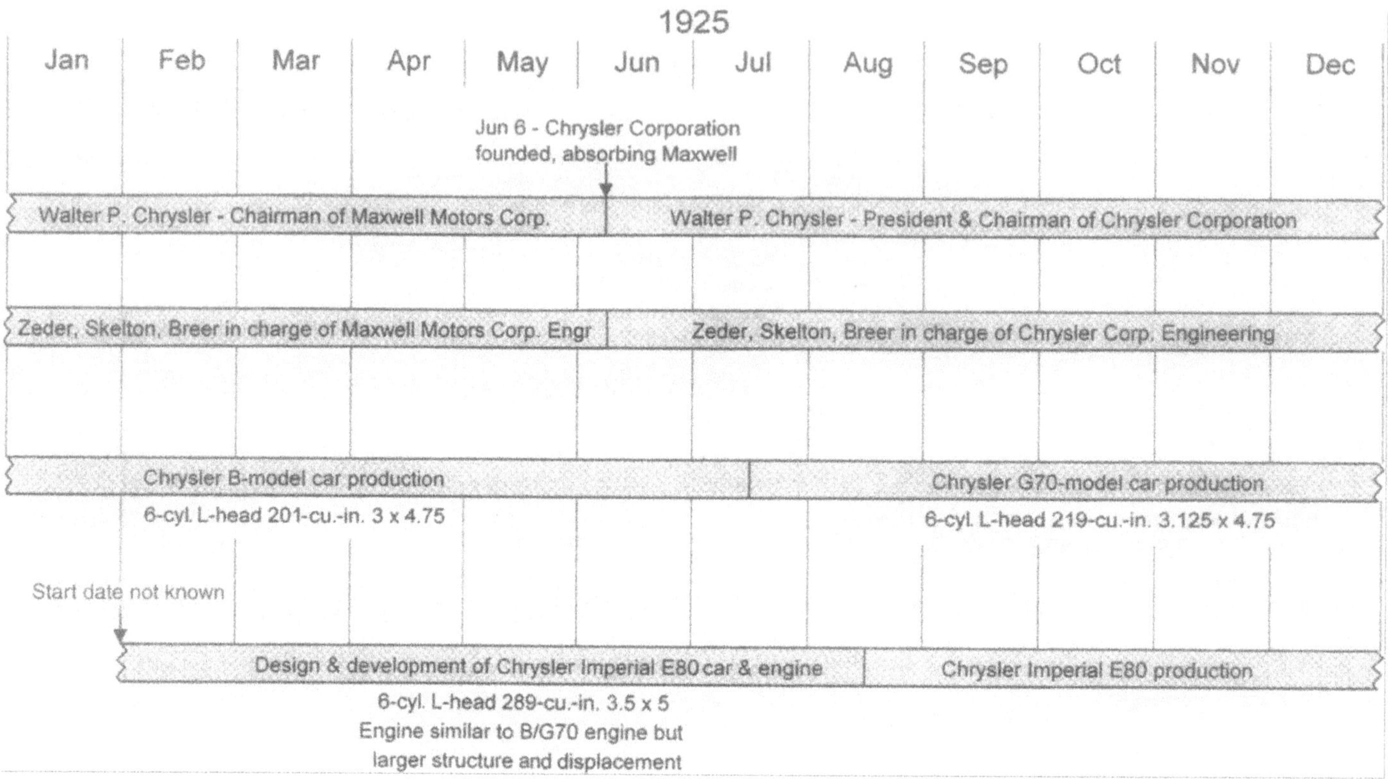

CHAPTER 3

Zeder, Skelton, and Breer's Three Engineering Masterpieces—Engines B, E, and H

More About the ZSB B-Model Engine for Chrysler

The November 11, 1922 meeting in New York City, mentioned in Chapter 2, between Walter P. Chrysler, Fred Zeder, Owen Skelton, and Carl Breer, which resulted in the launch of the B-model car and engine, can be regarded as a pivotal event leading to the formation of Chrysler Corporation two and a half years later. With the go-ahead and encouragement of Walter Chrysler, design work at Zeder, Skelton, and Breer (ZSB) Engineering Company at 24–26 Mechanic Street in Newark, New Jersey started with fresh enthusiasm. In the intervening months since they completed the design and development of the six-cylinder engine for the first Chrysler-Willys car and the follow-on C-model engine, probably for William C. "Billy" Durant, many new technical improvements had arrived that now were ready for use. One improvement was a timing chain instead of gears for driving the camshaft. Another was the use of aluminum for pistons, replacing cast iron. One of the most significant improvements, already mentioned, was the reshaping of the combustion chamber to promote turbulence of the fuel/air mixture before ignition. In the early to mid-1920s, development work on combustion chambers could be described as a technical free-for-all. Among the designs was one championed by Ricardo, cited in Chapter 2, which had a very small gap between the top of the piston and the bottom of the cylinder head at top dead center and positioned the spark plug in the center of the chamber volume above the valves. By 1923, Herschell-Spillman, Waukesha, Lycoming, and Continental used this design with minor variations.[1] Another similar chamber was the Tyler chamber of 1921, which also had a small gap and a chamber volume shaped like a hemisphere.[2] The theory was that when the piston came to the top of its stroke, the fuel/air mixture in the gap between the piston and the cylinder head would be squeezed out into the main chamber, causing great turbulence and assisting combustion to the extent that higher compression ratios could be tolerated and the engine could run at higher speeds, with the resultant increase in power and improvement in fuel economy.

During the formative stage of the B-engine, according to Carl Breer, ZSB Engineering "did a lot of development work on cylinder head [i.e., combustion chamber] design."[3] The first numbered B-engine layout no. BCL-1, drawn by Phil Kent probably in December 1922, was for a turbulent combustion chamber with the chamber roof above the piston pulled down to within 0.16 in. of the machined face of the cylinder head. Over the next several months of 1923, many variations of this type of turbulent chamber were drawn, indicating a vigorous dynamometer investigation by ZSB.

Although these layouts of combustion chambers show that ZSB had abandoned the old open combustion chamber in favor of a turbulent type of chamber, other early BCL-numbered layouts, made in December 1922, show that the general design arrangement of the initial B-engine was a virtual copy of the Chrysler-Willys engine, including the exact same graduated cylinder bore spacing of 4-4.25-4.5 from the end to the center cylinders, and the placement of the distributor on the gear case cover with its vertical shaft gear-driven from the camshaft. However, after about a month with only a few layouts made, a different design arrangement appeared, resulting in a much shorter engine length. This new arrangement located the oil pump on the right side of the cylinder block and attached the distributor, driven by a diagonal cross shaft, on the left side of the cylinder block. This design, together with other basic changes, transformed the

B-engine into a totally different engine, shown in the views of layout BCL-51, dated January 24, 1923, in Figs. 3.1 through 3.4. This second-design B-engine, with a displacement of 191 cu. in. and a bore and stroke of 3.00 × 4.5 in., was quite a bit smaller than the 224 cu. in. of the Chrysler-Willys 3.25 × 4.5 in. engine both in displacement and structural size. The front view of the engine is displayed in Fig. 3.1, a longitudinal section is shown in Fig. 3.2, a cross section through the #1 cylinder is shown in Fig. 3.3, and a cross section through the center of the engine is displayed in Fig. 3.4. Unlike the graduated cylinder spacing of the first design, the cylinders of the engine were spaced 3.875 in. apart except for the center two, which were 4.5 in. apart. A chain drive for the camshaft and a high-mounted water pump were part of the design, as were the characteristic ZSB seven-main-bearing crankshaft and full-pressure oiling to all major bearings. The layout obviously is incomplete because the cylinder head was not drawn.

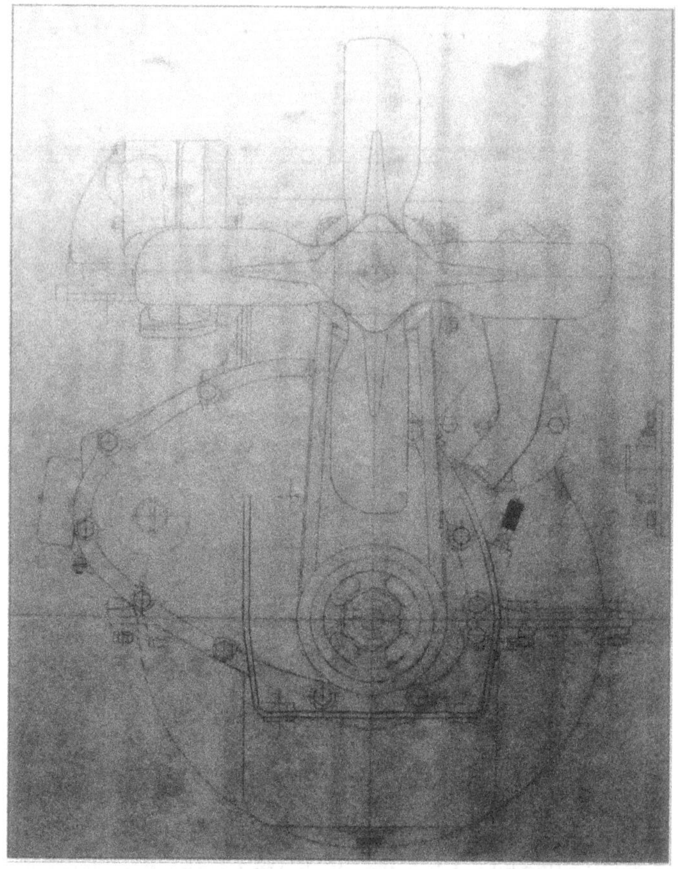

Fig. 3.1 *A front view of the second-design B-model 191-cu.-in. L-head six-cylinder engine. From Zeder, Skelton, and Breer Engineering Co. layout BCL-51, dated January 24, 1923.*

Approximately three months later, a third and what turned out to be the final design superseded this second design. The final design for the B-engine increased its displacement from the prior 191 cu. in. to 201.45 cu. in. by a stroke increase to 4.75 in. As part of the redesign, the cylinder spacing was increased to 4.0625 in. from 3.875 in., with the space between the center cylinders left unchanged at 4.5 in. In keeping with ZSB design practice, the camshaft, valves, and intake and exhaust manifolds were placed on the right side of the engine.

Similar to the second design, the camshaft was driven by a timing chain, visible in the front view of the engine shown in Fig. 3.5. Writers at the time often used the adjective "silent" in describing the chain because its multiple side-by-side link construction was quieter than gears or roller chains. These early chains did need provisions for adjustment because wear and stretch could lead to excessive slack. ZSB used the chain to drive the generator in addition to the camshaft and then designed a mounting for the generator so it could be moved to take up slack in the chain. This could be done with the engine running. The owner's manual of the original 1924 Chrysler advised the owner to tighten the chain tension adjusting screw until the chain began to hum, then "When the chain begins to hum, slowly release the adjusting screw until the noise is stopped, then securely tighten the cap screws and adjusting screw lock nut. Be careful not to leave the chain too tight or it will overload and burn out the generator bearings."

The crankshaft design of the new B-engine, shown in the longitudinal section in Fig. 3.6, was a ZSB hallmark: seven main bearings with full-pressure lubrication to all main bearings, connecting rod big-end bearings, and camshaft bearings. Also shown in Fig. 3.6 is the high-mounted water pump, derived from earlier designs previously described, which shared its shaft with the cooling fan. Driving the shaft was a V-belt that easily transmitted the power required for both the fan and the water pump. However, similar to the prior flat belts, these new V-belts needed adjustment for stretch. Taking advantage of the belt shape, the water pump/fan pulley was designed with a movable front flange that had a threaded hub. By rotating the flange, the pulley groove became narrower or wider. When the flange was adjusted to give the correct belt tension (described in the owner's manual as "snug, not tight"), a lock screw was tightened to prevent further movement. This also was a carry-over from the second-design B-engine.

The cross section through the #1 cylinder shown in Fig. 3.7 displays the crowning glory of the Chrysler B-model engine—its combustion chamber with a turbulence-inducing quench gap above the piston and a dome-shaped volume below the spark plug. This chamber allowed a relatively high 4.6:1 compression ratio to be used and moved the revolutions per minute at which its maximum 68 hp was produced up to 3200 rpm, a speed level higher than all other contemporary passenger car engines except the Duesenberg overhead camshaft engine.[4] In a departure from the second design, along the upper right side of the cylinder block were two tiers of ports rather than one, a design that also appeared on the D-engine. Six exhaust ports, one for each exhaust valve, composed the top tier. This arrangement provided plenty of cooling water around each exhaust valve seat—an excellent durability feature that can be seen partially on

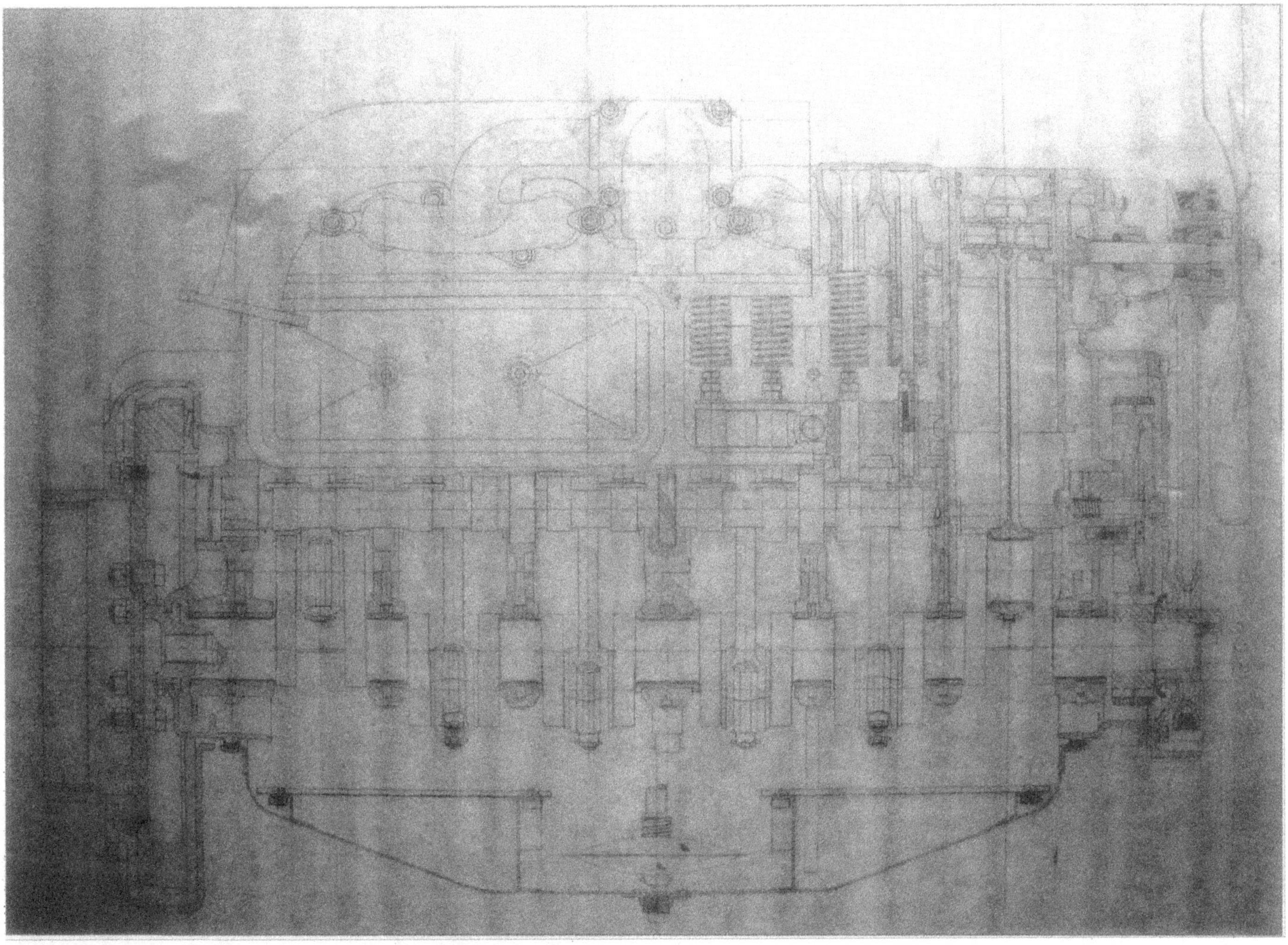

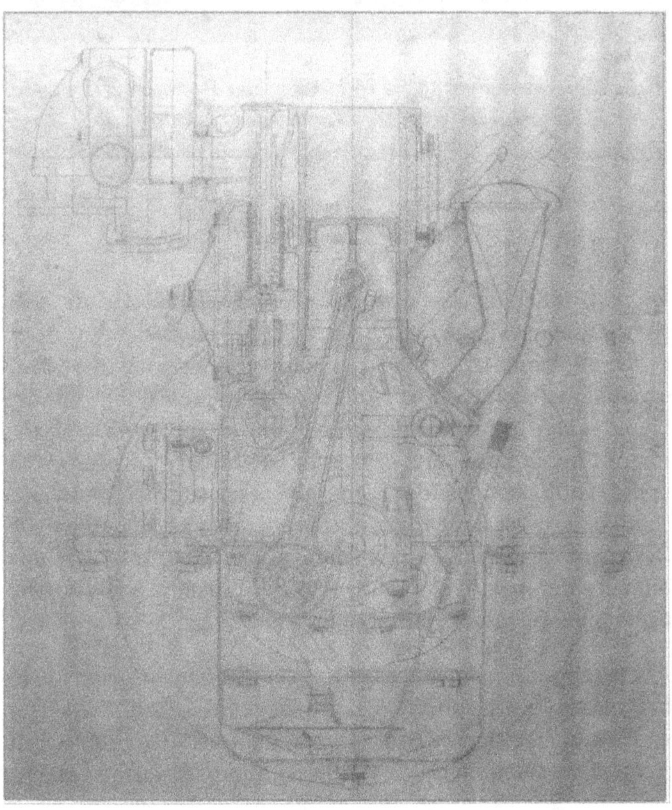

Fig. 3.2 *(Above) A longitudinal section of the second-design B-model L-head six-cylinder engine. A seven-main-bearing crankshaft, a chain drive for the camshaft, and a high-mounted water pump are included in the design. From Zeder, Skelton, and Breer Engineering Co. layout BCL-51, dated January 24, 1923.*

Fig. 3.3 *(Left) A cross section through the #2 cylinder of the second-design B-model L-head six-cylinder engine. The shallow skirt of the cylinder block and the aluminum piston are evident. The cylinder head has not yet been drawn. From Zeder, Skelton, and Breer Engineering Co. layout BCL-51, dated January 24, 1923.*

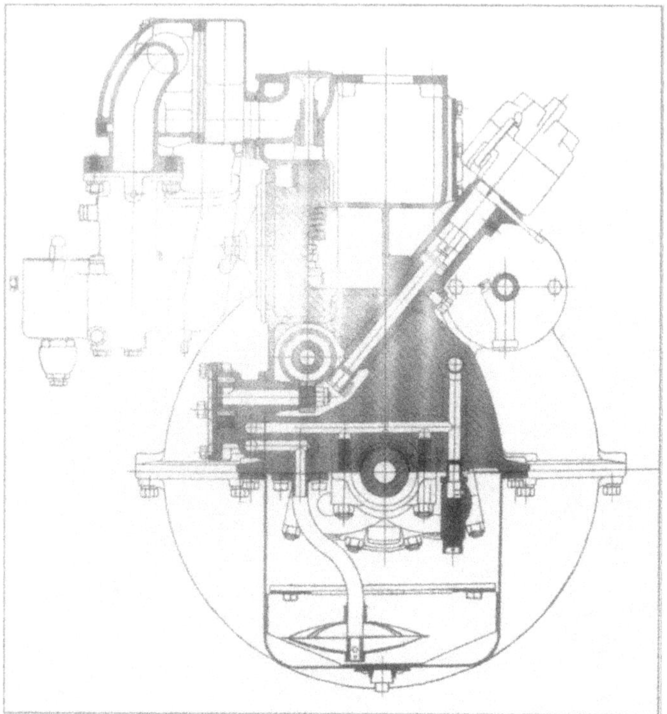

Fig. 3.4 *A cross section through the center of the second-design B-model L-head six-cylinder engine. The oil pump and distributor locations and drives are evident. From Zeder, Skelton, and Breer Engineering Co. layout BCL-51, dated January 24, 1923. Layout line clarity computer enhanced by W.L. Weertman.*

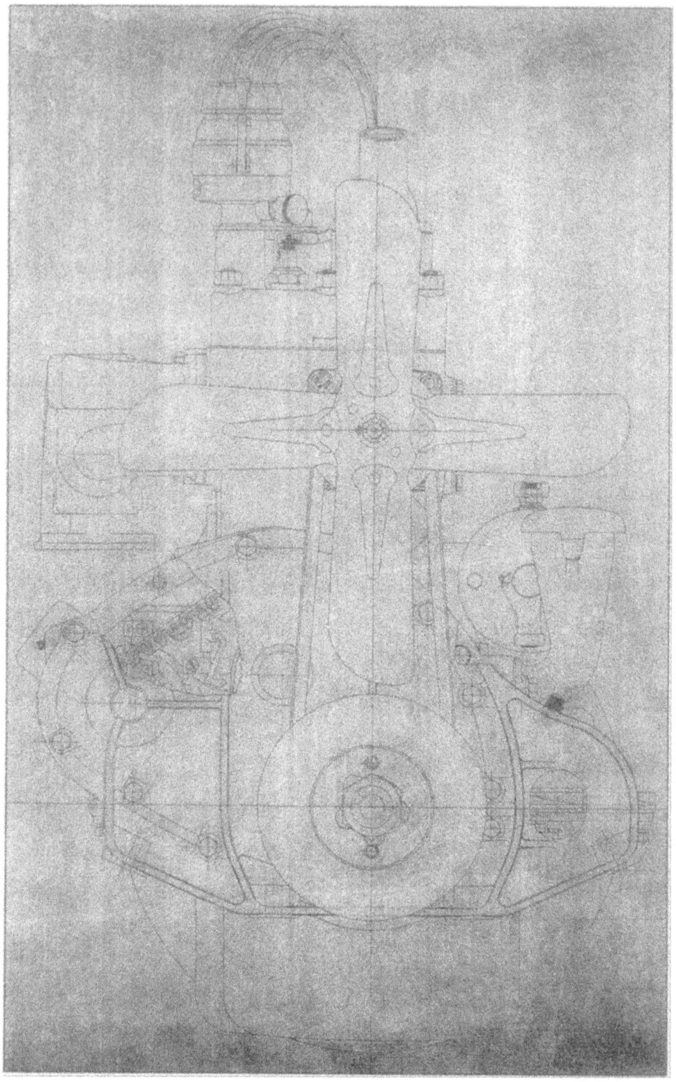

Fig. 3.5 *A front view of the 1924 Chrysler B-model 201-cu.-in. L-head six-cylinder engine. From Zeder, Skelton, and Breer Engineering Co. "Motor Assembly" drawing #50100.*

Fig. 3.7. The lower tier had only two ports that were internally divided to feed three intake valves apiece.

The cross section through the center of the engine, displayed on Fig. 3.8, shows the results of engine development work done by Carl Breer in obtaining equal fuel/air mixtures to all cylinders. The intake manifold riser is surrounded by an exhaust gas jacket (all part of the same intake and exhaust manifold casting), so that the intake manifold walls were heated quickly to vaporize any liquid fuel that could upset the distribution of fuel to the cylinders. The oil pump design approach of the Chrysler-Willys engine was applied to this new B-engine, resulting in the arrangement also shown on Fig. 3.8. Bolted to the bottom of the cylinder block and positioned inside the oil pan at the center of the right side of the engine, the vertical shaft of the pump was gear-driven from the camshaft. This location in the sump of the oil pan was excellent for the oil pump; being immersed in engine oil, it would not lose prime, and the possibility of external oil leaks was minimized. The part of the pump system that sometimes needed servicing—the pressure relief valve—was located on the left side of the crankcase where it was accessible from outside the engine. A new mounting location was found for the ignition distributor. Instead of mounting the distributor at the front of the engine as had been done on most of their earlier engines, or on the left side of the block as with the second B-engine design, ZSB gave it a home similar to the one on the D-engine at the center of the engine on top of the cylinder head. Its long driveshaft extended all the way down to the camshaft area of the cylinder block, where a tongue-and-groove joint connected it with the top of the oil pump and distributor driveshaft. In addition to being easily serviced, better protection for the distributor and spark plug wires from road splash and airborne water from the fan while driving in the rain could be expected from this new higher, drier location. Figures 3.9 and 3.10 show views of the right and left sides of the engine, respectively.

At the suggestion of Walter P. Chrysler, ZSB investigated filters being used in the mining industry for refining gold. Working with a supplier, they came up with a dash-mounted filter with a replaceable element that continuously removed particles from a

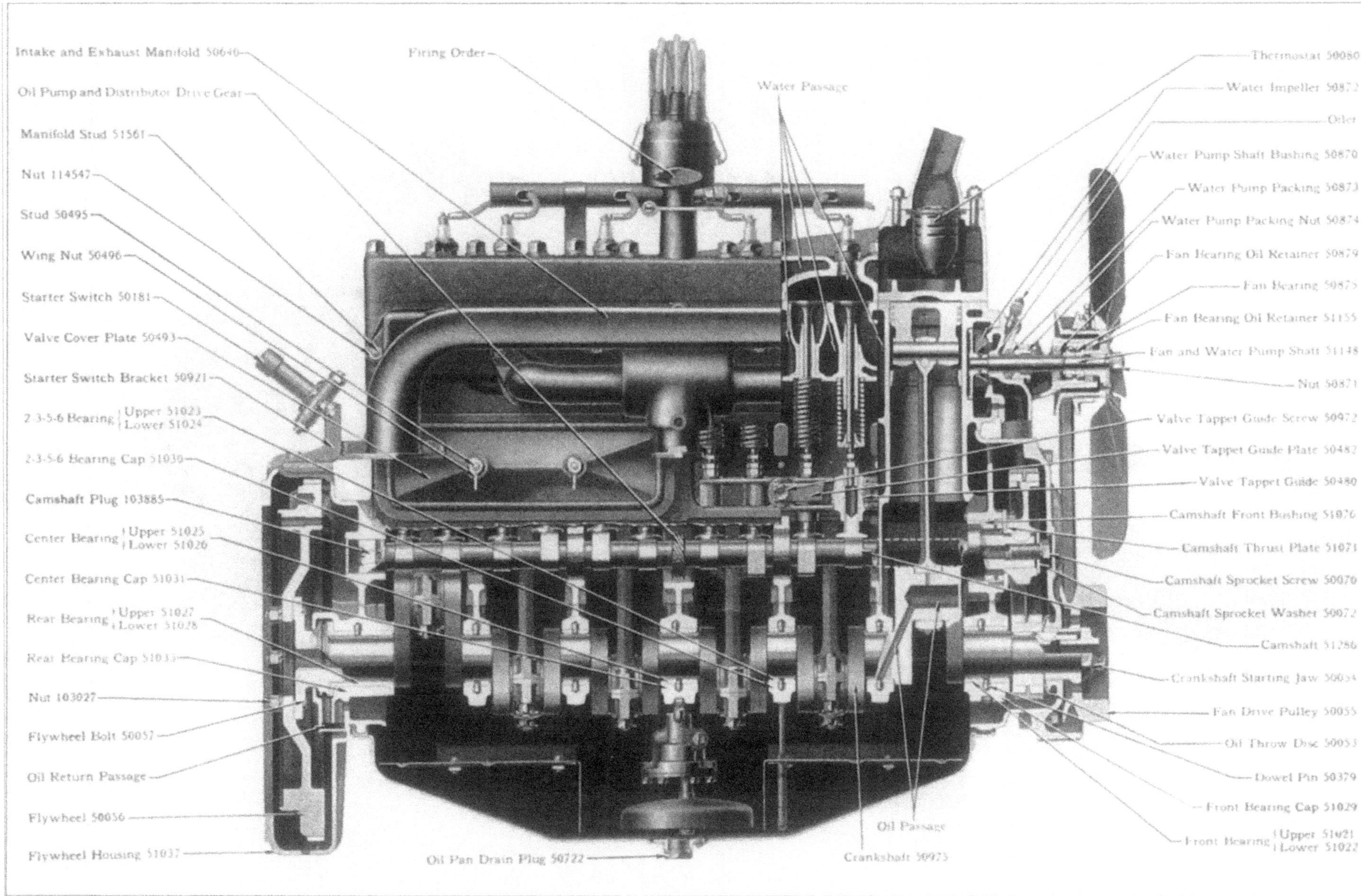

Fig. 3.6 *A longitudinal section of the 1924 Chrysler B-model 201-cu.-in. six-cylinder engine.*

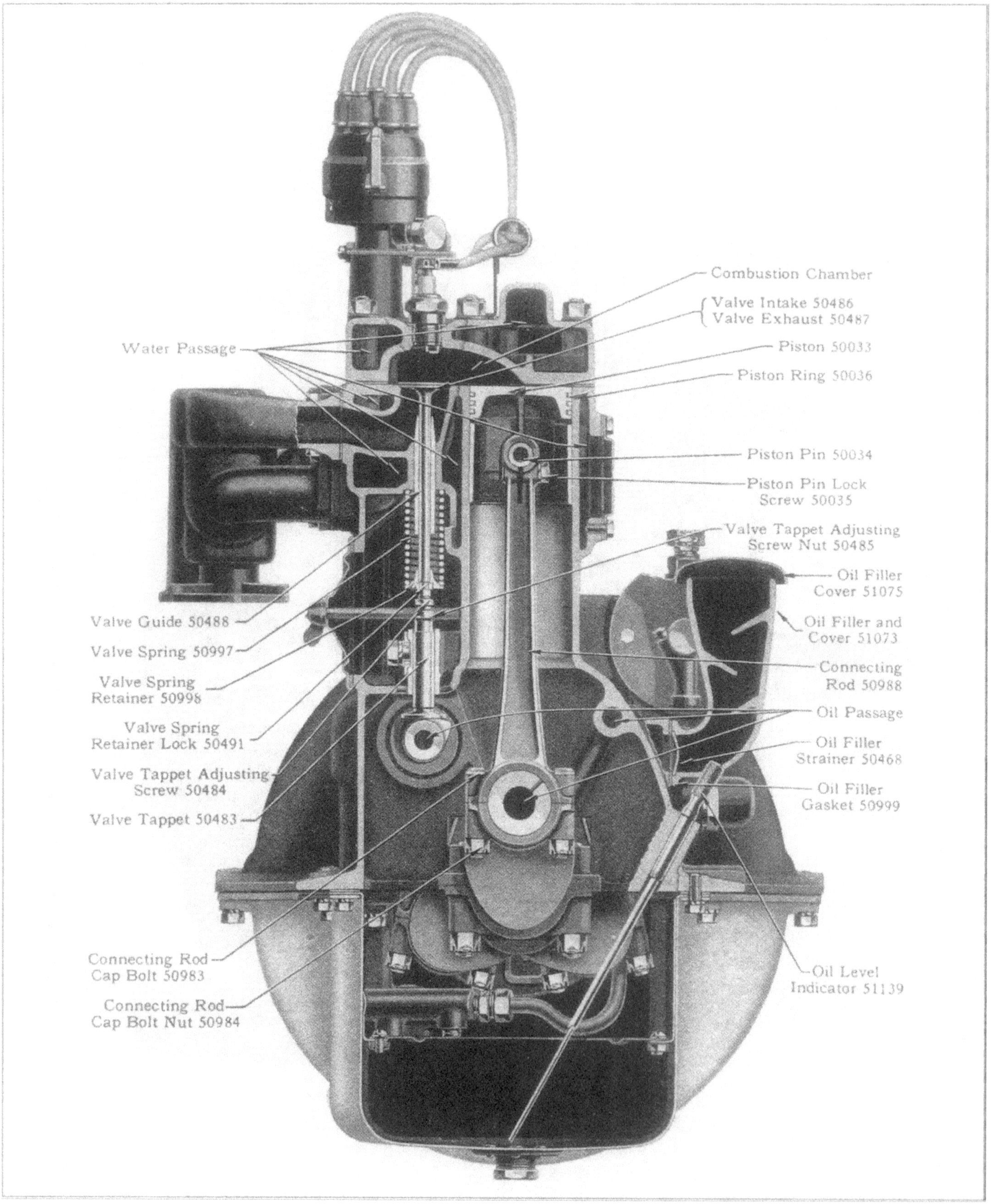

Fig. 3.7 *A cross section through the #1 cylinder and the #1 exhaust port of the 1924 Chrysler B-model 201-cu.-in. six-cylinder engine.*

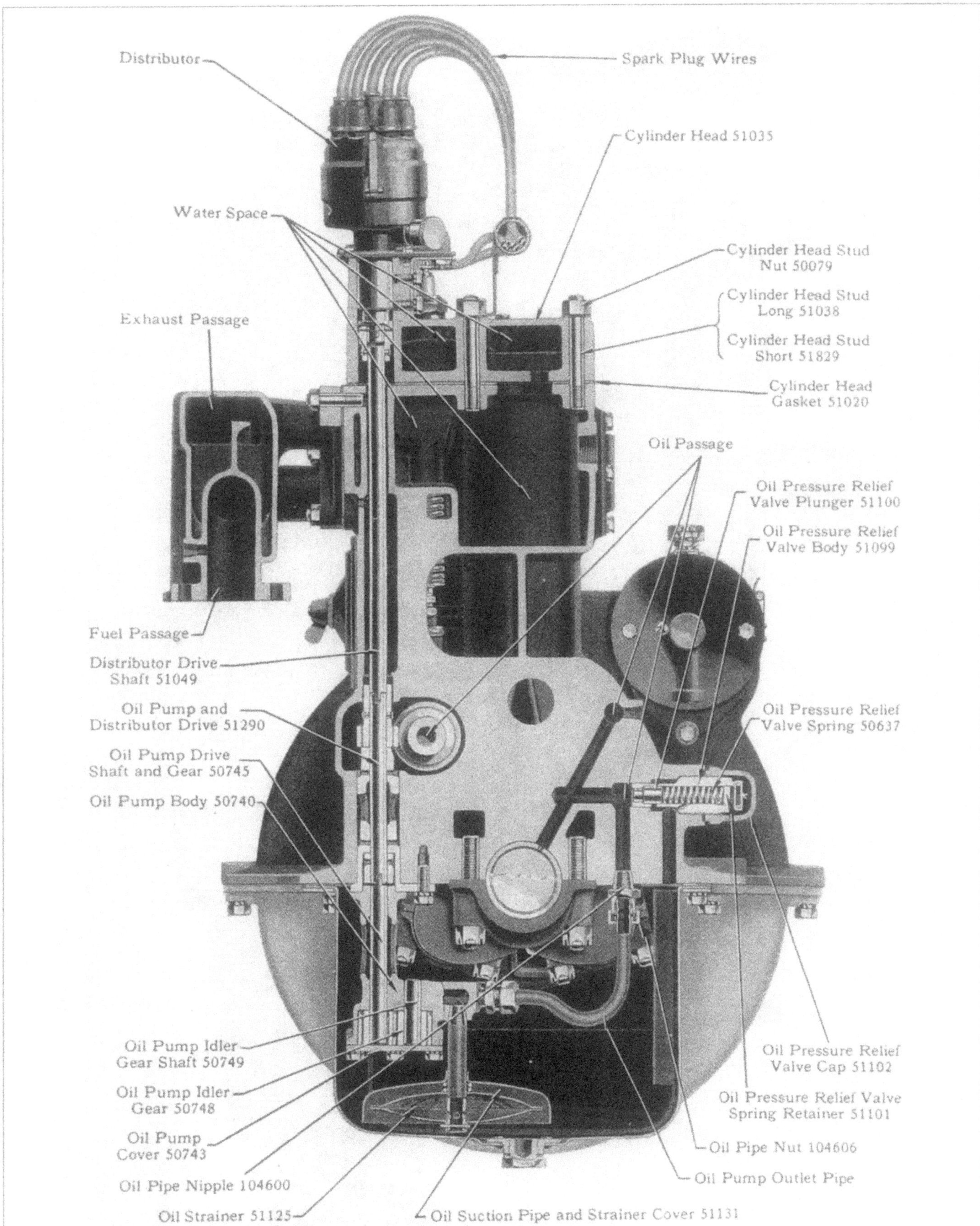

Fig. 3.8 *A cross section through the center of the 1924 Chrysler B-model six-cylinder engine, showing the exhaust heat jacket for the intake passage and the vertical drive for the oil pump and the distributor.*

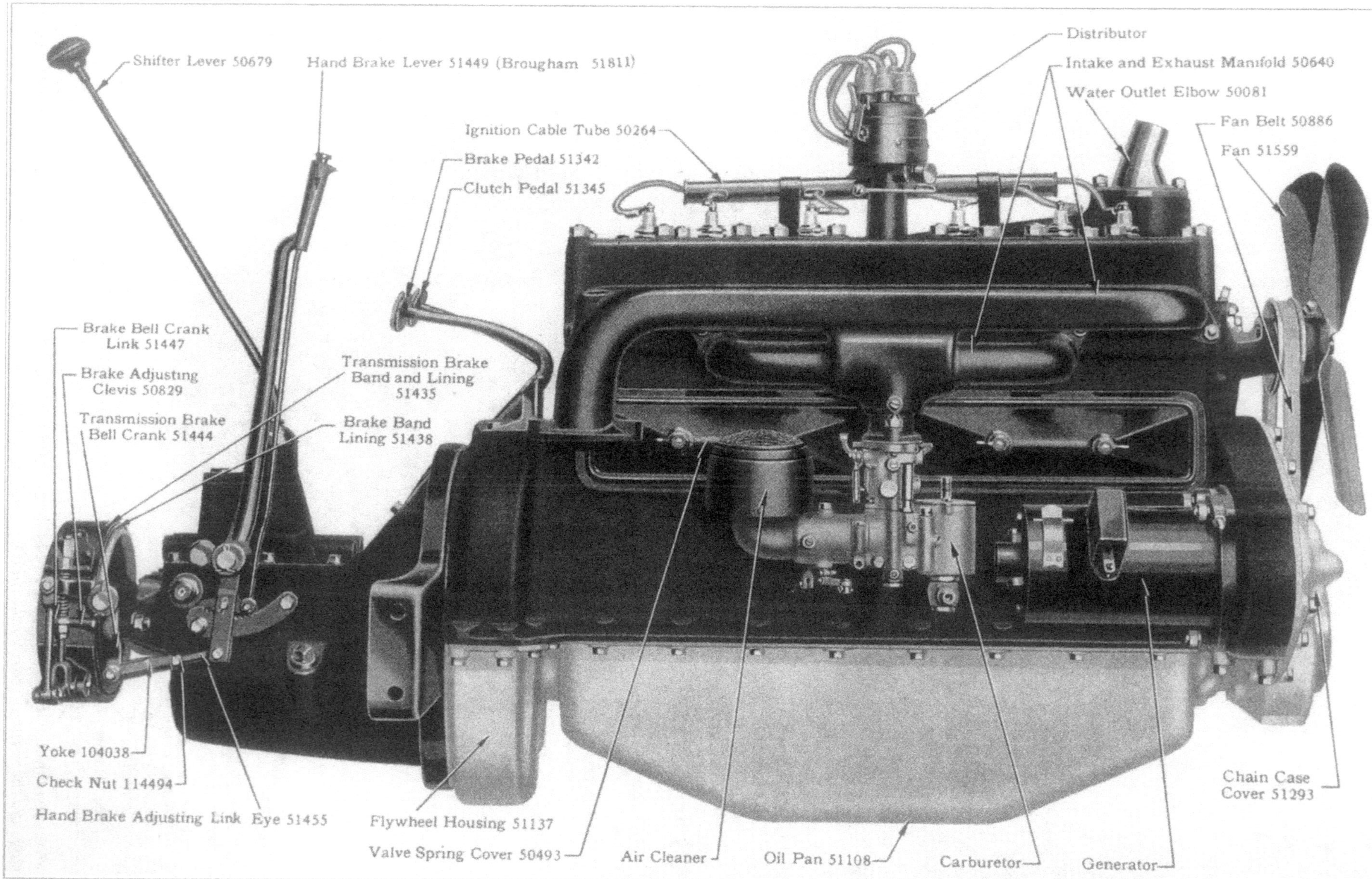

Fig. 3.9 *The right side of the 1924 Chrysler B-model engine assembly. To reduce weight, cast aluminum was used for the flywheel housing.*

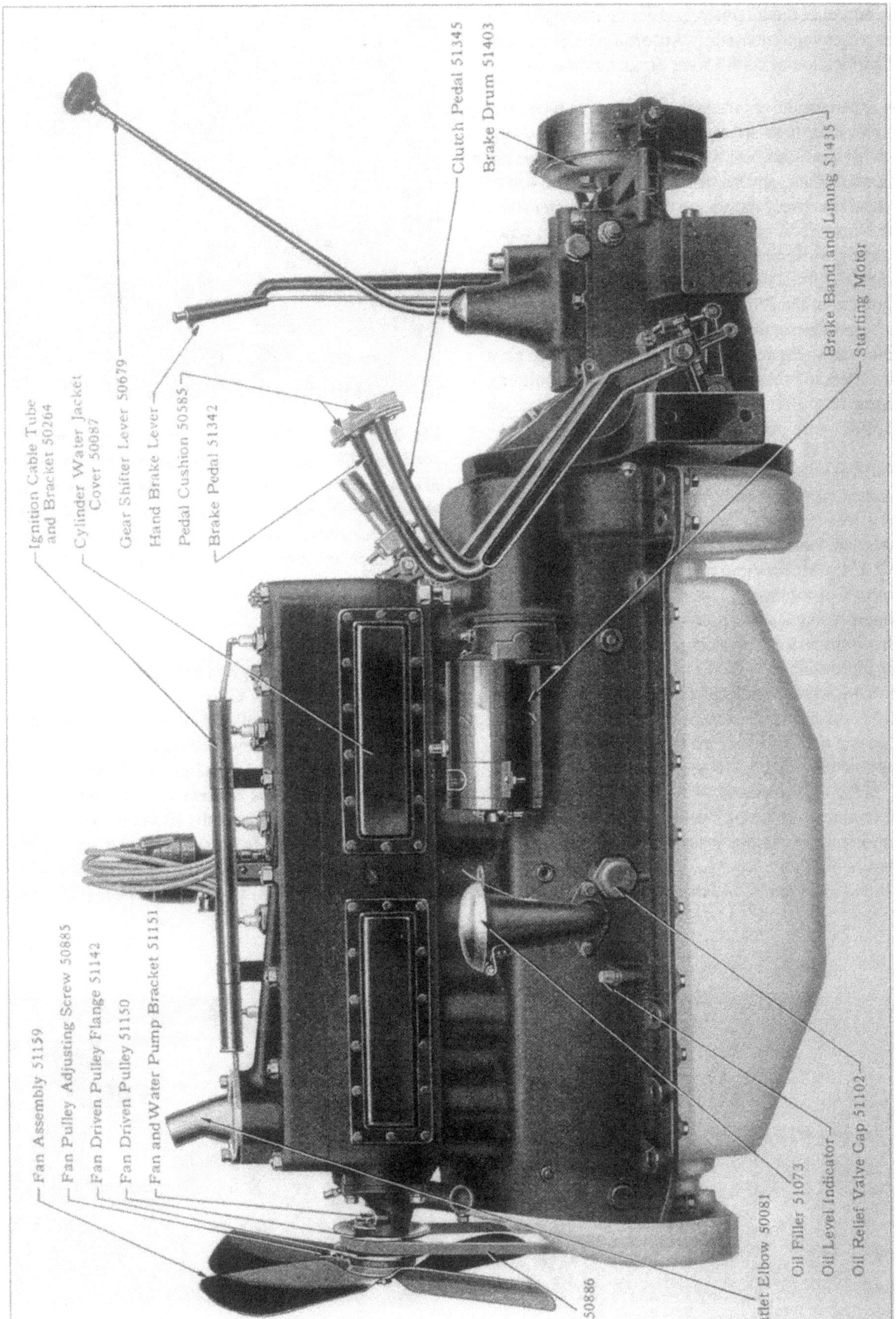

Fig. 3.10 *The left side of the 1924 Chrysler B-model engine assembly. The location of the distributor high on top of the cylinder head made it easily serviced.*

portion of the output of the oil pump, protecting the engine and extending the oil-change intervals.[5] According to Breer, this was the first application of such a filter on an automobile.[3]

A 201-cu.-in. experimental engine in this final design form with its high-compression, high-speed, and turbulent combustion chambers, its high-pressure oil to the bearings of its seven-main-bearing crankshaft, and its distributor located on top of the cylinder head was likely the one on the dynamometer stand at the Elizabeth plant facilities that Walter Chrysler was invited to see on a Saturday in April 1923.[3] Carl Breer, visiting friends and relatives in California, was not there, but Fred Zeder, Owen Skelton, and others of the ZSB Engineering staff were present to show off the smooth-running, powerful, and compact engine. Walter Chrysler was so impressed with what he saw that he gave the engineers orders to proceed immediately with the building of five prototype cars—all to be finished and ready to run by September 1, 1923, only four and a half months away![3]

I speculate that this successful demonstration of the engine on the dynamometer may have been the trigger for what Zeder, Skelton, and Breer had been hoping ever since the Willys bankruptcy turned their world upside down—an invitation from Walter P. Chrysler for them and their staff to join him at Maxwell Motors Corporation in Detroit. On June 6, 1923, the entire ZSB Engineering Company arrived in Detroit and set up shop in available space at a Chalmers production plant (owned at the time by Maxwell) on Jefferson Avenue.[6] As promised, the prototype cars were completed on schedule. One of the cars still exists, on proud display at the Walter P. Chrysler Museum, as shown on Fig. 3.11. The car's engine, pictured on Fig. 3.12, has a single number "6" steel-stenciled in the chain case housing of the cylinder block. An extensive test program for cars and engines, including 50-hour wide-open-throttle engine dynamometer durability tests, rolled up miles of car testing and hours of engine testing.[3] Meanwhile, production tooling was being installed in the Jefferson Avenue plant. By

Fig. 3.12 *The right side of the engine compartment of the 1924 prototype Chrysler B-model phaeton at the Walter P. Chrysler Museum. The combined intake and exhaust manifolds present a clean appearance. A louvered air filter for the updraft carburetor and a cylindrical vacuum fuel supply tank mounted to the dash are visible. Photo by W.L. Weertman.*

December 20, 1923, the first production cars came off the end of the line.[7] The new Chrysler car was made public at the January 1924 Automobile Show in New York and became an instant sales success. Seventeen months later, on June 6, 1925, Walter Chrysler, shown standing beside the car on Fig. 3.13, with the backing of his bankers and the approval of the Maxwell Motors Co. stockholders, incorporated a new company, Chrysler Corporation, in the state of Delaware, which absorbed all the assets of Maxwell Motors. A photo taken several years later, displayed on Fig. 3.14, shows Walter P. Chrysler with his

Fig. 3.11 *A 1924 prototype Chrysler B-model phaeton on display at the Walter P. Chrysler Museum. Photo by W.L. Weertman.*

Fig. 3.13 *Walter P. Chrysler stands with his namesake 1924 Chrysler B-model car.*

Fig. 3.14 *(From left to right) Walter P. Chrysler, Carl Breer, Fred Zeder, and Owen Skelton pose with a 1928 Chrysler model J72 car. For this model year, the displacement of the original engine was increased to 249 cu. in., and the compression ratio was increased to 5.1:1, giving the car 75 hp.*

three ace engineers, Fred Zeder, Owen Skelton, and Carl Breer, all posing with a 1928 Chrysler sedan.

Subsequent Changes

With its late 1924 arrival, the engine was not changed for 1925. For 1926, ZSB made a bore change from 3.0 to 3.125 in. to raise the displacement to 218.59 cu. in. At the same time, a torsional vibration damper, adapted by ZSB from the British Lancaster dampener concept, was added to the nose of the crankshaft to reduce audible and possibly even destructive torsional vibrations. Concurrent with these changes, both the main bearings and the connecting rod big-end bearings were increased in diameter to 2.25 in. and 2.0 in., respectively. The 1926 car and engine were coded G70, with the "70" implying its top speed.

In 1928, with another displacement increase to 248.9 cu. in. obtained by increasing both the bore and stroke to 3.25 × 5.0 in., bolted-on counterweights were added to the crankshaft to reduce the main bearing loads. For cold-starting improvement, a driver-controlled electrical heating element was added to the updraft carburetor to heat and vaporize liquid fuel that had accumulated.

The year 1928 also saw an increase in compression ratio from 4.8:1 to 5.1:1 for the standard engine, which, together with the displacement change, boosted the power to 75 hp. That same year was the start of the practice of offering an optional higher-compression cylinder head to take advantage of the premium higher-octane leaded "ethyl" gasoline becoming available. The standard compression cylinder head was called "Silver Dome," reflecting the shape of the combustion chamber above the valves. The high-compression cylinder head was painted red enamel and was advertised as the "Red Head." The 1928 J72 Red Head engine had a compression ratio of 6.2:1 and produced 85 hp.

For 1930, the displacement was further increased to a final 268.4 cu. in. by a bore change to 3.375 in. A number of other changes were made, including the use of a non-adjustable timing chain that drove only the camshaft. The generator was relocated to the left side of the engine and now was driven by the fan belt. Importantly, a lever-operated mechanical fuel pump was added to the right side of the cylinder block. With the carburetor no longer dependent on a gravity fuel feed from a vacuum tank in the engine compartment, the low-mounted updraft carburetor, which had been used for years, was replaced with a high-mounted downdraft carburetor—another "first" for Chrysler Corporation. The right side and left side exterior views are shown on Figs. 3.15 and 3.16, respectively. The longitudinal and cross-section views are shown on Figs. 3.17 and 3.18, respectively. These figures illustrate the 1930 version of the engine. The standard "Silver Dome" produced 93 hp @ 2000 rpm, topped by the optional high-compression "Red Head" that had a rating of 102 hp @ 3200 rpm.

This innovative, well-proportioned, and expertly designed engine—embodying the latest in engineering developments, which, upon its introduction in 1924 as the Chrysler B-model production engine, paved the way for the success of Chrysler Corporation—was discontinued at the end of the 1930 model year.

Table 3.1 shows the specifications for this engine.

Larger ZSB Seven-Main-Bearing L-Head Six-Cylinder Engine—The Chrysler Imperial E-Model Engine

Buoyed by the smashing sales success of the 1924 Chrysler B-model six, a new car line soon was added to broaden the corporate vehicle lineup. The new car, designed to be larger, more powerful, and more luxurious than the 1926 standard Chrysler G70 model, was introduced in 1926 as the "Chrysler Imperial." The Imperial name had been used in 1925 for the high-end Chryslers, but now it would be a separate model. This 1926 Chrysler Imperial car and engine was given an E-model code by ZSB Engineering. With "80" added, indicating capability speed of the vehicle, its model code became E80. As illustrated on Fig. 3.19, the car had the look of luxury and power.

A reasonably accurate description of the new engine designed and developed by ZSB for this new Chrysler Imperial car would be to describe it as an enlarged version of their original Chrysler six-cylinder engine. Figures 3.20 and 3.21 illustrate right and left outside views of the engine, respectively. Longitudinal and cross-section views appear on Figs. 3.22 and 3.23,

Fig. 3.15 *The right side of the 1930 Chrysler Model 77 six-cylinder engine. Compared to the 1924 version shown on Fig. 3.4, the generator has been moved to the left side of the engine, and a mechanical fuel pump has been added, which paved the way for the downdraft carburetor shown here.*

Fig. 3.16 *The left side of the 1930 Chrysler Model 77 six-cylinder engine. With the generator no longer used for timing chain adjustment, it could now be V-belt driven.*

Fig. 3.17 *A longitudinal section of the 1930 Chrysler Model 77 six-cylinder engine. The bore and stroke were 3.375 × 5.0 in., giving the engine a displacement of 268 cu. in. A multipiece torsional vibration damper is attached to the nose of the crankshaft.*

respectively. With a bore and stroke of 3.5 × 5.0 in., the engine had a displacement of 288.6 cu. in.—a hefty increase over the 201.45 cu. in. of the original B-engine. This required a new cylinder block with cylinder centers placed 4.5 in. apart, except for the center two, which were 5.1875 in. apart, and with a deck height of 15.505 in., almost an inch higher than the smaller engine. The overall design approach was the same: an L-head valve arrangement, shallow-skirt cylinder block, seven-main-bearing crankshaft, full-pressure lubrication, vertical oil pump and distributor drive, two-tier arrangement for intake and exhaust ports, water pump mounted high in front of the engine, a single-piece intake and exhaust manifold, and an updraft carburetor. The crankshaft had bolted-on counterweights to reduce the main bearing loads and was equipped with a vibration damper to control torsional vibration. The carburetor had an electric heater for vaporizing liquid fuel for improved cold starting.

A nice upscale touch was the use of cast aluminum in place of stamped steel for the valve spring covers on the right side of the cylinder block. Every time the owner or mechanic made a valve lash adjustment, he would be reminded that this was a quality engine. The aluminum covers also appear thick enough to help reduce valve noise. Another quality detail was the use of acorn nuts for the cylinder head studs instead of plain nuts. On the left side of the engine, the two cylinder block water jacket covers were noticeably farther apart than the covers used on the smaller G70 engine. Other than these features, the two engines appeared very similar.

For its 1926 launch year, the engine produced 92 hp @ 3000 rpm and 192 lb-ft torque @ 1000 rpm. With this engine under the hood, the 1926 E80 Chrysler Imperial had no trouble living up to the 80 designation assigned to it. An accolade for its performance came in the form of an invitation to participate in the May 1926 Indianapolis 500 race as the pace car—the first of many Chrysler cars to be selected for this prestigious task. Figure 3.24 shows a smiling Walter P. Chrysler with the pace car. Interestingly, for its race duties, the car was driven by Louis Chevrolet.[8]

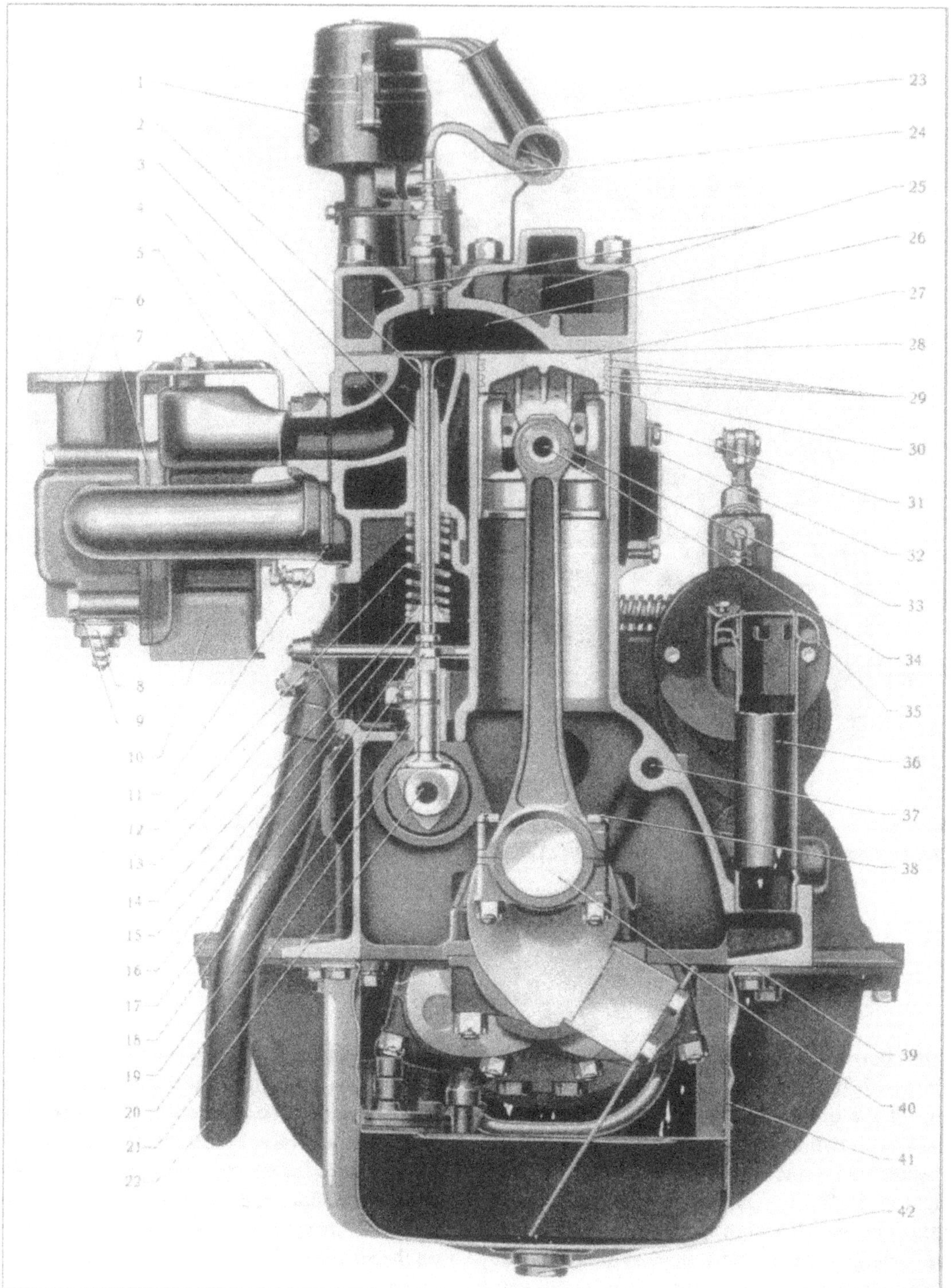

Fig 3.18 *A cross section through the #1 cylinder of the 1930 Chrysler Model 77 six-cylinder engine. Bolted-on counterweights that were added in 1928 are clearly visible here. The high-turbulence combustion chamber appears unchanged since its first appearance in 1924.*

TABLE 3.1
SPECIFICATIONS FOR THE B-MODEL ZSB L-HEAD SIX-CYLINDER ENGINE

Year	1922	1923	1924
Model	B-model—1st design	B-model—2nd design	B-model—Final design
Displacement (cu. in.)		191	201.45
Bore		3.0	3.0
Stroke		4.5	4.75
Compression ratio			4.6:1
BHP @ rpm			68 @ 3200
Torque (lb-ft) @ rpm			136 @ 1000
Combustion chamber	Turbulent	Turbulent	Turbulent
Cyl. block mat'l.—skirt type		Cast iron—shallow	Cast iron—shallow
Cyl. block deck height		14.0	14.38
Cyl. block length			32.0
Cyl. bore spacing	1-2, 5-6 = 4 2-3, 4-5 = 4.25 3-4 = 4.5	All except 3-4 = 3.875 3-4 = 4.5	All except 3-4 = 4.0625 3-4 = 4.5
Ck/shaft—no. of main brgs.	7	7	7
Ck/shaft main journal dia.		1.75	1.875
Crankpin journal dia.		1.75	1.875
Piston mat'l.—no. of rings		Aluminum—3	Aluminum—3
Connecting rod centers		9.75	10
Valve system		L-head	L-head
Valve events	220-230-2		220-230-2
Camshaft drive		Chain	Chain
Valve head dia.		Intake and exhaust 1.625	Intake and exhaust 1.59
Valve lash adjustment	Mech. tappets	Mech. tappets	Mech. tappets
Oil pump and distributor drive	Distr: vert. mount on gear case cover.	Distr. mtd. to left, ctr. of block; diag. drive from camshaft. Oil pump mtd. to right, ctr. of block, gear drive from camshaft.	Vertical at right-center of engine. Distr. on top of cyl. head. Oil pump at bottom of block. Full pressure.
Fuel system		Updraft carburetor	Updraft carburetor
Other			

(Continued on next page)

TABLE 3.1
(Continued)

Year	1926	1928	1930
Model	Chrysler G70	Chrysler J72	Chrysler W77
Displacement (cu. in.)	218.59	248.9	268.4
Bore	3.125	3.25	3.375
Stroke	4.75	5.0	5.0
Compression ratio	4.8:1	5.1:1 std.; 6.2:1 opt.	5.0:1 std.; 5.8:1 opt.
BHP@ rpm	68 @ 3000	75 @ 3200 std.; 85 @ 3200 opt.	93 @ 2000 std.; 102 @ 3200 opt.
Torque (lb-ft) @ rpm	145 @ 1000	160 @ 1000 std.; 174 @ 1000 opt.	180 @ 1000 std.; 196 @ 1000 opt.
Combustion chamber	Turbulent	Turbulent	Turbulent
Cyl. block mat'l.—skirt type	Cast iron—shallow	Cast iron—shallow	Cast iron—shallow
Cyl. block deck height	14.505	15.375	15.26
Cyl. block length	32.0	32.0	30.75
Cyl. bore spacing	All except 3-4 = 4.0625 3-4 = 4.5	All except 3-4 = 4.0625 3-4 = 4.5	All except 3-4 = 4.0625 3-4 = 4.5
Ck/shaft—no. of main brgs.	7	7	7
Ck/shaft main journal dia.	2.25	2.25	2.25
Crankpin journal dia.	2.0	2.0	2.0
Piston mat'l.—no. of rings	Aluminum—3	Aluminum—5	Aluminum—5
Connecting rod centers	10	10.4375	10.4375
Valve system	L-head	L-head	L-head
Valve events	220-230-2	220-230-2	220-230-2
Camshaft drive	Chain	Chain	Chain
Valve head dia.	Intake and exhaust 1.59	Intake 1.72; exhaust 1.59	Intake 1.72; exhaust 1.66
Valve lash adjustment	Mech. tappets	Mech. tappets	Mech. tappets
Oil pump and distributor drive	Vertical at right-center of engine. Distr. on top of cyl. head. Oil pump at bottom of block. Full pressure.	Vertical at right-center of engine. Distr. on top of cyl. head. Oil pump at bottom of block. Full pressure.	Vertical at right-center of engine. Distr. on top of cyl. head. Oil pump at bottom of block. Full pressure.
Fuel system	Updraft carburetor	Updraft carburetor	Downdraft carburetor
Other	Start of ck/s vib. damper	Start of bolted-on ck/s cwts.	

Fig. 3.19 *The 1926 Chrysler Imperial E80 two-door roadster. Under its hood was a new 288.6-cu.-in. six-cylinder engine, larger in structure and displacement than the original Chrysler engine.*

Subsequent Changes

The E80 engine remained unchanged throughout 1927. For the 1928 model, designated the L80, the displacement of the engine was increased to 309.6 cu. in. by a bore increase to 3.625 in. Two cylinder heads were offered: Silverdome, with a 4.75:1 compression ratio; and Red Head, with a 6.0:1 compression ratio. The engine versions produced 100 and 112 hp, respectively. From 1926 through 1929, the timing chain was adjusted by means of a pivoting generator located on the right side of the engine. In 1930, the last year the engine would power any passenger car, a non-adjustable, two-point (i.e., crankshaft-camshaft) timing chain was used, and the generator, relocated to the left side of the engine, was driven by a V-belt that also drove the fan and water pump.

Despite being dropped for passenger cars, the engine lived on. In 1930, the engine made its debut in Dodge trucks, powering the heavier models. The Dodge Truck Division continued using this "Big-Six" engine until the end of 1936 when, for 1937, the engine was replaced by a much larger, heavy-duty, L-head, truck/industrial/marine engine that is described in Chapters 15 and 16.

Table 3.2 shows specifications for this large Imperial ZSB engine.

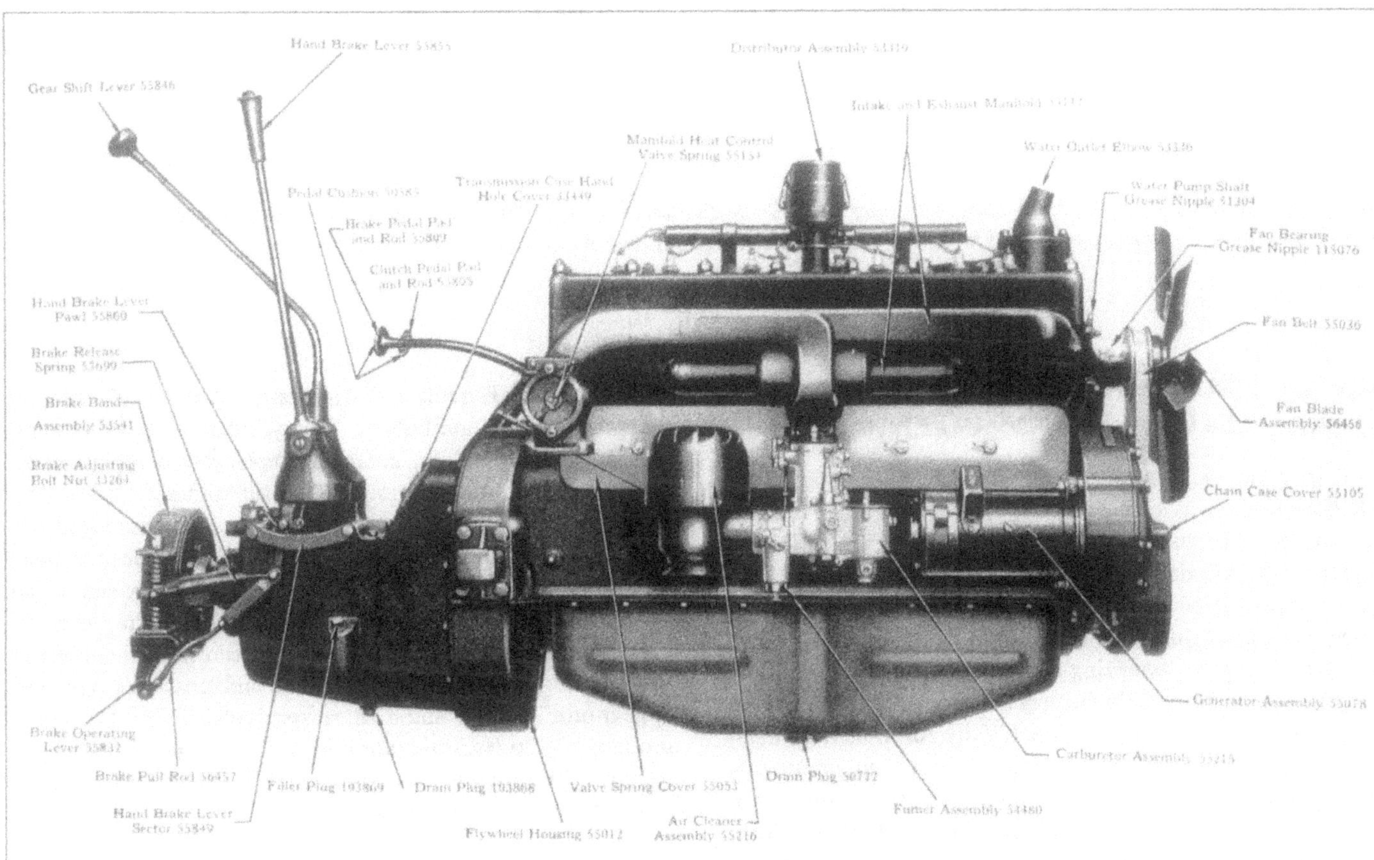

Fig. 3.20 *The right side of the 1926 Chrysler Imperial E80 six-cylinder engine. The engine was similar in appearance to the original Chrysler B-engine but larger throughout, including farther-apart cylinder bore centers.*

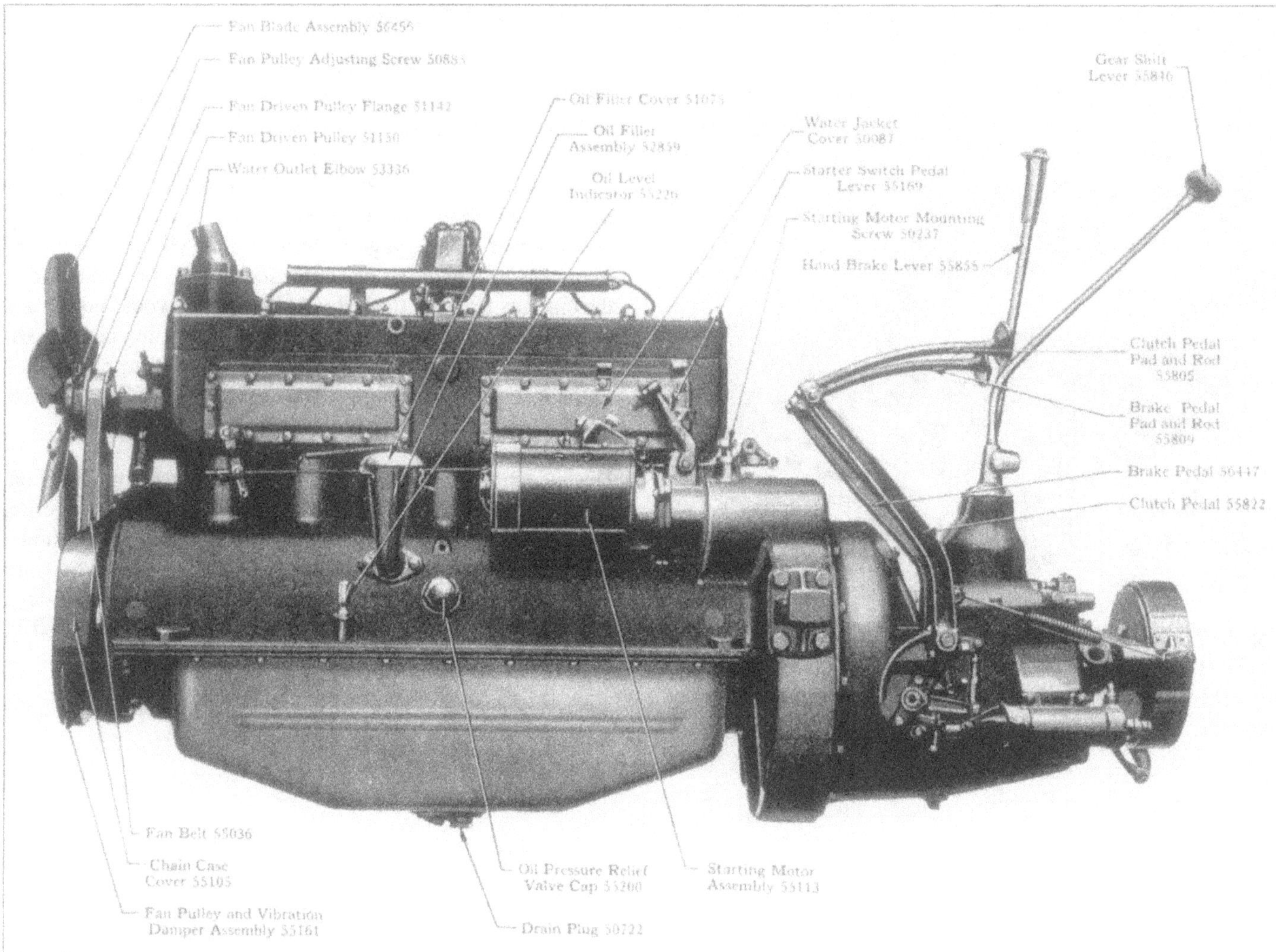

Fig. 3.21 *The left side of the 1926 Chrysler Imperial E80 six-cylinder engine. The two rectangular plates on the side of the cylinder block covered openings needed for water jacket core sand removal during the casting process.*

Smaller ZSB Seven-Main-Bearing L-Head Engine—The Chrysler H-Model Engine

Following the addition of the E80 Chrysler Imperial in 1926, another car line was added for 1927—a car somewhat smaller than the original Chrysler B (now the Chrysler G70), but larger than the four-cylinder car that was in the corporate lineup. The new car line carried the model code H60. Faced with creating a new engine for this car, ZSB certainly must have felt comfortable in simply downsizing their well-established excellent design. And that is exactly the way the engine turned out. The bore and stroke were set at 3.0 × 4.25 in. for a displacement of 180.24 cu. in.—well below the 218.59-cu.-in. engine of the Chrysler G70. The spacing of the cylinders was reduced to 3.875 in., with the center cylinders placed 4.4375 in. apart. This reduced cylinder spacing was reflected in the design of the cylinder head, crankshaft, camshaft, and manifolds, making them also shorter and lighter. The cylinder block deck height was reduced to 13.63 in., making the block considerably smaller and lighter. Although the engine was full of detail differences, its general layout and outward appearance, including the distributor on top of the cylinder head, unmistakably showed that the engine was a ZSB creation, giving assurance to the owner that he was getting the latest in engineering technology. Figures 3.25 and 3.26 show right and left side views of the engine, respectively. Longitudinal and cross-section views appear on Figs. 3.27 and 3.28, respectively. This H60 engine produced 54 hp @ 3000 rpm and 112 lb-ft torque @ 1000 rpm. Figure 3.29 shows a trim-looking 1927 Chrysler 60 two-door sedan, powered by the H60 engine.

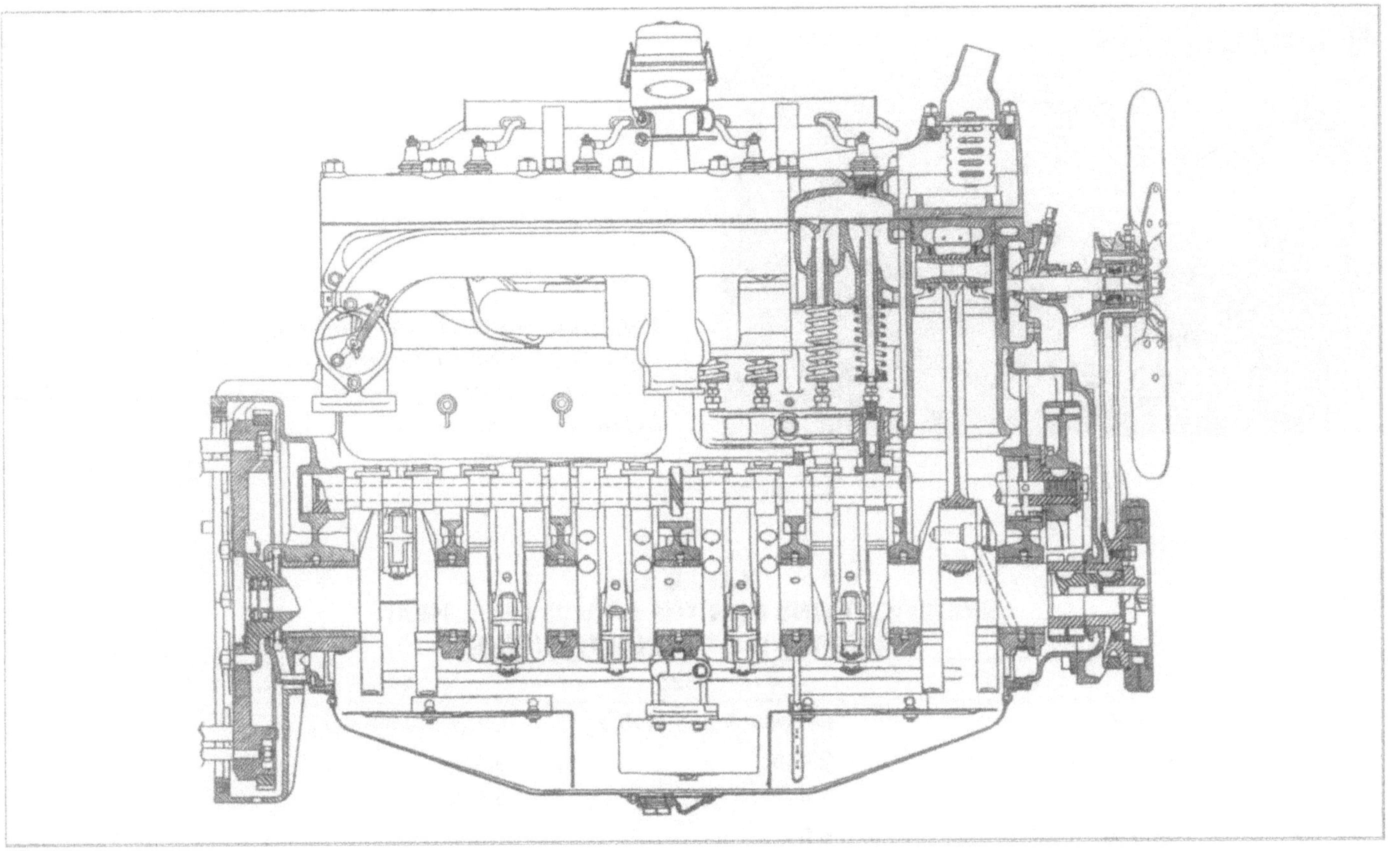

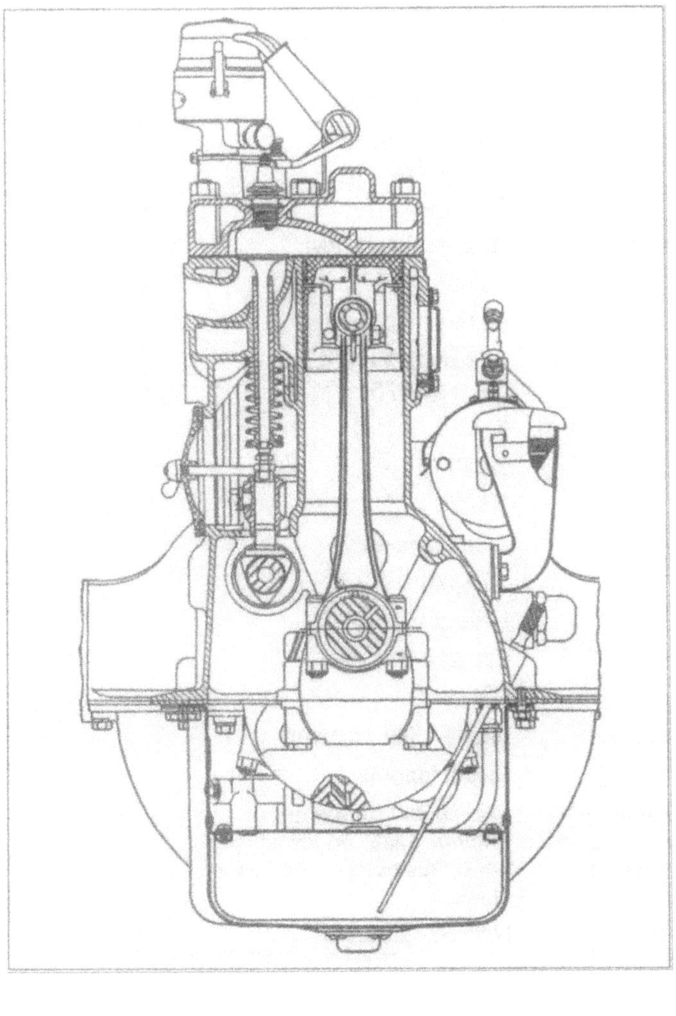

Fig. 3.22 *(Above) A longitudinal section of the 1926 Chrysler Imperial E80 six-cylinder engine. Visible are the seven main bearings of the engine, the vibration damper on the nose of the crankshaft, and the adjustable fan pulley front flange.*

Fig. 3.23 *(Left) A cross section through the #1 cylinder of the 1926 Chrysler Imperial E80 six-cylinder engine. At the top of the engine, the turbulent combustion chamber can easily be identified. In the crankcase, a crankshaft counterweight is sectioned to show one of its two attaching bolts.*

Fig. 3.24 *Walter P. Chrysler (right) and local Chrysler dealer Carl Wallerich (left) pose with the 1926 Chrysler Imperial roadster that was the pace car for the May 1926 Indianapolis 500 race.*

TABLE 3.2
SPECIFICATIONS FOR THE LARGE ZSB L-HEAD IMPERIAL SIX-CYLINDER ENGINE

Year	1926	1929–1930
Model	Chrysler Imperial E80	Chrysler Imperial L
Displacement (cu. in.)	288.6	309.6
Bore	3.5	3.625
Stroke	5.0	5.0
Compression ratio	4.7:1	4.75:1 std.; 6.0:1 opt.
BHP@ rpm	92 @ 3000	100 @ 3000 std.; 112 @ 3000 opt.
Torque (lb-ft) @ rpm	192 @ 1000	203 @1000 std.; 218 @ 1000 opt.
Combustion chamber	Turbulent	Turbulent
Cyl. block mat'l.–skirt type	Cast iron—shallow	Cast iron—shallow
Cyl. block deck height	15.505	15.625
Cyl. block length	35.688	35.688
Cyl. bore spacing	All except 3-4 = 4.5 3-4 = 5.1875	All except 3-4 = 4.5 3-4 = 5.1875
Ck/shaft–no. of main brgs.	7	7
Ck/shaft main journal dia.	2.625	2.625
Crankpin journal dia.	2.1875	2.1875
Piston mat'l.—no. of rings	Aluminum—3	Aluminum—3
Connecting rod centers	10.875	10.4375
Valve system	L-head	L-head
Valve events	220-230-2	220-230-2
Camshaft drive	Chain	Chain
Valve head dia.	Intake and exhaust 1.81	Intake 1.938; exhaust 1.81
Valve lash adjustment	Mech. tappets	Mech. tappets
Oil pump and distributor drive	Vertical at right-center of engine. Distr. on top of cyl. head. Oil pump at bottom of block.	Vertical at right-center of engine. Distr. on top of cyl. head. Oil pump at bottom of block.
Fuel system	Updraft carburetor	Updraft carburetor

Subsequent Changes

The 1928 engine, coded M62, was changed little from 1927 except for a change in standard compression ratio from 4.9:1 to 5.2:1, the addition of an optional high-compression (6.2:1) Red Head, and the addition of a road draft tube for crankcase ventilation. The standard engine produced 54 hp @ 3200 rpm and 115 lb-ft torque @ 1000 rpm. With the Red Head, the engine produced 60 hp @ 3200 rpm and 125 lb-ft torque @ 1000 rpm.

The displacement of the 1929 engine, coded P65, was increased to 195.6 cu. in. by a bore size increase to 3.125 in. The standard engine produced 65 hp @ 3200 rpm and 127 lb-ft torque @ 1000 rpm. With the Red Head, the engine produced 70 hp @ 3200 rpm and 135 lb-ft torque @ 1000 rpm. In addition to the displacement change, several other major changes were made to the engine: the timing chain was changed from a three-point to a two-point drive, with the generator moving to the left side of the engine, driven by the same V-belt used for the fan and water pump; massive, bolted-on steel counterweights were added to the crankshaft; and the exhaust manifold outlet was relocated from the rear to the front of the engine.

The 1930 model, coded CC66, started with a carryover 195.6-cu.-in. engine. During the model year, its displacement was increased to 218.6 cu. in. by a stroke change to 4.75 in. In addition, a mechanical fuel pump was added, and the carburetor and manifolds changed to a downdraft system. Power increased to 68 hp and 75 hp for the standard and Red Head versions, respectively, with torque increasing to 142 lb-ft and 153 lb-ft, respectively.

Table 3.3 shows the specifications for this small ZSB six-cylinder engine.

Chart 3.1 displays time lines for all three ZSB engines described in this chapter.

Fig. 3.25 *The right side of the 1927 Chrysler H60 180-cu.-in. six-cylinder engine. Looking outwardly similar to the original Chrysler B-model engine, the H60 engine was smaller in structure and displacement.*

Fig. 3.26 *The left side of the 1927 Chrysler H60 six-cylinder engine. Checking the oil level and servicing the oil pressure relief valve and starter motor were done easily from this side of the engine.*

Fig. 3.27 *A longitudinal section of the 1927 Chrysler H60 six-cylinder engine. With the cylinder bores spaced closer together than those of the larger ZSB-designed engines, the H60 engine shared the quality features of seven main bearings and a vibration damper on the nose of the crankshaft.*

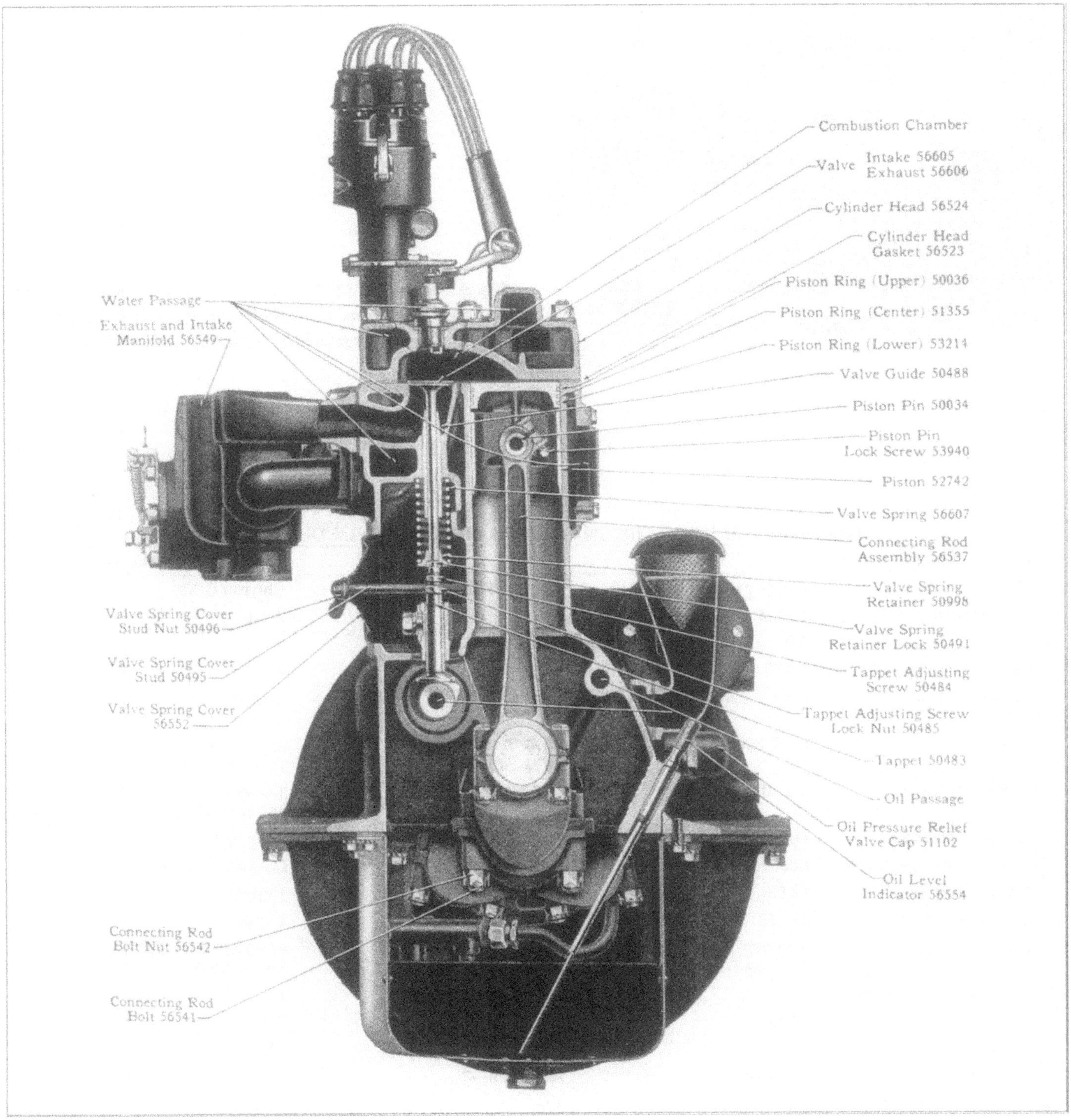

Fig. 3.28 *A cross section through the #1 cylinder of the 1927 Chrysler H60 six-cylinder engine. The sectioned port is an exhaust; below it, unsectioned, is one of two intake ports that divided inside the block to feed three cylinders each.*

Fig. 3.29 *The 1927 Chrysler H60 two-door sedan. The ZSB-designed 180-cu.-in. six-cylinder engine with its seven-main-bearing crankshaft provided smooth power for the car.*

TABLE 3.3
SPECIFICATIONS FOR THE SMALL ZSB L-HEAD SIX-CYLINDER ENGINE

Year	1927	1929	1930
Model	Chrysler H60	Chrysler P65	Chrysler CC66
Displacement (cu. in.)	180.24	195.6	218.6
Bore	3.0	3.125	3.125
Stroke	4.25	4.25	4.75
Compression ratio	4.9:1	5.2:1 std.; 6.0:1 opt.	5.1:1 std.; 6.0:1 opt.
BHP@ rpm	54 @ 3000	65 @ 3200 std.; 70 @ 3200 opt.	68 @ 3000 std.; 75 @ 3200 opt.
Torque (lb-ft) @ rpm	112 @ 1000	127 @1000 std.; 135 @ 1000 opt.	142 @1000 std.; 153 @ 1000 opt.
Combustion chamber	Turbulent	Turbulent	Turbulent
Cyl. block mat'l.—skirt type	Cast iron—shallow	Cast iron—shallow	Cast iron—shallow
Cyl. block deck height	13.63	13.63	13.925
Cyl. block length	31.188	31.188	31.188
Cyl. bore spacing	all except 3-4 = 3.875 3-4 = 4.4375	all except 3-4 = 3.875 3-4 = 4.4375	all except 3-4 = 3.875 3-4 = 4.4375
Ck/shaft—no. of main brgs.	7	7	7
Ck/shaft main journal dia.	1.875	2.0	2.25
Crankpin journal dia.	1.875	1.875	2.0
Piston mat'l.—no. of rings	Aluminum—3	Aluminum—3	Aluminum—3
Connecting rod centers	9.5	9.5	9.5
Valve system	L-head	L-head	L-head
Valve events	220-230-2	220-230-2	220-230-2
Camshaft drive	Chain	Chain	Chain
Valve head dia.	Intake and exhaust 1.31	Intake 1.44; exhaust 1.31	Intake 1.72; exhaust 1.66
Valve lash adjustment	Mech. tappets	Mech. tappets	Mech. tappets
Oil pump and distributor drive	Vertical at right-center of engine. Distr. on top of cyl. head. Oil pump at bottom of block.	Vertical at right-center of engine. Distr. on top of cyl. head. Oil pump at bottom of block.	Vertical at right-center of engine. Distr. on top of cyl. head. Oil pump at bottom of block.
Fuel system	Updraft carburetor	Updraft carburetor	Updraft carburetor
Other			Bolted-on ck/s cwts.

Chart 3.1
Time Lines—Zeder, Skelton, and Breer's Three Masterpiece Engines

Engines, Car Lines, and Displacements (cu. in.)

	Car	- - Model Years - -						
Engine	Line	1924	1925	1926	1927	1928	1929	1930
Original ZSB	Chrysler	201	201	219	249	249	249	268
7 main bearings	Model	B	B	Six - G70	G70	J72	R75	V*70, W77
Large ZSB	Imperial			289	289	310	310	310
7 main bearings	Model			E80	E80	L80	L	L
Small ZSB	Chrysler				180	180	196	219
7 main bearings	Model				H60	M62	P65	C66, V70

CHAPTER 4

Maxwell, Chrysler, and Plymouth Four-Cylinder Engines

Maxwell—A Car and a Company in Trouble

As related in Chapter 1, Walter P. Chrysler joined the Maxwell Motor Company in August 1920 at the request of the company's bankers, with a directive to bring it back to profitability from a near-bankruptcy state. Maxwell was suffering from a costly business deal with the Chalmers Motor Car Company, a huge stock of unsold cars, large inventories of unassembled parts, and, perhaps worst of all, a bad reputation because of field failures that had choked off sales of new cars.[1] The task was daunting, but Walter Chrysler was up to it. Stopping production of the solitary Maxwell product, the Maxwell 25 model, he saw that fixes were made to the troublesome areas—gas tanks, axles, and transmissions—before any more cars were shipped to the dealers. Although the sole engine of the Maxwell 25, which was a four-cylinder L-head, was not mentioned as a source of field problems, its design, which was virtually unchanged since 1916 and possibly even 1914, was hopelessly out of date, appearing very similar to the 1918 engine pictured on Fig. 4.1. Especially outdated were its crankshaft, with only two main bearings, and its plunger-type oil pump. Both of these features are shown on the lubrication diagram on Fig. 4.2 and longitudinal section on Fig. 4.3.

Fig. 4.1 *The right side of the engine compartment of the 1918 Maxwell car at the Chrysler Historical Collection. The engine is a 186-cu.-in. L-head four-cylinder with two main bearings. Photo by W.L. Weertman.*

Following Chrysler's arrival, a former Willys-Overland engineer named Harry T. Woolson made significant improvements that were incorporated during the 1922 model year run of the Maxwell 25 engine. He redesigned the crankshaft and crankcase to add a third main bearing at the center of the engine[1] and provided full-pressure oil to all main bearings and connecting rod big-end bearings. To boost the oil supply needed for the full-pressure oiling, he replaced the plunger pump with a cylinder-block-mounted rotary-gear oil pump driven by a gear on the camshaft. However, adding the third bearing was not easy. An examination of the detail drawings of the cylinder block, the center bearing cap, and the thick babbitt-coated bronze main bearings reveals that the design task was almost impossible to accomplish. The horizontal joint face between the new crankshaft bearing cap and the crankcase bulkhead, normally designed to be exactly at the crankshaft centerline, actually was raised 0.25 in. above the crankshaft centerline! A most unusual condition! I believe this unique design was the solution to a tough problem of assembling the three-main-bearing crankshaft into the crankcase. The design of the rear bearing and bearing support of the crankshaft, carried over from the original two-main-bearing crankshaft, required that the crankshaft be threaded through a 5-in. diameter hole at the rear of the crankcase before being swung into its final position. Doing this with the original two-main-bearing crankshaft was not a problem because the crankcase area of the cylinder

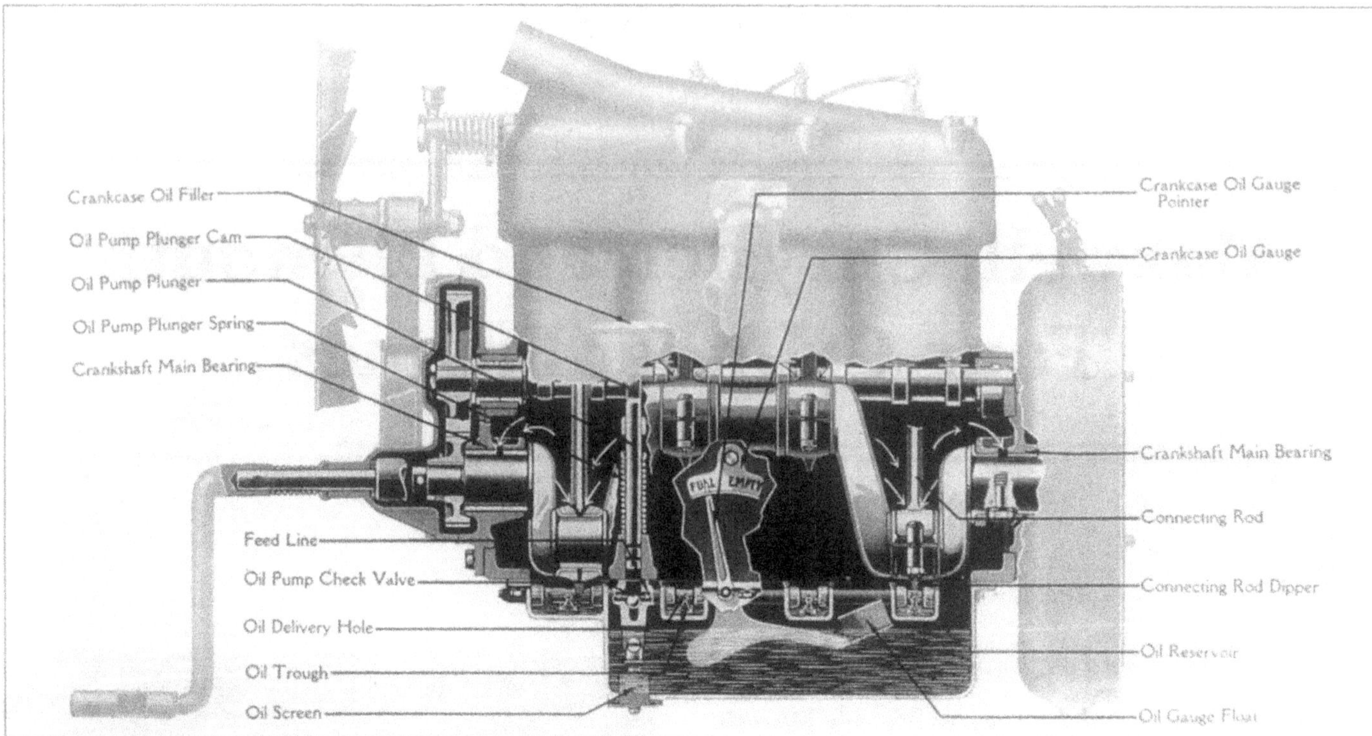

Fig. 4.2 *A lubrication diagram of the 1916 Maxwell 186-cu.-in. four-cylinder engine, showing the plunger oil pump, the splash troughs, and the two-main-bearing crankshaft.*

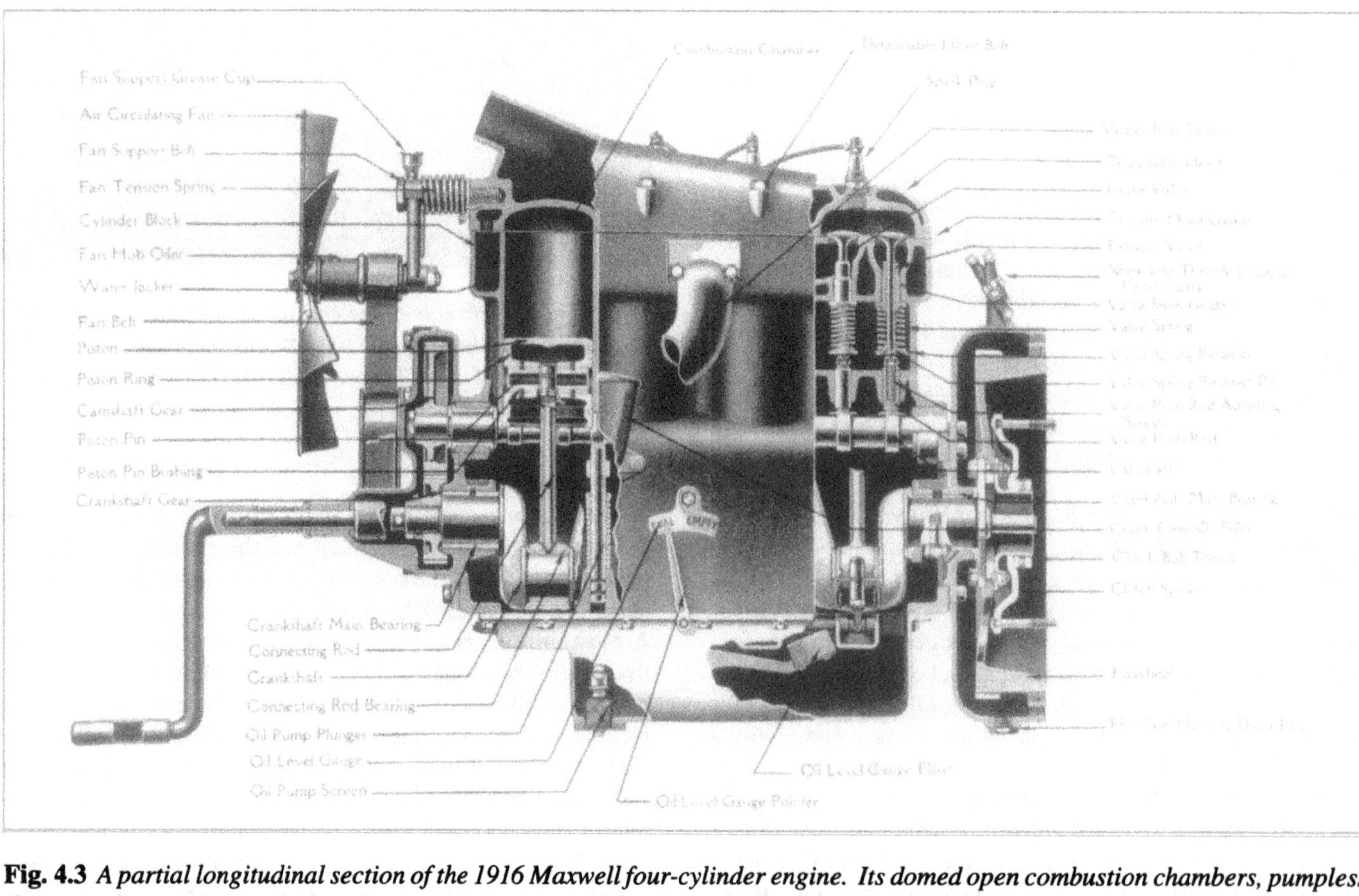

Fig. 4.3 *A partial longitudinal section of the 1916 Maxwell four-cylinder engine. Its domed open combustion chambers, pumpless thermosyphon cooling, splash and spray lubrication, and hand-crank starting were typical of many engines at the time.*

block was a wide-open space. When the additional cast iron bulkhead was added at the center of the crankcase to support the third bearing, the assembly hit a snag. Raising the joint face 0.25 in. for added assembly clearance evidently did the trick, and the design became acceptable, although not perfect. With the third bearing, which was larger in diameter than the other two—2.25 in. versus 1.875 in.—the engine was reported to run more smoothly.[1]

The 1922 Maxwell 25 engine is shown in the right and left outside views on Figs. 4.4 and 4.5, the longitudinal section on Fig. 4.6, and the lubrication system diagram on Fig. 4.7. With its camshaft, valves, and intake and exhaust manifolds located on the right side of the cylinder block, the engine had a bore and stroke of 3.625 × 4.5 in. for a displacement of 185.78 cu. in. Helical-toothed gears having a 1:2 speed ratio drove the camshaft from the crankshaft. The cast iron cylinder block had the centerlines of the #1 and #2 cylinders spaced 4.0 in. apart, those of cylinders #2 and #3 were spaced 4.625 in. apart, and those of cylinders #3 and #4 were spaced 4.0 in. apart. The block skirt was a deep 3.625 in. to accommodate the supports needed for the end main bearings of the crankshaft. Just below the top deck of the block, on its right side, was a row of five ports. Intake fuel/air mixture flowed into two of these ports that were divided inside the block to feed two cylinders each. The remaining three ports were for exhaust gases. Each end cylinder had its own exhaust port, and the two center cylinders shared a common exhaust port at the center of the block. Dome-shaped open combustion chambers provided a generous volume above the entire valve and piston head area of each cylinder. The gear case cover at the front of the engine was home for the ignition distributor, which was gear-driven from the nose of the camshaft. A gear cut into the camshaft between cylinders #3 and #4 drove the rotary-gear oil pump that was mounted on the right side of the cylinder block. Metal tubing transported engine oil in and out of the oil pump and to the main bearings. There was no water pump. Instead, the engine used a simple thermosyphon system (described in the 1916 owner's manual as "highly efficient and free from complications"), which relied on hot water rising through the engine and cooler water descending down the radiator to encourage circulation through the engine and radiator. High at the front of the engine were the generator and fan, sharing a common shaft and V-belt drive. The stylish 1922 Maxwell 25 touring car shown on Fig. 4.8 was powered with this engine, which had an output of 30 hp @ 2150 rpm.

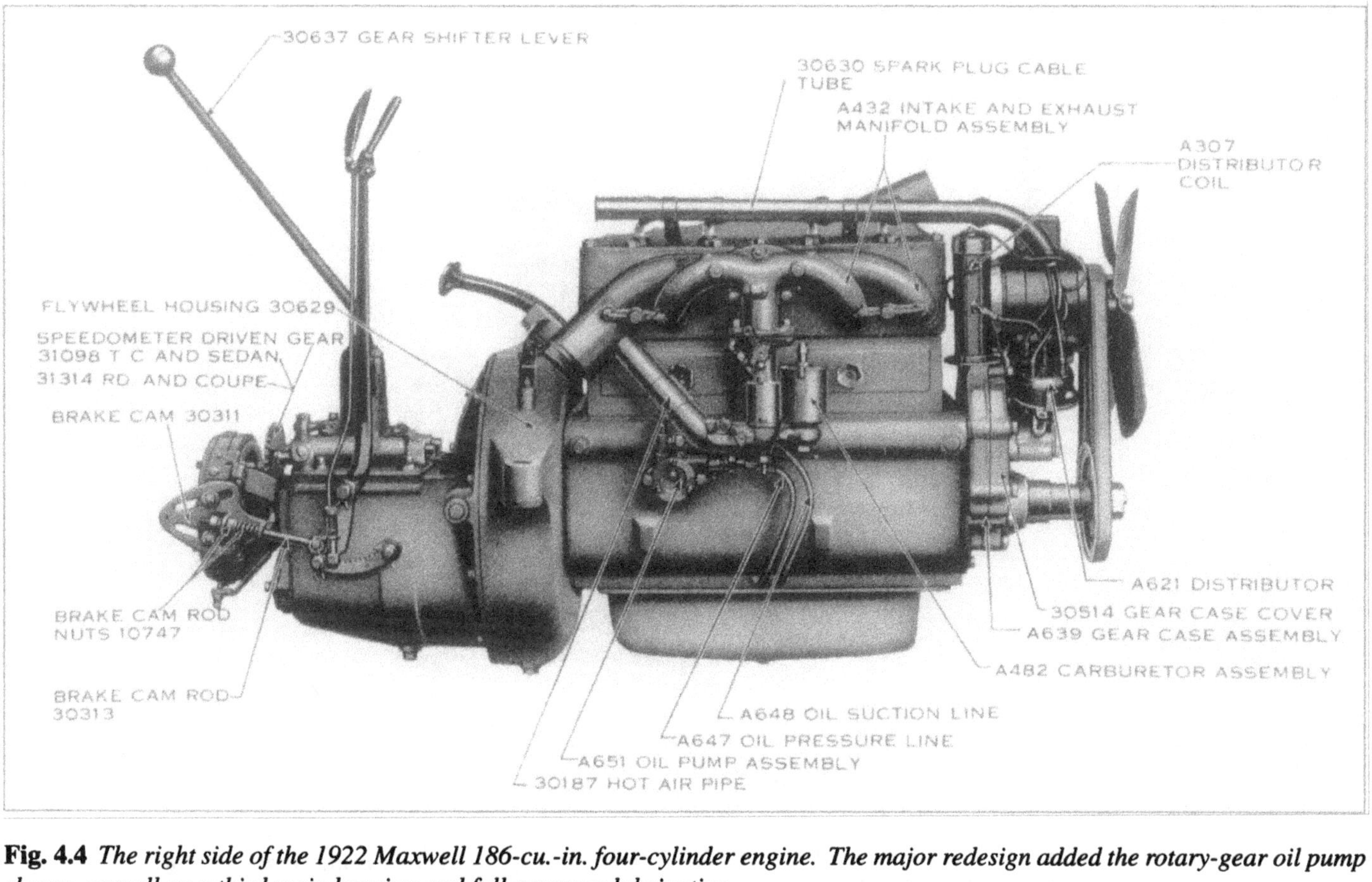

Fig. 4.4 *The right side of the 1922 Maxwell 186-cu.-in. four-cylinder engine. The major redesign added the rotary-gear oil pump shown, as well as a third main bearing and full-pressure lubrication.*

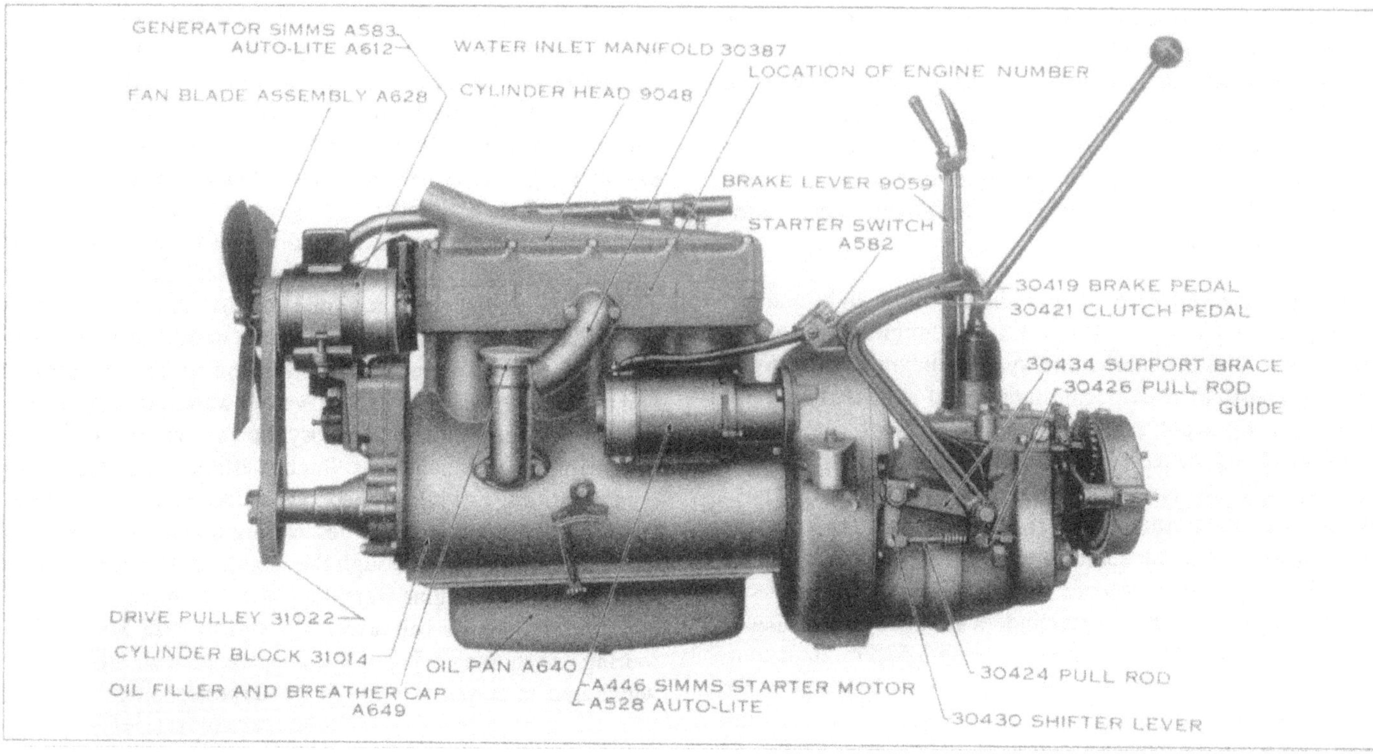

Fig. 4.5 *The left side of the 1922 Maxwell four-cylinder engine. The large diameter water inlet is evidence of its thermosyphon cooling system. The starter motor and belt-driven generator indicate that a hand crank is no longer the primary means of starting the engine.*

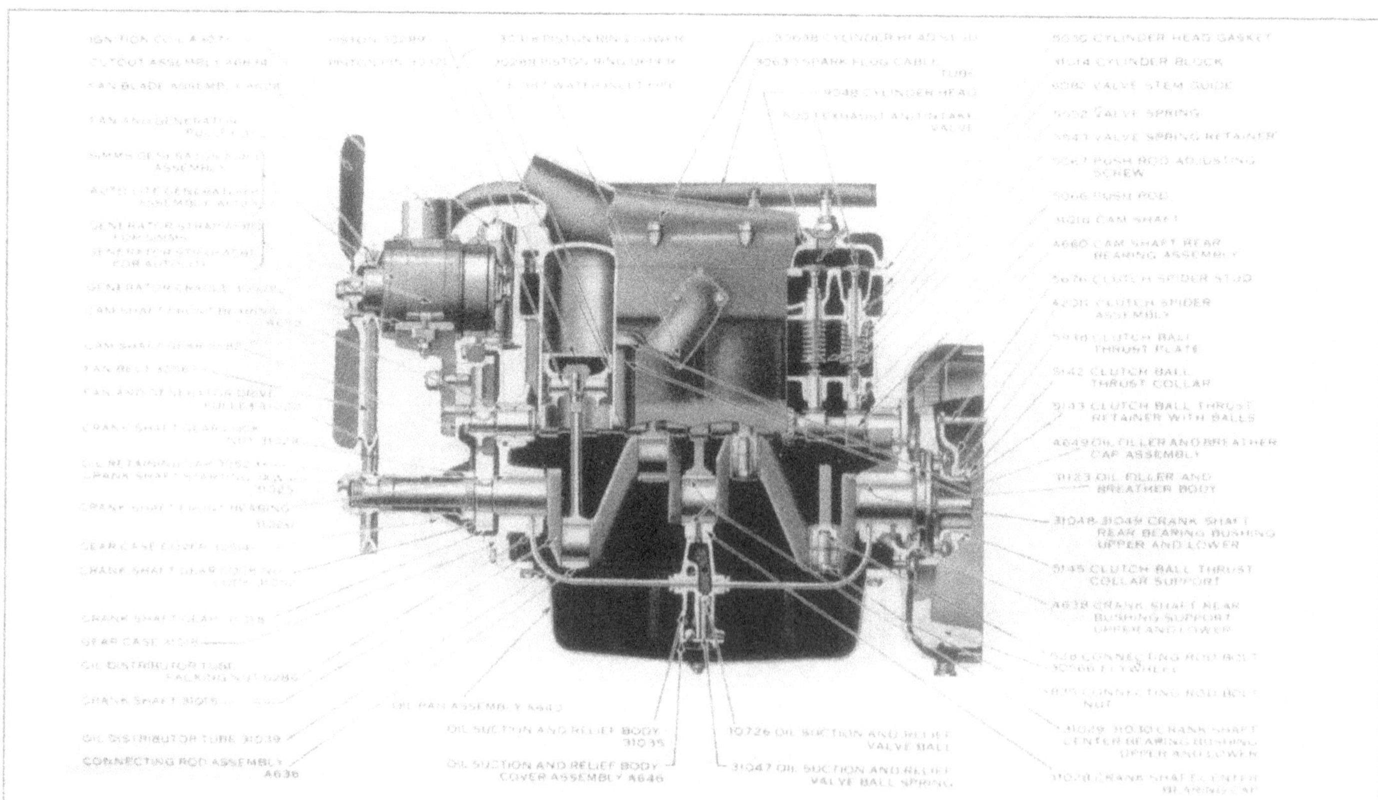

Fig. 4.6 *A longitudinal section of the 1922 Maxwell four-cylinder engine. The cap for the new center main bearing was attached with two vertical bolts. Supports for the end main bearings were unchanged from the prior two-main-bearing designs.*

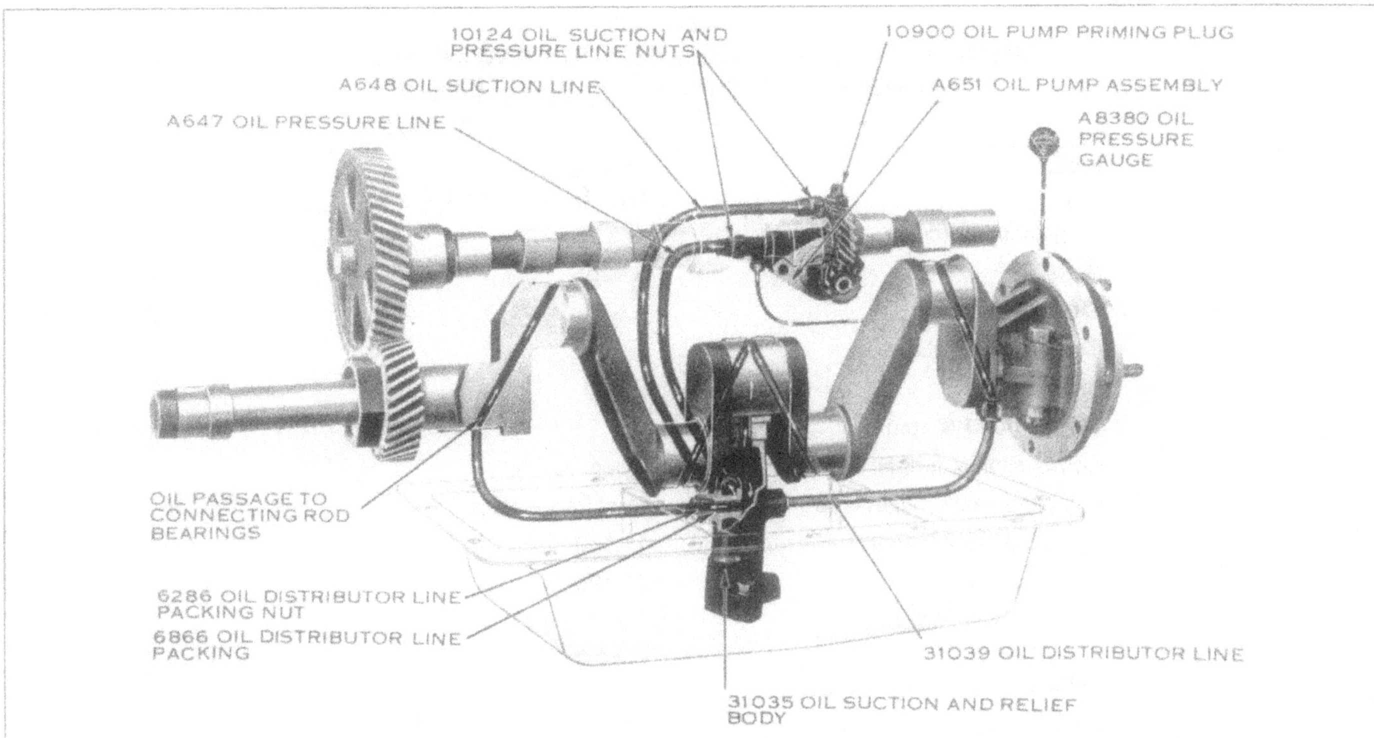

Fig. 4.7 *The lubrication system of the 1922 Maxwell four-cylinder engine. Tubing was used extensively to transport oil to and from the new oil pump mounted on the right side of the block.*

Fig. 4.8 *The 1922 Maxwell five-passenger touring car. The warmly clad driver and her passengers would have felt reassured to know that her 30-hp, 186-cu.-in. Maxwell four-cylinder engine was equipped with full-pressure lubrication to all its major bearings.*

Table 4.1 shows specifications of the 1921–1922 Maxwell 25 four-cylinder engine.

Some time after the arrival of Zeder, Skelton, and Breer at Maxwell with the ZSB Engineering Co. team in June 1923, followed by their appointment as head of the entire Maxwell engineering organization and completion of their work on the Chrysler B-model six-cylinder engine, ZSB turned their attention to the Maxwell 25 engine. The first power runs they made with the engine on the dynamometer told them something was wrong. After only a short time at wide-open throttle, the engine power would take a dive; however, following a cool-down period, the power would return.[2] What was happening? Exhaust valve seat distortion at the center of the engine turned out to be the culprit. The distortion was traced to inadequate cooling of the two center exhaust valve seats that warped when the engine got hot, losing their seal and, on cool-down, regaining it.[2] Two changes—incorporated in the 1925 Maxwell 25C engine—were made to solve the problem. One change raised the deck of the block from 11.5625 to 11.937 in., providing room for added coolant passages above the ports. The second change separated the two exhaust ports at the center of the engine so they were no longer "siamesed" into a single outlet port. With the ports now separated, a coolant passage was added between the two ports. The row of ports along the top of the right side of the cylinder block went from five for

TABLE 4.1
SPECIFICATIONS FOR THE LATE 1921 AND INITIAL 1922 MAXWELL L-HEAD FOUR-CYLINDER ENGINE

Year	1921–1922
Model	Maxwell 25 New Series, Good Maxwell
Displacement (cu. in.)	185.78
Bore	3.625
Stroke	4.5
Compression ratio	
BHP @ rpm	
Torque (lb-ft) @ rpm	
Combustion chamber	Open
Cyl. block mat'l.—skirt type	Cast iron—deep
Cyl. block deck height	11.563
Cyl. block length	20.125
Cyl. bore spacing	1-2, 3-4 = 4.0 2-3 = 4.625
Ck/shaft—no. of main brgs.	2
Ck/shaft main journal dia.	1.875
Crankpin journal dia.	1.875
Piston mat'l.—no. of rings	Aluminum—3
Connecting rod centers	7.875
Valve system	L-head
Valve events	215-215-neg. 5
Camshaft drive	Gear
Valve head dia.	Intake and exhaust 1.5
Valve lash adjustment	Mech. tappets
Oil pump	Plunger type, driven from camshaft. Splash and spray lubrication.
Distributor	Mtd. to rear of chain case cover
Cooling	Thermosyphon
Fuel system	Updraft carburetor

the prior design to six—four exhausts and two inlets. The old-style thermosyphon, pumpless cooling system, and open combustion chambers of the engine remained unchanged. With a conservative 3.9:1 compression ratio, output was rated at a modest 38 hp @ 2200 rpm, but now the engine could deliver it dependably to the driver.

The Four-Cylinder Maxwell Car Becomes a Chrysler

The 1925 Maxwell 25C was the end of the line for cars named Maxwell. For 1926, the car was rebadged as a Chrysler Four and was given an alphabetic F-model code. Later in the model year, the name was changed to Chrysler 58, and the model code was expanded to F58. This 1926 Chrysler four-cylinder engine actually was a direct carryover from the 1925 Maxwell 25C four-cylinder engine.

In what seemed to be a retrograde move in the following model year, 1927, the displacement of the engine was reduced from 185.78 to 170.3 cu. in. by a stroke reduction from 4.5 to 4.125 in. Model code letter "I" was assigned, with 1927 cars and engines labeled Chrysler I50. Despite a compression ratio boost to 4.1:1, the higher compression could not overcome the displacement reduction, and the power of this Chrysler I50 four-cylinder engine dropped to 35 hp. Figure 4.9 shows the upright styling of the car.

Fig. 4.9 *The 1927 Chrysler I50 model coupe. Its former Maxwell engine had a displacement of 170 cu. in., a compression ratio of 4.1:1, and 35 hp—a nice match for the compact coupe with its 106-in. wheelbase.*

Fig. 4.10 *The 1928 Chrysler-Plymouth Q deluxe rumble seat coupe. The Plymouth Q's 170-cu.-in. four-cylinder engine was the same 45-hp turbulent-combustion-chamber version as introduced mid-model on the Chrysler I52.*

The engine carried over to 1928, with the car and engine model code changing to I52. A higher-compression Red Head engine, still open chambered, was added to the product offering as standard on the roadster and optional on all others. Then, sometime after launch, two cylinder heads with turbulent chambers were released, with either a standard 4.6:1 ratio or a Red Head with a higher ratio. The engine responded to the change, with standard engine power jumping to 45 hp @ 2800 rpm.

With production starting in June 1928, Chrysler Corporation introduced a new car line to the motoring public. Carrying the name Chrysler Plymouth, it replaced the Chrysler I52 as the entry-level four-cylinder car line.

The Chrysler Four Becomes a Plymouth Four

Assigned a letter "Q" as its model code, the four-cylinder Chrysler Plymouth (shown on Fig. 4.10, with an engine compartment view on Fig. 4.11) was built from June 1928 to February 1929. Although its model year appears as 1928, 1929, or 1928–1929 in various publications and state registrations, when the 1938 model started production in October 1937, Plymouth described the 1938 car as the "Jubilee Plymouth, Climaxing Ten Years of Great Cars," effectively tagging the Plymouth Q as a 1928 model. The "Plymouth Chronological Succession of Models" charts published by Chrysler Engineering Technical Information Services are consistent with this position, showing the Plymouth Q as a 1928 model, a model year also shown in the information accompanying the Plymouth Q car on display at the Walter P. Chrysler Museum. With the Plymouth Q assigned to the 1928 model year, the three Plymouths that followed—the Plymouth U, 30U, and PA—were assigned to the 1929, 1930, and 1931 model years, respectively.

Fig. 4.11 *The left side of the engine compartment of the 1928 Plymouth Q coupe on display at the Walter P. Chrysler Museum. The rounded corners of the cylinder head of the engine are a legacy from its years as a Maxwell engine. Photo by W.L. Weertman.*

Carried over directly from the Chrysler I52 engine, the only significant change involving the engine during its production run was the dropping of the high-compression Red Head version. However, for the next model year, the engine of the 1929 Plymouth U was not a carryover or even a modification. It was almost a totally new four-cylinder engine.

The 1929 Plymouth (the Chrysler part of the name was dropped[3]) car and engine were assigned the model code letter "U." The U engine went through a remarkable redesign to make it stronger, with higher quality and easier manufacturing. Gone was the old cylinder block, with its awkward way of assembling the crankshaft through a hole in the rear bulkhead. The new block had a conventional crankcase with three husky main bearing caps that allowed the crankshaft to be given a normal, drop-in assembly into the crankcase. Taking advantage of the new design freedom, the designers gave the crankshaft greatly strengthened connectors between adjacent crankpins 1 and 2, and between crankpins 3 and 4. The cylinder block deck was raised almost an inch to 12.943 in., while at the bottom of the block, all vestiges of the ancient two-bearing design were removed, and the oil pan rail was raised 1 in. to be 2.625 in. below the crankshaft centerline rather than the prior 3.625 in. The new cylinder block had internal drilled holes to transport high-pressure oil to the bearings in place of the tubing that had been used. The cylinder head studs were relocated closer to the corners of the cylinder head, and the cylinder head corners were squared off. This gave a crisper appearance to the top of the engine. Through all this redesign, the engine's 3.625 in. bore and the cylinder spacing of 4.0-4.625-4.0 in. remained unchanged, as did the thermosyphon cooling system. The engine did receive a small displacement increase to 175.4 cu. in. by a stroke increase from 4.125 to 4.25 in. The compression ratio of the turbulent-chambered cylinder head remained at 4.6:1, and the engine output remained unchanged at 45 hp @ 2800 rpm. Figure 4.12 shows a front view of the engine. Figures 4.13 and 4.14 show longitudinal and cross-section views, respectively.

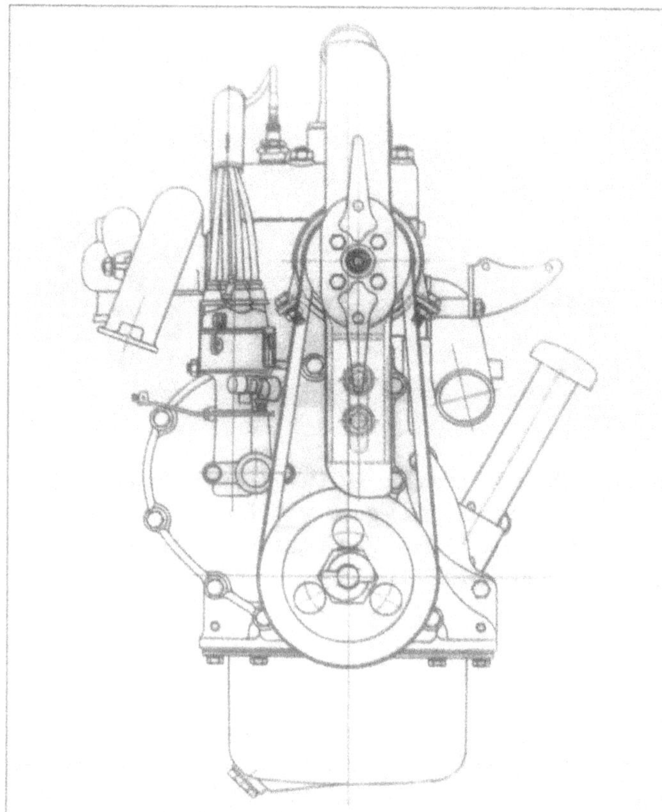

Fig. 4.12 *A front view of the 1929 Plymouth U 175.4-cu.-in. four-cylinder engine. A major redesign of the Maxwell-Chrysler-Plymouth Q engine that preceded it, the Plymouth U engine retained the thermosyphon cooling feature, requiring the large-diameter inlet pipe shown on the left side of the engine. From Engine Assembly Drawing 301000.*

The 1930 Plymouth model retained the U code but added the numeral "30" to make the model code 30U. April 1930 to June 1931 was the time span of its production. A longer stroke was among several significant changes made to the engine. With the taller block released in 1929, the designers had no trouble adding a half inch to the stroke, bringing it up to 4.75 in., which increased the displacement to 196.1 cu. in. This was almost exactly the same displacement as the ZSB six-cylinder 195.6-cu.-in. engine that powered the initial 1930 Chrysler 66. One important change for the 1930 Plymouth 30U engine was the addition of a V-belt-driven water pump, mounted at the top-front of the cylinder block, for positive coolant flow. This arrangement was somewhat similar to the design used on the three ZSB six-cylinder engines described in Chapter 3. As will be discussed in Chapter 15, light-duty Dodge Brothers trucks started using this engine in 1929, and for that year, a low-mounted water pump was provided for truck service. This 1930 30U engine had another improvement—the addition of a mechanical fuel pump. Although pressure fuel was now available, the carburetor remained as an updraft type and was not changed to a downdraft type. However, the vacuum tank for the fuel supply was no longer needed and became a thing of the past.

PA was the model code assigned to the 1931 Plymouth—the first of a series of double-letter codes that would end in 1935. The PA engine received a compression ratio increase to 4.9:1, bringing the engine power up to 56 hp @ 2800 rpm and the torque up to 130 lb-ft @ 1600 rpm. The engine also received a feature that made the 1931 Plymouth PA a uniquely desirable buy and, with application to all Chrysler Corporation engines soon afterward, gave all Chrysler Corporation cars and engines a marked improvement. The feature was called Floating Power®. Invented by Carl Breer, this new way to mount engines was developed after careful consideration of how to isolate the engine, with its shaking forces and torque pulses, from the rest of the car.[2] It really worked. Plymouth ad copy glowingly used the phrase "Smoothness of an Eight, Economy of a Four" over and over again.

Despite 1932 being its final year of production, which might have occasioned a slackening of engineering effort, the 1932 car and engine, which were model coded PB, saw a number

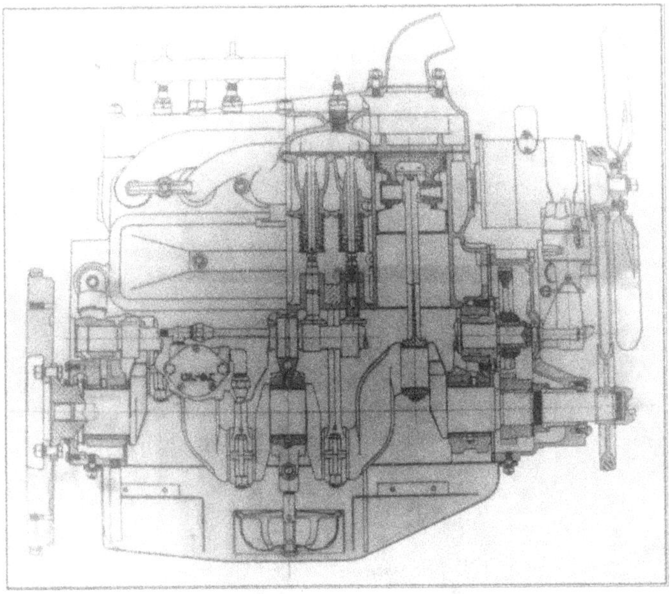

Fig. 4.13 *A longitudinal section of the 1929 Plymouth U four-cylinder engine. The redesign of the engine gave it a conventional crankcase, with three main bearing caps attached with vertical bolts and a much stronger crankshaft. From Engine Assembly Drawing 301000.*

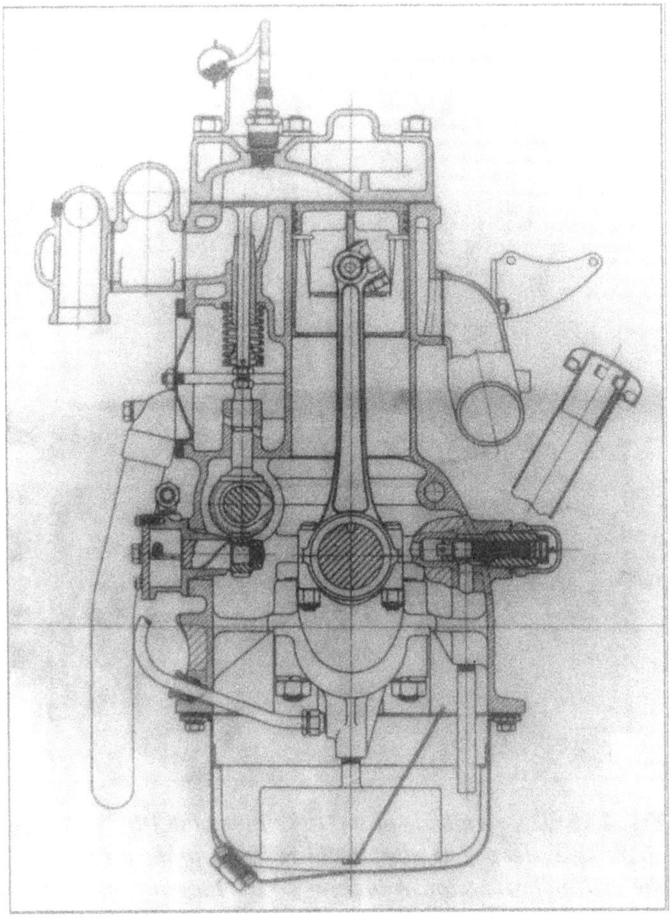

Fig. 4.14 *A cross section through the #1 cylinder of the 1929 Plymouth U four-cylinder engine. The turbulent combustion chamber with its minimal gap between the top of the piston and the bottom of the cylinder head has become a design standard for Chrysler L-head engines. From Engine Assembly Drawing 301000.*

of substantial changes. One of these changes was the return of the Red Head extra-cost option with its power and fuel economy-boosting high-compression-ratio (5.6:1) cylinder head. Another change was the addition of bolted-on counterweights for the crankshaft, which certainly reduced the main bearing loads and lowered engine noise. Another change was the opening of the exhaust manifold passages, which probably accounts for the substantial power and torque boost of the 1932 Plymouth PB engine, equipped with the standard 4.9:1 compression ratio Silver Dome cylinder head, to 65 hp @ 3400 rpm and 136 lb-ft @ 1600 rpm. A welcome quality feature for the Chrysler entry-level, price-leading car was a replaceable oil filter for the engine. Chrysler pioneered the use of engine oil filters on the 1924 Chrysler B-model six-cylinder engine, as noted in Chapter 3. Another quality move was a new air cleaner/silencer that had its entry point of inlet air high above the engine rather than low, almost at the bottom of the engine, as the prior air cleaner/silencer had. Figures 4.15 and 4.16 show right and left outside views of the Plymouth PB engine, respectively. A longitudinal section and a cross section appear on Figs. 4.17 and 4.18, respectively.

At the end of the 1932 model year, machining and assembly of the four-cylinder engine at the Chrysler Highland Park complex ceased, spelling the end of four-cylinder L-head engine manufacture by Chrysler. The building where the four-cylinder engines were made then became the Chrysler Engineering experimental road test garage for cars and trucks.[4] Plymouth and Dodge cars next would be powered by an all-new, six-cylinder L-head engine coming on-line at both the Dodge Main plant in Hamtramck and the new half-mile-long Plymouth assembly plant on Lynch Road in Detroit for the 1933 model year cars. Chapter 6 will tell the story of this new engine.

Table 4.2 shows specifications of the 1924 through 1928 Maxwell/Chrysler/Plymouth four-cylinder engines. Table 4.3 shows specifications of the 1929 through 1932 Plymouth four-cylinder engines.

Chart 4.1 shows time lines for all the four-cylinder engines described in this chapter.

Fig. 4.15 *The right side of the 1932 Plymouth PB 196-cu.-in. four-cylinder engine. Note the elaborate intake air silencing and cleaning system. The glass sediment bowl of the mechanical fuel pump, added in 1930, is pictured below and slightly ahead of the updraft carburetor. A bracket for the Floating Power® front engine mount can be seen directly behind the fan belt.*

Fig. 4.16 *The left side of the 1932 Plymouth PB four-cylinder engine. The oil filter, mounted on the side of the cylinder head, was made standard equipment to enhance engine protection.*

Fig. 4.17 *A longitudinal section of the 1932 Plymouth PB four-cylinder engine. The high-mounted water pump replaced the obsolete thermosyphon system.*

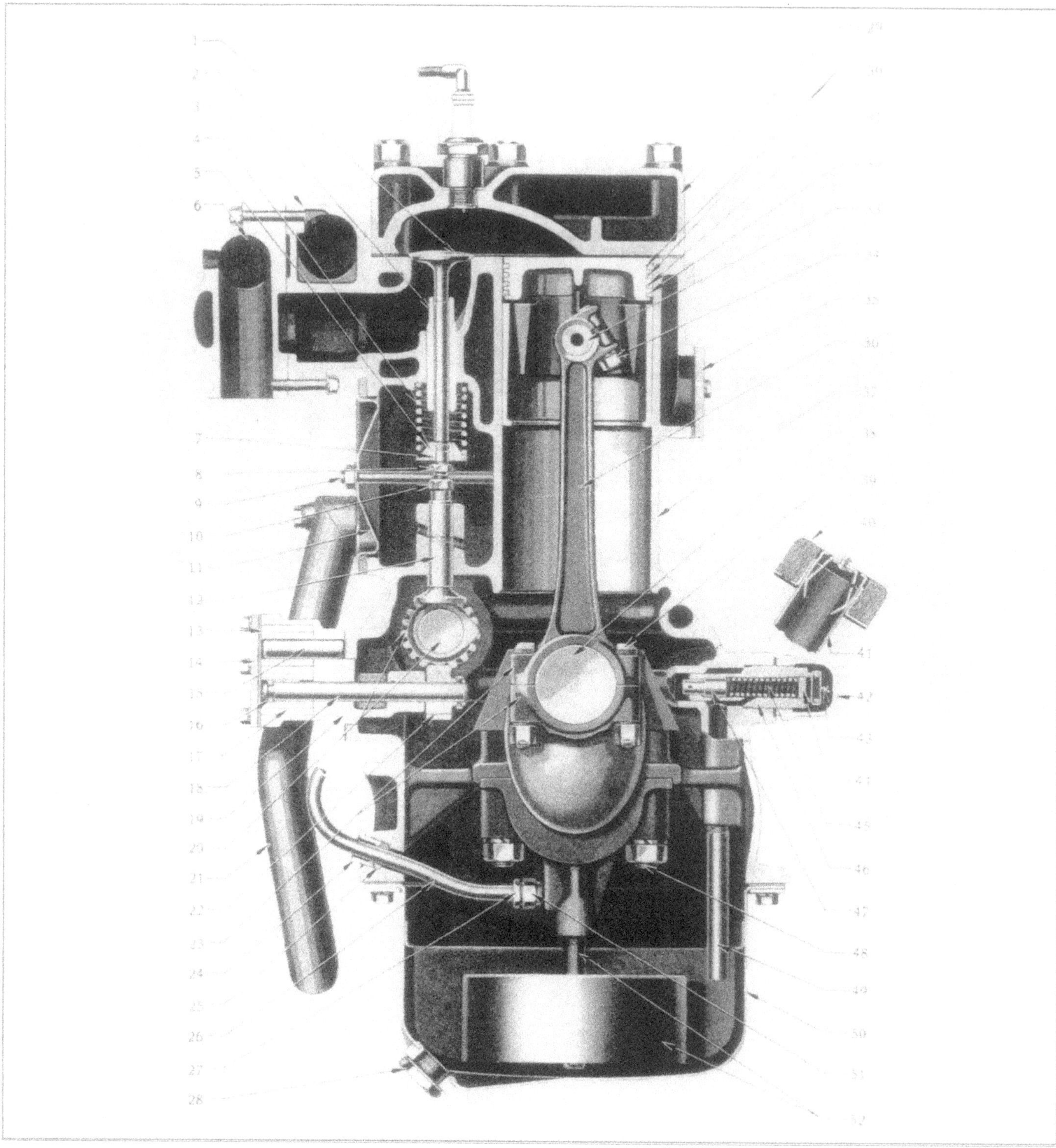

Fig. 4.18 *A cross section through the #1 cylinder of the 1932 Plymouth PB four-cylinder engine. The no-longer-needed hole for thermosyphon coolant has been capped with the thick plate located on the left side of the cylinder block near the top deck.*

TABLE 4.2
SPECIFICATIONS FOR MAXWELL, CHRYSLER, AND PLYMOUTH FOUR-CYLINDER ENGINES 1924–1928

Year	1924	1926	1927	1928
Model	Maxwell 25	Chrysler Four (later F58)	Chrysler I50	Chrysler I52 "after" and initial Plymouth Q
Displacement (cu. in.)	185.78	185.78	170.3	170.3
Bore	3.625	3.625	3.625	3.625
Stroke	4.5	4.5	4.125	4.125
Compression ratio	3.9:1	4.0:1	4.1:1	4.6
BHP@ rpm	34 @ 2200	38 @ 2200	35 @ 2200	45 @ 2800
Torque (lb-ft) @ rpm				
Combustion chamber	Open	Open	Open	Turbulent
Cyl. block mat'l.—skirt type	Cast iron—deep	Cast iron—deep	Cast iron—deep	Cast iron—deep
Cyl. block deck height	11.563	11.937	11.948	11.948
Cyl. block length	20.125	20.125	20.125	20.125
Cyl. bore spacing	1-2, 3-4 = 4.0 2-3 = 4.625	1-2, 3-4 = 4.0 2-3 = 4.625	1-2, 3-4 = 4.0 2-3 = 4.625	1-2, 3-4 = 4.0 2-3 = 4.625
Ck/shaft—no. of main brgs.	3	3	3	3
Ck/shaft main journal dia.	1 and 3 =1.875; 2 = 2.25	1 and 3 =1.875; 2 = 2.25	1 and 3 =1.875; 2 = 2.25	1 and 3 =1.875; 2 = 2.25
Crankpin journal dia.	1.875	1.875	1.875	1.875
Piston mat'l.—no. of rings	Aluminum—3	Aluminum—3	Aluminum—3	Aluminum—3
Connecting rod centers	7.875	7.875	7.875	7.875
Valve system	L-head	L-head	L-head	L-head
Valve events	220-232-neg 2	220-230-2	220-230-2	220-230-neg 2
Camshaft drive	Gear	Gear	Gear	Gear
Valve head dia.	Intake and exhaust 1.5	Intake and exhaust 1.59	Intake and exhaust 1.44	Intake and exhaust 1.44
Valve lash adjustment	Mech. tappets	Mech. tappets	Mech. tappets	Mech. tappets
Oil pump	Mtd to right side of block driven from camshaft. Full pressure lube.	Mtd. to right side of block driven from camshaft. Full pressure lube.	Mtd. to right side of block driven from camshaft. Full pressure lube.	Mtd. to right side of block driven from camshaft. Full pressure lube.
Distributor location	Front of gear case cover	Front of gear case cover	Front of gear case cover	Front of gear case cover
Cooling	Thermosyphon	Thermosyphon	Thermosyphon	Thermosyphon
Fuel system	Updraft carburetor	Updraft carburetor	Updraft carburetor	Updraft carburetor

TABLE 4.3
SPECIFICATIONS FOR 1929–1932 PLYMOUTH FOUR-CYLINDER ENGINES

Year	1929	1930	1931	1932
Model	Plymouth U	Plymouth 30U	Plymouth PA	Plymouth PB
Displacement (cu. in.)	175.4	196.1	196.1	196.1
Bore	3.625	3.625	3.625	3.625
Stroke	4.25	4.75	4.75	4.75
Compression ratio	4.6:1	4.6:1	4.9:1	4.9:1 std.; 5.5:1 opt.
BHP@ rpm	45 @ 2800	48 @ 2800	56 @ 2800	56 @ 2800 std.; 65 @ 3400 opt.
Torque (lb-ft) @ rpm		120 @ 1200	130 @ 1600	130 @ 1600 std.; 136 @ 1600 opt.
Combustion chamber	Turbulent	Turbulent	Turbulent	Turbulent
Cyl. block mat'l.—skirt type	Cast iron—deep	Cast iron—deep	Cast iron—deep	Cast iron—deep
Cyl. block deck height	12.943	12.943	12.943	12.943
Cyl. block length	21.00	21.00	21.00	21.00
Cyl. bore spacing	1-2, 3-4 = 4.0 2-3 = 4.625	1-2, 3-4 = 4.0 2-3 = 4.625	1-2, 3-4 = 4.0 2-3 = 4.625	1-2, 3-4 = 4.0 2-3 = 4.625
Ck/shaft—no. of main brgs.	3	3	3	3
Ck/shaft main journal dia.	2.25	2.25	2.25	2.25
Crankpin journal dia.	2.0	2.0	2.0	2.0
Piston mat'l.—no. of rings	Aluminum—3	Aluminum—3	Aluminum—3	Aluminum—3
Connecting rod centers	8.81	8.56	8.56	8.56
Valve system	L-head	L-head	L-head	L-head
Valve events	220-232-neg 2	220-232-neg 2	220-230-2	220-230-2
Camshaft drive	Gear	Gear	Gear	Gear
Valve head dia.	Intake 1.72; exhaust 1.59	Intake 1.84; exhaust 1.59	Intake 1.84; exhaust 1.59	Intake 1.84; exhaust 1.84
Valve lash adjustment	Mech. tappets	Mech. tappets	Mech. tappets	Mech. tappets
Oil pump	Mtd. to right side of block driven from camshaft. Full pressure lube.	Mtd. to right side of block driven from camshaft. Full pressure lube.	Mtd. to right side of block driven from camshaft. Full pressure lube.	Mtd. to right side of block driven from camshaft. Full pressure lube.
Distributor location	Front of gear case cover	Front of gear case cover	Front of gear case cover	Front of gear case cover
Cooling	Thermosyphon	Belt-driven water pump	Belt-driven water pump	Belt-driven water pump
Fuel system	Updraft carburetor	Updraft carburetor	Updraft carburetor	Updraft carburetor
Other			Start of Floating Power®	Ck/s has bolted-on cwts.

Chart 4.1
Time Lines—Maxwell, Chrysler, and Plymouth Four-Cylinder Engines

Engines, Car Lines, and Displacements (cu. in.)

Engine	Car Line	1921	1922-1925	1926	1927	1928	1929	1930	1931	1932
		- - Model Years - -								
Maxwell 4-cyl.	Maxwell	186								
2 main bearings	Model	25								
Maxwell-Chry-Ply	Maxwell		186							
4-cyl., 3 main bearings	Model		25 - 25C							
1 main bearing cap	Chrysler			186	170	170				
	Model			Four, F58	I-50	I-52				
	Plymouth					170				
	Model					Q				
Plymouth 4-cyl.	Plymouth						175	196	196	196
3 main brgs & caps	Model						U	30U	PA	PB

CHAPTER 5

Dodge Power Joins Chrysler

Dodge Brothers in 1928—A Company in Transition

At the time Chrysler purchased Dodge Brothers, Inc. on July 30, 1928,[1] Dodge Brothers had recently completed its program to leave the four-cylinder market and upscale its product line to only six-cylinder engines. The last four-cylinder powered car, a Model 128 Fast Four, had finished its production run on July 27, 1928,[2] only three days prior to the signing, leaving Dodge, as planned, with three car lines: Standard Six, Victory Six, and Senior Six. Two engines sufficed—a smaller 208.0-cu.-in. six-cylinder L-head that powered the Standard Six and the Victory Six, and a larger 224-cu.-in. six-cylinder L-head that powered the Senior Six model. Despite the recent demise of the four-cylinder engine, its importance to the formation and prosperity of Dodge Brothers gives it a rightful place in this account of Chrysler engines.

The Dependable Dodge Four

The year 1914 was the launch of the Dodge Brothers car and engine, five years after Henry Ford's first Model T rolled out of his modest factory to take its chances with the American car-buying public. Figure 5.1 portrays one of the Dodge cars built that year. Armed with experience from building engines, transmissions, and other major components for the successful Model T, Horace and John Dodge with engineer George Kiernan designed the engine for their own car.[3] In common with the Model T, the Dodge Brothers engine was an in-line four-cylinder with an L-head valve arrangement, open combustion chambers, a forged steel crankshaft with three main bearings, and a gear-driven camshaft. Unlike the Model T design that had a one-piece oil pan for both the engine and the planetary-gear transmission, allowing a common oil supply, the Dodge Brothers engine and conventional-gear transmission were completely separate assemblies. Displacing 212 cu. in. (upscale from the 176.7 cu. in. of the Model T), the Dodge engine had a bore and stroke of 3.875 × 4.5 in.; these dimensions would never change during the entire life of the engine. From front to rear, the spacings between the four cylinders were 4.625, 5.75, and 4.625 in. Figures 5.2 and 5.3 show views of the right and left sides of the 1919 engine, respectively. Figures 5.4 and 5.5 show engineering drawings of a longitudinal section and a cross section of the engine, respectively. For lubrication, the engine had a rotary eccentric-vaned oil pump, gear-driven from the crankshaft, located at the front of the engine below the crankshaft, together with an oil distribution tube in the crankcase and oil grooves cast into the crankcase walls. Gravity brought oil to the camshaft and crankshaft main bearings, while the connecting rods lubricated themselves by passing through splash troughs at the bottom of their strokes.

Fig. 5.1 *A 1914 Dodge Brothers touring car. Powered by a Dodge-built 212-cu.-in. L-head four-cylinder engine, this was the leadoff vehicle for Dodge Brothers as a producer of complete cars.*

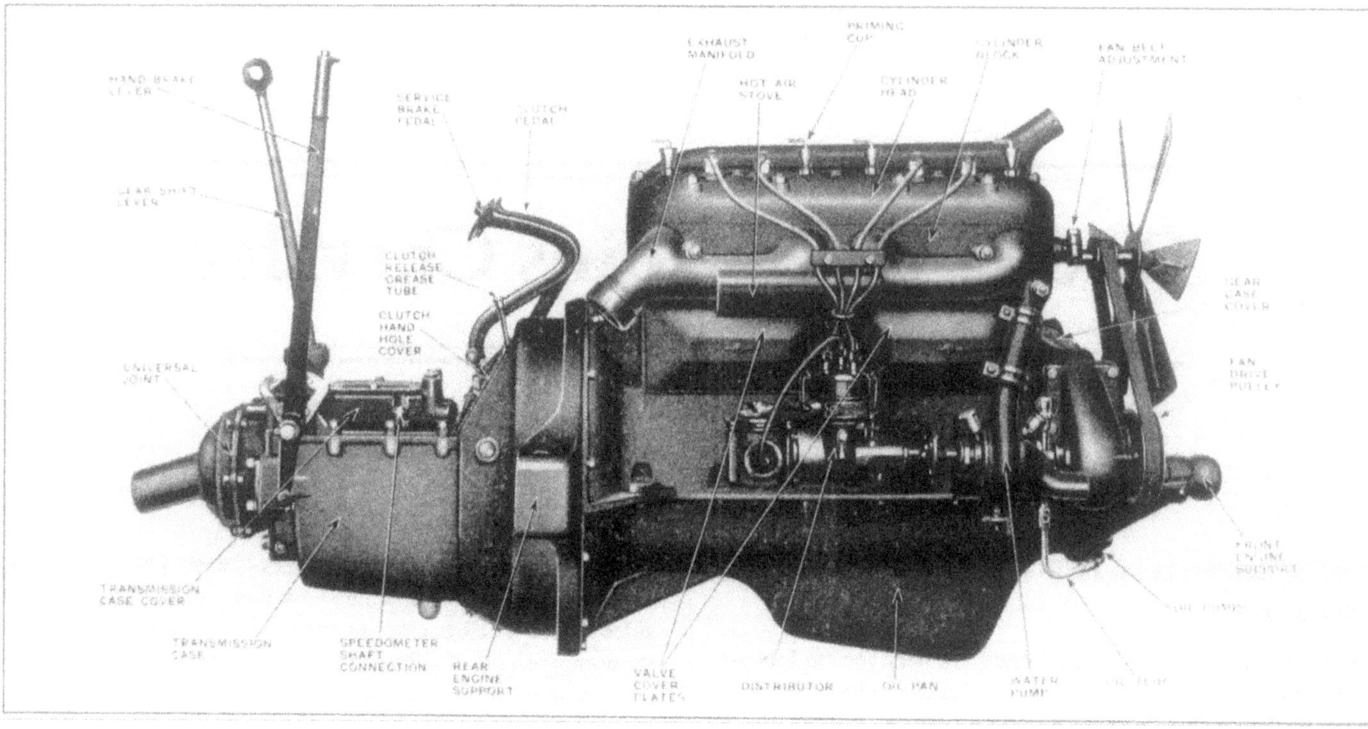

Fig. 5.2 *The right side of the 1919 Dodge Brothers 212-cu.-in. four-cylinder engine. A flat belt drove the fan, while the distributor was driven by a shaft system that extended through the water pump. Heat for the intake air of the engine was provided by a hot air stove, cast integrally with the exhaust manifold.*

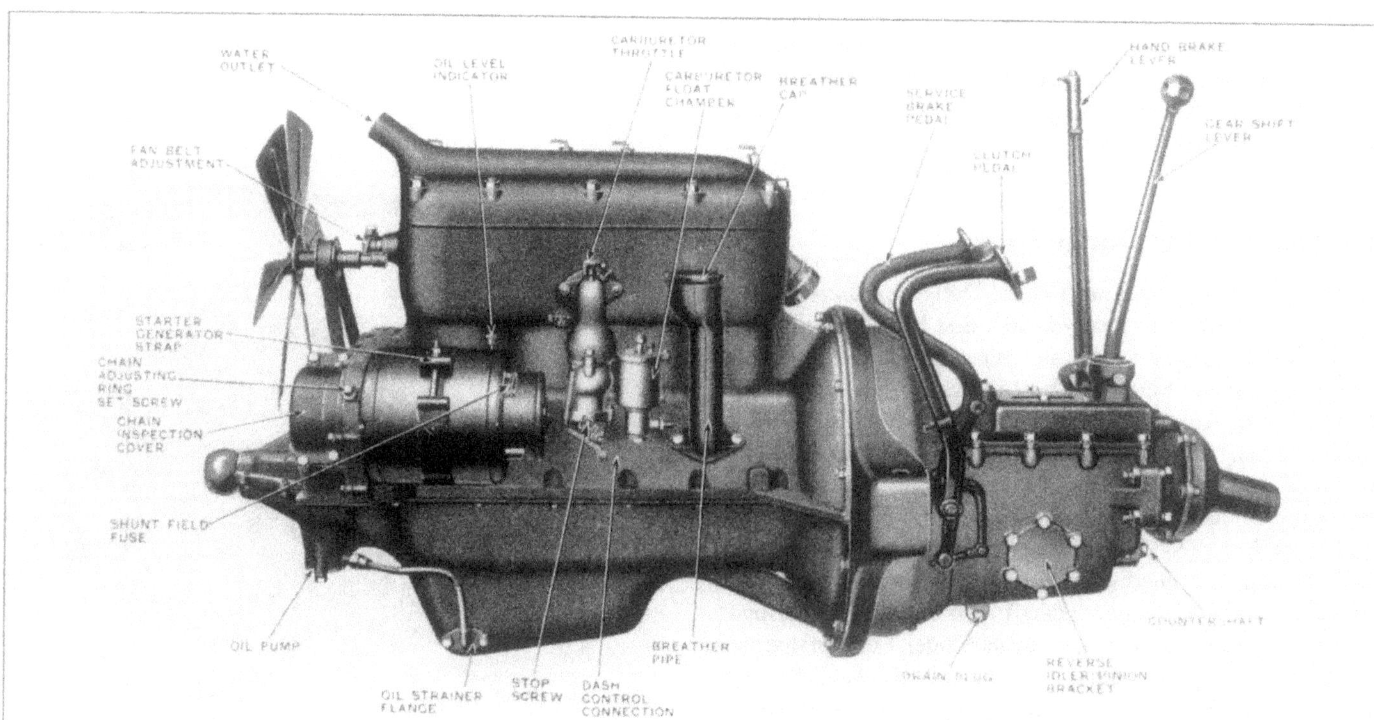

Fig. 5.3 *The left side of the 1919 Dodge Brothers four-cylinder engine. The large, cylindrical starter-generator provided the luxury of electric starting. The engine did not have an intake manifold; instead, passages inside the cylinder block distributed fuel/air mixture to the four intake valves. The external tubing shown at the bottom of the engine transported oil from the sump to the suction side of the oil pump.*

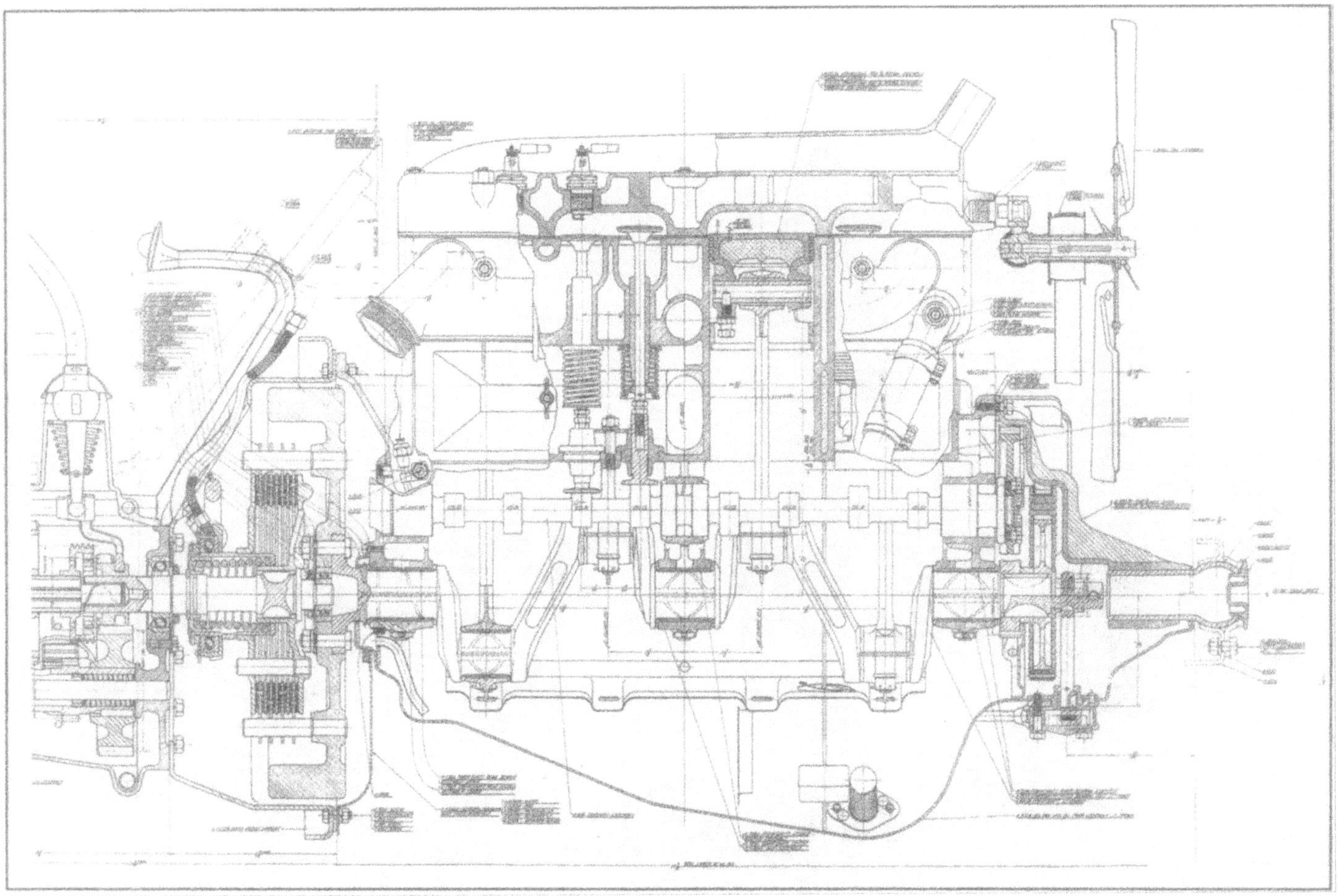

Fig. 5.4 *The longitudinal section of the 1925 Dodge Brothers 212-cu.-in. four-cylinder engine. Details as described in the text can be seen, including the oil pump at the front of the engine, splash troughs for the connecting rods, the racetrack-shaped hole through the center of the block for heated inlet air, and the round passage above it for fuel/air mixture to the intake valves. From Dodge Brothers Power Plant Assembly Drawing #22762.*

Unlike all Model Ts since the first 2,500 of them were built, the Dodge Brothers engine had a centrifugal water pump for positive circulation of coolant through the engine and the radiator. This was a quality feature that helped place it where the Dodge Brothers wanted to be—in the market segment above the Ford Model T. Another good feature of the Dodge engine was its provision for heating the intake fuel/air mixture during cold weather. Essentially two stages were used. First, the intake air (without benefit of any cleaning devices) entered an air jacket cast as part of the exhaust manifold. A short piece of sheet-steel tubing brought the heated air to a racetrack-shaped hole in the cylinder block, which went through the center of the block to emerge on the left side. A short, dual-walled tube next routed the air to the updraft carburetor, where it picked up fuel. Then, from the top of the carburetor, an elbow brought the fuel/air mixture back into the left side of the block, below the top deck. Entering a cast passage in the block, the fuel/air mixture gained additional heat as it traveled across the block to the right side, where, still inside the block, the mixture was routed to feed the four intake valves. Of interest, with this design the engine did not need an intake manifold; the passages inside the cylinder block served this purpose. The double-walled tube previously mentioned performed a weather-related function. The inner and outer segments of the tube each had four rectangular holes. The inner segment was fixed; the outer segment was movable. By rotating the outer segment, the holes could be lined up or blanked off. During cold-weather operation, with the holes blanked off, the carburetor drew heated air from the right side of the engine, as already described. For hot-weather driving, the outer segment could be rotated manually, lining up the holes and allowing unheated air to enter the carburetor, thereby bypassing the exhaust manifold heating system. The cooler air would help performance and avoid boiling the fuel in the carburetor during hot weather. The photo of the 1920 engine shown on Fig. 5.6 and a close-up photo of its carburetor area on Fig. 5.7 illustrate this system.

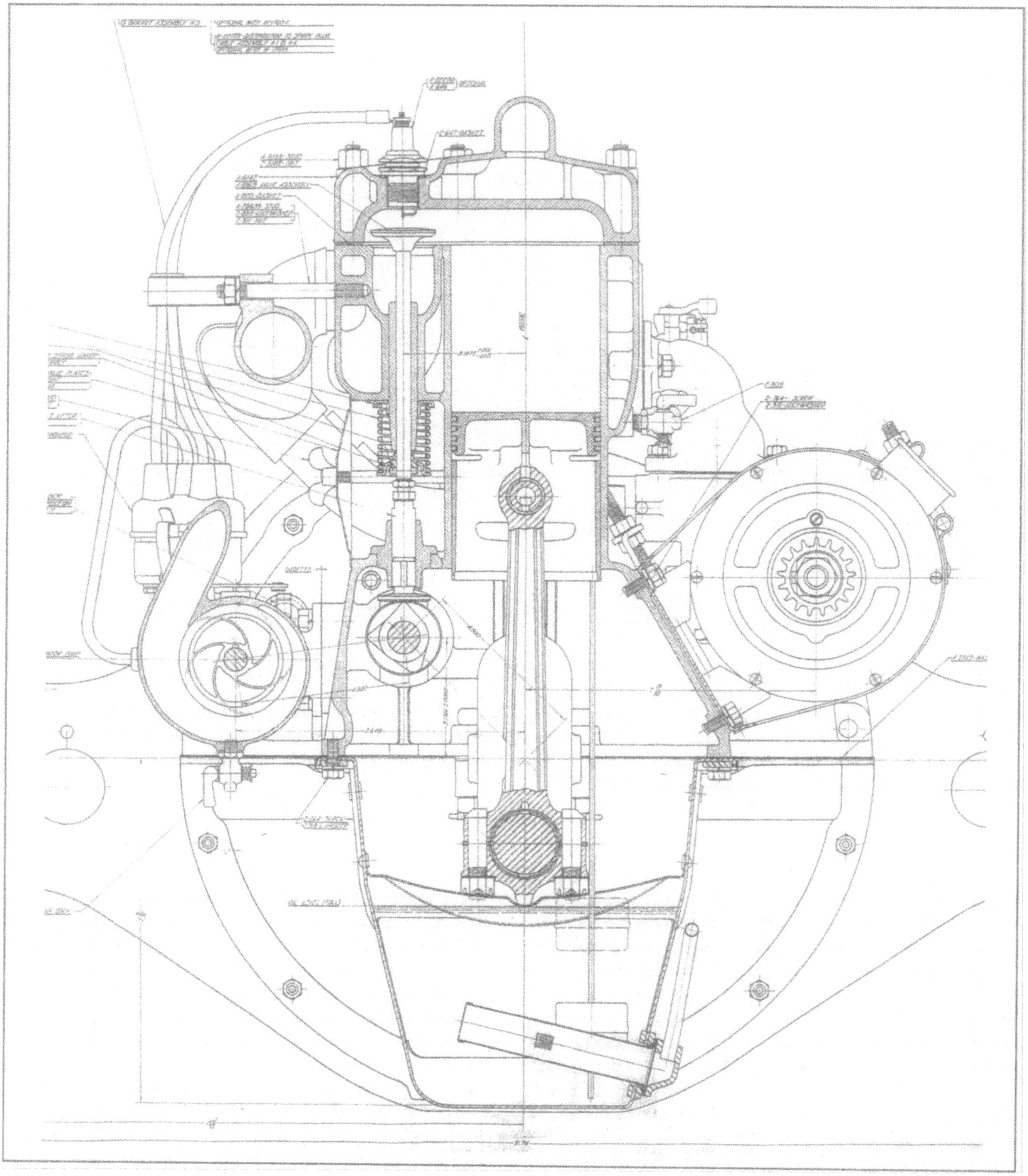

Fig. 5.5 *A cross section of the 1925 Dodge Brothers four-cylinder engine. Evident are features of the open combustion chamber, the water pump on the right side of the engine, and the large starter-generator on the left side. From Dodge Brothers Power Plant Assembly Drawing #22762.*

Fig. 5.6 *The left side of the engine compartment of the 1920 Dodge Brothers Screenside truck powered by a 212-cu.-in. Dodge four-cylinder engine. The elbow attached to the center of the cylinder block connects the updraft carburetor to the intake passage system inside the block. Vehicle in the Chrysler Historical Collection. Photo by W.L. Weertman.*

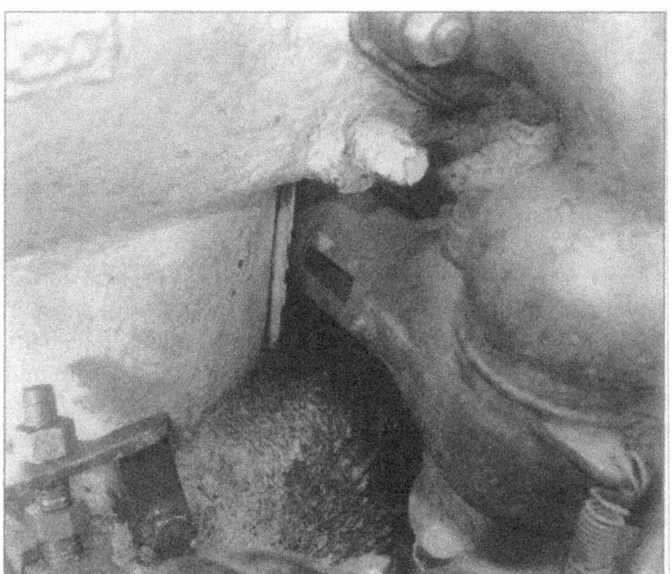

Fig. 5.7 *A close-up view of the carburetor area of the 1920 Dodge Brothers Screenside truck engine. The open rectangular hole (one of four) allows entry of unheated air into the carburetor. Rotating the sleeve shuts off the holes. Photo by W.L. Weertman.*

Probably the most important feature of the 1914 Dodge Brothers engine in its bid to occupy a place in the market above the entry-level Model T was its twelve-volt combined starter-generator bolted to the left-front corner of the engine. Driven by a chain from the crankshaft and spinning at three times the engine speed, the bulky cylindrical unit, which can be seen on Fig. 5.6, gave every owner the convenience of starting the engine of the car in a thoroughly modern and effortless manner.

Although a few changes were made, such as replacing the iron pistons with aluminum pistons in 1922, the 1914 Dodge Brothers engine remained pretty much the same for the next twelve years. Its somewhat unconventional but solid design, together with excellent manufacturing facilities, evidently led to the car having relatively few field problems. The result was a widespread reputation for genuine dependability, strengthened considerably by the successful use of Dodge Brothers cars by General John J. Pershing during the Mexican border campaign of 1916.[4]

This slow pace of new innovations changed in March 1926 when the starter-generator was replaced with two units: a starter motor, mounted on the left side of the rear flange of the cylinder block, and a six-volt generator, mounted on the right side of the engine and driven from the rear of the water pump shaft.[5] Another change was made at the same time. The distributor that had been located behind the water pump and driven through a flexible coupling was relocated to a much more solid position on the accessory housing where it was geared directly to the camshaft. In June 1926, a new bulbous-appearing sheet-metal air cleaner with radial, louver-like openings took engine inlet air from above the right side of the engine before sending it through the exhaust manifold heating jacket. For warm-weather operation, a bypass valve eliminated the exhaust manifold heater from the path of the inlet air.[5] Only four months later, in July 1926, another major change went into production—the addition of two main bearings, making a total of five for the engine.[5] However, as significant as these changes were, none affected the power-producing elements of the engine; thus, the output of the engine remained steady at 35 hp @ 2000 rpm.

Following these major changes to the original engine, in an even greater burst of four-cylinder creativity, the entire engine was replaced with a virtually all-new engine in March 1927 as power for the new Model 124 car. Its bore, stroke, displacement, cylinder spacing, and five-main-bearing crankshaft were retained, but everything else was new. Gone was the tortuous path for the inlet air. Instead, the right side of the cylinder block mounted conventional intake and exhaust manifolds. A vertical drive was located at the center-right side of the engine for driving the oil pump at the bottom of the block inside the oil pan and the distributor on top of the cylinder head. A timing chain and sprockets replaced the prior timing gears at the front of the engine. The timing chain also drove the generator, whose movable attachment was the means for chain tension adjustment. Outwardly, the engine, shown on Fig. 5.8, appeared similar to

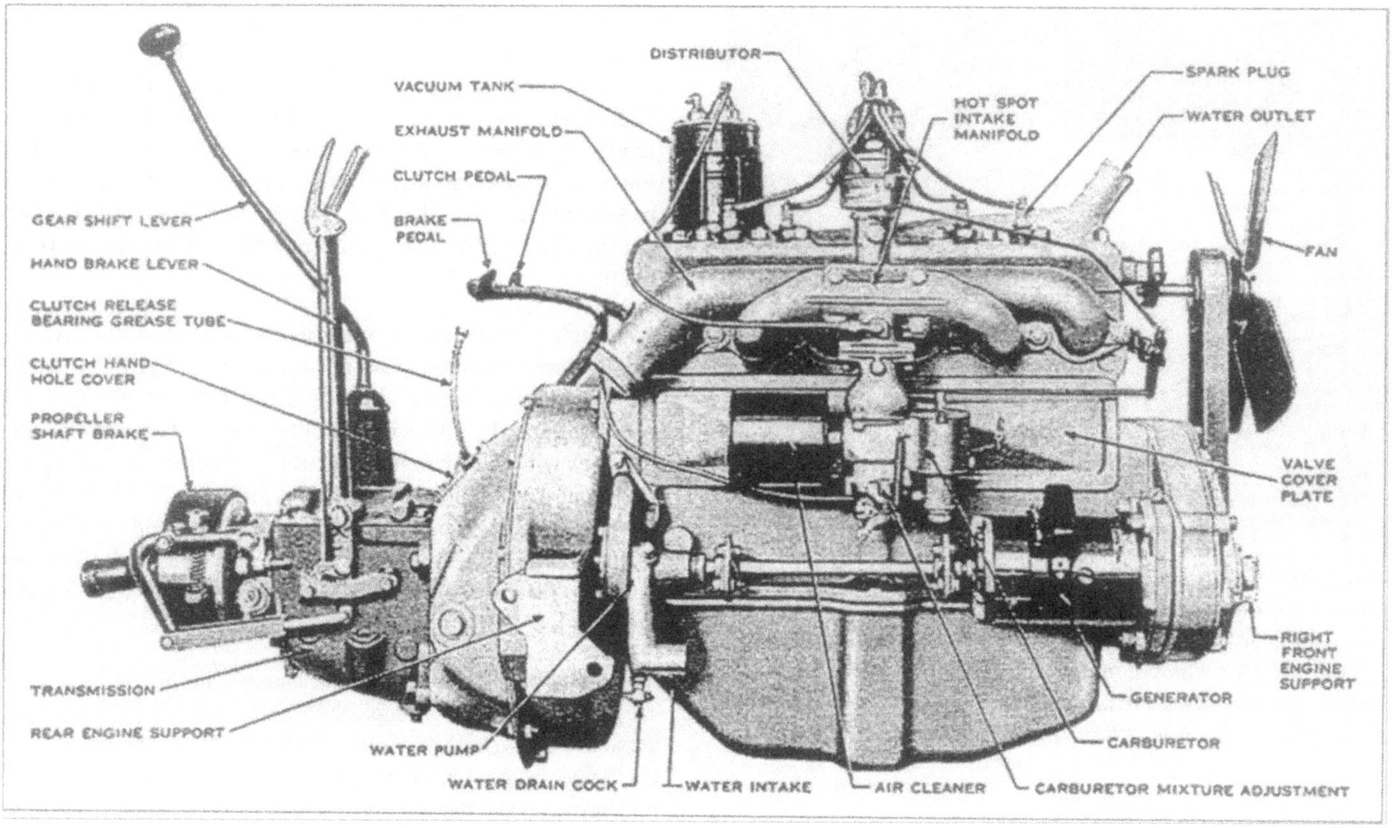

Fig. 5.8 *The right side of the 1927 Dodge Brothers 212-cu.-in. four-cylinder engine, as released for the Model 124 car. Although the cylinder bore center spacing and displacement were retained, the engine was all new.*

a four-cylinder version of the excellent Chrysler six-cylinder engines designed by Zeder, Skelton, and Breer (ZSB).

Somewhat puzzling was the Dodge Brothers Model 124 engine designers' reluctance to embrace several new technologies that would have considerably improved the engine. One of these was the V-type accessory drive belt, with its higher power transmittal capacity. Zeder, Skelton, and Breer took advantage of a V-belt to combine the drives of the water pump and the fan on the engines they designed for Chrysler cars. By contrast, Dodge Brothers designers stuck with the older flat belt to drive the fan, with the result that their water pump ended up at the back of the engine on the right side, driven by an exposed extension shaft, with universal joints. Another technology upgrade omission was the continuation of the splash and spray system of engine lubrication, while ignoring the high-pressure technology that was used universally on the ZSB-designed engines. Finally, and perhaps most importantly, the Dodge Brothers snubbed the turbulent combustion chamber that was well established and well publicized as an excellent way to boost engine performance and fuel economy. The new Model 124 engine was given a slight compression ratio increase to 4.1:1 from the prior 4.0:1 and, coupled with fewer restrictions in the inlet air path, provided a power increase to 40 hp @ 2400 rpm. Design details of the engine are shown in the longitudinal section on Fig. 5.9, and Fig. 5.10 shows the cross sections.

The car model run of the Model 124 ended in stages during 1927, depending on body type,[2] superseded by a shorter-wheelbase (108 versus 116 in.) Model 128. Powered with the same engine, the Model 128, advertised as the Fast Four and shown on Fig. 5.11, started production in July 1927.[2] Unfortunately, the Model 128 Fast Four car and engine became the victim of the Dodge Brothers plan to offer only cars with six-cylinder engines. When the new, low-priced Dodge Brothers Standard Six started production in February 1928, sales of the Model 128 Fast Four dropped, leading to its end of production, after only two model years, on July 27, 1928,[2] as noted at the beginning of this chapter.

Table 5.1 shows specifications for the 1925 through 1928 Dodge Brothers four-cylinder engines.

Dodge Brothers Senior Six Engine

In May 1927, Dodge Brothers started production of its first six-cylinder car and engine.[6] The stylish 1928 model year car,

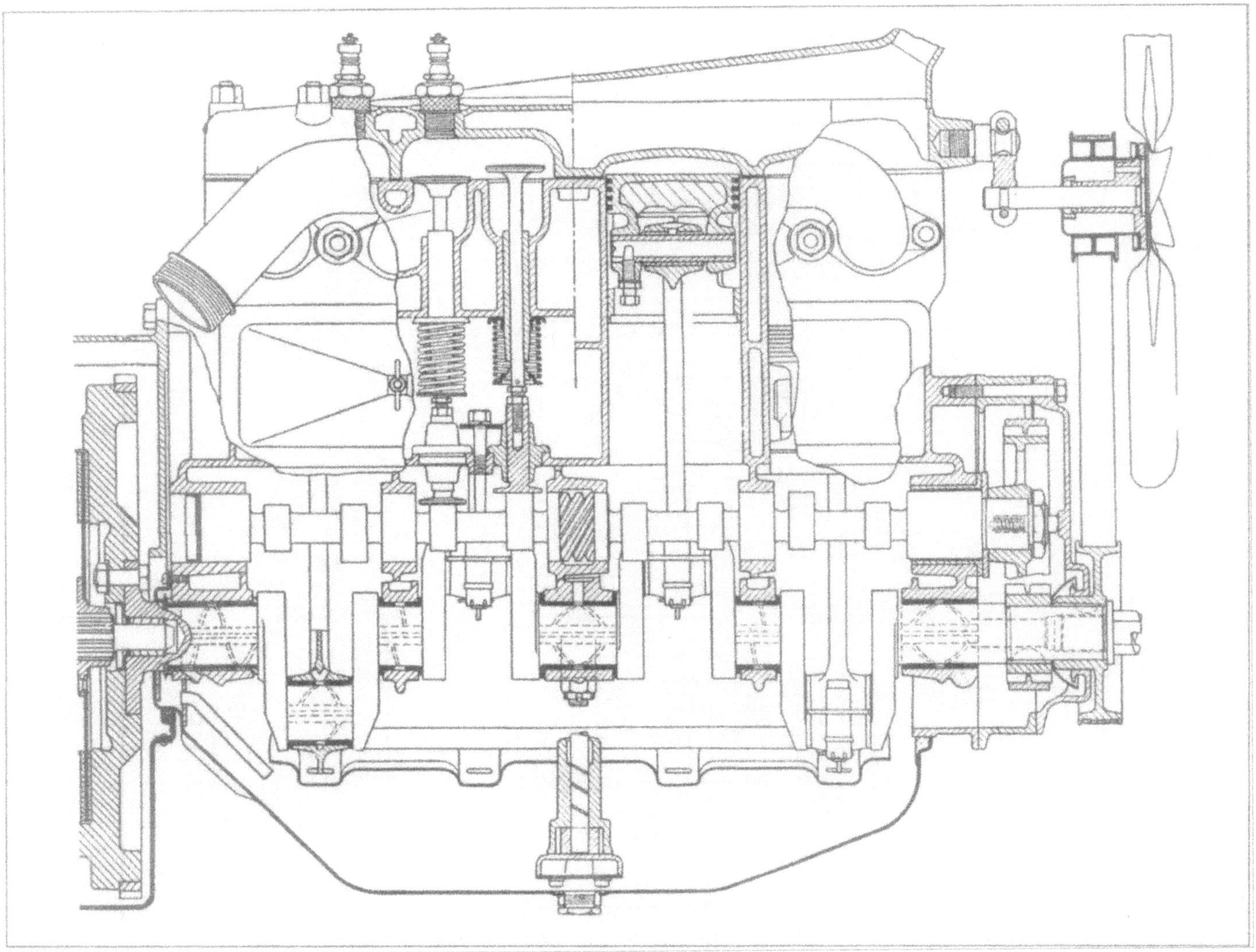

Fig. 5.9 *A longitudinal section of the 1927 Dodge Brothers 212-cu.-in. four-cylinder engine used in the Model 124 car. Note the connecting rod splash troughs and the five-main-bearing crankshaft. Despite having a rotary-gear oil pump, the splash and spray lubrication system was retained from the prior engine.*

shown in the advertising brochure picture on Fig. 5.12, was coded 2249 and bore the name Senior Six. A well-designed L-head engine provided 60 hp @ 2800 rpm from a displacement of 223.98 cu. in. with a typical bore and stroke for the era of 3.25 × 4.5 in. The words "well designed" are appropriate because the overall design of the engine was a dead ringer for the outstanding ZSB-designed Chrysler family of engines. Common features included turbulent combustion chambers, a seven-main-bearing crankshaft with full-pressure lubrication for all major bearings, a V-belt drive for the high-mounted water pump and fan at the front of the engine, a timing chain drive for the camshaft and generator with chain adjustment accomplished by moving the generator, and a vertical drive at the center of the engine for the oil pump at the bottom of the block and the distributor on top of the cylinder head. The hub of the front flange of the water pump/fan pulley was threaded for fan belt adjustment, similar to that of the ZSB-designed engines. A cylinder spacing of 4.25 in., except for the center two cylinders that were 4.9375 in. apart, showed the engine to be larger than the original ZSB Chrysler 70 series engine but not as large as the ZSB Chrysler Imperial 80 engine. Unlike the ZSB Chrysler engines, the cylinder block of the Dodge Brothers Senior Six engine had a deep skirt, giving the powerplant assembly greater stiffness for noise reduction. Figures 5.13 and 5.14 show a longitudinal section and cross section of the engine, respectively. Although designed by Dodge Brothers, the engine was not manufactured at the Dodge Main plant but was purchased from Continental Motors, the same source that William C. "Billy" Durant used for the Flint engine, as related in Chapter 2.

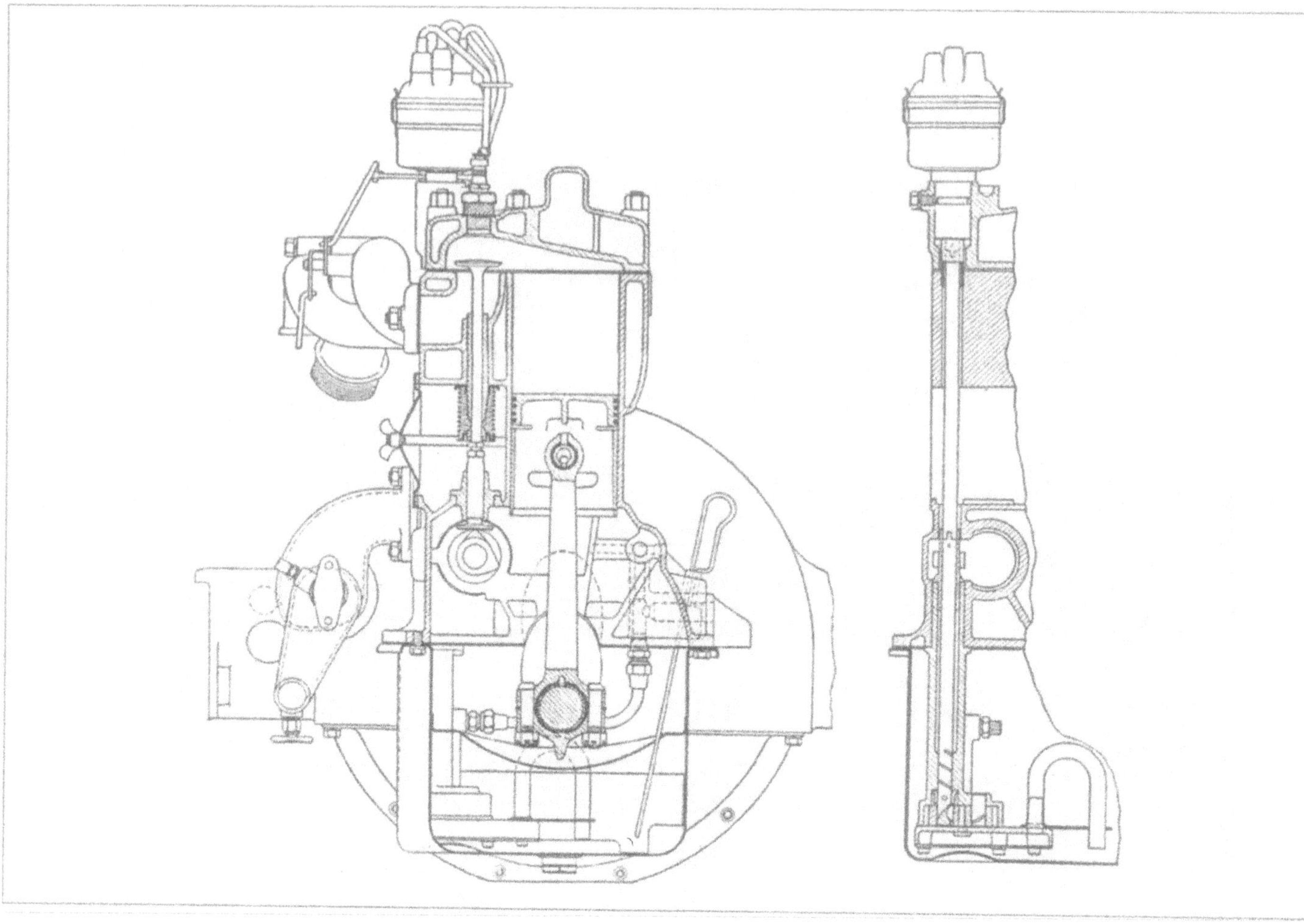

Fig. 5.10 *Cross sections of the 1927 Dodge Brothers four-cylinder engine. The general layout bears a striking similarity to the Chrysler ZSB-designed engines. However, the open combustion chamber is a puzzling carryover from the prior Dodge four-cylinder engine.*

Fig. 5.11 *This right-hand-drive export model 1928 Dodge Fast Four touring car was powered by the new Model 124 four-cylinder engine. The engine had a short production life; it was discontinued at the end of the 1928 model year.*

Midway in the model year, in January 1928, the car model designators became 2251 and 2252. For the 1929 model year, the car was coded as DB, and the car name "Dodge Brothers Senior" continued. This DB Senior engine was given changes that had a positive effect on its performance—the displacement was increased to 241.5 cu. in. by a bore increase to 3.375 in., and the compression ratio was bumped up to 5.6:1 from the prior 5.26:1, which resulted in an impressive output improvement to 78 hp @ 3000 rpm.

In addition to powering Dodge Brothers passenger cars, the six-cylinder Senior engine provided power for trucks built by Graham Brothers using Dodge chassis and drivelines, as will be described in Chapter 15. As good as the engine was, it did not survive the massive powerplant streamlining that took place two years after the purchase of Dodge Brothers, making its life only three model years long.

Table 5.2 shows specifications for the Dodge Brothers' Senior engine.

TABLE 5.1
SPECIFICATIONS FOR DODGE BROTHERS FOUR-CYLINDER ENGINES

Year	1925	Early 1927	Late 1927 Series 124, 1928 Series 128,129
Model	Dodge Brothers	Dodge Brothers	Dodge Brothers
Displacement (cu. in.)	212.28	212.28	212.28
Bore	3.875	3.875	3.875
Stroke	4.125	4.125	4.125
Compression ratio	4.0:1	4.0:1	4.1:1
BHP@ rpm	35 @ 2000	35 @ 2000	40 @ 2400
Torque (lb-ft) @ rpm			
Combustion chamber	Open	Open	Open
Cyl. block mat'l.—skirt type	Cast iron—shallow	Cast iron—shallow	Cast iron—shallow
Cyl. block deck height	13.375	13.375	13.375
Cyl. block length	24.719	23.469	23.469
Cyl. bore spacing	1-2, 3-4 = 4.625 2-3 = 5.75	1-2, 3-4 = 4.625 2-3 = 5.75	1-2, 3-4 = 4.625 2-3 = 5.75
Ck/shaft—no. of main brgs.	3	5	5
Ck/shaft main journal dia.	1.75	1.75	1.75
Crankpin journal dia.	1.375	1.375	1.375
Piston mat'l.—no. of rings	Aluminum—4	Aluminum—4	Aluminum—4
Connecting rod centers	9.125	9.125	9.125
Valve system	L-head	L-head	L-head
Valve events	223-223-neg. 2		223-223-neg. 2
Camshaft drive	Gear	Gear	Chain
Valve head dia.			Intake 1.72; exhaust 1.72
Valve lash adjustment	Mech. tappets	Mech. tappets	Mech. tappets
Oil pump	Gear type driven from nose of crankshaft	Gear type driven from nose of crankshaft	Bottom of block; vertical drive from camshaft
Distributor location	Rear of water pump shaft	Rear of water pump shaft	Top of cylinder head
Cooling	Aux. shaft driven water pump	Aux. shaft driven water pump	Water pump driven from generator
Fuel system	Updraft carburetor	Updraft carburetor	Updraft carburetor
Other	Combined starter/generator; no intake manifold	Separate starter and generator; no intake manifold	Separate starter and generator; normal intake manifold

Fig. 5.12 *The car model that was planned as the entry for Dodge Brothers into an upscale market before the purchase of the company by Chrysler Corporation was this 1928 Dodge Senior, powered by an all-new 224-cu.-in. L-head, six-cylinder engine.*

Fig. 5.13 *A longitudinal section of the 1928 Dodge Senior 224-cu.-in. six-cylinder engine. A seven-main-bearing crankshaft, full-pressure lubrication, and turbulent combustion chambers were modern features of this engine.*

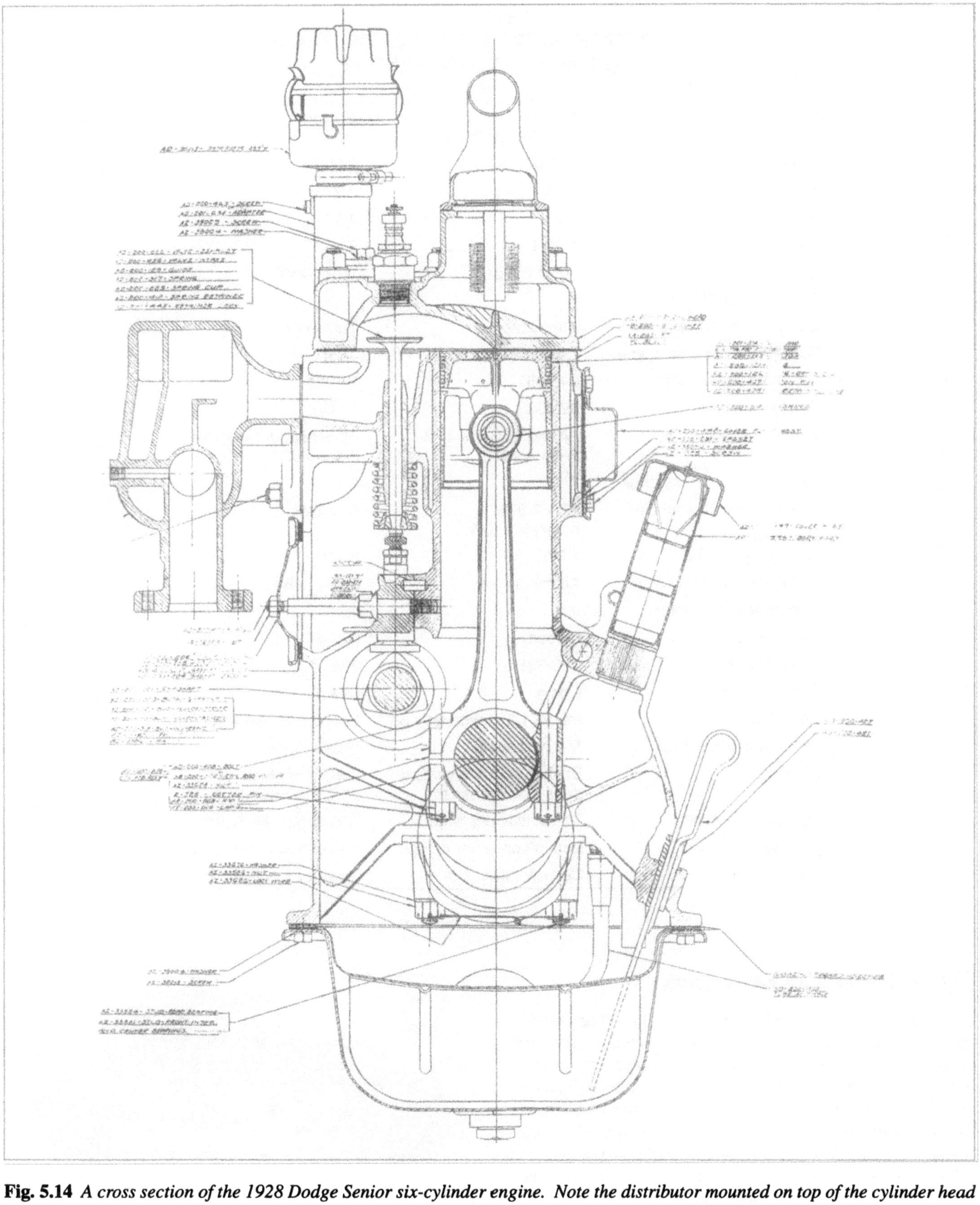

Fig. 5.14 *A cross section of the 1928 Dodge Senior six-cylinder engine. Note the distributor mounted on top of the cylinder head and the turbulent combustion chamber. From Dodge Brothers Engine Assembly Drawing #200772.*

TABLE 5.2
SPECIFICATIONS FOR THE DODGE BROTHERS SENIOR SIX-CYLINDER ENGINE

Year	1928	1930
Model	Senior 2249, 2251, 2252	Senior DB
Displacement (cu. in.)	223.98	241.5
Bore	3.25	3.375
Stroke	4.5	4.5
Compression ratio	5.26:1	5.56:1
BHP@ rpm	60 @ 2800	78 @ 3000
Torque (lb-ft) @ rpm		
Combustion chamber	Turbulent	Turbulent
Cyl. block mat'l.—skirt type	Cast iron—deep	Cast iron—deep
Cyl. block deck height	14.065	14.065
Cyl. block length	29.688	29.688
Cyl. bore spacing	All except 3-4 = 4.25 3-4 = 4.9375	All except 3-4 = 4.25 3-4 = 4.9375
Ck/shaft—no. of main brgs.	7	7
Ck/shaft main journal dia.	#1 = 2.375; others = 2.625	All 2.625
Crankpin journal dia.	2.375	2.375
Piston mat'l.—no. of rings	Aluminum—4	Aluminum—4
Connecting rod centers	9.5	9.5
Valve system	L-head	L-head
Valve events	228-236-6	228-236-6
Camshaft drive	Chain	Chain
Valve head dia.	Intake 1.66; exhaust 1.59	Intake 1.66; exhaust 1.59
Valve lash adjustment	Mech. tappets	Mech. tappets
Oil pump	Bottom of block; vertical drive from camshaft	Bottom of block; vertical drive from camshaft
Distributor location	Top of cylinder head	Top of cylinder head
Cooling	Belt-driven water pump	Belt-driven water pump
Fuel system	Updraft carburetor	Updraft carburetor

Dodge Brothers Victory Engine

In January 1928, about eight months after the Dodge Brothers Senior Six made its debut, a second six-cylinder-powered Dodge Brothers car was introduced. Figure 5.15 shows a coupe model of this new car line. With numerical designators of 130 and 131, the car was named Dodge Brothers Victory Six. Pegged below the Senior Six and essentially an upscale replacement for the long line of Dodge Brothers four-cylinder engines, the engine was designed with a smaller structure and some differences from the Senior Six.

Fig. 5.15 *The 1928 Dodge Victory Six coupe. Powered by a new 208-cu.-in. six-cylinder engine and somewhat smaller in structure and displacement than the Senior engine, the car was considered a hot performer in its class.*

Figures 5.16 and 5.17 show outside views of the right and left sides, respectively, of the 208-cu.-in. engine with a bore and stroke of 3.375 × 3.875 in. The cylinder spacing, a key element in basic engine size, for the Victory engine was set at 3.9375 in. for all but the center two cylinders that were 4.5 in. apart. With these numbers, the engine was, in terms of structural size, about halfway between the ZSB-designed original Chrysler six-cylinder engine and the later, smaller ZSB engine that powered the 60 series Chrysler cars. The familiar L-head valve arrangement, seven-main-bearing crankshaft, full-pressure lubrication, turbulent combustion chambers, vertical drive at the center of the engine for the oil pump at the bottom of the block and distributor on top of the cylinder head, and V-belt driven fan and water pump at the front of the engine were incorporated in the design. One part of the design was quite different from the normal L-head layout. Instead of being dead vertical, the valves were tipped at a 3 deg. 52 min. angle toward the cylinders. Regarding this valve inclination, the Dodge Brothers Handbook of Sales and Technical Information for the Victory Six advised its readers: "The valves are slightly inclined, being closer to the cylinder bore at the top than at the guides. This gives a more compact combustion chamber." The design was never repeated, indicating that the gains, if any, were less than noticeable. Of interest, this was not the only L-head engine with inclined valves. Around 1920, Studebaker produced an L-head six-cylinder engine with its valves inclined at a greater angle than the Dodge Victory engine. A longitudinal section, a cross section through the #1 cylinder, and a cross section through the center of the engine are shown on Figs. 5.18, 5.19, and 5.20, respectively. The modest output of 58 hp @ 3000 rpm indicated that the performance of the engine was not as well developed as the smaller ZSB P65 six-cylinder engine of 1929, which, with an identical compression ratio of 5.2:1, was getting 68 hp @ 3200 rpm out of its 195.6 cu. in.

Similar to the Dodge Brothers Senior Six engine, the Victory engine found good use in powering Dodge Brothers trucks, as will be described in Chapter 15. Also similar to the Senior Six, it did not survive the powerplant rationalization program of 1930 and, at the beginning of 1931, after powering Dodge cars for three years, was among the missing.

Table 5.3 shows specifications for the Dodge Brothers Victory and Standard Six engines.

Chart 5.1 shows time lines for the Dodge Brothers engines described in this chapter.

Fig. 5.16 *The right side of the 1929 Dodge Brothers DA 208-cu.-in. Victory six-cylinder engine. Released for the Dodge Victory and Standard Six models, the engine was close in size and appearance to the ZSB-designed engine used in the 60-series Chrysler cars. The flex hose connecting the rear tappet chamber cover to the air cleaner resembles an early form of closed crankcase ventilation.*

Fig. 5.17 *The left side of the 1929 Dodge Brothers DA Victory six-cylinder engine. An impressive oil filter is mounted near the center of the engine assembly.*

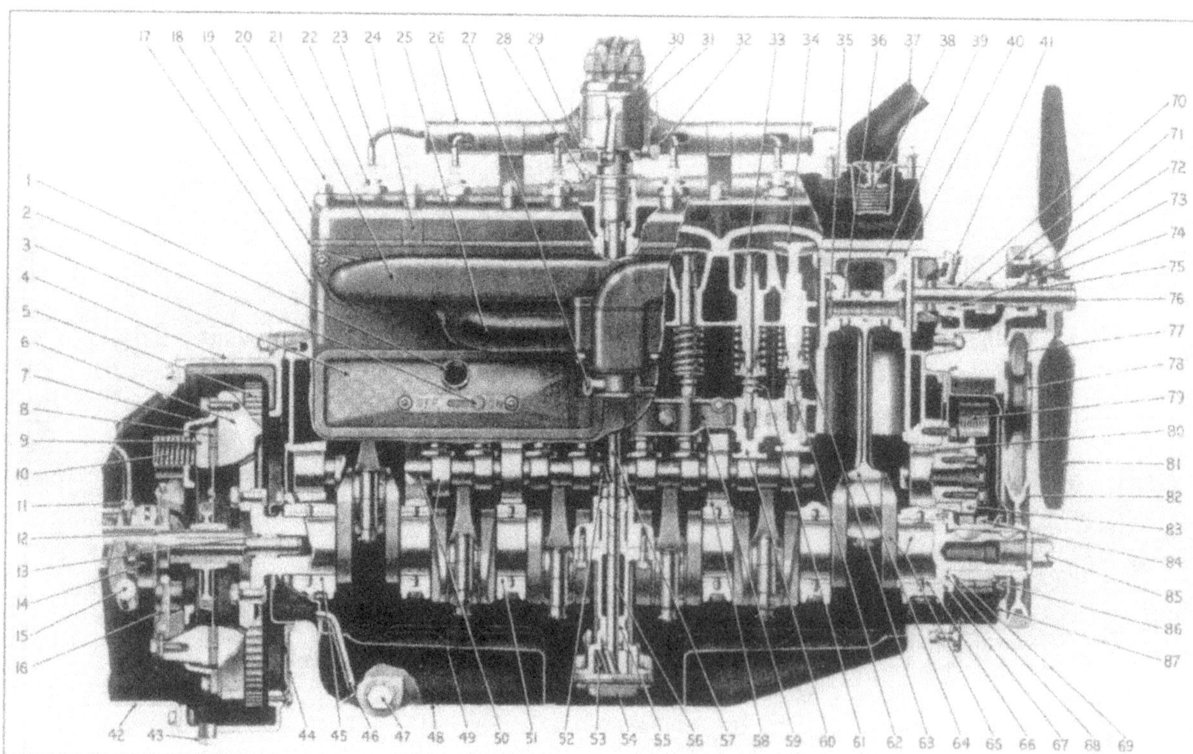

Fig. 5.18 *A longitudinal section of the 1929 Dodge Brothers DA Victory six-cylinder engine. Its seven-main-bearing crankshaft and vertical oil pump and distributor drive added to its similarity to the Chrysler ZSB-designed engines.*

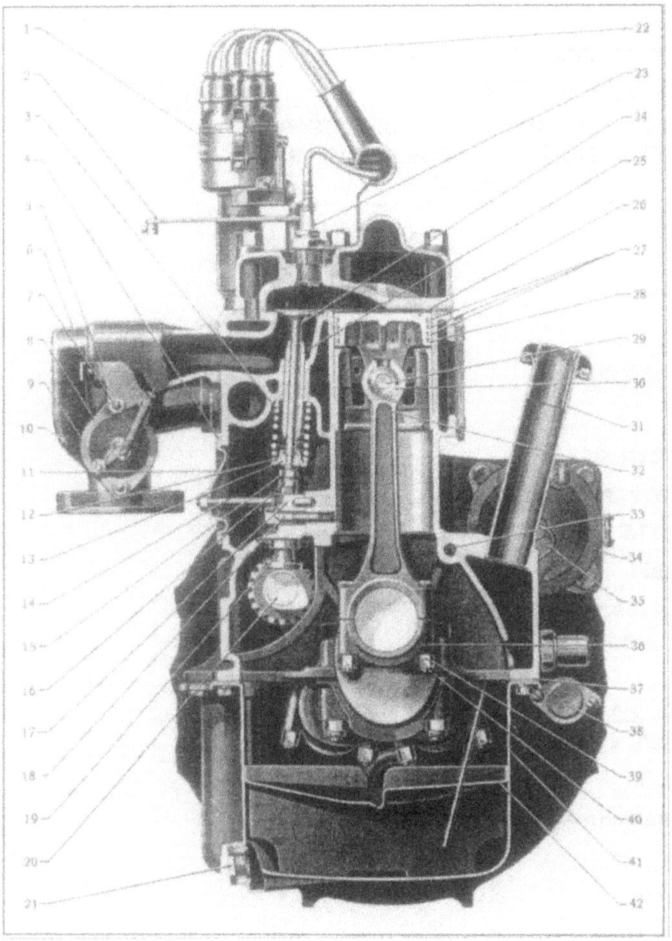

Fig. 5.19 *A cross section through the #1 cylinder of the 1929 Dodge Brothers DA Victory six-cylinder engine. An interesting aspect of the design is the tipping of the valves toward the cylinders. An exhaust port is sectioned; the round port below it is for intake fuel/air mixture.*

Fig. 5.20 *A cross section through the center of the 1929 Dodge Brothers DA Victory six-cylinder engine. The short skirt of the cylinder block was another feature in common with the Chrysler ZSB-designed engines. Note the large oil filter bracketed to the side of the cylinder block.*

TABLE 5.3
SPECIFICATIONS FOR THE DODGE BROTHERS VICTORY SIX-CYLINDER ENGINE

Year and Model	1928–1929 Victory Six 130M, 131M 1928–1929 Standard Six 140J, 141J 1929 DA
Displacement (cu. in.)	208
Bore	3.375
Stroke	3.875
Compression ratio	5.2:1
BHP@ rpm	58 @ 3000
Torque (lb-ft) @ rpm	
Combustion chamber	Turbulent
Cyl. block mat'l.—skirt type	Cast iron—Shallow
Cyl. block deck height	12.318
Cyl. block length	26.875
Cyl. bore spacing	All except 3-4 = 3.9375 3-4 = 4.5
Ck/shaft—no. of main brgs.	7
Ck/shaft main journal dia.	#1 = 2.25; others = 2.375
Crankpin journal dia.	2.125
Piston mat'l.—no. of rings	Aluminum—4
Connecting rod centers	8.0
Valve system	L-head
Valve events	228-236-6
Camshaft drive	Chain
Valve head dia.	Intake 1.65; exhaust 1.53
Valve lash adjustment	Mech. tappets
Oil pump	Bottom of block; vertical drive from camshaft
Distributor location	Top of cylinder head
Cooling	Belt-driven water pump
Fuel system	Updraft carburetor
Other	Valves are angled 3 deg, 52 min

Chart 5.1
Time Lines—Dodge Brothers Engines

Engines, Car Lines, and Displacements (cu. in.)

Engine	Car Line	1926	1927	1928	1929	Early 1930
Dodge Bros 4-cyl. Original design - 3 main bearings	Dodge Model	212				
Dodge Bros 4-cyl. Original design with 5 main brgs	Dodge Model		212			
Dodge Bros 4-cyl. New design with 5 main brgs	Dodge Model		212 124	212 Fast Four		
Dodge Bros Senior Six 7 main brgs, deep-skirt	Dodge Model			224 Senior	242 DB Senior	242 DB Sr
Dodge Bros Victory Six 7 main brgs, shallow-skirt	Dodge Model			208 Std 6, Victory 6	208 Std 6, Vict 6, DA	208 DA Six

CHAPTER 6

A New L-Head Six-Cylinder Design Takes Over

Following the purchase of Dodge Brothers in 1928, Chrysler cars and trucks were powered by no fewer than six different six-cylinder engines—the three engines designed by Zeder, Skelton, and Breer (ZSB) and described in Chapter 3; a recently released DeSoto engine described below; and the two Dodge sixes described in Chapter 5—with displacements ranging from 175 to 310 cu. in. Parts interchangeability was nil, and each engine had its own set of casting, forging, and machining lines. Meanwhile, in the late 1920s, a new marketing challenge had risen—the luxury field was turning to straight-eights as a minimum entry requirement.

Chrysler responded to the situation by creating not one but two new straight-eights to replace its luxury-market sixes and calling upon the somewhat diminutive new DeSoto six to replace all the remaining sixes so their production lines could be shut down. The story of the eight-cylinder engines will be told in Chapter 7. The story of the six-cylinder engine that started as a new engine for the new 1929 DeSoto car is told in the following paragraphs.

An Engine for DeSoto

The decision on the engine for the new medium-priced DeSoto car must have been a tough call. The 1928 Chrysler M62 was powered with the smallest of the three ZSB-designed engines that, at 180 cu. in. and 54 hp, would seem a leading candidate for the job. However, Harry T. Woolson, who was in charge of engine design,[1] came up with a new engine that matched its power (actually 1 hp better) with five fewer cubic inches in a more compact arrangement that was much cheaper to build. Woolson's engine understandably got the nod as the engine for DeSoto. Of interest, Harry Woolson's son, L. Irving Woolson, newly arrived at Chrysler two years after receiving his engineering degree, had his first job assignment in the Engine Design department and evidently made drawings for the Desoto engine.[2] Highly talented, he later would become president of the DeSoto Division. Although described as a DeSoto engine, the engine was designed and developed by the centralized Engineering Division and was manufactured at the Chrysler Jefferson Avenue plant because DeSoto had no manufacturing facilities at the time.

Assigned a single alphabetic designator K, the 1929 DeSoto was built in a number of body styles, including the sporty, wire-wheeled roadster shown on Fig. 6.1. Figure 6.2 gives a look under the hood of the 1929 DeSoto on display at the Walter P. Chrysler Museum. Additional outside views of the engine, showing the right and left sides, appear on Figs. 6.3 and 6.4, respectively.

Fig. 6.1 *The K-model roadster of the newly created 1929 DeSoto line of cars. A look under the hood of this sharp-looking, wire-wheeled car would have revealed an all-new Chrysler-built 175-cu.-in., 61.5-hp L-head six-cylinder engine.*

Fig. 6.2 *The right side of the engine compartment of the 1929 DeSoto K-model roadster on display at the Walter P. Chrysler Museum. The cylinder head of this 175-cu.-in. six-cylinder engine is painted red, indicating that it is the high-compression-ratio, 61.5-hp version that was standard on the roadster. Photo by W.L. Weertman.*

The in-line six-cylinder engine was laid out with the usual L-head valve arrangement having all valves located along the right side of the cylinder block. The bore and stroke were 3.0 × 4.125 in., giving a displacement of 174.9 cu. in. As shown on the longitudinal and cross-section views of the engine on Figs. 6.5. and 6.6, respectively, a turbulence-inducing narrow gap between the piston head and the bottom of the cylinder head permitted the engine to burn regular fuel with its high-compression (5.2:1) standard head. A higher-compression (6.0:1) Red Head, which was standard on the roadster and optional on other models, provided more performance but required high-test ethyl gas. Unlike the ZSB trio of engines with their two tiers of intake and exhaust ports, the intake and exhaust ports were compacted into a single row just below the top deck. The water pump was placed in its customary ZSB position, high at the front of the engine where it could share a shaft and V-belt drive system with the cooling fan.

Inside the crankcase, a major design shift took place. Instead of the ZSB signature of seven main bearings for the crankshaft, there were now only four. The crankshaft was reshaped to have husky connectors in place of the three missing bearings.

Fig. 6.3 *The right side of the Chrysler-built 1929 DeSoto K-model 175-cu.-in. six-cylinder engine. The oil pump has been moved from inside the oil pan to outside the cylinder block, just above the oil pan rail at the center of the engine. Two horizontal ribs reinforce the deep skirt of the new block.*

Fig. 6.4 *The left side of the 1929 DeSoto K-model six-cylinder engine. The distributor has been relocated from its former position prominent on top of the cylinder head to the center of this side of the cylinder block.*

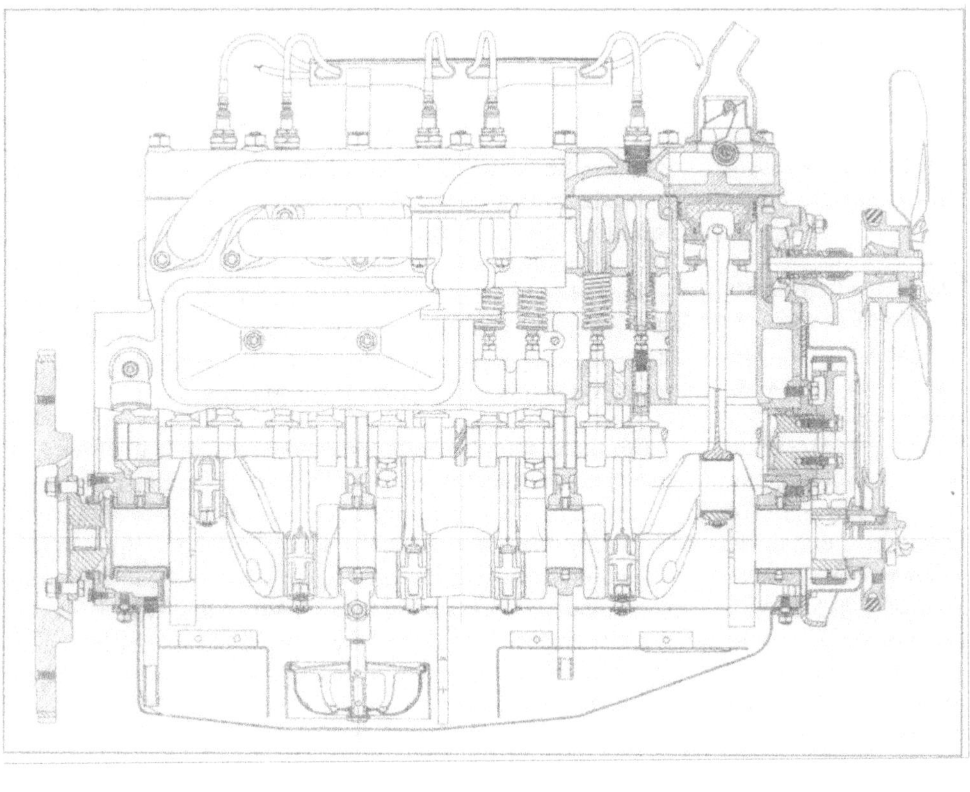

Fig. 6.5 *A longitudinal section of the 1929 DeSoto K-model six-cylinder engine. The use of four main bearings in place of the ZSB-customary seven represented a considerable departure in design approach. Stout connectors between the crankpins were provided.*

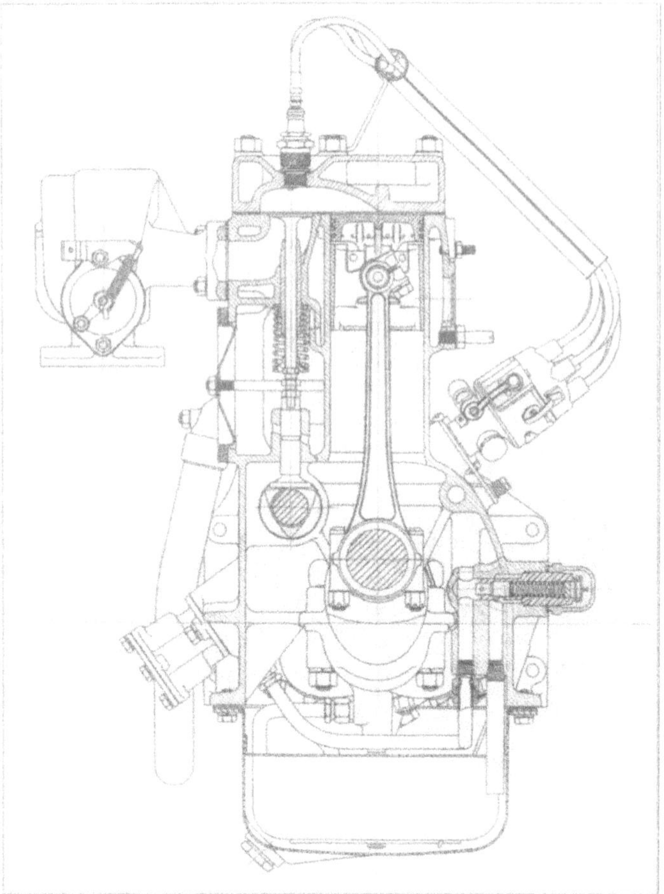

Fig. 6.6 *A cross section through the #1 cylinder of the 1929 DeSoto K-model six-cylinder engine. The favored turbulent combustion chamber with its tight quench gap was incorporated in the design of this new engine.*

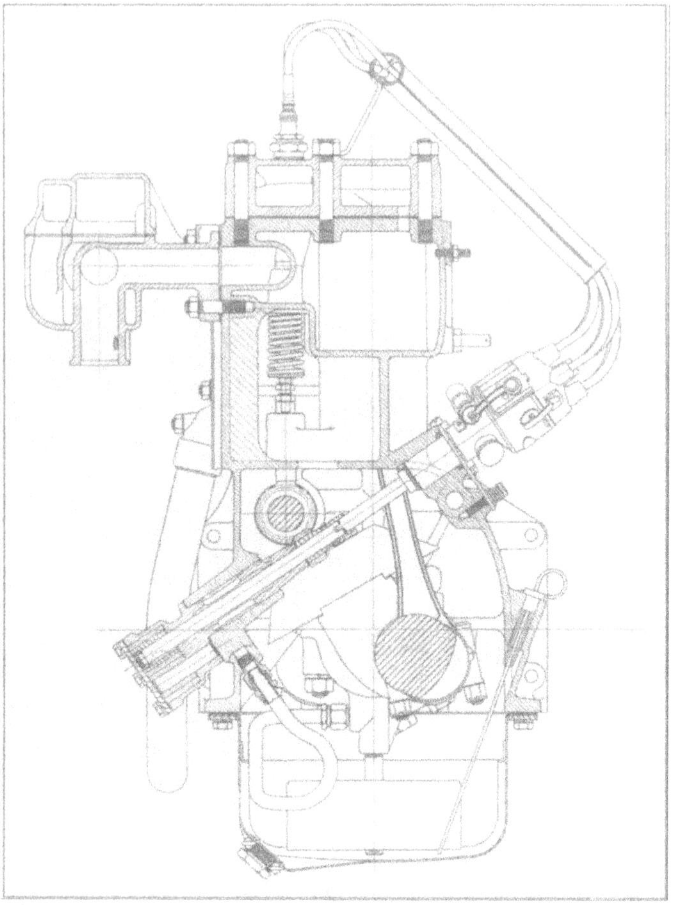

Fig. 6.7 *A cross section through the center of the 1929 DeSoto K-model six-cylinder engine. As depicted, a diagonal shaft, crossing the block below the cylinders, was used for driving the oil pump and the distributor. One of four bolted-on crankshaft counterweights can be seen beyond the connecting rod.*

Not only did this change make the engine less costly and simpler to build, it also permitted cylinders to be crowded closer together.

The distance between the centerlines of the first two cylinders (#1 and #2), the middle two cylinders (#3 and #4), and the last two cylinders (#5 and #6) was reduced to a tight 3.5625 in. The cylinder spacing between cylinders #2 and #3 and between cylinders #4 and #5 was considerably larger at 4.3125 in., giving room for the main bearings located below each pair. Gone was the elegantly designed vertical drive for the oil pump at the bottom of the cylinder block and distributor high on top of the cylinder head. In its place was a shorter drive that made a diagonal traverse at the center of the cylinder block, which can be seen on the cross section displayed on Fig. 6.7, with the distributor located on the upper-left side of the block and the oil pump on the lower-right side. A close look at the cross section discloses that the distributor driveshaft was positioned below the adjacent cylinder barrels, which allowed the designer to place the center cylinders, #3 and #4, at the snug 3.5625 in. spacing previously noted. Changing the location of the distributor from the top of the head to the side of the block had a substantial benefit for service. The cylinder head now could be removed and replaced without having to disturb the distributor. There was no need to retime the engine after scraping carbon from the combustion chamber and valve heads, which was a common maintenance procedure at the time.

With three main bearings removed, the torsional stiffness of the crankshaft increased so much that the torsional damper, a necessity on ZSB seven-main-bearing engines, was eliminated. A quality upgrade given to the engine was the addition of four bolted-on crankshaft counterweights to reduce the main bearing loads and to impart smoothness. Another quality item was the heavily ribbed, deep skirt of the cylinder block, which reduced powerplant flexing and provided a firm foundation for mounting of the oil pump.

Chrysler management evidently liked this new engine to the extent that, in December 1929 when a new 1930 Dodge DD model replaced the Dodge DA, instead of carrying over the Dodge Victory engine, this four-main-bearing, deep-skirt, six-cylinder L-head engine—with its bore enlarged to 3.125 in.—was chosen as the powerplant for the Dodge DD. With the larger bore, the displacement went up to 189.8 cu. in., and 5 hp were added to its power, for a total of 60 hp. A month later, in January 1930, a new and smaller Chrysler Six model, code lettered CJ, started production, powered by this engine with the displacement further enlarged to 195.6 cu. in. by a stroke increase to 4.25 in. Figure 6.8 shows a Chrysler CJ two-door coupe, one of several body styles offered with the engine. In a somewhat retro move, possibly to reduce cost, the Chrysler version was equipped with helical gears for driving its camshaft and did not use the bolted-on crankshaft counterweights. The results evidently were not satisfactory, because the timing chain and counterweights were reinstated the following year. In a catch-up move, DeSoto increased the displacement of its version in May 1930 to be the same as that of the 1930 Dodge DD. With the larger 189.8-cu.-in., 60-hp engine, the DeSoto was coded CK and was called the Finer DeSoto.

A new development for these 1930 engines was the addition of a mechanical fuel pump to the right side of the cylinder block where an eccentric on the camshaft between cylinders #1 and #2 depressed and released the spring-loaded operating lever of the pump. With the pump supplying pressurized fuel to the carburetor, the firewall of the car was cleared of the bulky vacuum tank that had been used for fuel transfer from the gas tank.

This sharing of the same basic engine by three different car lines—Dodge, DeSoto, and Chrysler—was an interesting event for Chrysler Corporation. In a way, it was a forerunner of the shift of engine production from the individual car divisions to centralized engine manufacturing operations, which took place in the mid-1950s, about forty years later.

Fig. 6.8 *The 1930 Chrysler CJ two-door coupe. Two levels of power were available—62 and 70 hp, both obtained from a 196-cu.-in. version of the four-main-bearing, deep-skirt, six-cylinder engine launched with the 1929 DeSoto K-model.*

Then, for the 1931 model year, a profound and unusual change was made to the engine—the bores were moved! I can only surmise that Engineering was faced with the challenge of increasing the displacement of the engine to provide more power and torque for improved vehicle performance, a basic move to help sales appeal in times of economic depression. However, when the new four-main-bearing, deep-skirt, six-cylinder engine was scrutinized for possible displacement increases beyond the 1930 levels, problems were revealed. Bore increases beyond the 3.125-in. diameter of the 1930 engine could lead to connecting rod big-end bearing overload and possibly could jeopardize the cylinder head gasket sealing between the narrowly spaced cylinders. Certainly many design studies must have been made before Engineering came to the unpleasant conclusion that increasing the bore size to 3.25 in. required moving the cylinder bores and lengthening the connecting rod big-end bearings.

Doing the design job without changing the overall length of the cylinder block required a deft reapportioning of the elements of the crankshaft and the placement of the cylinders. Each odd-numbered cylinder was moved 0.0625 in. toward the front of the cylinder block, and each even-numbered cylinder was moved 0.0625 in. to the rear. To increase the bearing area of the connecting rod big-ends, metal was added not symmetrically but on only one side, making the connecting rods nonsymmetrical. Thus, the offset connecting rod was born. The odd-numbered rods had the big-end extension facing toward the rear, and the even-numbered rods had it facing forward. The push–pull load transmitted down the shank of the rod was no longer in the center of the big-end bearing, but it was an acceptable solution, albeit with added complexity for production and service.

With this solution in hand, the bores of the 1931 engines—the DeSoto SA, the Dodge DH, and the Chrysler CM—were increased to 3.25 in. The DeSoto SA version with a 4.125-in. stroke had 205.3 cu. in. The Dodge DH version with a longer stroke of 4.25 in. gave it 211.5 cu. in., and the Chrysler CM version with an even longer stroke of 4.375 in. gave it the largest displacement of 217.8 cu. in. Power levels of the three versions were significantly higher at 72, 74, and 78 hp, respectively, and all at 3400 rpm. The engines could no longer get by without crankshaft vibration dampers, so dampers were added to the noses of all three crankshafts.

The demise of the other five six-cylinder engines came swiftly at the end of 1930. This four-main-bearing, deep-skirt six-cylinder engine single-handedly carried the whole power load for Chrysler Corporation six-cylinder passenger cars in 1931 and 1932. Displacements for 1931 are as previously noted. In 1932, the displacements were notched upward, with DeSoto receiving the 3.25 × 4.25 in., 211.5-cu.-in. size. Dodge received the 3.25 × 4.375 in., 217.8-cu.-in. size, and Chrysler received a new, longer-stroke, 3.25 × 4.5 in., 223.98-cu.-in. engine. As usual, the Chrysler version was the power leader, with 82 hp @

3400 rpm and torque of 156 lb-ft @ 1200 rpm. In a quality upgrade, all the 1932 six-cylinder engines received the benefit of the new Chrysler first—Floating Power® mounts, originally installed on 1931 Plymouth four-cylinder cars—for a marked improvement in keeping engine vibrations from being transmitted to the chassis.

Dodge Truck engineering also had to scramble to replace the Dodge Victory and Senior engines it had been using but which were no longer available after 1930. Therefore, this four-main-bearing, deep-skirt six-cylinder engine, at a displacement of 211.5 cu. in., was pressed into truck service and did well as a medium-duty engine, starting in 1931.

The year 1933 saw a reduction in the volume requirements of this four-main-bearing, deep-skirt, six-cylinder engine. Its usage in the Dodge cars dried up when the Dodge Main plant came on-line with a new, smaller six-cylinder engine (described in the next section), and, with a straight-eight engine covering the higher end of its market, Dodge no longer needed the engine. DeSoto continued using the engine at 217.8 cu. in. as power for its 1933 Model SD cars and, at 224 cu. in., it continued as power for the 1933 Chrysler Six CO cars. This 1933 realignment of the Dodge and DeSoto engines, with Dodge tooling a smaller-block engine and DeSoto sharing the larger Chrysler engine, signaled a shift in market position for the two car lines. The shift placed DeSoto between Chrysler and Dodge,[3] and Dodge between DeSoto and Plymouth, and this ranking would endure for decades. Nothing was new about the 1933 displacements, but a change was incorporated in the cylinder block that represented a watershed improvement in engine durability. The change amounted to six steel rings that were shrink-fitted into counterbores in the top deck of the cylinder block at the locations of the exhaust valve seats of the cylinders. The steel rings, or seat inserts as they were called, resisted the pounding of the exhaust valves, heated by exhaust gases, much better than did the cast iron of the cylinder block as conventionally used. The persistent problems of exhaust seat wear, warping, and erosion finally were overcome, thanks to this Chrysler "first."[4]

The year 1933 also saw another important change to this engine—its carburetion was changed from updraft to downdraft. Following the lead of the 1930 ZSB-designed Chrysler 65, 70, and 77 engines, the carburetor, manifolds, and air cleaner were redesigned to mount the carburetor above rather than below the intake manifold for improved air flow and fuel/air preparation. Advantage was taken of the uncrowded high location to greatly enlarge the inlet-silencing chamber of the air cleaner, as can be seen in the right- and left-side views of the engine shown on Figs. 6.9 and 6.10, respectively.

The year 1934 saw yet another upward move in displacement—this time, by a bore change to 3.375 in. With a carryover stroke of 4.5 in., the displacement increased to 241.5 cu. in., a size shared by both Chrysler and DeSoto. Further displacement changes were never made, and the 241.5-cu.-in. size remained intact through the last year of production for this engine in 1936. Engine improvement changes continued, however, including a crankshaft change in 1934 that resulted from a change in the forging process. The forging plant found a way to move red-hot steel to the locations along the crankshaft needed for counterweights. Thanks to the new forging process, counterweights became integral with the crankshaft forging; thus, the bolted-on counterweights were discontinued at considerable savings in both cost and complexity.

In 1934, the new models of cars carrying DeSoto, Chrysler, and Chrysler Imperial nameplates, with appearance influenced heavily by wind-tunnel tests and advanced structural analysis, were introduced to the car market. Of course, the new models were the Airflows. The 241.5-cu.-in. engine prepared for the 1934 DeSoto Airflow SE, the only car DeSoto sold in 1934, featured a high-compression (6.0:1) aluminum cylinder head that boosted output to a lively 100 hp @ 3400 rpm and 185 lb-ft torque @ 1200 rpm. The unique opening hood, hinged at the firewall, with its sloping "waterfall" grille required a low-mounted radiator and fan. To adapt the engine to the space available, the cooling fan of the DeSoto Airflow engine was relocated to the nose of the crankshaft, as were the fans on all Airflow engines. A disadvantage of this location was that the fan speed was reduced to the same speed as the engine revolutions per minute. Underscoring the luxury level of the 1934 Chrysler Airflow CU model, Chrysler powered it with only a straight-eight engine, confining 1934 use of the six-cylinder engine to the more conventional-appearing Chrysler Six CA and CB models. For the CA and CB models, two versions of the 241.5-cu.-in. engine were available—a standard 5.4:1 compression ratio, 93-hp version having a cast iron cylinder head, and an optional 6.2:1 compression ratio version with an aluminum head, similar to DeSoto, giving the same 100 hp and 185 lb-ft torque. In 1935, DeSoto added a more conventional Airstream model to its offerings.

More quality upgrades arrived in 1936 for the Chrysler and DeSoto engines. The cylinder water jackets in the cylinder block were made full length for improved cooling and reduced bore distortion, especially under heavy load conditions. For years, industry practice had been to water jacket only the upper portion of the cylinder barrels where most of the combustion heating occurred, and let the engine oil and outside air carry away heat from the lower portions of the barrels. However, temperature differences inevitably led to bore distortion. With ever-increasing compression ratios and combustion temperatures and pressures, improvements in bore temperature uniformity became necessary so the piston rings could do a good job of cylinder sealing and oil control. Full-length water jackets were the answer.

This water-jacket change was not the only improvement to the cooling system. With the water pump mounted at the front of the cylinder block, the front cylinders and the exhaust valve seats had an abundance of cooling water, while the rearmost exhaust valve seat areas were noticeably shortchanged. A sheet-steel (later changed to brass) tube running the length of

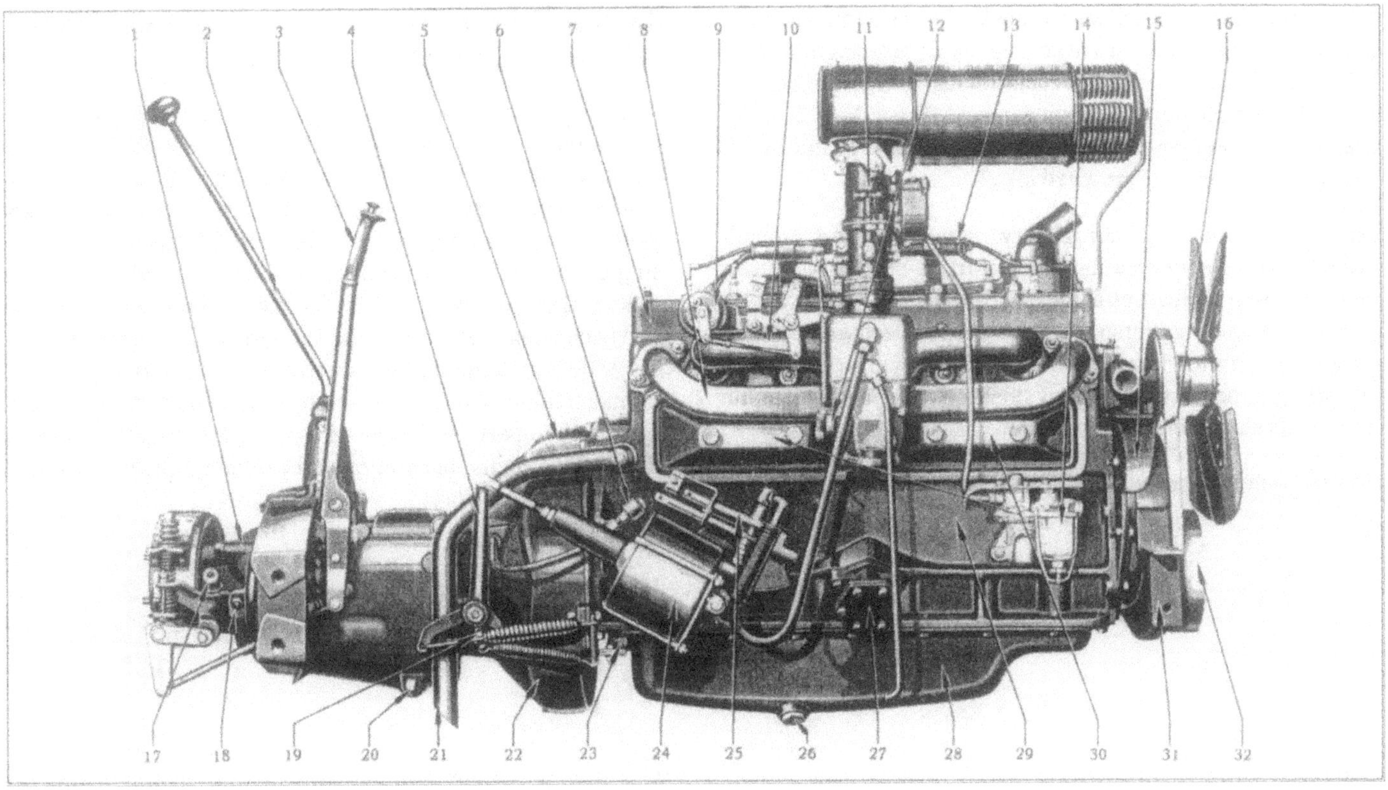

Fig. 6.9 *The right side of the 1933 Chrysler CO six-cylinder engine. The downdraft carburetor mounted on the top of the intake manifold was new. The mechanical fuel pump had been added in 1930. Built at two power levels, 83 and 89 hp, for the 1933 model year, the displacement of the engine was 224 cu. in.*

Fig. 6.10 *The left side of the 1933 Chrysler CO six-cylinder engine. The large air cleaner/silencer and the oil filter were quality assets of the engine.*

the cylinder block was added to equalize the cooling. The tube essentially bypassed the front-to-rear water jacket coolant flow and, with a hole in the tube at each exhaust port, brought cooling water directly to those critical areas. These 1936 features are shown on the longitudinal and cross sections of the engine on Figs. 6.11 and 6.12, respectively.

After nine years of powering Chrysler vehicles, this groundbreaking, four-main-bearing, deep skirt, L-head six-cylinder engine ceased production at the end of 1936. Its successor was a larger Chrysler Jefferson Avenue-built L-head six-cylinder engine that started production in 1937. It will be described following a look at the new and smaller Dodge and Plymouth engine, already mentioned, which is discussed next.

Table 6.1 shows the specifications for this engine.

Dodge and Plymouth Sixes—A Matched Pair of Classic L-Heads

When the market positions of Dodge and DeSoto finally were sorted out in 1933, with DeSoto emerging as the car line to compete in the market just below Chrysler, DeSoto, without an engine manufacturing facility of its own, was allocated a portion of the Chrysler engine output. This left Dodge and Plymouth in great need of a new, compact, low-cost six-cylinder engine to keep their sales rolling. The Plymouth four-cylinder engine had become no match for either the aggressive Chevrolet in-line six or the Ford flathead V-8, and Dodge had to have a good six-cylinder engine to anchor its product line. Both Plymouth and Dodge anticipated sufficient sales volume to justify tooling up a complete engine machining and assembly line of its own.

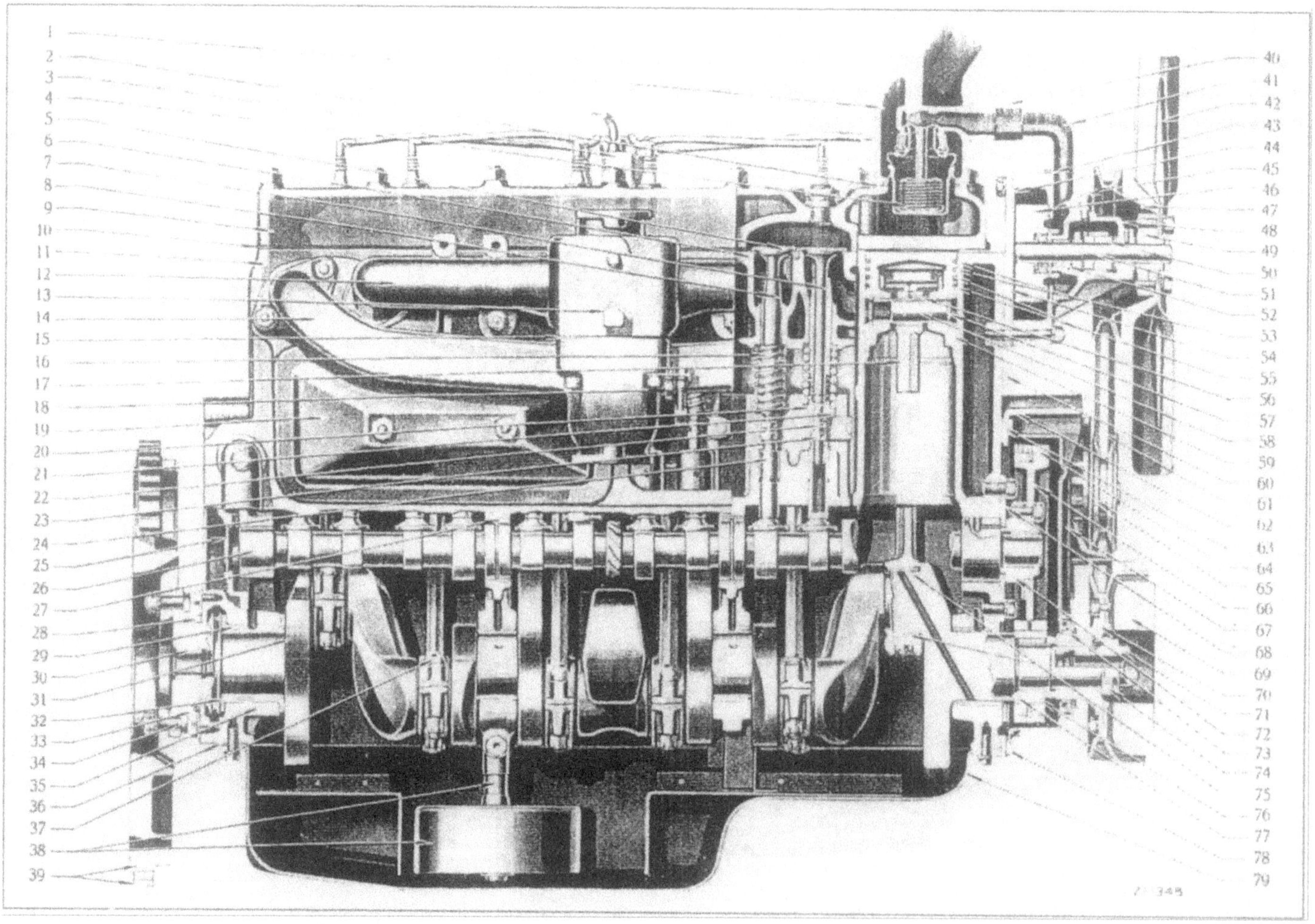

Fig. 6.11 *A longitudinal section of the 1936 Chrysler C7 and DeSoto S1 six-cylinder engine. In this, its last year of production, the engine was built at 241.5 cu. in., a displacement that was started in 1934.*

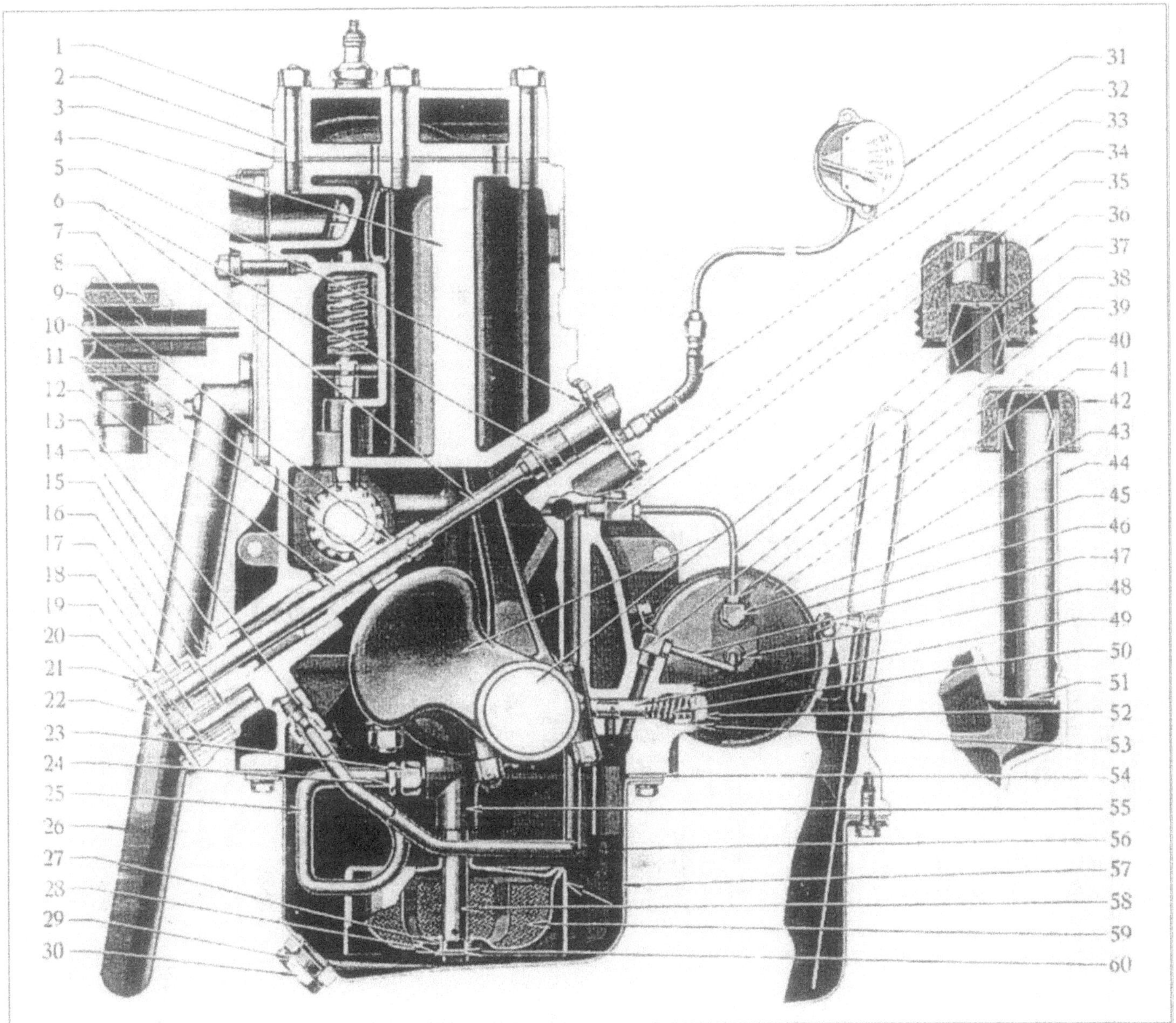

Fig. 6.12 *A cross section through the center of the 1936 Chrysler C7 and DeSoto S1 six-cylinder engine. Illustrated are the integrally forged crankshaft counterweights and the full-length water jacket. The water distribution tube is shown in the section just below the top deck.*

Engineering responded to the need with the design of a single new engine that could be fitted with a range of bores and strokes, allowing Dodge to have a larger-displacement version than Plymouth—a requirement for emphasizing the marketplace pecking order of the car divisions. The L-head valve arrangement, turbulent combustion chambers, single tier of intake and exhaust ports, diagonal drive for the oil pump and distributor at the center of the engine, timing chain drive for the camshaft, water pump mounted high at the front of the engine, crankshaft with four main bearings and integrally forged counterweights, torsional vibration damper, exhaust valve seat inserts, down-draft carburetor, mechanical fuel pump, and Floating Power® engine mounts all were part of the initial 1933 engine design and were similar to corresponding features on the 1933 DeSoto and Chrysler larger deep-skirt six-cylinder engine previously described. The engine designers equipped the engine with offset connecting rods right from the start, ensuring that the cylinder centers would never need to be moved. Cylinder center spacing of this new and smaller six-cylinder engine was set at 3.5625 in. for the front pair of cylinders, #1 and #2; the middle pair, #3 and #4; and the rear pair, #5 and #6. The pairs of cylinders straddling the two intermediate main bearings—#2 and #3, and

TABLE 6.1
SPECIFICATIONS FOR THE CHRYSLER-BUILT, FOUR-MAIN BEARING, DEEP-SKIRT, SIX-CYLINDER ENGINE

Model Year	1929	1930	1930	1931
Car line	DeSoto	Dodge	Chrysler	Dodge
Model	K	DD	CJ	DH
Displacement (cu. in.)	174.9	189.8	195.6	211.5
Bore	3.0	3.125	3.125	3.25
Stroke	4.125	4.125	4.25	4.25
Compression ratio	5.2:1 std.; 6.0:1 opt.	5.2:1	5.2:1 std.; 6.2:1 opt.	5.35:1
BHP@ rpm	55 @ 3000 std.; 61.5 opt.	60 @ 3400	62 @ 3200 std.; 70 @ 3200 opt.	74 @ 3400
Torque (lb-ft) @ rpm	110 @ 1200 std.	120 @ 1200	123 @ 10 std.; 35 @ 10 opt.	140 @ 1400
Combustion chamber	Turbulent	Turbulent	Turbulent	Turbulent
Cyl. block mat'l.—skirt type	Cast iron—deep	Cast iron—deep	Cast iron—deep	Cast iron—deep
Cyl. block deck height	13.005	13.005	13.005	13.005
Cyl. block length	26.75	26.75	26.75	26.75
Cyl. bore spacing	1-2, 3-4, 5-6 = 3.5625 2-3, 4-5 = 4.3125	1-2, 3-4, 5-6 = 3.5625 2-3, 4-5 = 4.3125	1-2, 3-4, 5-6 = 3.5625 2-3, 4-5 = 4.3125	1-2, 3-4, 5-6 = 3.6875 2-3, 4-5 = 4.1875
Ck/shaft—no. of main brgs.	4	4	4	4
Ck/shaft main journal dia.	2.25	2.25	2.25	2.25
Crankpin journal dia.	1.9375	1.9375	2.0	1.9375
Piston mat'l.—no. of rings	Aluminum—3	Aluminum—3	Aluminum—3	Aluminum—4
Connecting rod centers	8.9375	8.9375	8.875	8.8125—offset shank
Valve system	L-head	L-head	L-head	L-head
Valve events	220-230-2	220-230-2	220-230-2	220-230-2
Camshaft drive	Chain	Chain	Gear	Chain
Valve head dia.	Intake 1.47; exhaust 1.47	Intake 1.47; exhaust 1.47	Intake 1.47; exhaust 1.47	Intake 1.53: exhaust 1.47
Valve lash adjustment	Mechanical tappets	Mechanical tappets	Mechanical tappets	Mechanical tappets
Oil pump/distributor drive	Diagonal at center of block	Diagonal at center of block	Diagonal at center of block	Diagonal at center of block
Fuel system	Updraft carburetor	Updraft carburetor	Updraft carburetor	Updraft carburetor
Other	Bolted-on ck/s cwts.	Bolted-on ck/s cwts.		Bolted-on ck/s cwts.

(Continued on next page)

TABLE 6.1
(Continued)

Model Year	1931	1932	1934	1936
Car line	Chrysler	Chrysler	DeSoto	Chrysler
Model	CM	C1	Airflow SE	C7
Displacement (cu. in.)	217.8	223.98	241.5	241.5
Bore	3.25	3.25	3.375	3.375
Stroke	4.375	4.5	4.5	4.5
Compression ratio	5.35:1 std.; 6.2:1 opt.	5.35:1 std.; 6.2:1 opt.	6.2:1	6.0:1 std.; 6.5:1 opt.
BHP@ rpm	78 @ 3400 std.	82 @ 3400 std.	100 @ 3400	93 @ 3400 std.; 100 @ 3400 opt.
Torque (lb-ft) @ rpm	150 @ 1200 std.	156 @ 1200 std.	185 @ 1200	180 @ 1200 std.; 185 @ 1200 opt.
Combustion chamber	Turbulent	Turbulent	Turbulent	Turbulent
Cyl. block mat'l.—skirt type	Cast iron—deep	Cast iron—deep	Cast iron—deep	Cast iron—deep
Cyl. block deck height	15.005	15.005	15.005	15.005
Cyl. block length	26.75	26.75	26.75	26.75
Cyl. bore spacing	1-2, 3-4, 5-6 = 3.6875 2-3, 4-5 = 4.1875	1-2, 3-4, 5-6 = 3.6875 2-3, 4-5 = 4.1875	1-2, 3-4, 5-6 = 3.6875 2-3, 4-5 = 4.1875	1-2, 3-4, 5-6 = 3.6875 2-3, 4-5 = 4.1875
Ck/shaft—no. of main brgs.	4	4	4	4
Ck/shaft main journal dia.	2.25	2.25	2.5	2.5
Crankpin journal dia.	1.9375	1.9375	2.125	2.125
Piston mat'l.—no. of rings	Aluminum—4	Aluminum—4	Aluminum—4	Aluminum—4
Connecting rod centers	8.8125—offset shank	8.75—offset shank	8.75—offset shank	8.75—offset shank
Valve system	L-head	L-head	L-head	L-head
Valve events	220-230-2	220-230-2	230-230-2	230-230-2
Camshaft drive	Chain	Chain	Chain	Chain
Valve head dia.	Intake 1.53; exhaust 1.47	Intake 1.53; exhaust 1.47	Intake 1.53; exhaust 1.47	Intake 1.53; exhaust 1.47
Valve lash adjustment	Mechanical tappets	Mechanical tappets	Mechanical tappets	Mechanical tappets
Oil pump/distributor drive	Diagonal at center of block	Diagonal at center of block	Diagonal at center of block	Diagonal at center of block
Fuel system	Updraft carburetor	Updraft carburetor	1-bbl downdraft	1-bbl downdraft
Other	Bolted-on ck/s cwts.	Bolted-on ck/s cwts.	Integally forged ck/s cwts.; fan mtd. on nose of ck/s	Integally forged ck/s cwts.

#4 and #5—were placed 3.875 in. apart. The machined cylinder block was only 24.781 in. long, less than 4 in. longer than the block of the four-cylinder Plymouth engine it replaced. In addition to its closer-spaced cylinders, this new Plymouth and Dodge six-cylinder engine had another major difference from the DeSoto and Chrysler deep-skirt engine. The block skirt (i.e., the outer walls of the crankcase measured vertically from the crankshaft centerline to the oil pan rail at the bottom of the block) was shortened to 0.125 in., making it a shallow skirt and further reducing the weight and cost of the engine.

With both the Dodge four-cylinder and the Victory six-cylinder engines out of production, Dodge had plenty of room at its huge Dodge Main plant to install the machine tools and assembly line for building the engine. The Dodge foundry was located in the same complex of buildings, and its forge plant was not far away. At the Plymouth assembly plant on Lynch Road in Detroit, an extension was built at the south end of the plant in 1932[3] that quickly was occupied with the production tools and equipment needed to build its version of the new engine. This occurred in time for the 1933 Plymouth cars that started rolling off the end of the line in October 1932, one month ahead of Dodge vehicle production.

An engine displacement of 189.8 cu. in. with a bore and stroke of 3.125 × 4.125 in. sufficed for the Plymouth version, whereas the Dodge version had a longer stroke of 4.375 in. and displaced 201.3 cu. in. Views of the right and left sides of the 1933 Dodge engine are shown on Figs. 6.13 and 6.14, respectively, while a longitudinal section, a cross section through the #1 cylinder, and a cross section through the center of the engine are displayed on Figs. 6.15, 6.16, and 6.17, respectively. The Plymouth engine was available in two power levels: 70 hp @ 3600 rpm with the standard 5.5:1 compression ratio, and 76 hp @ 3600 rpm with an extra-cost Red Head having a higher compression ratio of 6.5:1. Dodge had similar offerings, with its larger engine delivering 75 hp @ 3600 rpm and 81 bhp @ 3600 rpm for the standard and Red Head, respectively. Figure 6.18 shows a stylish 1933 Dodge convertible powered by a 201-cu.-in. Dodge-built engine. Figure 6.19 shows a 1933 Plymouth two-door business coupe powered by a Plymouth-built 190-cu.-in. engine.

Aluminum was used as the cylinder head material for the 1934 high-compression-ratio (6.5:1) Red Head engine. That same year, the Dodge version was boosted to 217.8 cu. in. by a bore increase to 3.25 in., and the Plymouth version took the former size of the Dodge engine, which was 201.3 cu. in. The 1935

Fig. 6.13 *The right side of the 1933 Dodge DP 201-cu.-in. six-cylinder engine. As originally released, the engine was up to date with a downdraft carburetor, mechanical fuel pump, and Floating Power® engine mounts.*

Fig. 6.14 *The left side of the 1933 Dodge DP six-cylinder engine. The engine was protected with the quality features of an air cleaner/silencer and an oil filter.*

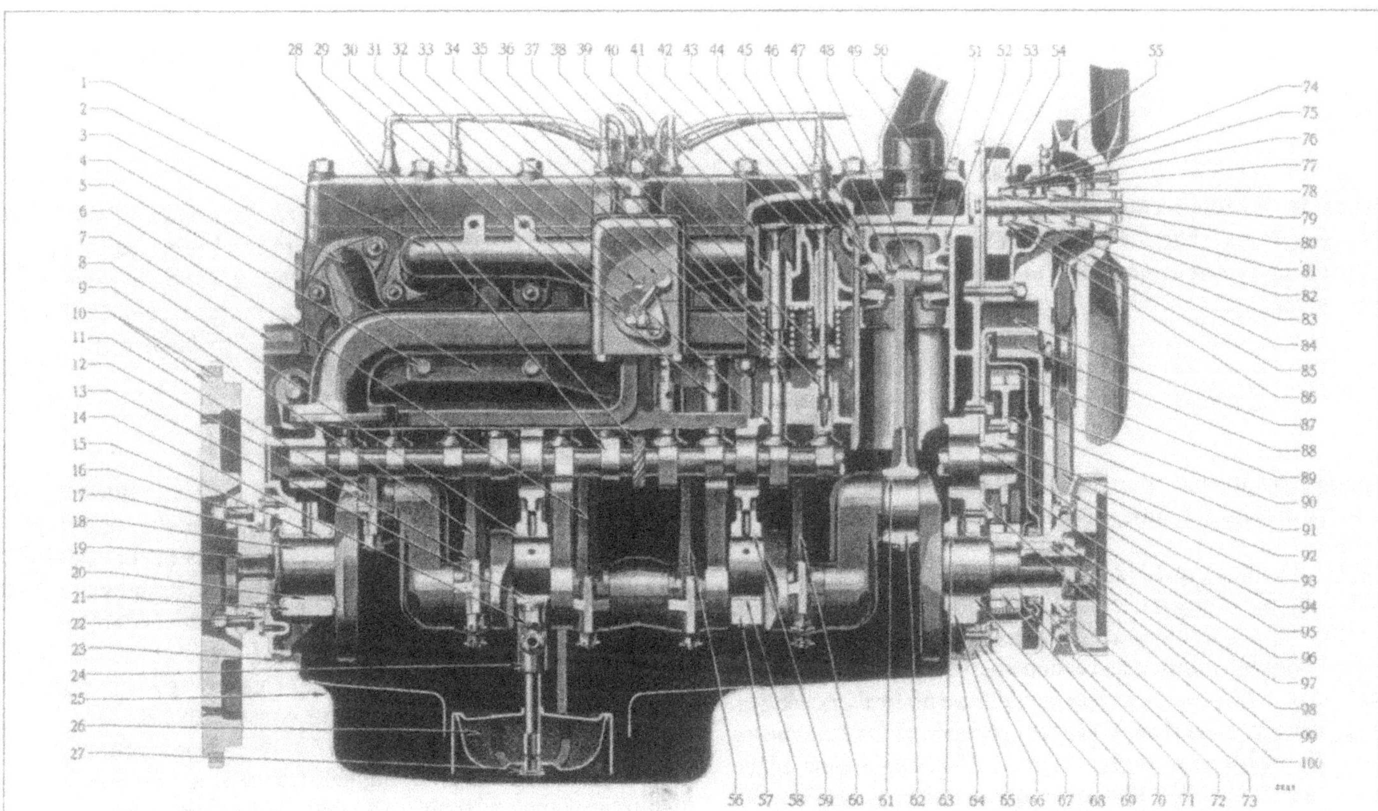

Fig. 6.15 *A longitudinal section of the 1933 Dodge DP six-cylinder engine. The four-main-bearing crankshaft with its sturdy connectors between crankpins was designed with four integral counterweights.*

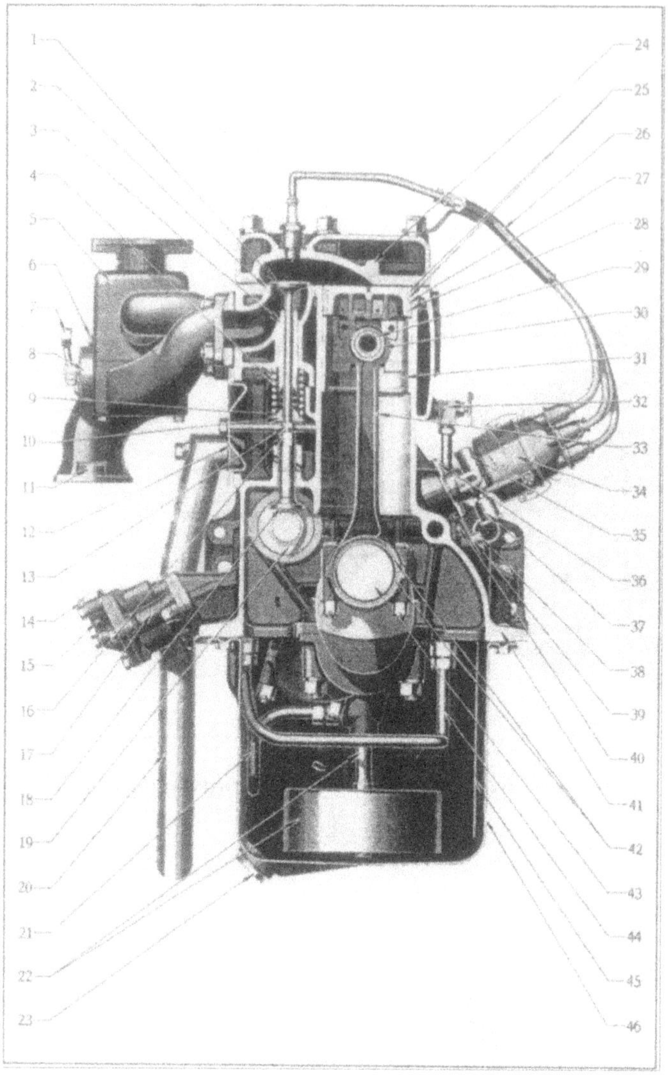

Fig. 6.16 *A cross section through the #1 cylinder of the 1933 Dodge DP six-cylinder engine. The shrink-fitted exhaust valve seat inserts contributed to the excellent durability of the engine.*

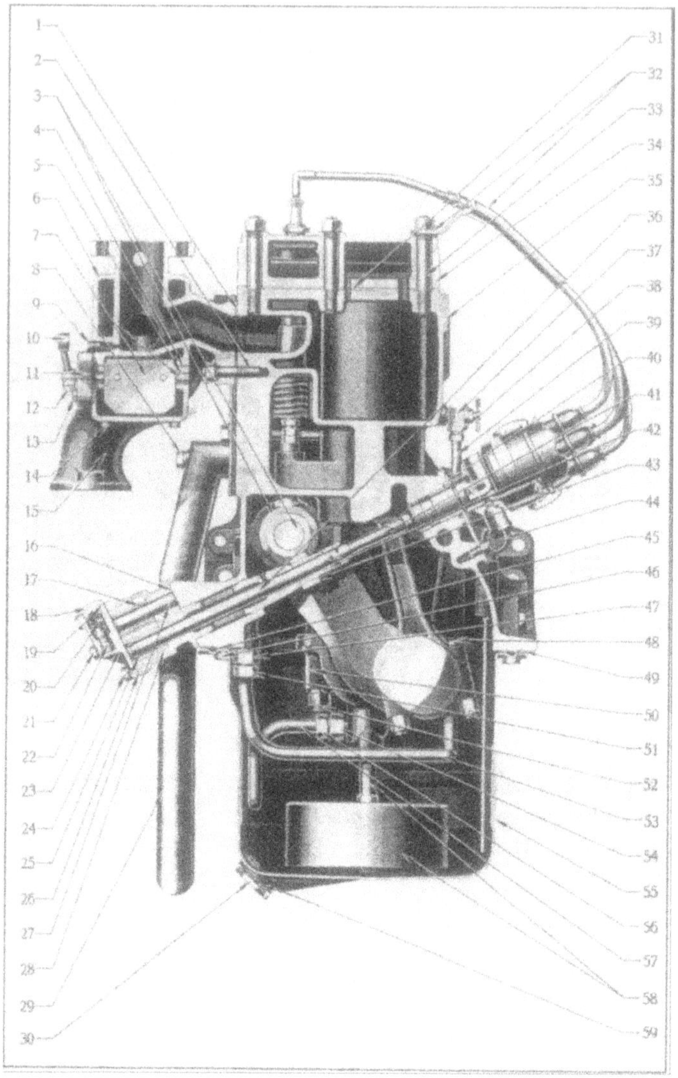

Fig. 6.17 *A cross section through the center of the 1933 Dodge DP six-cylinder engine. With its shallow skirt and cylinders spaced closer together, this engine was lighter and more compact than the deep-skirt Chrysler-built six-cylinder engine having the same type of oil pump and distributor drive.*

engines saw the incorporation of full-length water jackets and a water distribution tube, as described in the prior section.

Eight years later, in 1942, the Dodge engine was increased to 230.2 cu. in. with a 4.625 stroke, and Plymouth moved up to 217.8 cu. in. Twelve years after that, during the 1954 model year, Plymouth brought its version up to the same 230.2 cu. in. as Dodge, a displacement that would have no further changes. The year 1959 was the last year of production of the engine by both Plymouth and Dodge. Figure 6.20 shows a view of the right side of the 1959 Plymouth MP1 engine. A longitudinal section, a cross section through the #1 cylinder, and a cross section through the center of the engine appear on Figs. 6.21, 6.22, and 6.23, respectively.

The importance of this engine to the general well-being of Chrysler Corporation during the 1930s, 1940s, and 1950s cannot be overstated. Every year, hundreds of thousands of Plymouths and Dodges equipped with this engine were sold to car and truck buyers who appreciated the dependable power under the hood. A review of the 1952 Plymouth in the April 1952 issue of *Motor Trend* stated, "This power plant is highly developed, highly reliable and of conservative design." Not bad for an engine that at the time had been in production longer than any other Chrysler engine. Car model names changed, styling changed, bodies changed, and wheelbases changed. But this engine just kept soldiering along.

Fig. 6.18 *A 1933 Dodge DP convertible. For this car, which was especially sporty-looking with its folded-down windshield and wire wheels, the power came from the new 201-cu.-in., four-main-bearing, six-cylinder engine tooled up at the huge Dodge Main plant in Hamtramck, Michigan.*

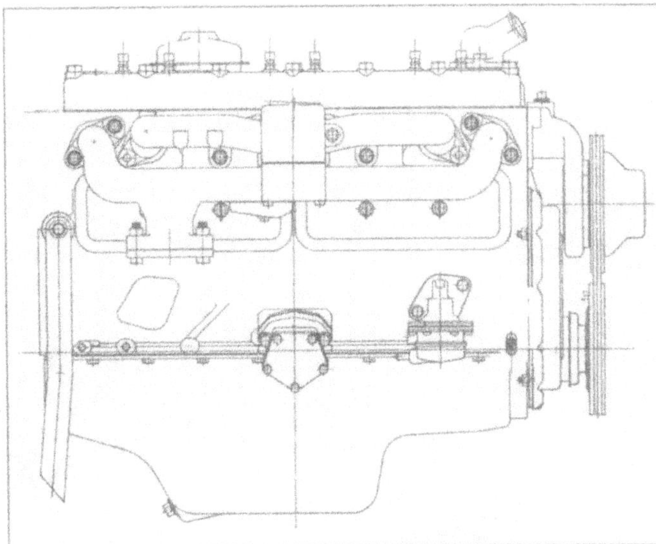

Fig. 6.20 *The right side of the 1959 Plymouth MP1 230-cu.-in. six-cylinder engine. In this, its final year of passenger car production, the appearance of the engine was little changed since its 1933 introduction.*

Fig. 6.19 *A 1933 Plymouth PD two-door business coupe. Its 190-cu.-in. six-cylinder engine, which was the first Plymouth six, was built at the Plymouth assembly plant on Lynch Road in Detroit. The engine was a near-duplicate of the Dodge 201-cu.-in. six-cylinder engine.*

But the days of this engine were numbered. By the date of the article, both Ford and Chevrolet were fielding cars with overhead valve six-cylinder engines that could outperform the Plymouth and Dodge L-heads. Finally, in 1960, a new engine emerged that allowed the Plymouth and Dodge L-head to retire with appreciation of a job well done. The new engine was the Slant Six, whose story is told in Chapter 11.

Table 6.2 shows the specifications for this long-lived, L-head engine.

Four-Main-Bearing, Shallow-Skirt, Six-Cylinder Engine for DeSoto and Chrysler

The four-main-bearing, shallow-skirt six-cylinder Plymouth and Dodge engine described in the preceding section must have been viewed as a smashing success in terms of power output, cost to manufacture, size, and weight, causing DeSoto and Chrysler to consider a similar engine—adjusted, of course—for their large-car needs. The 1937 Chrysler Royal C16 shown on Fig. 6.24 and the 1937 DeSoto S3 model were launched with such an engine as a successor to the four-main-bearing, deep-skirt, six-cylinder engine previously described, which dated from 1929. The engine was indeed a look-alike to the Plymouth and Dodge engine except for its larger size.

Cylinder spacing between the front pair of cylinders (#1 and #2), the middle pair (#3 and #4), and the rear pair (#5 and #6) was 3.75 in. Between cylinder pair #2and #3 and cylinder pair #4 and #5, the spacing was a hefty 4.5 in. With that much room available, the crankshaft had plenty of strength, the connecting rod shank was nicely centered above the big-end bearing, and offset connecting rods were not required, allowing the same connecting rod assembly to be used in all six cylinders. Being released for the 1937 models, the engine incorporated from the start all the many quality features that had been developed on its predecessor engine: exhaust valve seat inserts, full-length water jackets, downdraft carburetor, and so forth. These have already been described. A longitudinal section and a cross section through the center of the engine are shown on Figs. 6.25 and 6.26, respectively.

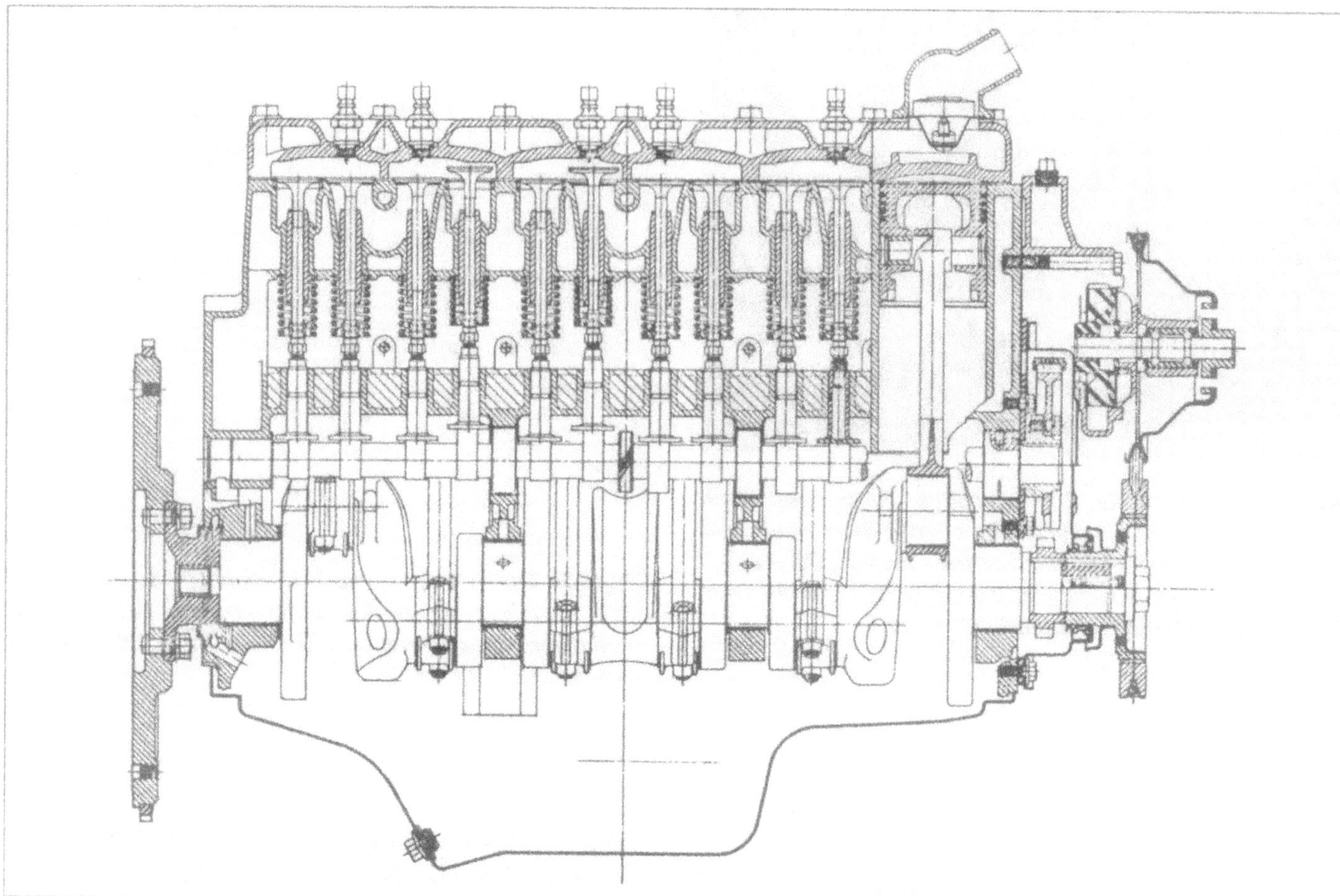

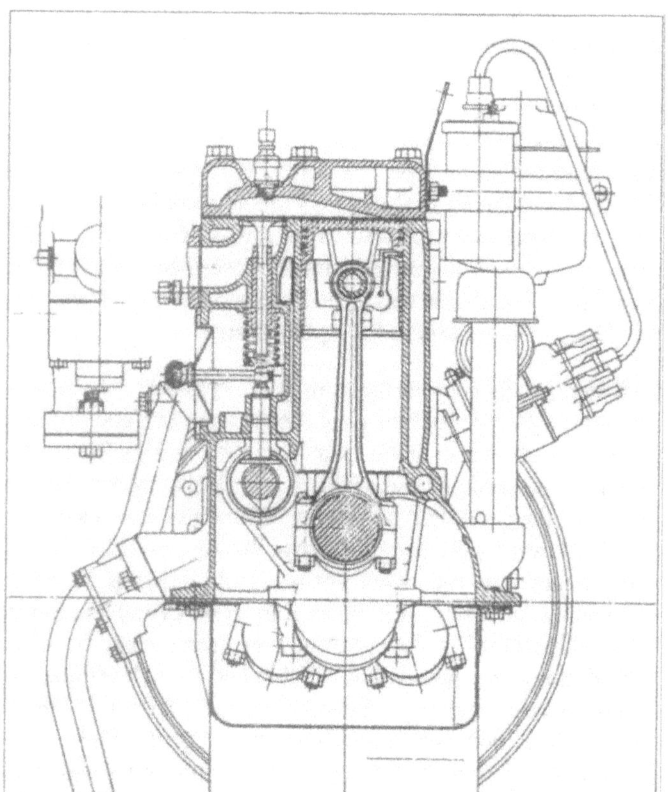

Fig. 6.21 *(Above) A longitudinal section of the 1959 Plymouth MP1 six-cylinder engine. The shallow arch sectioned in the combustion chamber of the #1 cylinder was the result of changes to improve engine breathing while retaining turbulence. A groove cut into the crankshaft vibration damper inertia ring was used to drive the fan belt.*

Fig. 6.22 *(Left) A cross section through the #1 cylinder of the 1959 Plymouth MP1 six-cylinder engine. The full-length water jacket and the water distribution tube were introduced in 1935.*

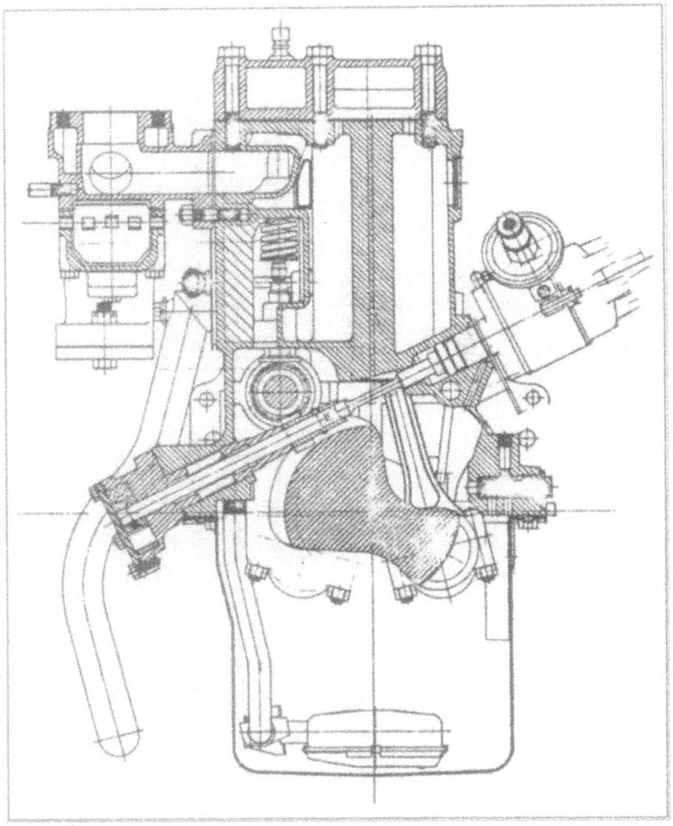

Fig. 6.23 *A cross section through the center of the 1959 Plymouth MP1 six-cylinder engine. The innovative pivoting oil strainer had a float chamber to keep the oil inlet opening just below the surface so the cleanest oil would be sent to the oil pump.*

Interestingly, the 1937 launch engine for both Chrysler and DeSoto had a considerably smaller displacement than the preceding deep-skirt engine—228.1 cu. in. versus 241.5 cu. in. However, the 1937 engine exactly matched it for power—93 hp with the standard cylinder head, and 100 hp with the optional (for Chrysler) aluminum high-compression-ratio head. Based on detailed specifications, this probably was due to the larger valves and the slightly higher compression ratios of the engine. The engine used a 3.375-in.-diameter bore and a 4.25-in. stroke to give it the 228.1 cu. in. previously noted. The next year, engines built for the 1938 DeSoto S5 stayed with the 228.1-cu.-in. displacement, but engines built for the 1938 Chrysler Royal C18 had a displacement increase to 241.5 cu. in. by a stroke change to 4.5 in. Starting with the 1942 DeSoto S10 and the Chrysler C34 models, the bores of both engines were increased to 3.4375 in., giving 236.7 and 250.6 cu. in. displacements, respectively. In 1951, Desoto and Chrysler shared a single engine size of 250.6 cu. in. Then in 1952, Chrysler increased the displacement of its engine to 264.5 cu. in., with a final stroke change to 4.75 in. The year 1954 was the last year of passenger car use of this excellent engine. In the following year, 1955, both DeSoto and Chrysler passenger cars were powered exclusively by V-8 engines.

Table 6.3 shows the specifications for this engine.

Dodge trucks, which started using the engine the first year it was available, found it to be great for medium-duty trucks and, with considerable effort, managed to keep it in production, obviously at a significantly reduced volume, through 1967. A major incentive for this extended duty was its use in the popular WM300 Dodge PowerWagon that needed the low silhouette of the L-head engine. Chapter 15 presents more information on the truck version of this four-main-bearing, shallow-skirt DeSoto and Chrysler engine.

Because the small Plymouth and Dodge and the larger Chrysler and Desoto shallow-skirt, L-head six-cylinder engines were similar in appearance, the length of the cylinder block at the head deck surface often was used for positive identification. With the Chrysler and DeSoto engine block head deck measuring 25.7 in., and the Plymouth and Dodge measuring 23.4 in., they were known as the 25-in. and 23-in. block (also as the long-block and short-block) engines, respectively. The actual overall machined lengths of the cylinder blocks were considerably longer at 27.06 and 24.78 in., but the head deck length was much easier to measure when the engine was installed in a vehicle.

The main manufacturing location of the Chrysler and DeSoto engine was the Chrysler Jefferson Avenue plant in Detroit, home to engine building since Chrysler was formed and perhaps even before that. A second location was across the Detroit River in neighboring Windsor, Ontario, the home of Chrysler Canada. A new engine plant was built adjacent to the Windsor assembly plant in time to start production of the 1937 models. Such a move was made necessary by Canadian local content laws, and this engine was the one that was tooled. Because Chrysler Canada favored equipping Canadian-built cars, including Plymouth and Dodge, with Canadian-built engines, Canadian-built Plymouths and Dodges ended up having these large Chrysler engines under the hoods. The production of this four-main-bearing, shallow-skirt engine ceased at the Windsor engine plant when the plant was readied to build Slant Six engines at the start of 1960 model year.

Chart 6.1 shows time lines for the three four-main-bearing, L-head, six-cylinder engines described in this chapter.

TABLE 6.2
SPECIFICATIONS FOR THE PLYMOUTH AND DODGE, FOUR-MAIN BEARING, SHALLOW-SKIRT, SIX-CYLINDER ENGINES

Model Year	1933	1933	1934	1942
Car line	Plymouth	Dodge	Dodge	Plymouth
Model	PC, PD	DP	DR	P14
Displacement (cu. in.)	189.8	201.3	217.8	217.8
Bore	3.125	3.125	3.25	3.25
Stroke	4.125	4.375	4.375	4.375
Compression ratio	5.5:1 std.; 6.5:1 opt.	5.5:1 std.; 6.2:1 opt.	5.6:1 std.; 6.5:1 opt.	6.8:1
BHP@ rpm	70 @ 3600 std.; 76 @ 3600 opt.	75 @ 3600 std.; 81 @ 3600 opt.	82 @ 3600 std.; 87 @ 3600 opt.	95 @ 3400
Torque (lb-ft) @ rpm	130 @ 1200 std.; 136 @ 1200 opt.	136 @ 1200 std.; 144 @ 1200 opt.	150 @ 1200 std.; 160 @ 1200 opt.	172 @ 1600
Combustion chamber	Turbulent	Turbulent	Turbulent	Turbulent
Cyl. block mat'l.—skirt type	Cast iron—shallow	Cast iron—shallow	Cast iron—shallow	Cast iron—shallow
Cyl. block deck height	12.125	12.13	12.13	12.13
Cyl. block length	24.781*	24.781	24.781	24.781
Cyl. bore spacing	1-2, 3-4, 5-6 = 3.5625 2-3, 4-5 = 3.875	1-2, 3-4, 5-6 = 3.5625 2-3, 4-5 = 3.875	1-2, 3-4, 5-6 = 3.5625 2-3, 4-5 = 3.875	1-2, 3-4, 5-6 = 3.5625 2-3, 4-5 = 3.875
Ck/shaft—no. of main brgs.	4	4	4	4
Ck/shaft main journal dia.	2.25	2.25	2.25	2.25
Crankpin journal dia.	1.9375	1.9375	2.0625	2.0625
Piston mat'l.—no. of rings	Aluminum—4	Aluminum—4	Aluminum—4	Aluminum—4
Connecting rod centers	8.0625	8.0625	7.9375	7.9375
Valve system	L-head	L-head	L-head	L-head
Valve events	220-230-2	220-230-2	220-230-2	236-236-18
Camshaft drive	Chain	Chain	Chain	Chain
Valve head dia.	Intake 1.47; exhaust 1.47	Intake 1.47; exhaust 1.47	Intake 1.47; exhaust 1.47	Intake 1.53; exhaust 1.41
Valve lash adjustment	Mechanical tappet	Mechanical tappet	Mechanical tappet	Mechanical tappet
Oil pump/distributor drive	Diagonal at center of block	Diagonal at center of block	Diagonal at center of block	Diagonal at center of block
Fuel system	1-barrel carburetor	1-barrel carburetor	1-barrel carburetor	1-barrel carburetor
Other			Opt. ratio cyl. head is alum.	

* 23.4 at head deck

(Continued on next page)

TABLE 6.2
(Continued)

Model Year	1942	1955	1958	1959
Car line	Dodge	Plymouth	Dodge	Plymouth
Model	D22	P26	LD1	MP1
Displacement (cu. in.)	230.2	230.2	230.2	230.2
Bore	3.25	3.25	3.25	3.25
Stroke	4.625	4.625	4.625	4.625
Compression ratio	6.7:1	7.4:1	8.0:1	8.0:1
BHP@ rpm	105 @ 3600	117 @ 4000	138 @ 4000	132 @ 3600
Torque (lb-ft) @ rpm	185 @ 1600	194 @ 1600	208 @ 1600	205 @ 1600
Combustion chamber	Turbulent	Turbulent	Turbulent	Turbulent
Cyl. block mat'l.—skirt type	Cast iron—shallow	Cast iron—shallow	Cast iron—shallow	Cast iron—shallow
Cyl. block deck height	12.13	12.13	12.13	12.13
Cyl. block length	24.781	24.781	24.781	24.781
Cyl. bore spacing	1-2, 3-4, 5-6 = 3.5625 2-3, 4-5 = 3.875	1-2, 3-4, 5-6 = 3.5625 2-3, 4-5 = 3.875	1-2, 3-4, 5-6 = 3.5625 2-3, 4-5 = 3.875	1-2, 3-4, 5-6 = 3.5625 2-3, 4-5 = 3.875
Ck/shaft—no. of main brgs.	4	4	4	4
Ck/shaft main journal dia.	2.5	2.5	2.5	2.5
Crankpin journal dia.	2.0625	2.0625	2.0625	2.0625
Piston mat'l.—no. of rings	Aluminum—4	Aluminum—4	Aluminum—4	Aluminum—4
Connecting rod centers	7.8125	7.8125	7.8125	7.8125
Valve system	L-head	L-head	L-head	L-head
Valve events	236-236-18	236-236-18	236-236-18	236-236-18
Camshaft drive	Chain	Chain	Chain	Chain
Valve head dia.	Intake 1.53; exhaust 1.41	Intake 1.53; exhaust 1.41	Intake 1.53; exhaust 1.41	Intake 1.53; exhaust 1.41
Valve lash adjustment	Mechanical tappet	Mechanical tappet	Mechanical tappet	Mechanical tappet
Oil pump/distributor drive	Diagonal at center of block	Diagonal at center of block	Diagonal at center of block	Diagonal at center of block
Fuel system	1-barrel downdraft	1-barrel downdraft	2-barrel downdraft	1-barrel downdraft
Other				

Fig. 6.24 *A 1937 Chrysler Royal C16 four-door, eight-passenger sedan. Power for this car and for the 1937 Desoto S3 model came from an all-new 228-cu.-in. six-cylinder engine recently tooled at the Chrysler Jefferson Avenue plant.*

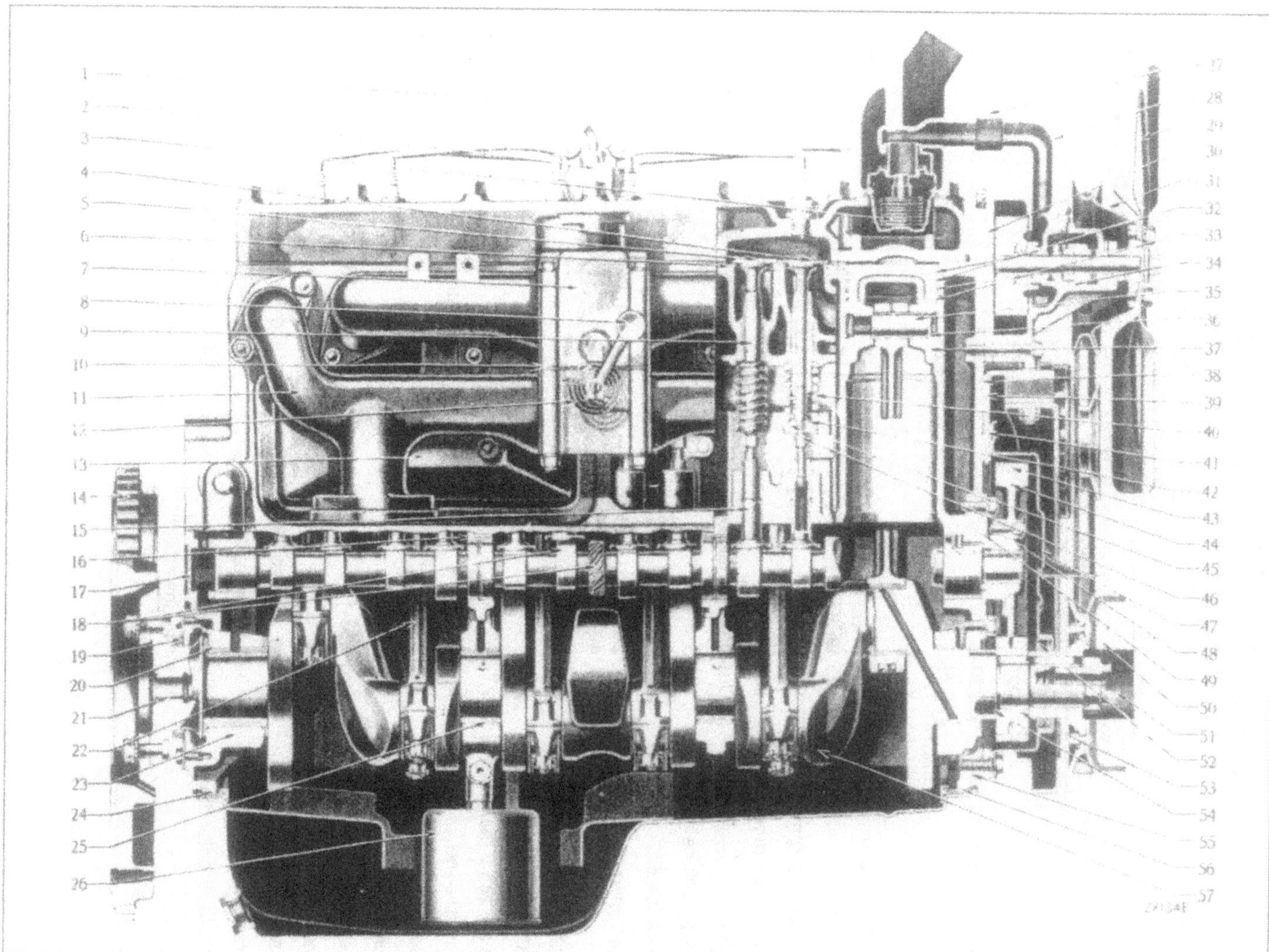

Fig. 6.25 *A longitudinal section of the 1937 Chrysler C16 and the DeSoto S3 228-cu.-in. six-cylinder engine. Looking much like a scaled-up version of the 1933 Plymouth and Dodge four-main-bearing, six-cylinder engine, the 1937 Chrysler engine had its cylinder bore centers farther apart, allowing it to eventually reach a displacement of 265 cu. in.*

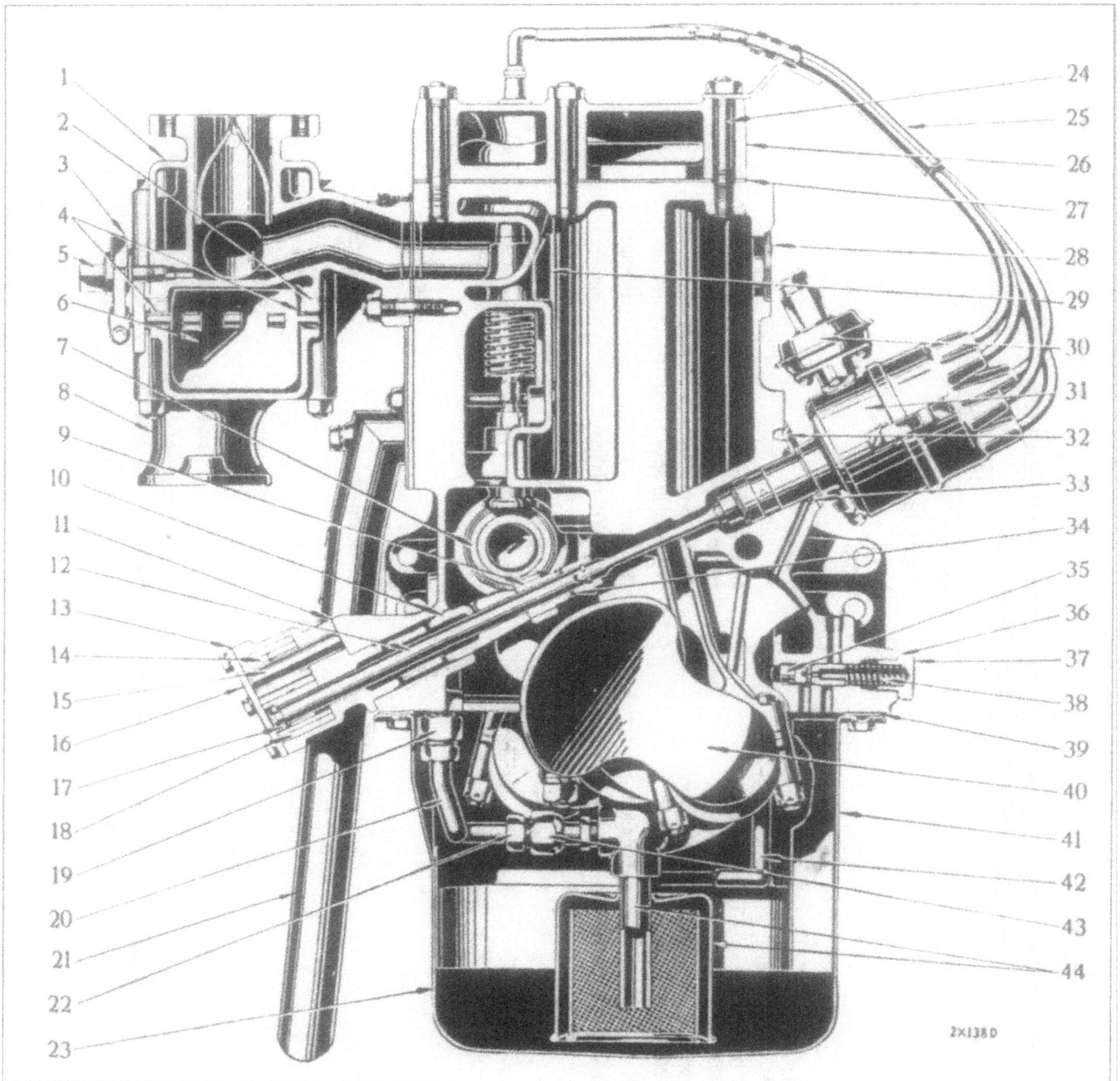

Fig. 6.26 *A cross section through the center of the 1937 Chrysler C16 and DeSoto S3 six-cylinder engine. The noticeable hump in the intake port was developed to retain liquid fuel on the heated floor for more thorough evaporation. Both the weight and the cost of the engine benefited from the use of a shallow skirt for the cylinder block.*

TABLE 6.3
SPECIFICATIONS FOR THE CHRYSLER-BUILT, FOUR-MAIN BEARING, SHALLOW-SKIRT, SIX-CYLINDER ENGINE

Model Year	1937	1937	1938	1942
Car line	Chrysler	DeSoto	Chrysler	DeSoto
Model	Royal C16	S3	Royal C18	S10
Displacement (cu. in.)	228.1	228.1	241.5	236.7
Bore	3.375	3.375	3.375	3.4375
Stroke	4.25	4.25	4.5	4.25
Compression ratio	6.5:1 std.; 7.0:1 opt.	6.5:1	6.2:1 std.; 7.0:1 opt.	6.6:1
BHP@ rpm	93 @ 3600 std.; 100 @ 3600 opt.	93 @ 3600	93 @ 3600 std.; 102 @ 3600 opt.	115 @ 3800
Torque (lb-ft) @ rpm	168 @ 1200 std.; 176 @ 1200 opt.	168 @ 1200	180 @ 1200 std.; 190 @ 1200 opt.	190 @ 1600
Combustion chamber	Turbulent	Turbulent	Turbulent	Turbulent
Cyl. block mat'l.—skirt type	Cast iron—shallow	Cast iron—shallow	Cast iron—shallow	Cast iron—shallow
Cyl. block deck height	12.13	12.125	12.125	12.13
Cyl. block length	27.0625*	27.0625	27.0625	27.0625
Cyl. bore spacing	1-2, 3-4, 5-6 = 3.75 2-3, 4-5 = 4.5	1-2, 3-4, 5-6 = 3.75 2-3, 4-5 = 4.5	1-2, 3-4, 5-6 = 3.75 2-3, 4-5 = 4.5	1-2, 3-4, 5-6 = 3.75 2-3, 4-5 = 4.5
Ck/shaft—no. of main brgs.	4	4	4	4
Ck/shaft main journal dia.	2.5	2.5	2.5	2.5
Crankpin journal dia.	2.125	2.125	2.125	2.125
Piston mat'l.—no. of rings	Aluminum—4	Aluminum—4	Aluminum—4	Aluminum—4
Connecting rod centers	8.0	8.0	7.875	8.0
Valve system	L-head	L-head	L-head	L-head
Valve events	230-230-2	230-230-2	230-230-10	236-236-18
Camshaft drive	Chain	Chain	Chain	Chain
Valve head dia.	Intake 1.66; exhaust 1.53	Intake 1.66; exhaust 1.53	Intake 1.66; exhaust 1.53	Intake 1.72; exhaust 1.53
Valve lash adjustment	Mechanical tappet	Mechanical tappet	Mechanical tappet	Mechanical tappet
Oil pump/distributor drive	Diagonal at center of block	Diagonal at center of block	Diagonal at center of block	Diagonal at center of block
Fuel system	1-barrel carburetor	1-barrel carburetor	1-barrel carburetor	1-barrel carburetor
Other				

* 25.7 at head deck

(Continued on next page)

TABLE 6.3
(Continued)

Model Year	1942	1952	1952–1953–1954
Car line	Chrysler	DeSoto	Chrysler
Model	C34	S15	Windsor C51-C60-C62
Displacement (cu. in.)	250.6	250.6	264.5
Bore	3.4375	3.4375	3.4375
Stroke	4.5	4.5	4.75
Compression ratio	6.6:1	7.0:1	7.0:1
BHP@ rpm	120 @ 3800	116 @ 3600	119 @ 3600
Torque (lb-ft) @ rpm	200 @ 1600	208 @ 1600	218 @ 1600
Combustion chamber	Turbulent	Turbulent	Turbulent
Cyl. block mat'l.—skirt type	Cast iron—shallow	Cast iron—shallow	Cast iron—shallow
Cyl. block deck height	12.13	12.13	12.13
Cyl. block length	27.0625	27.0625	27.0625
Cyl. bore spacing	1-2, 3-4, 5-6 = 3.75 2-3, 4-5 = 4.5	1-2, 3-4, 5-6 = 3.75 2-3, 4-5 = 4.5	1-2, 3-4, 5-6 = 3.75 2-3, 4-5 = 4.5
Ck/shaft—no. of main brgs.	4	4	4
Ck/shaft main journal dia.	2.5	2.5	2.5
Crankpin journal dia.	2.125	2.125	2.125
Piston mat'l.—no. of rings	Alum and cast iron**—4	Aluminum—4	Aluminum—4
Connecting rod centers	7.875	7.875	7.875
Valve system	L-head	L-head	L-head
Valve events	236-236-18	236-236-18	236-236-18
Camshaft drive	Chain	Chain	Chain
Valve head dia.	Intake 1.66, exhaust 1.53	Intake 1.72, exhaust 1.53	Intake 1.72, exhaust 1.5
Valve lash adjustment	Mechanical tappet	Mechanical tappet	Mechanical tappet
Oil pump/distributor drive	Diagonal at center of block.	Diagonal at center of block.	Diagonal at center of block.
Fuel system	1-barrel carburetor	1-barrel carburetor	1-barrel carburetor
Other			

** Cast iron on some wartime engines.

Chart 6.1
Time Lines—Six-Cylinder L-Head Engines with Four Main Bearings

Engines, Car Lines, and Displacements (cu. in.)

Engine	Car Line	1929	1930	1931	1932	1933	1934	1935	1936
Chrysler 6-cyl. Deep-skirt	Dodge		190	212	218				
	Model		DD	DH	DL				
	DeSoto	175	190	205	212	218	242	242	242
	Model	K	CK	SA	SC	SD	SE	SF, SG	S1, S2
	Chrysler		196	218	224	224	242	242	242
	Model		CJ	CM	C1	CO	CA, CB	C6	C7
Dodge/Ply 6-cyl. Shallow-skirt	Dodge					201	218	218	218
	Model					DP	DR, DS	DU	D2
	Plymouth					190	201	201	201
	Model					PC, PD	PE, PF, PG	PJ	P1, P2

Chart 6.1
(Continued)

Engines, Car Lines, and Displacements (cu. in.)

Engine	Car Line	1937	1938	1939	1940	1941	1942	1946–1948	1949
Dodge/Ply 6-cyl. Shallow-skirt	Dodge	218	218	218	218	218	230	230	230
	Model	D5	D8	D11	D14, D17	D19	D22	D24	D29, D30
	Plymouth	201	201	201	201	201	218	218	218
	Model	P3, P4	P5, P6	P7, P8	P9, P10	P11, P12	P14	P15	P17, P18
Chrysler 6-cyl. Shallow-skirt	DeSoto	228	228	228	228	228	237	237	237
	Model	S3	S5	S6	S7	S8	S10	S11	S13
	Chrysler	228	242	242	242	242	251	251	251
	Model	C16	C18	C22	C25	C28	C34	C38	C45

Chart 6.1
(Continued)

Engines, Car Lines, and Displacements (cu. in.)

Engine	Car Line	1950	1951–1952	1953	1954	1955	1956	1957	1958	1959
		- - Model Years - -								
Dodge/Ply 6-cyl. Shallow-skirt	Dodge	230	230	230	230	230	230	230	230	230
	Model	D33, D34	D41, D42	D46, D47	D51, D52	D56	D52	D72	LD1	MD1
	Plymouth	218	218	218	218 230	230	230	230	230	230
	Model	P19, P20	P22, P23	P24	P25	P26	P28	P30	LP1	MP1
Chrysler 6-cyl. Shallow-skirt	DeSoto	237	251	251	251					
	Model	S14	S15	S18	S20					
	Chrysler	251	251 1951 265 1952	265	265					
	Model	C48	C51	C60	C62					

CHAPTER 7

Engines of Luxury—The Straight-Eights

By the late 1920s, a must-have for entry into the luxury car market was an engine with at least eight cylinders. Chrysler Corporation six-cylinder engines were smooth, powerful, and quiet, but the market was demanding more than that—it wanted more cylinders. Chrysler Corporation launched its first straight-eight engine, surprisingly not in its Chrysler or Imperial car lines but in Dodge and DeSoto models. Dodge and DeSoto may have been looking at eight-cylinder engines as a way of competing better in the fiercely competitive mid-range car market against other makes that were already selling eight-cylinder cars.[1] Regardless, production of the first Chrysler straight-eight-powered cars, the 1930 model Dodge DC and DeSoto CF, started in December 1929. Figures 7.1 and 7.2 show a 1930 Dodge DC two-door coupe and a 1930 DeSoto CF convertible, respectively. Both cars have an eight-cylinder engine.

Eight Cylinders with Five Main Bearings—The First Chrysler Corporation Straight-Eight Engine

The Dodge sales staff must have been pleased at having a straight-eight engine, with its connotations of luxury, added to their offerings. The manufacturing staff probably was even more elated. The Dodge four-cylinder engine had ceased production several years prior, and production of the Dodge-built six-cylinder Victory engine was about to end, leaving the Dodge Main plant with no engines to build. The straight-eight was an opportunity to remain in the engine manufacturing business, and they grabbed it.

Fig. 7.1 *A 1930 Dodge DC coupe. Inside the engine compartment was a new 221-cu.-in. straight-eight manufactured at the Dodge Main plant. The 75 hp and 145 lb-ft torque of the engine gave the car a considerable boost over the 60 hp and 120 lb-ft torque of the six-cylinder model.*

Fig. 7.2 *A 1930 DeSoto CF convertible. In keeping with the corporate marketplace hierarchy of the time, its straight-eight engine had a displacement of 208 cu. in. This was 13 cu. in. less than the 221 cu. in. version installed in Dodge cars.*

Evidently well-satisfied with the outcome of the 1929 four-main-bearing, deep-skirt, six-cylinder engine as described in Chapter 6, Engineering produced a design for the straight-eight that, with some accuracy, could be described as the same concept but with two cylinders added. Carried over into this eight-cylinder engine were the 1929 six-cylinder engine features of a ribbed skirt, 2.625 in. deep, for the cylinder block; a two-cylinder span between the main bearings; and a diagonally angled oil pump and distributor drive at the center of the engine. A downdraft carburetor was part of the design right from the beginning, as shown in the right- and left-side views on Figs. 7.3 and 7.4. A longitudinal section, a cross section through the #1 cylinder, and a cross section through the center of the engine are displayed on Figs. 7.5, 7.6, and 7.7, respectively. With the eight-cylinder design, a significant adjustment had to be made at the center of the engine. A main bearing bulkhead, not present with the six-cylinder engine, vied for space with the diagonal oil pump and the distributor drive. The solution was to elevate the diagonal drive to a higher location between the cylinders, above the structure of the bulkhead, although this required the cylinder block to become longer. A comparison of the cross section of the six-cylinder engine on Fig. 6.7 of Chapter 6 and the eight-cylinder engine on Fig. 7.7 illustrates this difference.

Spacing of the cylinders of the straight-eight was set at a tight 3.5 in. for the four pairs of cylinders that did not have a main bearing between them, a more generous 3.9375 in. for the two pairs of cylinders that straddled the intermediate main bearings, and 4.125 in. for the center pair of cylinders. The 1930 launch-year Dodge and DeSoto engines shared the small bore diameter of 2.875 in., making the DeSoto version, with a 4.0-in. stroke, only 208 cu. in. At this time, Dodge still was positioned above DeSoto and received the longer stroke of 4.25 in., for a 221 cu. in. displacement. In 1931, the size of both versions was increased, the DeSoto model CF* (the asterisk does not indicate a footnote; it was part of the model designation) moving up to 221 cu. in. and the Dodge DG, given a larger 3.0-in. bore, becoming 240 cu. in.

Evidently not wanting to seem stodgy compared to Dodge and DeSoto, Chrysler started installing this five-main-bearing straight-eight engine in its 1931 CD model in June 1930, dropping the ZSB-designed sixes from its luxury market.

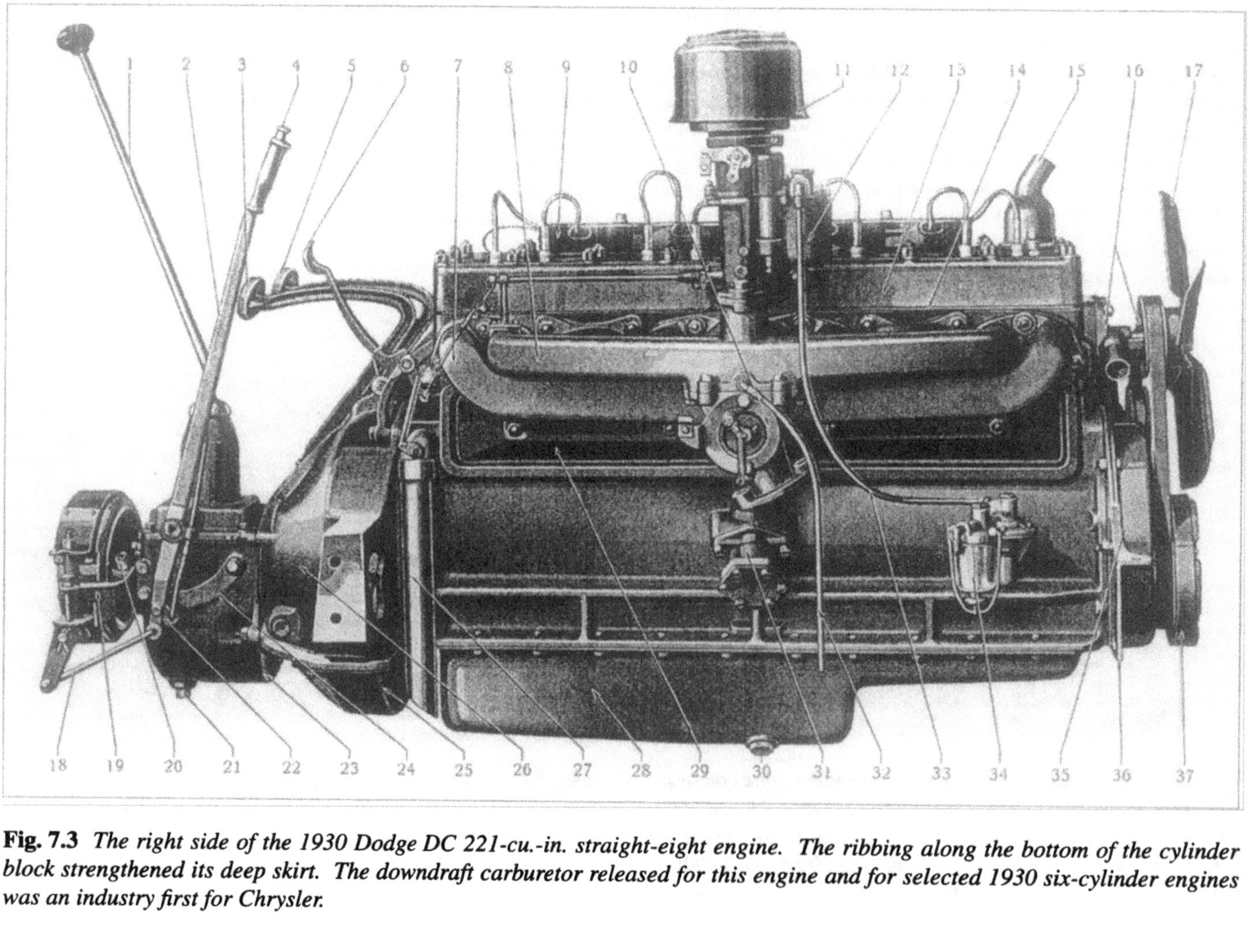

Fig. 7.3 *The right side of the 1930 Dodge DC 221-cu.-in. straight-eight engine. The ribbing along the bottom of the cylinder block strengthened its deep skirt. The downdraft carburetor released for this engine and for selected 1930 six-cylinder engines was an industry first for Chrysler.*

Fig. 7.4 *The left side of the 1930 Dodge DC straight-eight engine. The eight-cylinders-long cylinder block provided ample room for mounting the six-volt generator at the front, the oil filter near the center, and the starter motor at the rear of the block.*

Fig. 7.5 *A longitudinal section of the 1930 Dodge DC straight-eight engine. For 1930 and 1931, the five-main-bearing crankshaft, as shown, did not have counterweights. In 1932, bolted-on counterweights were added, followed by integrally forged counterweights in 1934.*

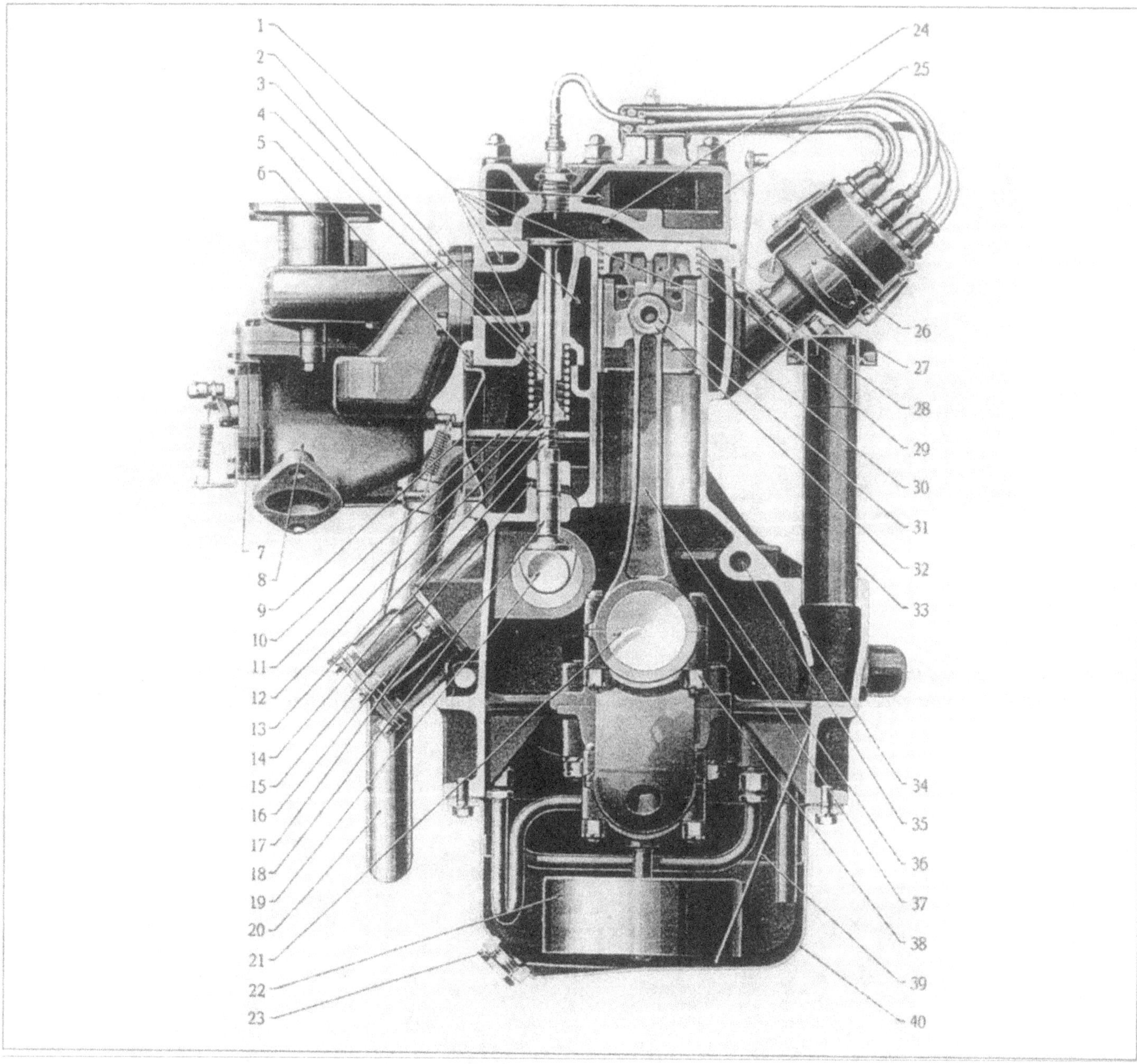

Fig. 7.6 *A cross section through the #1 cylinder of the 1930 Dodge DC straight-eight engine. The cylinder head has the classic turbulent combustion chamber developed by Zeder, Skelton, and Breer (ZSB). This view shows how the deep skirt was tied into the bulkhead structure by triangular webs in the crankcase.*

The first engines were identical to the Dodge DG engine at 240 cu. in. and probably were built at the Dodge Main plant. Sometime after launch, Chrysler moved above the 240-cu.-in. Dodge-size engine with a bore increase to 3.125 in., giving its engine 260.8 cu. in. Late in the model year, a larger and more luxurious Chrysler CD DeLuxe model was introduced that was powered by an even larger version of the engine, 282.1 cu. in., which produced an impressive 100 hp @ 3400 rpm and 190 lb-ft torque @ 1400 rpm. This last increase in displacement was not easy to achieve, however.

The displacement increase to 282.1 cu. in. was accomplished by a bore increase from 3.125 to 3.25 in., and present was the exact same problem described in Chapter 6 that afflicted the 1931 Dodge, DeSoto, and Chrysler four-main bearing, deep-skirt six-cylinder engine when its bore was increased to 3.25 in. Not surprisingly, the solution that worked on the six-cylinder engine was applied to this engine. The cylinder centers were judiciously moved, resulting in the narrow 3.5-in. spacing being increased to 3.625 in., the intermediate spacing being reduced to 3.8125 in., and 0.125 in. being removed from the center

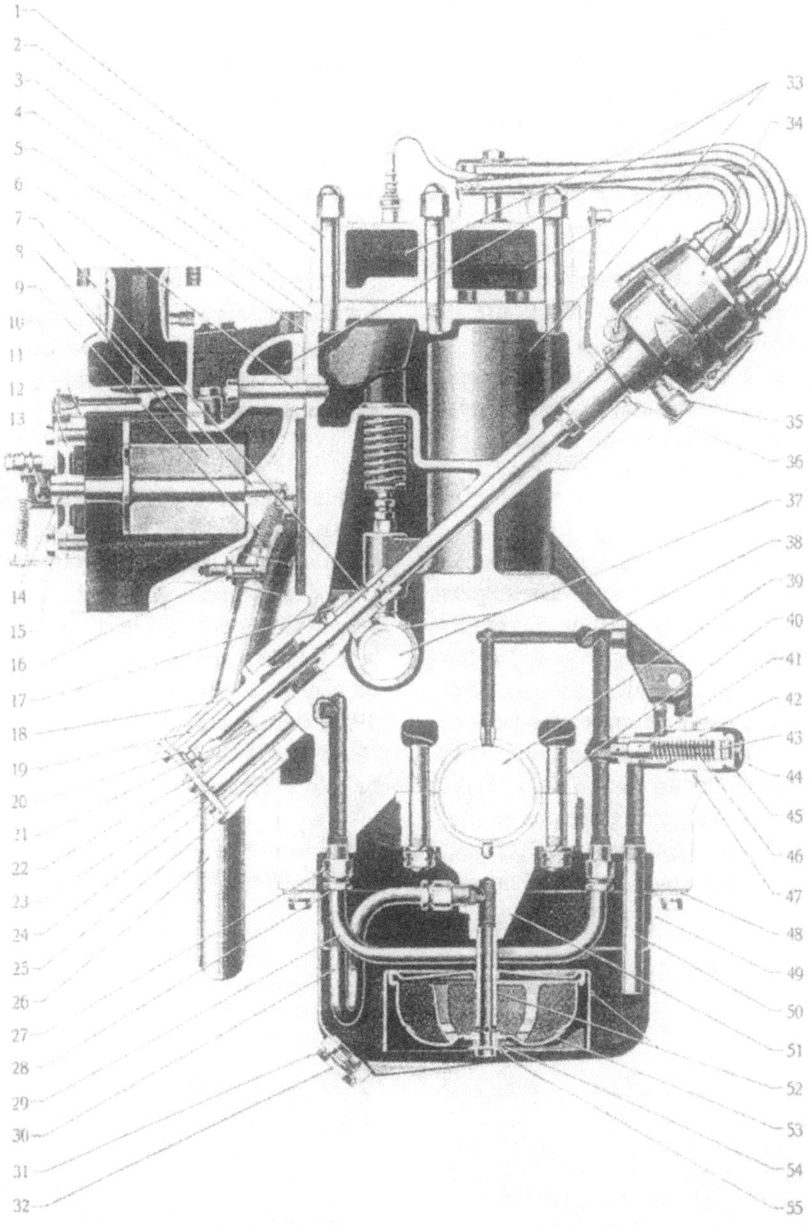

Fig. 7.7 *A cross section through the center of the 1930 Dodge DC straight-eight engine. The diagonal drive for the oil pump and the distributor had to be positioned above the camshaft to safeguard the structural integrity of the main bearing bulkhead.*

spacing to make it 4.0 in. As before, the new connecting rods had longer big ends and offset shanks—four offset one way and four the opposite way—requiring the engine assemblers to be careful in assembling one type of rod in the odd-numbered cylinders and the other type in the even-numbered cylinders. Similar to the six-cylinder engine, the solution was not ideal, but it worked.

The revised 4.25-in.-stroke crankshaft was fitted with bolted-on counterweights for main bearing load reduction. The 1931 282.1-cu.-in. Chrysler engine with its additional crankshaft counterweighting, modifications to the combustion chamber, and revised manifolding was reported to be smoother than any of its predecessors.[2]

At the end of 1931, after only two years, DeSoto cancelled its straight-eight engine. Dodge and Chrysler carried on with the 1932 Dodge DK engine taking over the 282-cu.-in. size and the 1932 Chrysler CP stretching the stroke to 4.5 in. for a 298.6-cu.-in. engine. Both the Dodge and Chrysler versions were upgraded with steel exhaust valve seat inserts and Floating Power® engine mounts. Replaceable connecting rod bearings appeared in the larger Chrysler engine but not in the Dodge, which continued with the long-used cast-babbitt bearings. This suggested that by now, Chrysler was independent of Dodge for eight-cylinder engine supply and was machining and assembling its version of the engine at the engine-savvy Chrysler Jefferson Avenue plant.

The 1933 Dodge DO, powered with a carryover 1932 Dodge DK engine, turned out to be the last Dodge use of the straight-eight engine. Chrysler expanded its 1933 use of the engine with the addition of a 273.8-cu.-in.-displacement, 4.125-in.-stroke version as power for the new Chrysler Royal Eight CT model. At the upper end of its model offerings, Chrysler used the 298.6-cu.-in. version as power for its 1933 Chrysler Imperial Eight CQ model.

In 1934, the unmistakable, eye-catching Airflow models began appearing on the nation's streets and highways. The engine for the smooth-looking 1934 Chrysler Airflow Eight CU shown on Fig. 7.8 was the 298.6-cu.-in. version with the fan moved down to the nose of the crankshaft for the reasons outlined in the section of Chapter 6 that described the DeSoto Airflow six-cylinder engine. Equipped with an aluminum, 6.5:1 compression-ratio cylinder head, and a two-barrel carburetor, the Chrysler 298.6-cu.-in. straight-eight engine produced an impressive 122 hp @ 3400 rpm and 225 lb-ft torque @ 1600 rpm. Powering the top-of-the-line Chrysler Airflows—the Chrysler Airflow Imperial CV and the 137.5-in.-wheelbase Chrysler Airflow Custom Imperial CX—was a larger 323.5-cu.-in. version of the engine, similar to the 298.6-cu.-in. engine, but with a longer 4.875-in. stroke. Equipped with a 6.5:1 high-compression-ratio aluminum cylinder head and a two-barrel carburetor as standard equipment, its power and torque were an even more impressive 130 hp @ 3400 rpm and 250 lb-ft @ 1600 rpm.

Starting in 1934, the crankshafts of the engines had integrally forged counterweights. Cylinder block water jackets were made full length in 1936, and the water distribution tube for improved cooling of the exhaust valve seat area was added at the same time.

This 323.5-cu.-in. engine became the standard power for Chrysler luxury cars for year after year—through the remainder of the 1930s, all of the 1940s, and the start of the 1950s. Included in the many cars were the one-of-a kind pre-World War II 1939 Chrysler Imperial parade car shown on Fig. 7.9, owned by retired Chrysler engineer Bruce Thomas, and the sporty wood-sided 1948 Chrysler Town and Country convertible shown on Fig. 7.10. Its reign of top-end power finally ended in late 1950, eclipsed by the arrival of a worthy replacement—the Hemi-head Chrysler FirePower V-8 of 1951, whose story will be told in Chapter 8.

Table 7.1 shows the specifications for this five-main-bearing, L-head, straight-eight engine.

Fig. 7.8 *The 1934 Chrysler CU Airflow four-door sedan on display at the Walter P. Chrysler Museum. Under the hinged-at-the-rear hood was a 122-hp 298.6-cu.-in. straight-eight engine with its cooling fan mounted on the nose of the crankshaft. Photo by W.L. Weertman.*

Fig. 7.9 *The 1939 Chrysler Imperial C24 Custom Parade Phaeton powered by a 323.5-cu.-in. straight-eight engine. Originally used to transport dignitaries at the 1939 World's Fair in New York on June 10, 1945, the car was ridden by General Dwight Eisenhower in the New York City Victory Parade celebrating the Allied victory in Europe near the end of World War II. Displayed at the Walter P. Chrysler Museum, the car was on temporary loan from owner Bruce R. Thomas. Photo by W.L. Weertman.*

Fig. 7.10 *This 1948 Chrysler C39N Town and Country convertible is on display at the Walter P. Chrysler Museum. Its 323.5-cu.-in. straight-eight engine gave the car smooth, abundant performance. Photo by W.L. Weertman.*

Regal Power for the 1931 Chrysler Imperial—A Straight-Eight with Nine Main Bearings

With swooping fenders, white-walled spare tires buckled securely in dual wheel wells, richly outfitted interiors, and a hood that seemed to stretch forever, the super-luxury cars of the late 1920s and early 1930s, including the 1931 Chrysler Imperial CG pictured on Fig. 7.11, were designed to provide personal transportation with style, comfort, and elegance. Views of the right and left sides of the engine compartment of the Imperial CG as shown on Figs. 7.12 and 7.13, respectively, illustrate that under its hood, the car had an engine as impressive as the exterior. With a cylinder block machined to a length of 37.125 in. and a crankshaft length of 41.78 in., it was the longest automotive engine ever manufactured by Chrysler. Likewise, with a bore and stroke of 3.5 × 5.0 in., the 384.8-cu.-in. displacement of this engine towered above all other Chrysler engines built at the time—and for many years to come. The general layout of the engine was classic ZSB—a main bearing between every cylinder, making a total of nine; a vertical drive at the right-center of the engine for the oil pump located at the bottom of the block, inside the oil pan, and the distributor located on top of the cylinder head; full pressure for all main bearings, connecting rod big-end bearings, and camshaft bearings; and turbulent combustion chambers and the water pump and fan at the top-front of the engine.

An interesting question that could be posed is: Why was the engine designed to be similar to the older ZSB-designed engines instead of the newer-style Dodge, DeSoto, and Chrysler five-main-bearing straight-eight engine with a diagonal drive for the oil pump and the distributor? With all the key participants long gone, I can only speculate that it was designed either with a sense of nostalgia for the glory days of the original Imperial or with a pragmatic eye toward utilizing some of the machining equipment that would become surplus once the three original ZSB-designed engines ended production. Perhaps it was both reasons. With bolted-on counterweights for its nine-main-bearing crankshaft, the engine had a 3-in. deep skirt for the cylinder block and a uniform cylinder spacing of 4.0 in., except at the center of the engine, where the cylinders were 4.625-in. apart. The long, clean lines of the engine are evident in the right- and left-side views shown on Figs. 7.14 and 7.15, respectively. Similar to the earlier ZSB-designed Imperial six-cylinder engine, acorn nuts capped the cylinder head studs in place of ordinary machine nuts. The views on Figs. 7.16, 7.17, and 7.18 show a longitudinal section, a cross section through the #1 cylinder, and a cross section through the center of the engine, respectively.

The bore, stroke, and displacement of the engine never changed during its six-year life from 1931 through 1936. Both a standard 5.2:1 compression ratio cylinder head and an optional Red Head having a compression ratio of 5.8:1 were available on the 1931 CG model, providing power levels of 125 hp @ 3200 rpm and 135 hp @ 3200 rpm, respectively. Torque levels were 260 lb-ft @ 1200 rpm and 280 lb-ft @ 1200 rpm, respectively. With that much power and torque available, the engine inevitably was pressed into service as a heavy-duty truck, marine, and industrial engine, duties that are outlined in Chapters 15 and 16.

The year 1932 brought change for this big straight-eight engine, now coded CL. The Floating Power® engine mount system was installed, and eight shrink-fitted steel exhaust valve seat inserts were added to the top of the cylinder block to eliminate the problem of exhaust valve seat sinkage.

In 1933, the corporation was invited by the Indianapolis Motor Speedway to furnish a Chrysler Custom Imperial powered by the engine as a pace car for the May 1933 Indianapolis 500 race, an invitation fulfilled by the handsome car shown on Fig. 7.19.

By 1934, the exhaust manifold, originally a single-piece design, had been redesigned to a three-piece construction, reducing thermal stress during heat-up and cool-down. As with the other engines installed in Chrysler Airflow models, when this engine powered the 1934–1936 Chrysler Airflow Custom Imperials, the fan was attached to the nose of the crankshaft.

During its last three years, 1934 through 1936, the engine powered the Chrysler Airflow Custom Imperial CW. Fitted with an aluminum cylinder head having a 6.5:1 compression ratio, the engine produced 150 hp @ 3200 rpm and an impressive 300 lb-ft torque @ 1600 rpm, which was sorely needed to propel the 5000-plus pound car.[3] All engines were equipped with a two-barrel downdraft carburetor.

By the end of 1936, sales volume of the only model using the engine, the Chrysler Airflow Custom Imperial CW, dropped

TABLE 7.1
SPECIFICATIONS FOR THE DODGE AND CHRYSLER-BUILT, FIVE-MAIN BEARING, STRAIGHT-EIGHT ENGINES

Model Year	1930	1930	Initial 1931	Later 1931
Car line	Dodge	DeSoto	Chrysler	Chrysler
Model	Eight DC	Eight CF	Eight CD	Eight CD
Displacement (cu. in.)	220.7	207.7	240.33	260.8
Bore	2.875	2.875	3.0	3.125
Stroke	4.25	4.0	4.25	4.25
Compression ratio	5.4:1	5.2:1	5.2:1 std.; 5.4:1 opt.	5.3:1 std.; 6.2:1 opt.
BHP@ rpm	75 @ 3400	70 @ 3400	80 @ 3400 std.; 82 @ 3400 opt.	88 @ 3400 std.
Torque (lb-ft) @ rpm	145 @ 1400	132 @ 1200	160 @ 1200 std.; 163 @ 1400 opt.	170 @ 1400 std.
Combustion chamber	Turbulent	Turbulent	Turbulent	Turbulent
Cyl. block mat'l.—skirt type	Cast iron—deep	Cast iron—deep	Cast iron—deep	Cast iron—deep
Cyl. block deck height	13.005	13.005	13.005	13.005
Cyl. block length	33.1875	33.1875	33.1875	33.1875
Cyl. bore spacing	1-2, 3-4, 5-6, 7-8 = 3.5 2-3, 6-7 = 3.9375 4-5 = 4.125	1-2, 3-4, 5-6, 7-8 = 3.5 2-3, 6-7 = 3.9375 4-5 = 4.125	1-2, 3-4, 5-6, 7-8 = 3.5 2-3, 6-7 = 3.9375 4-5 = 4.125	1-2, 3-4, 5-6, 7-8 = 3.5 2-3, 6-7 = 3.9375 4-5 = 4.125
Ck/shaft—no. of main brgs.	5	5	5	5
Ck/shaft main journal dia.	2.5	2.5	2.5	2.5
Crankpin journal dia.	2.25	2.25	2.25	2.25
Piston mat'l.—no. of rings	Aluminum—4	Aluminum—4	Aluminum—4	Aluminum—4
Connecting rod centers	8.875	8.875	8.875	8.875
Valve system	L-head	L-head	L-head	L-head
Valve events	220-230-2	220-230-2	220-230-2	220-230-2
Camshaft drive	Chain	Chain	Chain	Chain
Valve head dia.	Intake 1.41; exhaust 1.41	Intake 1.41; exhaust 1.41	Intake 1.41; exhaust 1.41	Intake 1.47; exhaust 1.41
Valve lash adjustment	Mechanical tappet	Mechanical tappet	Mechanical tappet	Mechanical tappet
Oil pump/distributor drive	Diagonal at center of block	Diagonal at center of block	Diagonal at center of block	Diagonal at center of block
Fuel system	Downdraft carburetor	Downdraft carburetor	Downdraft carburetor	Downdraft carburetor

(Continued on next page)

TABLE 7.1
(Continued)

Model Year	1931	1932	1933	1933
Car line	Chrysler	Chrysler	Dodge	Chrysler
Model	Deluxe Eight CD	Eight CP	DO	Royal Eight CT
Displacement (cu. in.)	282.1	298.6	282.1	273.8
Bore	3.25	3.25	3.25	3.25
Stroke	4.25	4.50	4.25	4.125
Compression ratio	5.2:1 std.; 6.2:1 opt.	5.2:1 std.; 6.2:1 opt.	6.2:1 std.; 5.2:1 opt.	5.4:1 std.; 6.2:1 opt.
BHP@ rpm	100 @ 3400 std.	100 @ 3400 std.	100 @ 3400 std.; 92 @ 3400 opt.	90 @ 3400 std.; 98 @ 3400 opt.
Torque (lb-ft@ rpm)	190 @ 1400 std.	200 @ 1200 std.	200 @ 1200 std.; 184 @ 1200 opt.	190 @ 1200 std.; 192 @ 1200 opt.
Combustion chamber	Turbulent	Turbulent	Turbulent	Turbulent
Cyl. block mat'l.—skirt type	Cast iron—deep	Cast iron—deep	Cast iron—deep	Cast iron—deep
Cyl. block deck height	13.005	13.005	13.005	13.005
Cyl. block length	33.1875	33.1875	33.1875	33.1875
Cyl. bore spacing	1-2, 3-4, 5-6, 7-8 = 3.625 2-3, 6-7 = 3.8125 4-5 = 4.0	1-2, 3-4, 5-6, 7-8 = 3.625 2-3, 6-7 = 3.8125 4-5 = 4.0	1-2, 3-4, 5-6, 7-8 = 3.625 2-3, 6-7 = 3.8125 4-5 = 4.0	1-2, 3-4, 5-6, 7-8 = 3.625 2-3, 6-7 = 3.8125 4-5 = 4.0
Ck/shaft—no. of main brgs.	5	5	5	5
Ck/shaft main journal dia.	2.5	2.5	2.5	2.5
Crankpin journal dia.	2.25	2.1875	2.1875	2.1875
Piston mat'l.—no. of rings	Aluminum—4	Aluminum—4	Aluminum—4	Aluminum—4
Connecting rod centers	8.8125—offset shank	9.0—offset shank	8.8125—offset shank	9.0—offset shank
Valve system	L-head	L-head	L-head	L-head
Valve events	220-230-2	220-230-2	220-230-2	220-230-2
Camshaft drive	Chain	Chain	Chain	Chain
Valve head dia.	Intake 1.47; exhaust 1.41	Intake 1.47; exhaust 1.41	Intake 1.47; exhaust 1.41	Intake 1.47; exhaust 1.41
Valve lash adjustment	Mechanical tappet	Mechanical tappet	Mechanical tappet	Mechanical tappet
Oil pump/distributor drive	Diagonal at center of block	Diagonal at center of block	Diagonal at center of block	Diagonal at center of block
Fuel system	Downdraft carburetor	Downdraft carburetor	Downdraft carburetor	Downdraft carburetor
Other	Bolted-on ck/s cwts.	Bolted-on ck/s cwts.	Bolted-on ck/s cwts.	Bolted-on ck/s cwts.

(Continued on next page)

TABLE 7.1
(Continued)

Model Year	1934	1946–1948	1949–1950
Car line	Chrysler	Chrysler, Crown Imperial	Chrysler, Imperial
Model	Airflow CV, CX	C39, C40	C46, C47, C49, C50
Displacement (cu. in.)	323.5	323.5	323.5
Bore	3.25	3.25	3.25
Stroke	4.875	4.875	4.875
Compression ratio	6.5:1	6.7:1	7.25:1
BHP@ rpm	130 @ 3400	135 @ 3400	135 @ 3200
Torque (lb-ft) @ rpm	250 @ 1600	270 @ 1600	270 @ 1600
Combustion chamber	Turbulent	Turbulent	Turbulent
Cyl. block mat'l.—skirt type	Cast iron—deep	Cast iron—deep	Cast iron—deep
Cyl. block deck height	13.005	13.005	13.005
Cyl. block length	33.1875	33.1875	33.1875
Cyl. bore spacing	1-2, 3-4, 5-6, 7-8 = 3.625 2-3, 6-7 = 3.8125 4-5 = 4.0	1-2, 3-4, 5-6, 7-8 = 3.625 2-3, 6-7 = 3.8125 4-5 = 4.0	1-2, 3-4, 5-6, 7-8 = 3.625 2-3, 6-7 = 3.8125 4-5 = 4.0
Ck/shaft—no. of main brgs.	5	5	5
Ck/shaft main journal dia.	2.703	2.703	2.703
Crankpin journal dia.	2.1875	2.1875	2.1875
Piston mat'l.—no. of rings	Aluminum—4	Aluminum—4	Aluminum—4
Connecting rod centers	9.0—offset shank	9.0—offset shank	9.0—offset shank
Valve system	L-head	L-head	L-head
Valve events	226-230-6	C39: 236-232-18 C40: 236-236-18	236-236-18
Camshaft drive	Chain	Chain	Chain
Valve head dia.	Intake 1.47; exhaust 1.41	Intake 1.51; exhaust 1.34	Intake 1.51; exhaust 1.34
Valve lash adjustment	Mechanical tappet	Mechanical tappet	Mechanical tappet
Oil pump/distributor drive	Diagonal at center of block	Diagonal at center of block	Diagonal at center of block
Fuel system	2-barrel carburetor	2-barrel carburetor	2-barrel carburetor
Other	Fan mtd. to nose of ck/s		

Fig. 7.11 *The 1931 Chrysler Imperial CG Dual Cowl Phaeton on display at the Walter P. Chrysler Museum. Its 385-cu.-in. straight-eight engine had nine main bearings and cylinder bores spaced farther apart than those of the five-main-bearing straight-eight already in production. Photo by W.L. Weertman.*

Fig. 7.12 *The right side of the engine compartment of the 1931 Chrysler Imperial CG Dual Cowl Phaeton on display at the Walter P. Chrysler Museum. The red-painted cylinder head identifies the engine as the optional 5.8:1 compression-ratio engine with an output of 135 hp and 280 lb-ft of torque. Photo by W.L. Weertman.*

Fig. 7.13 *The left side of the engine compartment of the 1931 Chrysler Imperial CG Dual Cowl Phaeton. The long, corrugated cover seals the water jacket opening of the cylinder block used for sand removal during the casting process. The acorn nuts for the cylinder head studs add a quality touch to the appearance of the engine. Photo by W.L. Weertman.*

Fig. 7.14 *The right side of the 1931 Chrysler Imperial 385-cu.-in. straight-eight engine. By 1934, the long, single-piece exhaust manifold shown had been changed to a three-piece design to reduce temperature stresses during engine operation.*

Fig. 7.15 *The left side of the 1931 Chrysler Imperial 385-cu.-in. straight-eight engine. The engine's coolant thermostat, mounted on the top-front of the cylinder head, was linked to radiator shutters. When installed in the 1934 Chrysler Airflow Custom Imperial CW model, the engine fan was mounted on the nose of the crankshaft.*

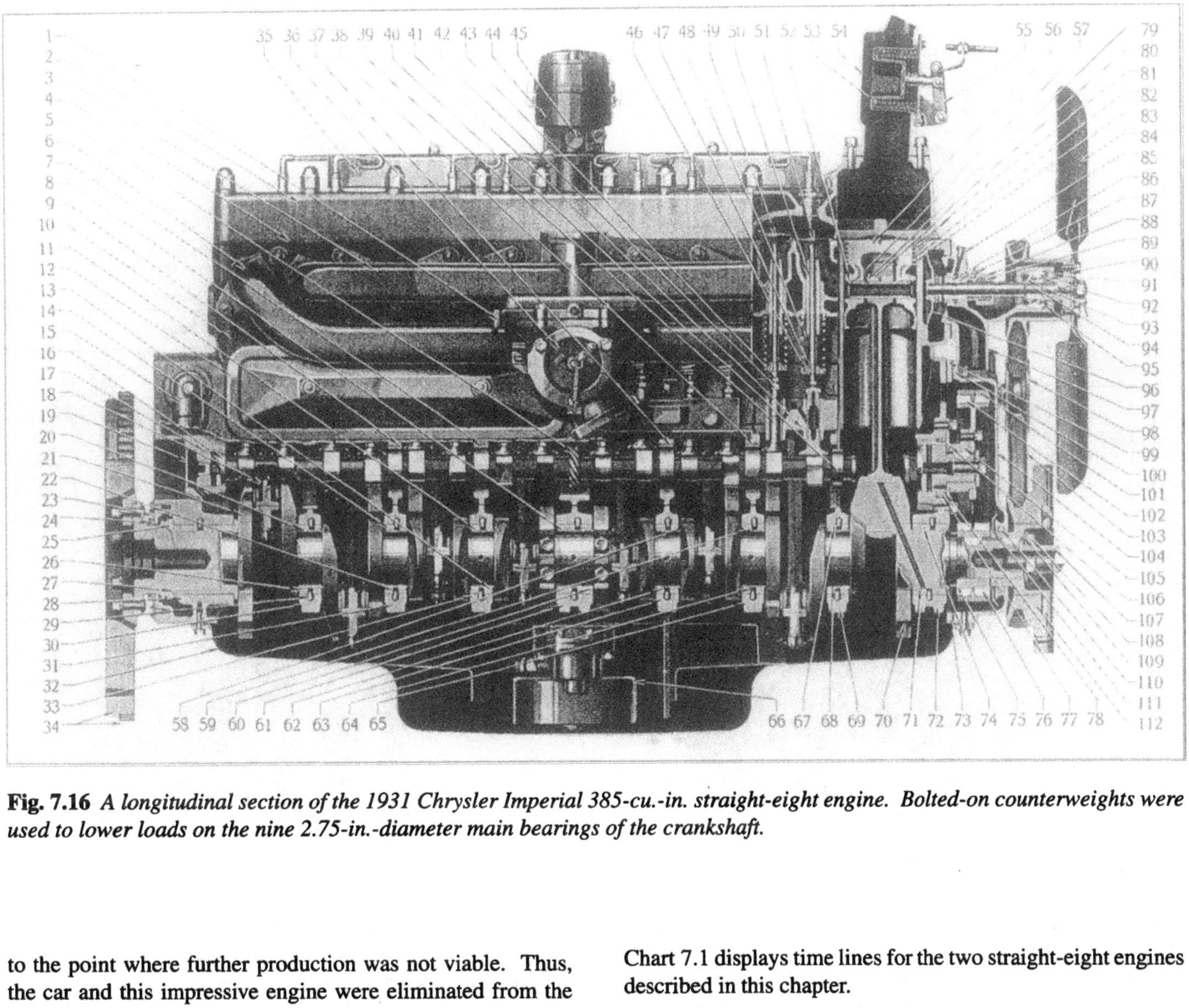

Fig. 7.16 *A longitudinal section of the 1931 Chrysler Imperial 385-cu.-in. straight-eight engine. Bolted-on counterweights were used to lower loads on the nine 2.75-in.-diameter main bearings of the crankshaft.*

to the point where further production was not viable. Thus, the car and this impressive engine were eliminated from the corporate lineup.

Table 7.2 shows the specifications for this Imperial nine-main-bearing, L-head, straight-eight engine.

Chart 7.1 displays time lines for the two straight-eight engines described in this chapter.

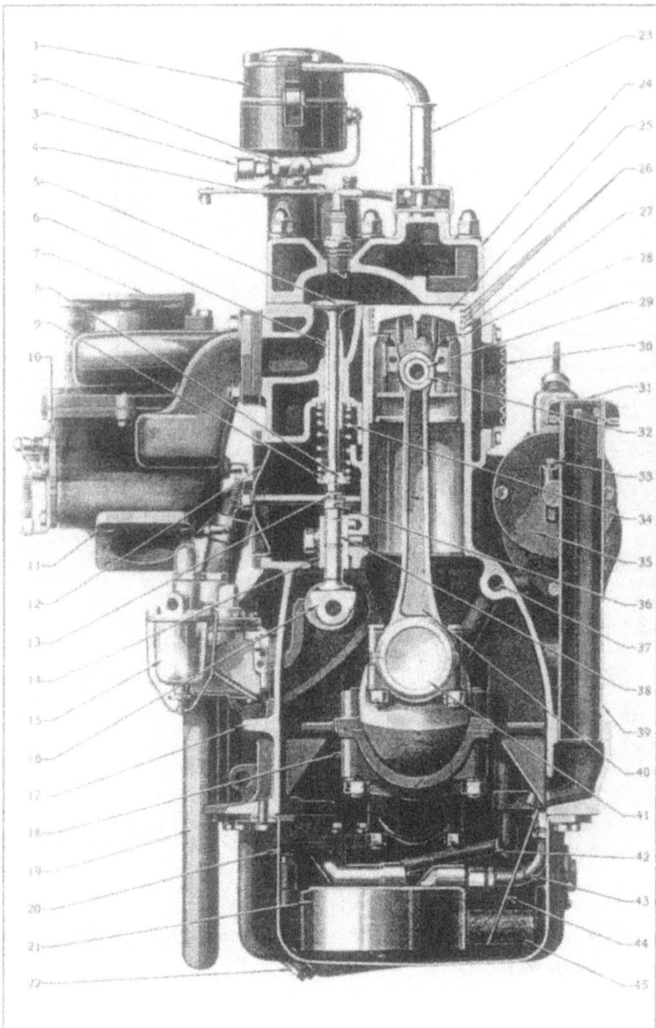

Fig. 7.17 *A cross section through the #1 cylinder of the 1931 Chrysler Imperial 385-cu.-in. straight-eight engine. The ZSB classic turbulent, domed combustion chamber can be seen. To accommodate the 5-in. stroke of the engine, the top deck of the cylinder block was located 14.1875 in. above the crankshaft centerline.*

Fig. 7.18 *A cross section through the center of the 1931 Chrysler Imperial 385-cu.-in. straight-eight engine. The design of the vertical shaft system for driving the oil pump and the distributor appears to be a carryover from the earlier ZSB-designed engines.*

Fig. 7.19 *(Left) This classy-looking 1933 Chrysler Custom Imperial CL convertible roadster was the pace car for the May 1933 Indianapolis 500 race. Its big 385-cu.-in. nine-main-bearing straight-eight engine gave Chrysler vice president Byron Foy plenty of power for his driving duties. Photo courtesy of Indianapolis Motor Speedway.*

TABLE 7.2
SPECIFICATIONS FOR THE IMPERIAL NINE-MAIN-BEARING, STRAIGHT-EIGHT ENGINE

Model Year	1931	1932	1933	1934–1936
Car line	Chrysler Imperial	Chrysler Imperial	Chrysler	Chrysler
Model	Eight CG	Eight CH, Custom Eight CL	Custom Imperial	Airflow Custom Imperial
Displacement (cu. in.)	384.8	384.8	384.8	384.8
Bore	3.5	3.5	3.5	3.5
Stroke	5.0	5.0	5.0	5.0
Compression ratio	5.2:1 std.; 5.8:1 opt.	5.2:1 std.; 5.8:1 opt.	5.8:1 std.; 5.2:1 opt.	6.5:1
BHP@ rpm	125 @ 3200 std.; 135 @ 3200 opt.	125 @ 3200 std.; 135 @ 3200 opt.	135 @ 3200 std.; 125 @ 3200 opt.	150 @ 3200
Torque (lb-ft) @ rpm	260 @ 1200 std.; 280 @ 1200 opt.	260 @ 1200 std.; 280 @ 1200 opt.	280 @ 1200 std.; 260 @ 1200 opt.	300 @ 1600
Combustion chamber	Turbulent	Turbulent	Turbulent	Turbulent
Cyl. block mat'l.—skirt type	Cast iron	Cast iron	Cast iron	Cast iron
Cyl. block deck height	14.1875	14.1875	14.1875	14.1875
Cyl. block length	37.125	37.125	37.125	37.125
Cyl. bore spacing	All except 4-5 = 4.0 4-5 = 4.625	All except 4-5 = 4.0 4-5 = 4.625	All except 4-5 = 4.0 4-5 = 4.625	All except 4-5 = 4.0 4-5 = 4.625
Ck/shaft—no. of main brgs.	9	9	9	9
Ck/shaft main journal dia.	2.75	2.75	2.75	2.75
Crankpin journal dia.	2.1875	2.1875	2.1875	2.1875
Piston mat'l.—no. of rings	Aluminum—5	Aluminum—5	Aluminum—5	Aluminum—5
Connecting rod centers	10	10	10	10
Valve system	L-head	L-head	L-head	L-head
Valve events	220-230-2	220-230-2	220-230-2	220-230-2
Camshaft drive	Chain	Chain	Chain	Chain
Valve head dia.	Intake 1.72; exhaust 1.66	Intake 1.72; exhaust 1.66	Intake 1.72; exhaust 1.66	Intake 1.72; exhaust 1.66
Valve lash adjustment	Mechanical tappet	Mechanical tappet	Mechanical tappet	Mechanical tappet
Oil pump/distributor drive	Vertical at center of engine	Vertical at center of engine	Vertical at center of engine	Vertical at center of engine
Fuel system	2-barrel downdraft	2-barrel downdraft	2-barrel downdraft	2-barrel downdraft

Chart 7.1
Time Lines—Straight-Eight L-Head Engines

Engines, Car Lines, and Displacements (cu. in.)

Engine	Car Line	1930	1931	1932	1933	1934	1935	1936	1937
Dodge/Chrysler, 5 main bearings	Dodge	221	240	282	282				
	Model	DC	DG	DK	DO				
	Desoto	208	221						
	Model	CF	CF*						
	Chrysler		240-261-282	299	274	299	274 - 323	274 - 323	323
	Model		CD - CD - CD*	CP	CT	CU	CZ - C1	C8 - C9	C17
	Imperial				299	323	323	323	274 - 323
	Model				CQ	CV, CX	C2, C3	C10, C11	C14 - C15
Imperial, 9 main bearings	Imperial		385	385	385	385	385	385	
	Model		CG	CH, CL	CL*	CW	CW*	CW**	

NOTE: The asterisks do not indicate a footnote; they are part of the model designator.

Chart 7.1
(Continued)

Engines, Car Lines, and Displacements (cu. in.)

Engine	Car Line	1938	1939	1940	1941	1942	1946–1948	1949	1950
Chrysler, 5 main bearings	Chrysler	299	323	323	323	323	323	323	323
	Model	C19	C23	C26	C30	C36	C39	C46	C49
	Imperial	299 - 323	323	323	323	323	323	323	323
	Model	C19 - C20	C23, C24	C27	C33	C37	C40	C46, C47	C49, C50

CHAPTER 8

Hemispheres and Polyspheres—The First Generation of V-8s

Throughout the 1930s and early 1940s, engine dynamometers in the Chrysler Engineering complex at Highland Park, Michigan, clocked hour after hour in searching for clues as to what might be better than the familiar, dependable, easy-to-build, in-line six-cylinder and eight-cylinder L-head engines that had been in production for years. Among the several L-head V-8 engines tested was the engine shown on Fig. 8.1, a 289-cu.-in., 3.5 × 3.75 in. V-8, Engineering coded A26, which produced a respectable 131 hp.

Also tested during this investigation period were four-, six-, and eight-cylinder horizontally opposed engines. For careful comparisons of different engine types, single-cylinder engines were used, one of the singles being a sleeve valve engine. Among the more unusual engines was a small 66.6-cu.-in. five-cylinder radial liquid-cooled L-head that was designed to power a front-wheel-drive compact car. Named the Star Car after the shape of the engine, the car and engine did not progress beyond the prototype stage.[1] However, despite all the vigorous testing, nothing surfaced with enough basic improvement to warrant replacing the solidly entrenched L-head engines being produced by the thousands day after day at the Chrysler Jefferson Avenue, Dodge Main, and Plymouth Lynch Road plants.

Fig. 8.1 *The left side of the experimental L-head V-8. Coded A26, the engine was one of many experimental engines designed, built, and tested to evaluate different sizes and arrangements. Photo dated October 1945.*

The priority of defense work during World War II caused civilian engine work to be largely discontinued. However, as the end of the war approached, the need for renewed effort to establish the course of future production engines became more urgent. The preface to an Engineering Technical Report dated October 9, 1947 stated: "Preparatory to the design of a new group of post-war engines, a proposal was made, on 12/7/44... to conduct a basic investigation of cylinder types, which might be applied to the design of multicylinder engines producing substantially increased performance and improved operating characteristics."[2] Co-authored by R. Dean Engle and John B. Platner, the report presented the encouraging results of tests of the single-cylinder engine equipped with double overhead camshafts, shown on Fig. 8.2, and its hemispherical combustion chamber that appears on Fig. 8.3. It was the start of a broad investigation program by the Engine Development Laboratory, Engineering Research, and Engine Design, which eventually led to the outstanding 1951 Chrysler FirePower Hemi® head V-8 engine. Chrysler Engineering already had favorable experience with hemispherical-shaped combustion chambers in developing the A45 World War II experimental XI-2220 inverted V-16 aircraft engine, described later in Chapter 17, and an experimental V-12 tank engine coded A65. Automobiles with hemispherical combustion chambers, long recognized as the best for output, had been built at least since 1904, as witnessed by a 1904 Welch SOHC four-cylinder engine and the Belgian Pipe Co. twin-camshaft OHV engine of the same year.

Fig. 8.2 *Front-top view of a single-cylinder laboratory test engine equipped with double overhead camshafts and a hemispherical combustion chamber. Photo dated January 1946.*

For the investigation program, a number of experimental single-cylinder engines, in-line sixes, and in-line eights with various types of valve arrangements and chamber shapes including overhead valves, overhead camshafts, hemispherical chambers, and F-heads were designed, built, and tested. Part of the effort was directed toward running high-output European engines on the dynamometer to obtain readings on their specific outputs, volumetric efficiency, thermal efficiency, detonation characteristics, and so forth. A double overhead camshaft Alfa Romeo hemi-chamber engine, a Healy overhead valve hemi using pushrods and two camshafts, an overhead valve Fiat with in-line valves, a Riley, and an F-head Rolls-Royce were among those procured and tested. Figure 8.4 shows an experimental double-overhead-camshaft Hemi® head 235-cu.-in. six-cylinder test engine built by Chrysler Engineering and coded A173.

Prior to the start of this investigation program, conventional engineering wisdom held that an open combustion chamber such as the hemi would require a lower compression ratio to avoid detonation or "spark knock" versus a chamber that had a turbulence-inducing feature such as the narrow gap above the piston of the L-head engines. However, the dynamometer results showed that the hemi chamber was not more knock sensitive than other chambers, suggesting that the hemispherical shape, with its low surface-to-volume ratio and with the central location of the spark plug, made up for any loss of the mixture turbulence. The hemi chamber had an inherent volumetric efficiency advantage, in that the incoming charge and exiting

Fig. 8.3 *Hemispherical combustion chamber of the double-overhead-camshaft single-cylinder test engine. Test results encouraged further investigation.*

Fig. 8.4 *Experimental double overhead camshaft hemi-chambered six-cylinder test engine. Coded A173, the cross-flow 235-cu.-in. engine added more data favoring the hemispherical combustion chamber.*

exhaust gas could easily flow past the heads of their respective valves without being restricted by the nearby chamber walls, as is the case with in-line overhead valves. With the knock sensitivity handicap substantially removed, the superior specific output, thermal efficiency, and volumetric efficiency of the hemi head design put it, in terms of performance, ahead of all contenders for the Chrysler powerplant of the future. The final challenge to its acceptance was the concern of production compatibility and cost. The dual-overhead-camshaft arrangements usually associated with hemi-chambered engines required many more expensive parts and careful adjustments of the camshaft drive system and valve lash settings that would result in severe cost penalties and procedures not adaptable to high-volume engine production.[3] To avoid these complications, both Powerplant Research and Engine Design came up with designs of in-line six-cylinder and V-8 engines combining the hemi chamber and valve placement with a single camshaft mounted in the cylinder block using tappets, pushrods, and shaft-mounted rocker arms to actuate the valves.

An integral part of the future engine direction was the dropping of the in-line type of eight-cylinder engine and embracing the V-8 arrangement. Overall, car layouts certainly would benefit from its compactness, and, most importantly, the V-8 would permit a major shift in the bore and stroke relationship of future engines, with the bore becoming larger and the stroke becoming shorter for any given displacement. The larger bores would provide room for larger valves that increased the airflow in and out of the cylinders, resulting in a desirable boost in power output per cubic inch. By its shape, the hemi chamber was particularly good at flowing air and minimizing heat loss, giving it the efficiency advantages previously noted.

Figure 8.5 shows an experimental in-line six-cylinder Hemi® head engine that was coded A221 and was designed by the Engine Design department under Mel Carpentier. Figure 8.6 shows an experimental V-8 engine that was coded A182 and was designed by the Powerplant Research department under Ray White.

With investigation testing completed and major problems resolved, a corporate decision was needed on whether or not to proceed with the Hemi® head V-8 for production. Steering away from the smooth, quiet, L-head, in-line eight-cylinder engine, with its tooling that was amortized years ago, toward a new engine type that would require a substantial investment of dollars was a difficult decision for corporate management. However, Bill Drinkard, head of Engine Development, knowing from the vast amount of test data that Chrysler had a breakthrough engine and with the support of senior development engineer John Platner, finally prevailed in his arguments that this was the way to go.[4] With a favorable decision to proceed with the Hemi® V-8 engine as a production program, Mel Carpentier assigned his ace lead designer, Fred Shrimpton, to make the design of the first production Hemi® head V-8 engine, the 331-cu.-in. A239.[5]

Chrysler FirePower Hemi® Head V-8

FirePower was the appropriate name created for the first Chrysler production V-8 engine, originally known by its Engineering A239 designator. Scheduled to power the 1951 Chrysler Saratoga, Chrysler New Yorker, and Chrysler Imperial cars, production started about December 1950 at the Chrysler Jefferson Avenue plant in Detroit. The V-8 arrived none too soon because Cadillac and Oldsmobile had introduced new high-performing overhead valve V-8 engines two years earlier with the 1949 model year—introductions that changed the course of powerplant development for decades. As a showcase for Chrysler

Fig. 8.5 *The right side of the experimental 231-cu.-in. hemi-chambered pushrod-operated, overhead valve six-cylinder engine. The Engine Design department started a layout for the engine, which was coded A221, in 1948.*

Fig. 8.6 *The right side of the experimental hemi-chambered pushrod-operated overhead valve V-8 engine. The 330-cu.-in. engine, coded A182, was the first Chrysler Hemi® head V-8 engine.*

Corporation in general and Chrysler Engineering in particular, the FirePower V-8 engine had to be not just good but outstanding—and it was. With a bore and stroke of 3.8125 × 3.625 in., its 331-cu.-in. displacement delivered 180 hp @ 4000 rpm and a hefty torque of 312 lb-ft @ 2000 rpm—both output figures far above the 135 hp @ 3400 rpm and 270 lb-ft @ 1600 rpm of the 323.5-cu.-in. straight-eight engine that it replaced. Figure 8.7 shows an outside view of the V-8 engine, and Figs. 8.8 and 8.9 show a longitudinal section and a cross section, respectively.

Certainly the most impressive feature of the Chrysler V-8 engine was its cylinder head. The broad, powerful-looking shape shown in the cross section was the result of the hemispherical chamber, the placement of the valves, and valve actuation. The intake and exhaust valves, located directly across from each other, had an included angle of 58.5 deg. Above the valves, two rocker shafts were mounted solidly in cast iron pedestals—an inboard shaft for the intake rocker arms, and an outboard shaft for the exhaust rocker arms. The spark plugs, located deep inside the cylinder head and near the center of the hemi chamber, were accessed through long tubes that provided an oil-tight seal. The word "FirePower" was embossed on the wide, formed-steel cylinder head covers, further emphasizing the look of power and performance. Replaceable guides for all valve stems, shrink-fitted alloy-steel seat inserts for the exhaust valves, and inherent excellent cooling of the exhaust valve seats due to the hemi chamber design enhanced engine durability. Hydraulic tappets provided automatic valve lash adjustment while minimizing valve noise.

Fig. 8.8 *A longitudinal section of the 1951 331-cu.-in. Chrysler FirePower Hemi® head V-8 engine. Typical of V-8 engines, two connecting rods from opposite banks share each crankpin of the five-main-bearing crankshaft. During 1953, the cylinder block was shortened by removal of the flywheel housing extension located at the left end of the section.*

Fig. 8.7 *A right-front view of the 1951 331-cu.-in. Chrysler FirePower Hemi® head V-8 engine. With the release of this engine, coded A239, Chrysler gained immediate prominence for its high-performing cars. This engine is on display at the Walter P. Chrysler Museum. Photo by W.L. Weertman.*

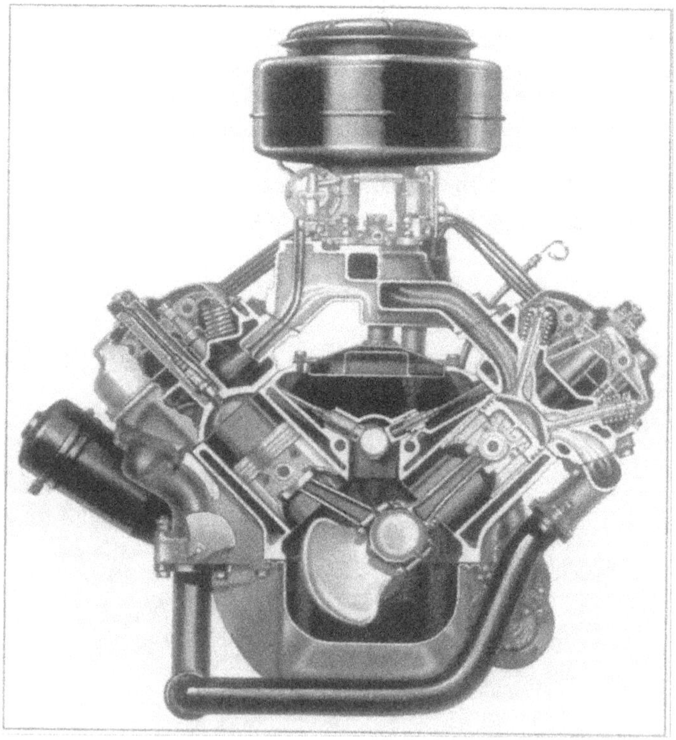

Fig. 8.9 *A cross section of the 1951 331-cu.-in. Chrysler FirePower Hemi® head V-8 engine. The hemispherical combustion chamber and double rocker arm shafts for each cylinder head are shown in both right and left banks of the engine. The right cylinder head shows the spark plug in its ideal location at the center of the combustion chamber.*

The cast iron cylinder block had the customary V-8 included bank angle of 90 deg, a deck height of 10.38 in., and a shallow skirt that extended only 0.125 in. below the crankshaft centerline. The upper half of the flywheel housing was integrated into the rear of the block (see Fig. 8.8), causing the machined length of the block to be a lengthy 27.78 in. A spacing of 4.5625 in. was used between all cylinder centers. Cylinders were numbered from front to rear, with the leading left bank having odd-numbered cylinders, and the right bank having even-numbered cylinders. This was a system that would be repeated on all Chrysler V-8s. Taking advantage of the leading left bank, a vertical shaft, gear-driven from the camshaft, was located at the left-rear of the engine. The ignition distributor, mounted on top of the block, picked up its drive from the top of the vertical shaft, while the oil pump of the engine, attached to the #5 main bearing cap, was driven from the lower end of the vertical shaft. The robust forged-steel crankshaft, with its five 2.5-in.-diameter main journals and four 2.25-in. crankpins, was designed with strength for infinite life, a necessity for power section parts. Forged-steel connecting rods with 6.625-in. centers and three-ring aluminum pistons completed the power section of the engine.

For cold weather operation, exhaust gas from the entire right bank was routed across the engine through a crossover passage in the bottom of the intake manifold. The hot exhaust gases heated the intake fuel/air passages directly above, helping to evaporate liquid fuel and to improve mixture distribution, as well as to warm the choke-sensing element.

A technical paper on the engine, co-authored by William E. Drinkard and Mel L. Carpentier, was presented to the Society of Automotive Engineers on March 6, 1951.[6] In December of the same year, in an article that appeared in *Motor Trend* magazine,[7] James C. Zeder outlined the reasons for the excellent performance of the hemi chamber.

Chrysler Division's experienced engine production people quickly gained the skills needed to build the new V-8 engine, and the sight of neat rows of engines at the end of the assembly line (as shown on Fig. 8.10) became commonplace.

The 1951 models powered by the engine were the Chrysler Saratoga, New Yorker, Imperial, and Crown Imperial. Although styling changes from the prior model year were limited to the grille, taillights, and other minor modifications, the cars nevertheless had clean and attractive lines, as shown on Fig. 8.11 on the photo of a 1951 Chrysler New Yorker convertible. With a rapidly growing reputation for top performance by new Chrysler V-8 powered cars, a similar New Yorker convertible, shown on Fig. 8.12, was chosen to pace the field of race cars competing in the May 1951 Indianapolis 500 race.

Although the performance of the engine certainly was its strong suit, its fuel efficiency also was remarkable with a Chrysler Imperial and Chrysler Crown Imperial, both equipped with the 180-hp 331-cu.-in. FirePower engine, taking home first-place trophies in the 1951 Mobilgas Economy Run. The two cars, entered in Class G (i.e., cars listing from $3,001 to $3,300) and Class I (cars listing from $4,001 to $6,000) averaged 21.174 and 19.208 mpg, respectively, on the 840-mile course that ran from Los Angeles to the Grand Canyon.

Altogether, the engine represented a bold and brilliant advancement of Chrysler automotive power.

Fig. 8.10 *An impressive view of Chrysler power—new Chrysler FirePower V-8 engines in shipping racks at the Chrysler Jefferson Avenue plant.*

Fig. 8.11 *The 1951 Chrysler C52 New Yorker convertible. The bold "V" on the front of the hood proclaimed that under the hood was the new Chrysler 180-hp FirePower Hemi® V-8. This vehicle is on display at the Walter P. Chrysler Museum. Photo by W.L. Weertman.*

Fig. 8.12 *Chrysler Division president Dave Wallace is at the wheel of the Hemi® engine-powered 1951 Chrysler New Yorker convertible selected as the pace car for the May 1951 Indianapolis 500 race.*

Subsequent Changes

Once production started, no significant changes were made to the engine through early 1953. Then, during 1953, the flywheel housing extension was removed from the cylinder block, reducing the machined length of the block by 6 in. to 21.78 in. Figure 8.13 shows an outside view of the shortened 1953 engine.

The October 1952 issue of *Motor Trend* magazine gave accolades to the engine, saying, "The Chrysler Corporation can be proud of its power-packed V-8 engine, a true engineering achievement...Power of the FirePower is astounding. Since the introduction of the sensational FirePower engine with the 1951 line of Chrysler automobiles, its fame among motor-minded car owners has steadily grown until today it is generally conceded that this powerful V-8 is among the foremost pace-setters."

The quest for more power surged ahead in 1954 when intake and exhaust valve head sizes were increased, bringing the rated power of the two-barrel carburetor version up to 195 hp @ 4400 rpm. The year 1954 also saw the start of a four-barrel carburetor version of the engine that produced an impressive 235 hp and 330 lb-ft torque. In 1955, a compression ratio increase to 8.5:1 from the original 7.5:1 brought the power of the 331-cu.-in. four-barrel version up to 250 hp. Figures 8.14 and 8.15 show a longitudinal view and a cross-section view, respectively, of the 1954–1955 engine. In 1956, a bore increase from 3.81 to 3.94 in. added 23 cu. in. to the displacement, for a total of 354 cu. in., boosting the output level to 280 hp and 380 lb-ft of torque. Only a year later, in 1957, an incremental bore change to 4.00 in., together with a substantial stroke increase to 3.90 in., further increased the displacement to a final 392 cu. in. The stroke increase was more than could be accommodated within the original 10.38-in. deck height of the cylinder block, requiring the deck to be raised to 10.865 in. A somewhat surprising ancillary change to the raised deck block was a change to the angle of the tappet bores. The tappet bores

Fig. 8.13 *The 1953 Chrysler FirePower 331-cu.-in. Hemi® V-8. Except for shortening the cylinder block during the model year, little was changed on this 180-hp engine, with carburetion limited to a two-barrel carburetor.*

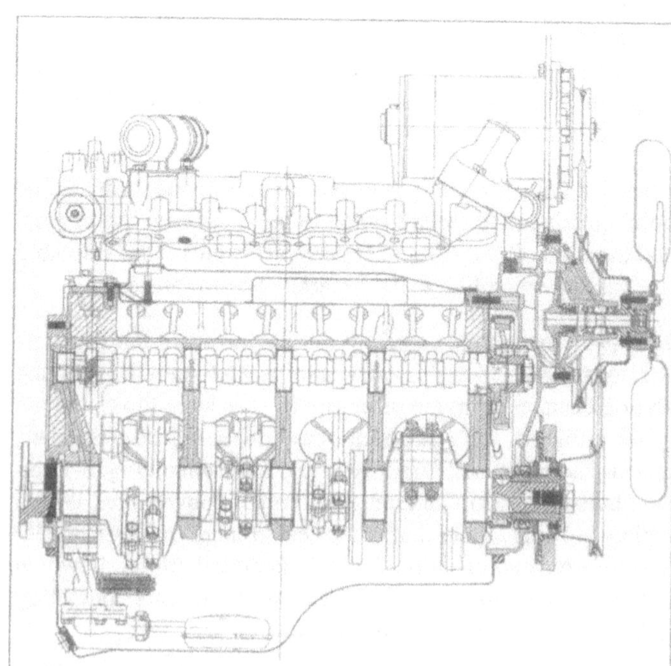

Fig. 8.14 *A longitudinal section of the 1954 Chrysler 331-cu.-in. Hemi® V-8 engine. The cylinder block was 6 in. shorter than that of the 1951 design.*

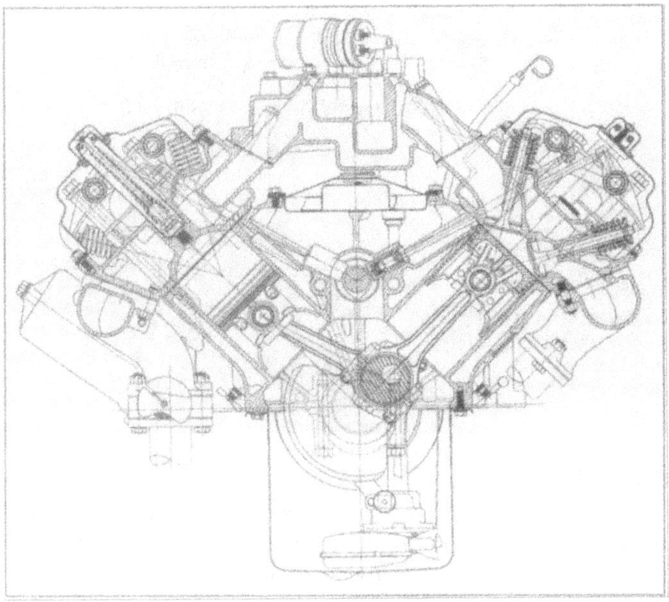

Fig. 8.15 *A cross section of the 1954 Chrysler 331-cu.-in. Hemi® V-8 engine. In this view, little appears to have been changed from the initial 1951 design.*

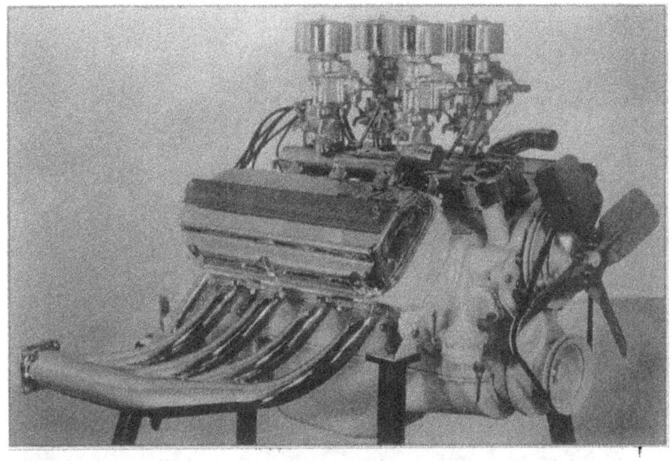

Fig. 8.16 *The experimental K-310 version of the 331-cu.-in. Chrysler Hemi® V-8 engine. The four single-barrel carburetors and swept-back exhaust headers shown were among many components tested in exploring the output potential of the Chrysler V-8.*

of the original block were not parallel to the 90-deg included angle of the cylinder bores but were inclined at a broader 123-deg included angle. When the deck was raised, this angle had to be reduced, or the drills and chamfering tools used for machining the tappet bores would have left too small a land at the top deck for the cylinder head gasket sealing ring. A 5-deg reduction of the angle to 118 deg was all that was required but did mean that, in addition to adjustments throughout the cylinder block machining lines, camshafts for the low deck and raised deck engines would not be interchangeable.

The longer stroke of the 392-cu.-in. engine required that its crankshaft journal diameters be increased to maintain sufficient crankshaft strength. The new diameters for the main journals and the crankpins were 2.6875 and 2.375 in., respectively.

Power Programs

When the huge task of preparing the 1951 331-cu.-in. Hemi® head V-8 engine for initial production was largely completed, Engine Development turned its attention to probing the power potential of the engine. Compression ratio increases, valve size increases, longer valve events, lower restriction exhaust manifolding, and multiple carburetion all were explored in a special program coded K-310. The engine, shown in Fig. 8.16 in one of its many versions, reached a peak power of 309 hp @ 5200 rpm, as reported by Engineering Vice President James C. Zeder in the October 1952 quarterly SAE *Transactions*, with the qualifier, "This work is in progress."[3] In the same report, Zeder noted that the engine was achieving specific outputs similar to those of four-valves-per-cylinder racing engines. Engine Development engineers John Platner and Don Moore, indeed, in 1951, started their dynamometer work with the 331-cu.-in. Hemi® head V-8 for serious racing as power for cars fielded by Briggs Cunningham for racing at LeMans and as possible power for an Indianapolis 500 race car. Assigned the program code A311, the engine was equipped with roller tappets, a gear-driven camshaft, Hillborn continuous-flow fuel injection, 12:1 high-compression pistons, tubular exhaust headers, tuned intake pipes, a Scintilla magneto ignition, and other power-inducing modifications. Its first on-track racing experience appears to have been in 1951, when it powered three Cunningham C2-R cars in the June 23–24, 1951 LeMans race. One of the cars finished the 24-hour race in eighteenth place.[8] At the end of the next race on June 14–15, 1952, one of the two Cunningham C4-R cars, pictured on Fig. 8.17, finished a respectable fourth, with Briggs Cunningham and Bill Spear doing the driving. A year later, on June 13–14, 1953, a Cunningham C5-R car powered with the A311 engine did even better, finishing third with Phil Walters and John Fitch co-driving.[9]

The program for running the A311 engine in an Indianapolis 500 race was off to an auspicious start but then took an unexpected turn. With the A311 331-cu.-in. engine well developed and turning out plenty of power, a Kurtis-Kraft car powered by the engine was sent to the famous two-and-a-half-mile track for tryouts in October 1952 in preparation for the Memorial Day 1953 event. Lap times showed that the engine, pictured on Fig. 8.18, and the car, shown on Fig. 8.19, had the potential not only to run at competitive speeds but to actually win.[10] However, development efforts suddenly were thwarted by a late change in the rules, which lowered the maximum "Stock Block" displacement from 335 to 275 cu. in. Hoping to salvage

Fig. 8.17 *The Cunningham C4-R LeMans race car. Power came from a race-prepared A311 Chrysler Hemi® V-8 engine.*

Fig. 8.19 *Race car #66 combined a Kurtis-Kraft chassis with an A311 Chrysler FirePower Hemi® V-8 engine. Although the car had the potential to win with its original 331-cu.-in. displacement, it was no longer competitive when the displacement was dropped to 271 cu. in.*

Fig. 8.18 *The high-output version of the Chrysler Hemi® V-8, coded A311, prepared for the May 1953 Indianapolis 500 race with fuel injection, tuned intake pipes, and other performance features.*

something, a destroked 271-cu.-in. engine, coded A311-6, was prepared quickly and installed in race car #66. Laps run in May 1953, preceding the Indianapolis 500 race, showed the car to be noncompetitive, and it did not attempt to qualify for the race. However, it was not the only car with an A311 Hemi® engine. A private entry car, #25, with a similar combination of a 271-cu.-in. A311 engine in a Kurtis-Kraft chassis did attempt to qualify; however, as predicted, that car did not make the cut. The car, now owned by enthusiast and race car driver Joe Freeman, still exists and is in immaculate condition.[11] The career of the A311 engine continued, however, with an assignment during the mid-1950s to power the Firestone Indianapolis tire test cars.[12]

All this high-performance work in pure racing cars certainly gained welcomed publicity, but what about a real car with plenty of power and road-handling capability for performance enthusiasts? That may have been the thought of Bob Roger, chief engineer of Chrysler Division. Bob was a true engine expert, having been in Engine Development at Central Engineering during the original test and development program of the original 331-cu.-in. Hemi® V-8.[4] After much effort, his possible thought on something special became a reality with the introduction, in 1955, of the first Chrysler 300 car, pictured on Fig. 8.20. Equipped with power-boosting dual four-barrel carburetors in a tandem arrangement, as shown on Fig. 8.21, and a long-duration camshaft with mechanical tappets for solid performance at high revolutions per minute, the output of the engine rose to the signature 300-hp level that gave the car its model designation. In addition to the outstanding performance of the engine, the chassis of the car received modifications to give it the road-handling capability of an exceptionally good sports car. When the base Chrysler C72 Hemi® V-8 moved up to 354 cu. in. and a 9:1 compression ratio in 1956, that year's Chrysler 300B incorporated the same changes, moving its power up to 340 hp. An even higher power level of 355 hp was available in the form of an optional version having a compression ratio of 10:1. Of interest, a few 1957 Dodge Coronet club sedans and convertibles were built with the 340-hp Chrysler 300B 354-cu.-in. engine. These cars, engine coded D501, apparently were intended for use in competitive events.

The arrival of the high-performance 1955 Chrysler 300 with its excellent road-handling ability caught the interest of Carl Kiekhaefer, the performance-minded owner of his own company, Kiekhaefer Corporation, a manufacturer of Mercury outboard engines. Kiekhaefer organized a multi-Chrysler 300 car racing team, hired top-notch drivers and crew members,

Fig. 8.20 *The 1955 Chrysler 300 two-door hardtop. Created by the Chrysler Division chief engineer and engine expert, Bob Roger, the car, powered by a high-performance 331-cu.-in. Chrysler Hemi® V-8 engine, quickly set a new standard of automotive performance excellence. The car is on display at the Walter P. Chrysler Museum. Photo by W.L. Weertman.*

Fig. 8.21 *A right-front view of the 1955 Chrysler 300 FirePower Hemi® V-8 engine. The tandem four-barrel carburetors on top of the intake manifold, a long-duration camshaft, mechanical valve gear, and other special parts contributed to the engine achieving its signature 300 hp.*

demanded perfection in preparation and running of the cars, entered the NASCAR Grand National series, and, to use the vernacular, blew everybody else off the tracks. Twenty-two wins were recorded for Kiekhaefer's Chrysler 300s for 1955, with Kiekhaefer's ace driver, Tim Flock, capturing eighteen of the total. The following year, 1956, a fleet of Chrysler 300Bs replaced the 300s, and the winnings continued. The other contestants ruefully knew that when the Kiekhaefer team trucks advertising Mercury outboard engines arrived at a racetrack and the big, immaculate, white Chryslers 300s came rolling down the ramps after the back doors of the trucks were opened, there was no question as to the winning make. The only question was which of Kiekhaefer's drivers would end up in the winner's circle. For 1956, with twenty-two wins for Kiekhaefer's Chryslers and eight wins for the Dodge cars that Kiekhaefer also fielded that year, Buck Baker got the checkered flag the most often, with nine behind the wheel of a Chrysler 300B and five piloting a Dodge. Following the 1956 season, Kiekhaefer disbanded the racing team. One reason cited was that many fans were outspokenly displeased with the total dominance of the Kiekhaefer cars on the long tracks, a view that could negate the intended objective of the entire program—to increase sales of Mercury outboard engines.

As noted, the displacement of the base Chrysler C76 Hemi® V-8 increased in 1957 to 392 cu. in., requiring the use of a raised-deck cylinder block. The displacement also applied to the 1957 Chrysler 300C engine that appears on Fig. 8.22 and was available in two compression ratios of 9.25:1 and 10:1. Power outputs were 375 and 390 hp, respectively. In 1958, the next Chrysler 300 series model, the 300D, was introduced. The standard 300D, equipped as usual with tandem four-barrel carburetors, was rated at 380 hp. A surprising new 390-hp

Fig. 8.22 *An outside view of the 1957 Chrysler 300C 392-cu.-in. FirePower Hemi® V-8 engine. The offset air cleaner was designed to provide sufficient clearance to the hood. With its optional 10:1 compression ratio, the engine output was 390 hp. The engine is on display at the Walter P. Chrysler Museum. Photo by W.L. Weertman.*

optional version, pictured on Fig. 8.23, appeared on the 1958 300D—the use of electronic fuel injection. Named the Bendix Electrojector, the system used solid-state electronics and eight individual fuel injectors. This electronic control of fuel injection (as contrasted to earlier electromechanical systems) was a Chrysler first in the industry. Unfortunately, its use was short-lived. After exposure to a greater range of driving conditions in customers' hands, occasions of control malfunction due to electromagnetic interference were encountered, and these were serious enough to cause the engines to be retrofitted with the standard 300D tandem four-barrel carburetor setup. (Historical note: a 1957½ Nash Rambler Rebel almost preceded Chrysler with this electronic fuel injection system; however, control unit problems caused it to be cancelled before any production cars were built.)[13]

Fig. 8.23 *An under-hood view of the optional 1958 Chrysler 300D 392-cu.-in. FirePower Hemi® V-8 engine equipped with Bendix multipoint fuel injection. An early application of electronically controlled fuel injection with an injector for each intake port, the engine had the same 390-hp rating as the tandem-carburetor version but an additional 5 lb-ft torque at 435 lb-ft.*

Production of this exemplary first-generation Chrysler Hemi® head V-8 engine ended at the close of the 1958 model year, its replacement being the big-block B and RB engines that will be described in Chapter 9. In addition to powering passenger cars, the engine was put to work as a heavy-duty truck, marine, and industrial engine. These applications are described in Chapters 15 and 16.

The design of the Chrysler Hemi® V-8 engine was so solid and so well received that it set the pattern for two reduced-size versions that quickly followed—the DeSoto and the Dodge Hemi® head V-8 engines.

Table 8.1 shows the specifications for this Chrysler Hemi® head V-8 engine.

DeSoto Firedome Hemi® Head V-8

The year 1952 was notable for DeSoto. The division finally had an engine it could call its own. Ever since the DeSoto car line was created in 1929, DeSoto had been using engines built by other divisions, mostly Chrysler-built sixes with a few Dodge-built straight-eights in the early 1930s. The marketplace pressure to provide DeSoto with an up-to-date overhead valve V-8 was the opportunity to release an engine that would be tooled and used exclusively by DeSoto. By using the same Hemi® head design approach that was so successful with the Chrysler V-8, it was sure to generate the same kind of customer enthusiasm. In keeping with the well-understood division marketing hierarchy, the DeSoto engine obviously would have to be somewhat smaller than the Chrysler Hemi® head V-8 already in production. Mel Carpentier again assigned lead designer Fred Shrimpton to the task, and, as might be expected, a powerful-looking engine emerged that was similar in appearance, as pictured on Figs. 8.24 and 8.25. Not obvious on either of these outside views or on the longitudinal and cross section views shown on Figs. 8.26 and 8.27 is the fact that the cylinder spacing and the deck height were reduced versus the Chrysler Hemi®; the cylinder spacing became 4.3125 in. (versus 4.5625 in.), and deck height became 9.569 in. (versus 10.38 in.). Engineering coded A275, the DeSoto engine's 2.375-in.-diameter main journals and 2.0625-in.-diameter crankpins were proportionately downsized from the 331 cu. in. of the Chrysler Hemi®, as were the 1.75-in. intake valves and the 1.406-in. exhaust valves. The angle between the two valves, which was 58.5 deg on the Chrysler Hemi®, was reduced to 53 deg on the DeSoto Hemi®. Initially powering the 1952 DeSoto S17 Firedome cars that included the convertible shown on Fig. 8.28, the bore and stroke of the DeSoto Hemi® V-8 was 3.625 × 3.375 in. and gave it a displacement of 276 cu. in. With a two-barrel carburetor and a 7.1:1 compression ratio, the engine produced 160 hp and 250 lb-ft of torque.

Subsequent Changes

The 276-cu.-in. engine went unchanged until 1954, when it gained 10 hp by a compression ratio increase from 7.1:1 to 7.5:1. For the 1955 S22 Firedome model, a bore increase to 3.72 in. and larger valves with head diameters of 1.84 and 1.50 in. for intakes and exhausts, respectively, moved the power upward to 185 hp for the now 291-cu.-in. engine. A companion 1955 S21 Fireflite model's engine version was similar but used a four-barrel carburetor instead of a two-barrel carburetor, giving it a rating of 200 hp and 274 lb-ft torque. This 291-cu.-in. displacement level lasted only one year. For the following 1956 year, the S23 Firedome and S24 Fireflite

TABLE 8.1
SPECIFICATIONS FOR THE CHRYSLER FIREPOWER HEMI® HEAD V-8 ENGINE

Model Year	1951–1953	1955	1956	1956
Car line	Chrysler and Imperial	Chrysler	Chrysler and Imperial	Chrysler
Model	C52, C53, C54, C55, C58, C59	C68-300	C70, C71, C72, C73	300B
Displacement (cu. in.)	331	331	354	354
Bore	3.8125	3.8125	3.94	3.94
Stroke	3.625	3.625	3.625	3.625
Compression ratio	7.5:1	8.5:1	9.0:1	9.0:1
BHP@ rpm	180 @ 4000	300 @ 5200	280 @ 4600	340 @ 5200
Torque (lb-ft) @ rpm	312 @ 2000	345 @ 3200	380 @ 2800	385 @ 3400
Combustion chamber	Hemispherical	Hemispherical	Hemispherical	Hemispherical
Angle between valves	58.5°	58.5°	58.5°	58.5°
Cyl. block mat'l.—skirt type	Cast iron—shallow skirt	Cast iron—shallow skirt	Cast iron—shallow skirt	Cast iron—shallow skirt
Cyl. block deck height	10.38	10.32	10.32	10.32
Cyl. block length	27.78	21.78	21.78	21.78
Cyl. bore spacing	4.5625	4.5625	4.5625	4.5625
Tappet angle (from vertical)	61.5°	61.5°	61.5°	61.5°
Ck/shaft—no. of main brgs.	5	5	5	5
Ck/shaft main journal dia.	2.5	2.5	2.5	2.5
Crankpin journal dia.	2.25	2.25	2.25	2.25
Piston mat'l.—no. of rings	Aluminum—3	Aluminum—3	Aluminum—3	Aluminum—3
Connecting rod centers	6.625	6.625	6.625	6.625
Valve system	OHV—double rocker shaft	OHV—double rocker shaft	OHV—double rocker shaft	OHV—double rocker shaft
Valve events	252-244-30	280-270-60	252-244-30	280-270-60
Camshaft drive	Chain	Chain	Chain	Chain
Valve head dia.	Intake 1.8125; exhaust 1.5	Intake 1.94; exhaust 1.75	Intake 1.94; exhaust 1.75	Intake 1.94; exhaust 1.75
Valve lash adjustment	Hydraulic tappet	Mech. —rocker arm screw	Hydraulic tappet	Mech. —rocker arm screw
Oil pump/distributor drive	Vertical at left-rear of block	Vertical at left-rear of block	Vertical at left-rear of block	Vertical at left-rear of block
Fuel system	2-barrel carburetor	Tandem 4-barrel carbs	4-barrel carburetor	Tandem 4-barrel carbs

(Continued on next page)

TABLE 8.1
(Continued)

Model Year	1957	1957	1958	1958
Car line	Dodge	Chrysler 300C	Chrysler	Chrysler
Model	D501	C76-300	300D	300D—Optional
Displacement (cu. in.)	354	392	392	392
Bore	3.94	4.0	4.00	4.00
Stroke	3.625	3.90	3.90	3.90
Compression ratio	10.0:1	9.25:1	10.0:1	10.0:1
BHP@ rpm	340 @ 5200	375 @ 5200	380 @ 5200	390 @ 5200
Torque (lb-ft) @ rpm	380 @ 3800	420 @ 4000	435 @ 3600	435 @ 3600
Combustion chamber	Hemispherical	Hemispherical	Hemispherical	Hemispherical
Angle between valves	58.5°	58.5°	58.5°	58.5°
Cyl. block mat'l.—skirt type	Cast iron— shallow skirt	Cast iron—shallow skirt	Cast iron—shallow skirt	Cast iron—shallow skirt
Cyl. block deck height	10.32	10.865	10.865	10.865
Cyl. block length	21.78	21.78	21.78	21.78
Cyl. bore spacing	4.5625	4.5625	4.5625	4.5625
Tappet angle (from vertical)	61.5°	59°	59°	59°
Ck/shaft—no. of main brgs.	5	5	5	5
Ck/shaft main journal dia.	2.5	2.6875	2.6875	2.6875
Crankpin journal dia.	2.25	2.375	2.375	2.375
Piston mat'l.—no. of rings	Aluminum—3	Aluminum—3	Aluminum—3	Aluminum—3
Connecting rod centers	6.625	6.951	6.951	6.951
Valve system	OHV—double rocker shaft	OHV—double rocker shaft	OHV—double rocker shaft	OHV—double rocker shaft
Valve events	280-270-60 Opt. 284-284-69	280-270-60 Opt. 300-300-95	276-276-55	276-276-55
Camshaft drive	Chain	Chain	Chain	Chain
Valve head dia.	Intake 1.94; exhaust 1.75	Intake 2.0; exhaust 1.75	Intake 2.0; exhaust 1.75	Intake 2.0; exhaust 1.75
Valve lash adjustment	Mech.—rocker arm screw	Mech.—rocker arm screw	Mech.—rocker arm screw	Mech.—rocker arm screw
Oil pump/distributor drive	Vertical at left-rear of block	Vertical at left-rear of block	Vertical at left-rear of block	Vertical at left-rear of block
Fuel system	Tandem 4-barrel carbs	Tandem 4-barrel carbs	Tandem 4-barrel carbs	MPI fuel injection

Fig. 8.24 *A right-front view of the1952 DeSoto S17 276-cu.-in. Firedome Hemi® V-8 engine. The DeSoto engine had cylinder bore centers closer together than those of the Chrysler Fire-Power Hemi® V-8. The engine is on display at the Walter P. Chrysler Museum. Photo by W.L. Weertman.*

Fig. 8.25 *A left-front view of the 1952 DeSoto S17 276-cu.-in. Firedome Hemi® V-8 engine. The broad cylinder head covers are indicative of the space needed for the double rocker shafts and the rocker arms involved in the Hemi® valve gear arrangement. The engine is on display at the Walter P. Chrysler Museum. Photo by W.L. Weertman.*

models were powered with 330-cu.-in. engines, the additional 39 cu. in. being accomplished by a stroke increase to 3.8 in. The longer stroke required the journals of the crankshaft to be increased to 2.5 and 2.25 in. for main bearings and crankpins, respectively, to maintain crankshaft strength and, most importantly, required that the cylinder block decks be raised. The new deck height was 10.37 in., and, similar to the Chrysler Hemi® V-8 already described, the tappet bank angle had to be changed. The original included tappet bank angle of 128 deg of the DeSoto Hemi® V-8 was reduced to 120 deg. These changes plus larger valves—1.94 in. intake and 1.75 in. exhaust—and a compression ratio increase to 8.5:1 brought the power level of the two-barrel S23 Firedome engine up to 230 hp and the four-barrel S24 Fireflite engine up to 255 hp. Shown on Fig. 8.29, an S24 Fireflite convertible, driven by DeSoto Division President L. Irving Woolson, was selected as the 1956 Indianapolis 500 pace car, possibly in recognition of DeSoto's emphasis on performance, having put only Hemi® V-8s under the hood since the start of 1955.

Further building its performance image, DeSoto launched a new model in 1956—the S24 DeSoto Adventurer. Extra power for the Adventurer came from a Firedome engine, slightly overbored to a 3.78-in. diameter, giving it 341 cu. in. with the additional power enhancements of high-ratio 9.25:1 pistons, a special camshaft, and, best of all, tandem four-barrel carburetors, similar in arrangement to those of Chrysler 300s. This potent engine delivered 320 hp and 356 lb-ft torque. For 1957, the two-barrel S25 Firedome and four-barrel S26 Fireflite models were upgraded with the 1956 Adventurer's 341-cu.-in. displacement and 9.25:1 compression ratio, while the 1957 S26 Adventurer's engine was bored out again to 3.80 in., giving it a 4-cu.-in. edge at 345 cu. in. The rated power levels of the Firedome and Fireflite engines became 270 and 295 hp, respectively. The engine of the 1957 DeSoto Adventurer, a car that looked fast even when standing still, as evident in Fig. 8.30, was rated at 345 hp, for an impressive specific output of 1 hp per cubic inch.

The DeSoto Hemi® V-8 engine, built by DeSoto Division at its Warren Avenue plant in Dearborn, Michigan, as shown on Fig. 8.31, ended its career at the close of 1957. For 1958, a new engine, the centrally manufactured big-block B engine V-8, would power all DeSotos.

Table 8.2 shows the specifications for this DeSoto Hemi® head V-8 engine.

Dodge Red Ram Hemi® Head V-8

Following DeSoto, it was Dodge Division's turn to get a new V-8 engine. Thus, a third Hemi® head engine—coded A274, with a cylinder spacing of 4.1875 in. and a deck height of 9.2865 in., somewhat smaller than that of the DeSoto engine—was designed, developed, and released for the 1953 Dodge

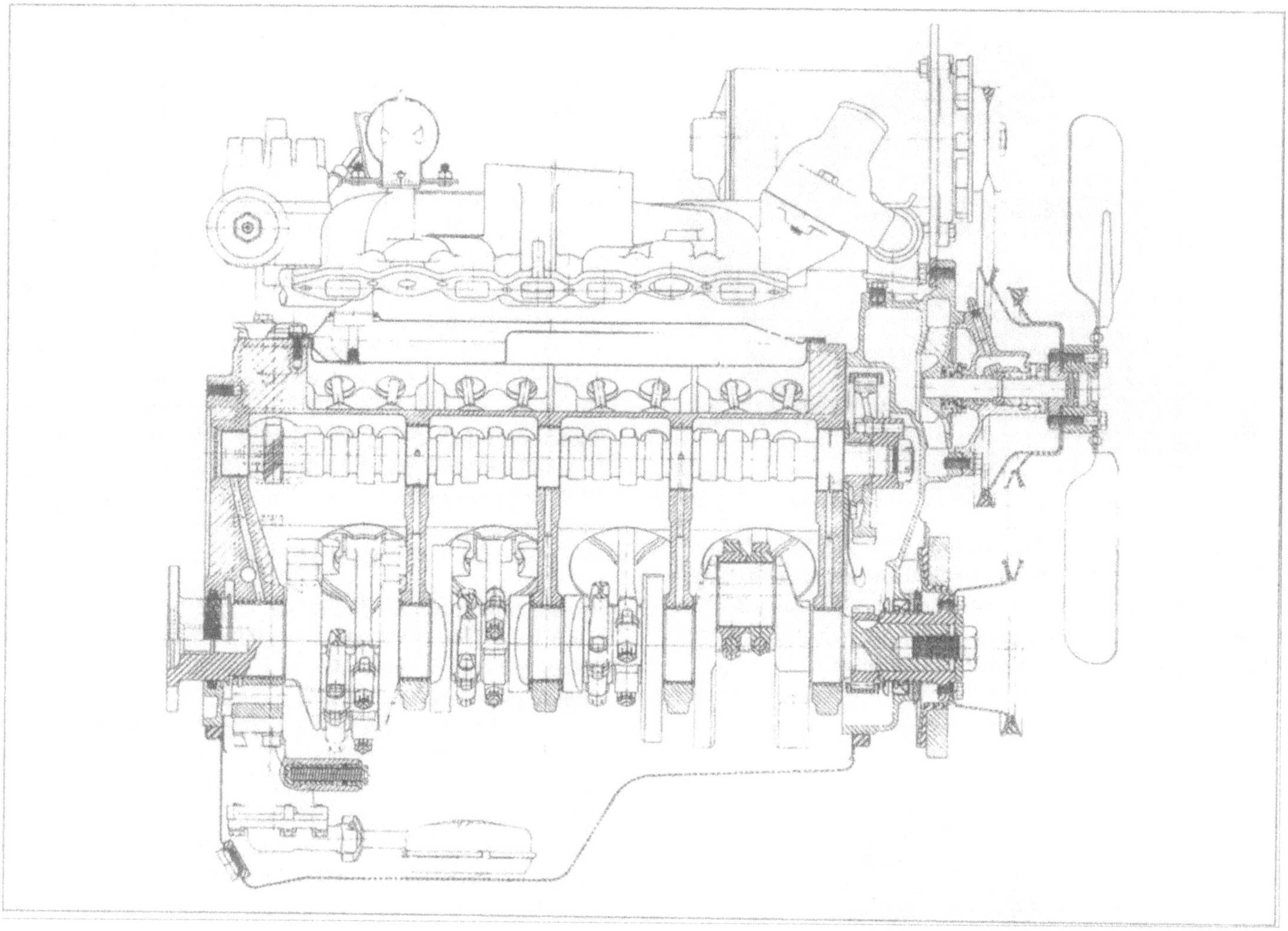

Fig. 8.26 *A longitudinal section of the 1953 DeSoto S16 276-cu.-in. Firedome Hemi® V-8 engine. The gear near the rear of the camshaft drove the distributor on top of the cylinder block and the oil pump below, bolted to the #5 main bearing cap.*

car line that included the D44 Coronet shown on Fig. 8.32. Figure 8.33 shows an exterior view of the new Dodge Hemi® V-8 engine. The vast Dodge Main plant in Hamtramck was the location for the new tooling required to produce the engine. With the precedence set by the Chrysler and DeSoto Hemi® V-8 engines, the design of the general arrangement and the cylinder head of the Dodge Hemi® V-8 had a strong similarity. Design details displayed on the longitudinal and cross-section views shown on Figs. 8.34 and 8.35, respectively, can be compared with the same sections of the earlier engines. The angle between the intake valve and the exhaust valve of the Dodge Hemi® was 52 deg, slightly less than the 53 deg of the DeSoto, as previously noted. The 3.4375-in. bore and 3.25-in. stroke of the initial 1953 Dodge Red Ram engine gave it a displacement of 241 cu. in. Equipped with a two-barrel carburetor, the only carburetor offered, and having a compression ratio of 7.1:1, the engine produced 140 hp and 220 lb-ft torque.

Because Mel Carpentier assigned the task of designing this engine to the same lead designer, Fred Shrimpton, the similarity of design between the three Hemi® engines—Chrysler, Desoto, and Dodge—is not surprising. What might be surprising is the lack of parts commonality among the three engines. Because of the difference in basic structural sizes of the engines as exemplified by the different cylinder spacings, all the major parts—blocks, heads, intake and exhaust manifolds, crankshafts, camshafts, and oil pans—were different. Connecting rods, valves, most gaskets, and many minor parts also were different. Each car division required a complete set of production tooling for its engine, and the three engines were filled with similar but different parts.

Named the Red Ram, the initial 1953 241-cu.-in. Hemi® V-8 had a straightforward model application—standard use on the Dodge Coronet Eight models. For 1954, this engine carried

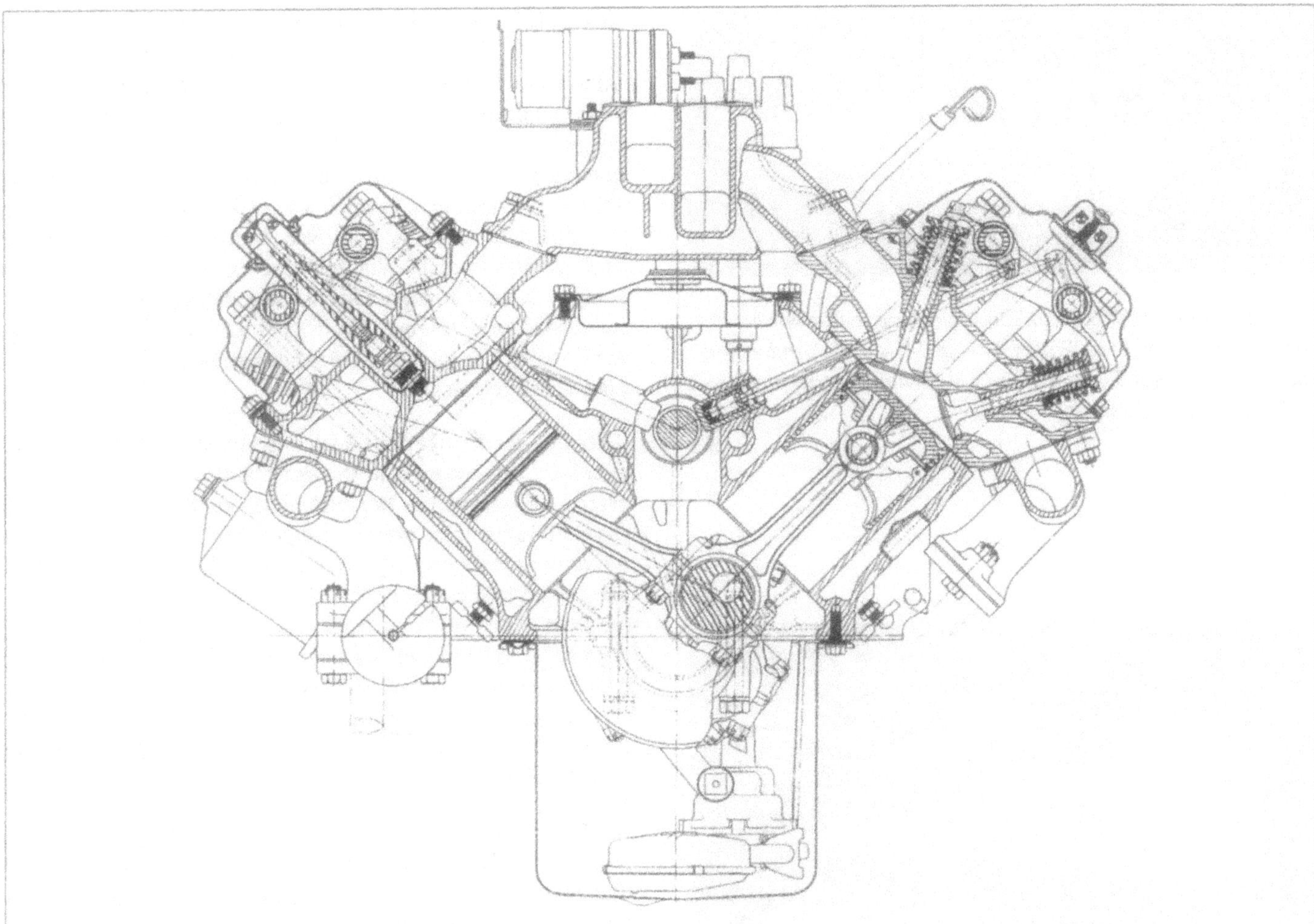

Fig. 8.27 *A cross section of the 1953 DeSoto S16 Hemi® V-8 engine. Although appearing similar to the cross section of the Chrysler Hemi® V-8 engine (Fig. 8.15), the valve angles and tappet angles of the DeSoto engine were slightly different.*

Fig. 8.28 *The 1952 DeSoto S17 Firedome convertible. The new 160-hp DeSoto Hemi® V-8 engine provided a noticeable performance gain over the previous year's engine: a 116-hp L-head six-cylinder engine.*

Fig. 8.29 *The 1956 Indianapolis 500 pace car. DeSoto president L. Irving Woolson, shown at the wheel, was the driver of the Hemi®-engine-powered 1956 DeSoto S24 Fireflite convertible during the May 1956 race. Photo courtesy of Indianapolis Motor Speedway.*

Fig. 8.30 *The 1957 DeSoto S26 Adventurer two-door hardtop was a fast-looking car. With a 345-hp 345-cu.-in. tandem four-barrel carburetor DeSoto Hemi® V-8 engine under the hood, it had both show and go.*

Fig. 8.31 *Desoto Hemi® V-8 engines are shown near the end of the production line at the DeSoto Warren Avenue plant in 1954.*

over as power for the Meadowbrook Eight model, while a high-compression-ratio (7.5:1) 150-hp version was installed as standard on the Coronet Eight and Royal models. Both versions of the engine were called Red Ram. A third version of the engine was available in the spring of 1954 on a limited production car. Dodge had been invited to provide the pace car for the 1954 Indianapolis 500 Memorial Day race and obliged by preparing a special 1954 Dodge Royal convertible, shown on Fig. 8.36, that was given a 500 designator. The engine of this car was modified to have a four-barrel carburetor and a high-performance Offenhauser aluminum intake manifold. The performance image generated for Dodge by the pace car triggered a decision to build 701 similar cars for sale to the general public.[14] The standard engine for these cars was the 150-hp Dodge Royal Hemi® V-8 with a two-barrel carburetor. However, a factory-approved, dealer-installed conversion package was available that replaced the two-barrel carburetor setup with an aluminum Offenhauser intake manifold mounting a four-barrel carburetor. Figure 8.37 shows a look under the hood of one of these Offenhauser-equipped 1954 Dodge pace car duplicates.

Although the Dodge Red Ram engine garnered praise for its excellent performance in the same way as did the Chrysler FirePower engine, the hemispherical combustion chamber of the Red Ram engine also provided remarkable fuel efficiency, which was demonstrated by capturing a first-place finish in the 1953 Mobilgas Economy Run. Entered in Class C (low-medium priced eight-cylinder cars), the Dodge Coronet four-door sedan with manual transmission and overdrive, powered by a 241-cu.-in. Dodge Red Ram 140-hp Hemi® engine, posted an average fuel consumption of 23.4189 mpg over the 1,206-mile course that had been lengthened to run from Los Angeles to Sun Valley, Idaho.

In 1955, Dodge transferred the Red Ram name to a newly released, lighter, and less expensive version of its V-8 engine that had a polysphere-shaped combustion chamber. This polysphere engine will be described in a later section of this chapter. The 1955 Dodge Hemi® V-8 was renamed the Super Red Ram. Displacement of the 1955 Super Red Ram Hemi® engine was upped to 270 cu. in. from the prior 241.3 cu. in. by a simple bore increase to 3.625 in. With the compression ratio increased to 7.6:1, the two-barrel carburetor version of the engine, which was standard on the Dodge D55-3 Custom Royal, saw its power climb to 183 hp and torque to 245 lb-ft. A second version of the engine equipped with a four-barrel carburetor and titled Super Red Ram with Power Pack produced 193 hp. This second version was available as an option on the entire line of 1955 Dodge C55 models.

In 1956, more dramatic changes were made to the Dodge Hemi® V-8 engine. The displacement of the engine was increased to 315 cu. in. by a stroke change from 3.25 to 3.8 in., which was long enough to require the cylinder block decks to be raised from their original 9.287 in. to 10.38 in. In turn, this caused the included tappet bank angle to be reduced from its original 130 deg to 116 deg. The compression ratio also took a substantial jump to 9.25:1. To maintain crankshaft strength, the main journals of the crankshaft went up from 2.375 to 2.5 in. and the crankpins from 1.9375 to 2.25 in. Both the intake and exhaust valve heads were increased in diameter and, for 1956, were 1.87 and 1.53 in., respectively. With more V-8 engine volume shifting to the polysphere version, the size of which also was increased to 315 cu. in. with the higher-deck cylinder

TABLE 8.2
SPECIFICATIONS FOR THE DESOTO HEMI® HEAD V-8 ENGINE

Model Year	1952–1953	1955	1956	1956
Car line	DeSoto	DeSoto	DeSoto	DeSoto
Model	Firedome S16–S17	Fireflite S21	Firedome S23	Adventurer S24
Displacement (cu. in.)	276	291	330	341
Bore	3.625	3.72	3.72	3.78
Stroke	3.344	3.344	3.8	3.8
Compression ratio	7.1:1	7.5:1	8.5:1	9.25:1
BHP @ rpm	160 @ 4000	200 @ 4400	230 @ 4400	320 @ 5200
Torque (lb-ft) @ rpm	250 @ 2000	274 @ 2800	305 @ 2800	356 @ 4000
Combustion chamber	Hemispherical	Hemispherical	Hemispherical	Hemispherical
Angle between valves	53°	53°	53°	53°
Cyl. block mat'l.—skirt type	Cast iron—shallow	Cast iron—shallow	Cast iron—shallow	Cast iron—shallow
Cyl. block deck height	9.54	9.54	10.37	10.37
Cyl. block length	20.844	20.844	20.844	20.844
Cyl. bore spacing	4.3125	4.3125	4.3125	4.3125
Tappet angle (from vertical)	64°	64°	60°	60°
Ck/shaft—no. of main brgs.	5	5	5	5
Ck/shaft main journal dia.	2.375	2.375	2.5	2.5
Crankpin journal dia.	2.0625	2.0625	2.25	2.25
Piston mat'l.—no. of rings	Aluminum—3	Aluminum—3	Aluminum—3	Aluminum—3
Connecting rod centers	6.0625	6.0625	6.623	6.623
Valve system	OHV-DRS	OHV-DRS	OHV-DRS	OHV-DRS
Valve events	244-244-26	244-244-26	252-244-6	280-270-60
Camshaft drive	Chain	Chain	Chain	Chain
Valve head dia.	Intake 1.75; exhaust 1.41	Intake 1.75; exhaust 1.41	Intake 1.94; exhaust 1.75	Intake 1.94; exhaust 1.75
Valve lash adjustment	Hydraulic tappet	Hydraulic tappet	Hydraulic tappet	Hydraulic tappet
Oil pump/distributor drive	Vertical at left-rear of block	Vertical at left-rear of block	Vertical at left-rear of block	Vertical at left-rear of block
Fuel system	2-barrel carburetor	4-barrel carburetor	2-barrel carburetor	Tandem 4-barrel carbs

(Continued on next page)

TABLE 8.2
(Continued)

Model Year	1957	1957	1957
Car line	DeSoto	DeSoto	DeSoto
Model	Firedome S25	Fireflite S26	Adventurer S26
Displacement (cu. in.)	341	341	345
Bore	3.78	3.78	3.80
Stroke	3.8	3.8	3.8
Compression ratio	9.25:1	9.25:1	9.25:1
BHP @ rpm	270 @ 4600	295 @ 4600	345 @ 5200
Torque (lb-ft) @ rpm	350 @ 2400	375 @ 2800	355 @ 3600
Combustion chamber	Hemispherical	Hemispherical	Hemispherical
Angle between valves	53°	53°	53°
Cyl. block mat'l.—skirt type	Cast iron—shallow	Cast iron—shallow	Cast iron—shallow
Cyl. block deck height	10.37	10.37	10.37
Cyl. block length	20.844	20.844	20.844
Cyl. bore spacing	4.3125	4.3125	4.3125
Tappet angle (from vertical)	60°	60°	60°
Ck/shaft—no. of main brgs.	5	5	5
Ck/shaft main journal dia.	2.5	2.5	2.5
Crankpin journal dia.	2.25	2.25	2.25
Piston mat'l.—no. of rings	Aluminum—3	Aluminum—3	Aluminum—3
Connecting rod centers	6.623	6.623	6.623
Valve system	OHV-DRS	OHV-DRS	OHV-DRS
Valve events	252-252-30	252-252-30	280-270-60
Camshaft drive	Chain	Chain	Chain
Valve head dia.	Intake 1.94; exhaust 1.75	Intake 1.94; exhaust 1.75	Intake 1.94; exhaust 1.75
Valve lash adjustment	Hydraulic tappet	Hydraulic tappet	Hydraulic tappet
Oil pump/distributor drive	Vertical at left-rear of block	Vertical at left-rear of block	Vertical at left-rear of block
Fuel system	2-barrel carburetor	4-barrel carburetor	Tandem 4-barrel carbs

Fig. 8.32 *The 1953 Dodge D44 Coronet two-door sedan. The "V" on the hood below the Ram ornament was a clue that the powerplant of the car was one of the new Dodge 140-hp 241-cu.-in. Hemi® V-8 engines.*

Fig. 8.33 *A left-front view of the 1953 Dodge Hemi® V-8 engine. Similar in general design to the preceding Chrysler and DeSoto Hemi® engines but on a smaller scale, the broad cylinder head covers are indicative of the Hemi® valve gear underneath. The engine is on display at the Walter P. Chrysler Museum. Photo by W.L. Weertman.*

block, the Hemi® engine became a high-performance-only engine named the D500.

Appropriate for the high-speed nature of the D500 engine, mechanical tappets were used in place of hydraulic tappets. Equipped with a four-barrel carburetor, the standard D500 engine produced 260 hp and 330 lb-ft torque. For even more performance-minded car buyers, a high-output version was created, the D500-1, with two four-barrel carburetors in a tandem arrangement. The rated output of the D500-1 engine was 295 hp @ 4400 rpm. The D500 engine was part of a chassis performance package applied to the D500 Custom Royal Lancer Two-Door, the D500 Custom Royal Convertible, and the D500 Coronet V-8 Club Sedan models.[15] The D500-1 tandem four-barrel engine was an option on these models.[16] It appears that later in the model year, the engines were made available as general options throughout the car line.

For 1957, the engine added another 10 cu. in. to its displacement by a small one-sixteenth-inch overbore, bringing the displacement to a final 325 cu. in. The single four-barrel Hemi® engine, now producing 285 hp and 345 lb-ft torque, retained its D500 designator. The tandem four-barrel version, renamed the Super D500, saw its output move up to 310 hp with 350 lb-ft torque—a high note for the last year of production of the Dodge Hemi® head V-8 engine.

Table 8.3 shows the specifications for this Dodge Hemi® head V-8 engine.

1955 Spitfire V-8—A Single-Rocker-Shaft Polysphere Head Engine for Chrysler

High-speed performance was the realm where the Hemi® V-8 engine was king, but with it came the burdens of extra weight and extra manufacturing cost. The marketplace also had a need for large quantities of engines with good low- to mid-speed performance (e.g., power for family four-door sedans and station wagons). In the case of Chrysler, this low- to mid-speed power demand slot had been filled for years with its Chrysler Jefferson Avenue-built six-cylinder L-head engine. A V-8 with lower cost than the Hemi® would be a desirable addition and would allow discontinuance of the aged L-head six-cylinder engine. Engine Design and Engine Development responded with the creation of a new engine, the A389 engine, having a much smaller (in cross section) cylinder head bolted to the block structure already well developed and tooled for the Chrysler FirePower V-8. By reorienting the valves to a diagonal arrangement in the plan view, placing the exhaust valve stem parallel to the cylinder axis, and leaving the intake valve tipped toward the intake manifold, both valves could be operated with a single rocker arm shaft. Also, the spark plugs were accessible from outside the cylinder head cover, while a semblance of the hemispherical chamber shape could be retained. This new chamber shape was

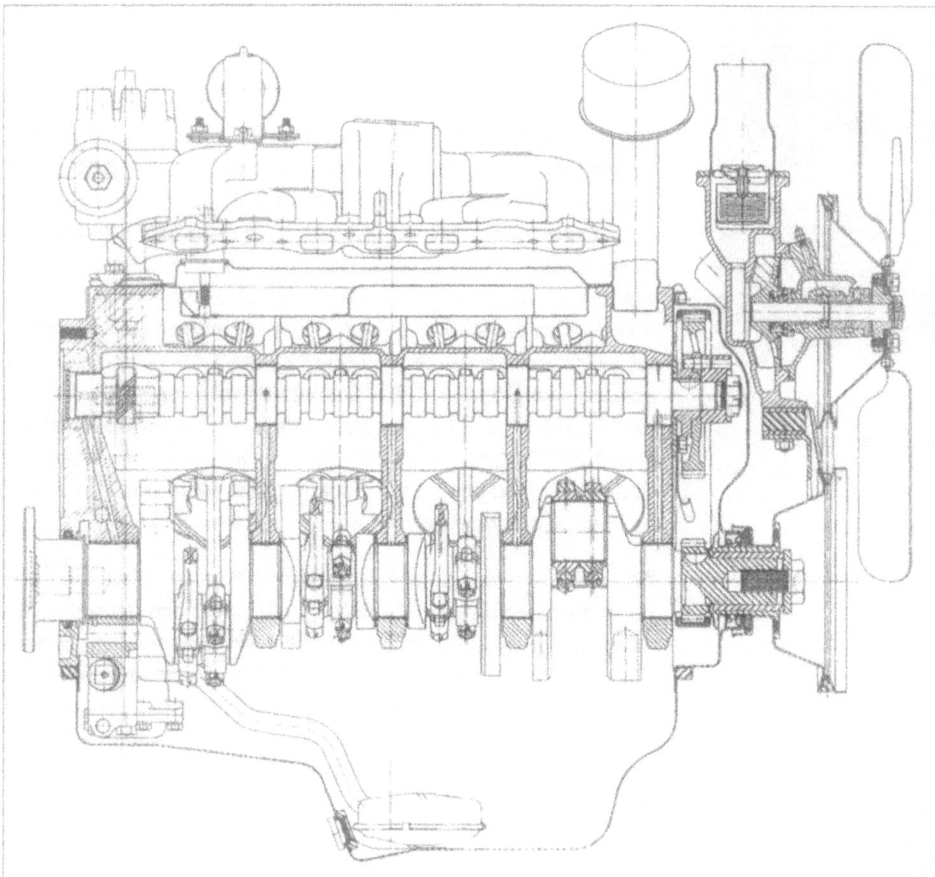

Fig. 8.34 *A longitudinal section of the 1953 Dodge 241-cu.-in. Hemi® V-8 engine. The drawing shows the similarity of this engine to the Chrysler and Desoto Hemi® V-8 engines. Because of its short stroke and husky crankshaft journals, the Dodge engine did not need a torsional vibration damper.*

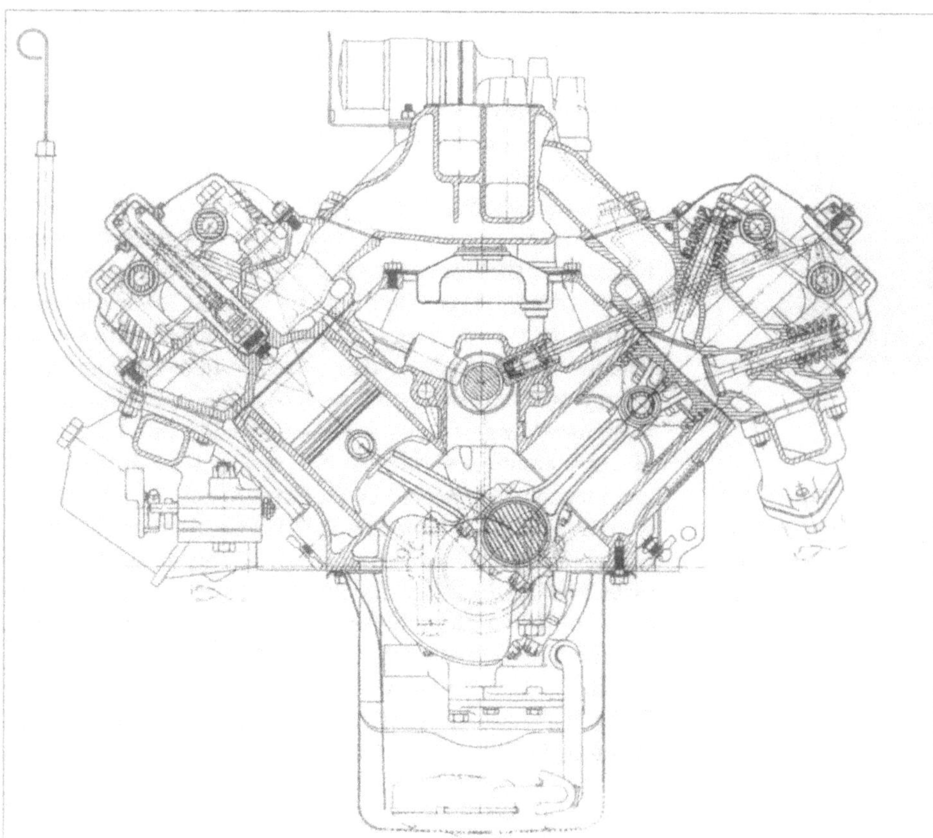

Fig. 8.35 *A cross section of the 1953 Dodge Hemi® V-8 engine. Although similar in general with the cross sections of the Chrysler and DeSoto Hemi® V-8s, the valve angles and tappet bore angles differ slightly.*

Fig. 8.36 *The 1954 Indianapolis 500 pace car. With a Hemi® V-8 under the hood, Dodge was recognized as a hot performing car. Dodge Division President William Newberg was the driver during the May 1954 race.*

Fig. 8.37 *The left side of the engine compartment of a 1954 Dodge Royal 500 pace car duplicate built and sold to the general public. Power upgrades for the 241-cu.-in. Dodge Hemi® V-8 engine included an aluminum Offenhauser intake manifold, which is visible in the photo, with a four-barrel carburetor. This vehicle was on temporary display at the Walter P. Chrysler Museum. Photo by W.L. Weertman.*

described as polyspherical. Figures 8.38 and 8.39 show a cross section and a longitudinal section, respectively, of the engine. The initial engine, used in the 1955 Chrysler Windsor models, including the station wagon shown on Fig. 8.40, had the same 3.625-in. stroke as the 1955 FirePower Hemi® but a reduced bore of 3.625 in. (versus 3.81 in.), giving it a displacement of 301 cu. in. Figure 8.41 shows an outside view of the engine, which generated 188 hp and 275 lb-ft torque. In the following year, 1956, the displacement was increased to 331 cu. in. by a bore increase to 3.81 in. In 1957, with another bore increase to 3.94 in., the displacement was brought up to 354 cu. in., where it remained through 1958 when the engine went out of production and was replaced by the big-block B V-8 engine. During its last year, a two-barrel carburetor version of this Chrysler poly-head single-rocker-shaft V-8, producing 290 hp and 385 lb-ft torque, powered the 1958 Chrysler Windsor, while the 1958 Chrysler Saratoga used a four-barrel carburetor version of the engine having an output of 310 hp and 405 lb-ft torque.

Table 8.4 shows the specifications for this Chrysler poly-head V-8 engine.

Dodge 1955 Single-Rocker-Shaft Poly-Head V-8 Engine

After releasing its Hemi® head V-8 engine in 1953 and enjoying its performance reputation, Dodge, in the same way as Chrysler, realized that many cost-conscious buyers in its market segment wanted V-8 power but would never utilize the high-speed horsepower capability of the Hemi®. The design solution for Dodge was the same as that for Chrysler, as previously described, and was introduced in the same 1955 model year as the Chrysler engine. Coded A388, the Dodge V-8 engine with its lower weight, smaller cross section, and single-rocker-shaft cylinder head used polysphere combustion chambers with the valves arranged similarly to those of the Chrysler Spitfire engine. Figures 8.42 and 8.43 show an outside view and a cross section of the engine, respectively. The block structure was the same as that of the Dodge Hemi® engine. Developed and released for the 1955 Dodge Coronet D55-1 model, the engine had a bore and stroke of 3.63 × 3.256 in., giving a displacement of 270 cu. in. Titled Red Ram, a name originally associated with the Hemi® head version of the Dodge-built V-8 as previously noted, and built with a two-barrel carburetor and 7.6:1 compression ratio, the output of the engine was 175 hp and 240 lb-ft torque.

In addition to building single-rocker-shaft poly-head engines for its own use, Dodge was asked to supply Plymouth with V-8 engines until the new Plymouth Division Mound Road engine plant came on line, which was a year away. Dodge agreed. Armed with blueprints from Engine Design for two lower-displacement versions—241 and 259 cu. in. (the latter often rounded up to 260 cu. in. in sales literature and other publications—of its poly-head engine, Dodge was able to ship engines to the Plymouth Lynch Road assembly plant in time for the start of 1955 production. The 241- and 259-cu.-in. sizes were easily accomplished by using smaller 3.4375- and 3.5625-in. bores, respectively, versus the 3.63-in. bore of the 270-cu.-in. Dodge poly-head engine. Plymouth named its Dodge-built engine the Hy-Fire. Similar to the 270-cu.-in. Dodge version, it was equipped with hydraulic tappets, a two-barrel carburetor, and a 7.6:1 compression ratio. The 241-cu.-in.

TABLE 8.3
SPECIFICATIONS FOR THE DODGE HEMI® HEAD V-8 ENGINE

Model Year	1953	1955	1956	1956
Car line	Dodge	Dodge	Dodge	Dodge
Model	Coronet D44, D48	Custom Royal D55-3	D500	D500-1
Displacement (cu. in.)	241	269	315	315
Bore	3.4375	3.625	3.625	3.625
Stroke	3.25	3.25	3.8	3.8
Compression ratio	7.0:1	7.6:1	9.25:1	9.25:1
BHP @ rpm	140 @ 4400	183 @ 4400	260 @ 4800	295 @ 4400
Torque (lb-ft) @ rpm	220 @ 2000	245 @ 2400	330 @ 3000	
Combustion chamber	Hemispherical	Hemispherical	Hemispherical	Hemispherical
Angle between valves	52°	52°	52°	52°
Cyl. block mat'l.—skirt type	Cast iron—shallow	Cast iron—shallow	Cast iron—shallow	Cast iron—shallow
Cyl. block deck height	9.29	9.29	10.38	10.38
Cyl. block length	20.844	20.844	20.844	20.844
Cyl. bore spacing	4.3125	4.3125	4.3125	4.3125
Tappet angle (from vertical)	65°	65°	58°	58°
Ck/shaft—no. of main brgs.	5	5	5	5
Ck/shaft main journal dia.	2.375	2.375	2.5	2.5
Crankpin journal dia.	1.9375	1.9375	2.25	2.25
Piston mat'l.—no. of rings	Aluminum—3	Aluminum—3	Aluminum—3	Aluminum—3
Connecting rod centers	5.9355	5.9355	6.618	6.618
Valve system	OHV-DRS	OHV-DRS	OHV-DRS	OHV-DRS
Valve events	244-244-26	244-244-26	252-252-30	280-270-60 Opt: 284-284-69
Camshaft drive	Chain	Chain	Chain	Chain
Valve head dia.	Intake 1.66; exhaust 1.41	Intake 1.66; exhaust 1.47	Intake 1.87; exhaust 1.53	Intake 1.87; exhaust 1.53
Valve lash adjustment	Hydraulic tappet	Hydraulic tappet	Rocker arm screw	Rocker arm screw
Oil pump/distributor drive	Vertical at left-rear of block	Vertical at left-rear of block	Vertical at left-rear of block	Vertical at left-rear of block
Fuel system	2-barrel carburetor	2-barrel carburetor	4-barrel carburetor	Tandem 4-barrel carbs

(Continued on next page)

TABLE 8.3
(Continued)

Model Year	1957	1957
Car line	Dodge	Dodge
Model	D500	Super D500
Displacement (cu. in.)	325	325
Bore	3.69	3.69
Stroke	3.8	3.8
Compression ratio	9.25:1	9.25:1
BHP @ rpm	285 @ 4800	310 @ 4800
Torque (lb-ft) @ rpm	345 @ 2800	350 @ 3200
Combustion chamber	Hemispherical	Hemispherical
Angle between valves	52°	52°
Cyl. block mat'l.—skirt type	Cast iron—shallow	Cast iron—shallow
Cyl. block deck height	10.38	10.38
Cyl. block length	20.344	20.344
Cyl. bore spacing	4.1875	4.1875
Tappet angle (from vertical)	58°	58°
Ck/shaft—no. of main brgs.	5	5
Ck/shaft main journal dia.	2.5	2.5
Crankpin journal dia.	2.25	2.25
Piston mat'l.—no. of rings	Aluminum—3	Aluminum—3
Connecting rod centers	6.618	6.618
Valve system	OHV-DRS	OHV-DRS
Valve events	256-256-38	256-256-38
Camshaft drive	Chain	Chain
Valve head dia.	Intake 1.87; exhaust 1.53	Intake 1.87; exhaust 1.53
Valve lash adjustment	Hydraulic tappet	Hydraulic tappet
Oil pump/distributor drive	Vertical at left-rear of block	Vertical at left-rear of block
Fuel system	4-barrel carburetor	Tandem 4-barrel carbs

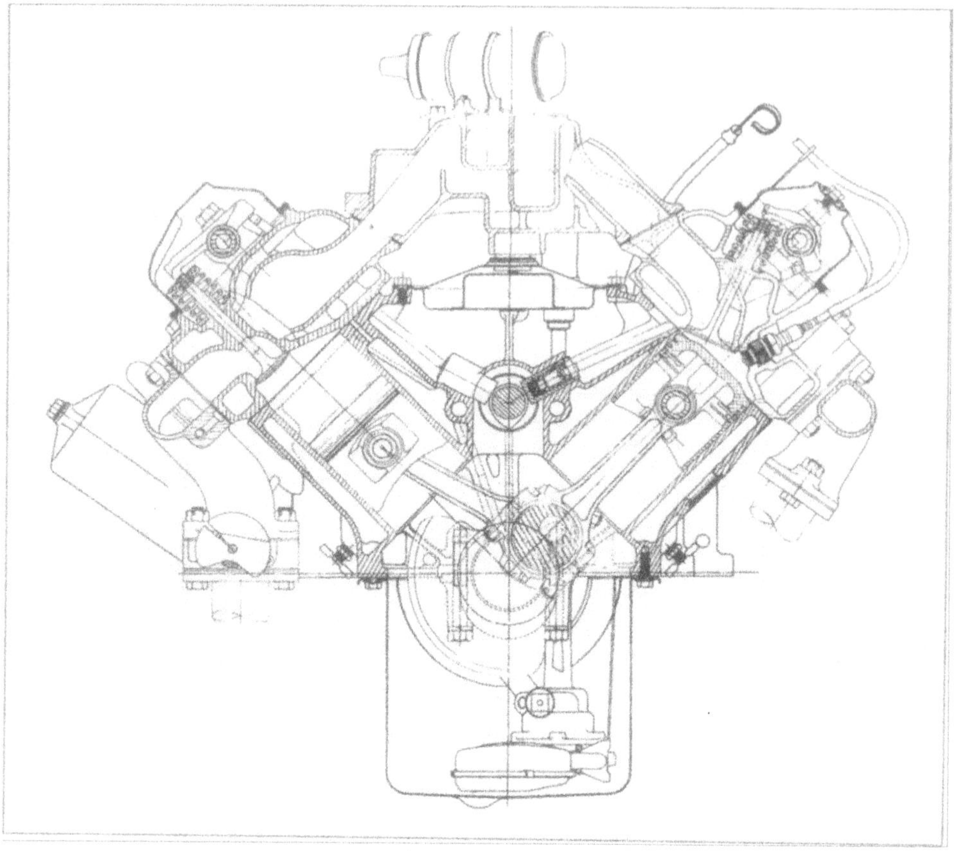

Fig. 8.38 *A cross section of the 1955 Chrysler C67 Spitfire 301-cu.-in. poly-head V-8 engine. Note the exhaust valve parallel to the cylinder axis, the single rocker shaft, and the spark plug accessible from outside the cylinder head cover.*

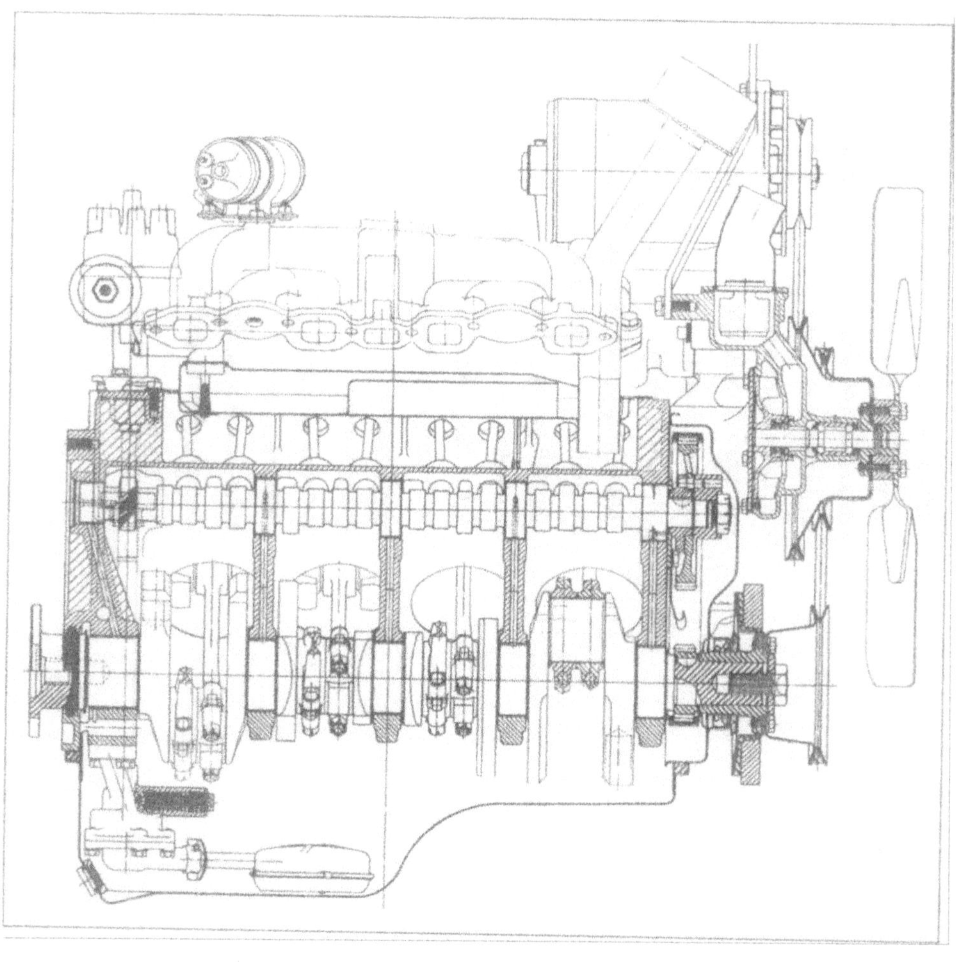

Fig. 8.39 *A longitudinal section of the 1955 Chrysler C67 Spitfire poly-head V-8 engine. The cylinder block design appears virtually identical to the Hemi® version of the engine.*

Fig. 8.40 *The 1955 Chrysler Windsor station wagon. Its power was a 301-cu.-in. Chrysler Spitfire V-8. This was the first year since the corporation was founded that Chrysler model cars were not available with a six-cylinder engine.*

Fig. 8.41 *A right-front view of the 1955 Chrysler 301-cu.-in. Spitfire poly-head V-8. The engine was lighter, more compact, and less expensive to build than the Hemi® head version.*

Plymouth engine produced 157 hp and 217 lb-ft torque; the larger 259-cu.-in. engine had an output of 167 hp and 231 lb-ft torque. Shortly after production started, the Plymouth dual V-8 engine offering was pared down to only one, the 259-cu.-in. 167-hp engine. Because the 1955 Chevrolet and Ford base V-8s both were rated at 162 hp, Plymouth could claim it had the highest horsepower standard V-8 in the low-priced field.[17] For additional performance, an optional four-barrel carburetor package was available on the 259-cu.-in. Hy-Fire engine that gave the driver command of 177 hp. Although Dodge Division clearly was the production source of the engine and the user of the largest displacement version, the 259-cu.-in. two-barrel version built for Plymouth was selected as the subject of an SAE paper, presented by Engine Development Assistant Chief Engineer Harold Welch on March 2, 1955, which outlined the design and development of the engine.[18]

Subsequent Changes

The single-rocker-shaft poly-head 270-cu.-in. Red Ram V-8 engine of the Dodge Coronet was carried over to 1956, except that its compression ratio was bumped up to 8:1, thereby bringing its power and torque up to 189 hp and 266 lb-ft, respectively. Plymouth still needed V-8s in 1956 while engine production at the Mound Road engine plant ramped up, and Dodge filled the need by shipping its 270-cu.-in. V-8 to the Plymouth Lynch Road assembly plant as optional power for the Plymouth Savoy and Plaza models. Still named Hy-Fire, the engine was rated at 180 hp and 260 lb-ft torque as installed in Plymouths.

With the 1956 longer-stroke, raised-deck 315-cu.-in. Dodge Hemi® V-8 previously described, Dodge also brought out a longer-stroke, raised-deck 315-cu.-in. poly-head V-8, making a two-barrel Super Red Ram version standard power on its Royal and Custom Royal. Power and torque ratings were 218 hp and 309 lb-ft, respectively. Sales popularity led this Super Red Ram engine to be made a generally available option throughout the car line, including station wagons. For those who wanted an engine with even more performance, a four-barrel version of the 315-cu.-in. poly-head V-8 engine was added to the Dodge optional engine lineup. Carrying the somewhat lengthy title of Super Powered Super Red Ram, it delivered 230 hp and 316 lb-ft torque. All these 1956 270- and 315-cu.-in. engines shared the same 8.0:1 compression ratio.

For 1957, Dodge made some changes and simplified the engine build. Dodge increased the bore size of its raised deck engine by one-sixteenth of an inch to 3.69 in., adding 10 cu. in. for a total of 325 cu. in. The compression ratio was increased to 8.5:1. Only two versions of this engine were built: a two-barrel-carburetor 245-hp version, and a four-barrel-carburetor 260-hp version. Dodge used both versions in its model lineup, with the names Red Ram and Super Red Ram, respectively. With plenty of engines available from the efficient Mound Road engine plant, Plymouth no longer used any Dodge-built engines, but the DeSoto Division became a customer. For 1957, DeSoto introduced a new model, the S27 Firesweep, into its lineup. The car was actually a Dodge-sized vehicle that was assembled at the Dodge Main plant, rode on a Dodge chassis, and was available with the same two- and four-barrel 325-cu.-in. single-rocker-shaft poly-head Dodge V-8 engines as used by Dodge. This was the only poly-head V-8 engine ever used by DeSoto; its own V-8 engine line produced Hemi® engines exclusively.

The 325-cu.-in. Dodge poly-head V-8 engine carried over to its final 1958 year in both its two- and four-barrel carburetor versions. Another step-up in compression ratio to 9:1 brought

TABLE 8.4
SPECIFICATIONS FOR THE CHRYSLER SPITFIRE POLY-HEAD V-8 ENGINE

Model Year	1955	1956	1958
Car line	Chrysler	Chrysler	Chrysler
Model	Windsor C67	Windsor C71	Saratoga LC2
Displacement (cu. in.)	301	331	354
Bore	3.625	3.8125	3.94
Stroke	3.625	3.625	3.625
Compression ratio	8.01	8.5:1	9.0:1
BHP @ rpm	188 @ 4400	225 @ 4400	310 @ 4600
Torque (lb-ft) @ rpm	275 @ 2400	310 @ 2800	405 @ 2800
Combustion chamber	Polyspherical	Polyspherical	Polyspherical
Cyl. block mat'l.—skirt type	Cast iron—shallow skirt	Cast iron—shallow skirt	Cast iron—shallow skirt
Cyl. block deck height	10.32	10.32	10.32
Cyl. block length	21.78	21.78	21.78
Cyl. bore spacing	4.5625	4.5625	4.5625
Crankshaft—no. of main brgs.	5	5	5
Crankshaft main journal dia.	2.5	2.5	2.5
Crankpin journal dia.	2.25	2.25	2.25
Piston mat'l.—no. of rings	Aluminum—3	Aluminum—3	Aluminum—3
Connecting rod centers	6.625	6.625	6.625
Valve system	OHV—single rocker shaft	OHV—single rocker shaft	OHV—single rocker shaft
Valve events	252-244-30	252-244-30	252-256-30
Camshaft drive	Chain	Chain	Chain
Valve head dia.	Intake 1.81; exhaust 1.5	Intake 1.81; exhaust 1.5	Intake 1.94; exhaust 1.75
Valve lash adjustment	Hydraulic tappet	Hydraulic tappet	Hydraulic tappet
Oil pump and distributor drive	Vertical at left-rear of engine	Vertical at left-rear of engine	Vertical at left-rear of engine
Fuel system	2-barrel carburetor	2-barrel carburetor	4-barrel carburetor

Fig. 8.42 *A left-front view of the 1955 Dodge 270-cu.-in. poly-head V-8 engine. Similar to the Chrysler poly-head V-8, the Dodge poly-head V-8 presents a more compact appearance than its Hemi® head counterpart.*

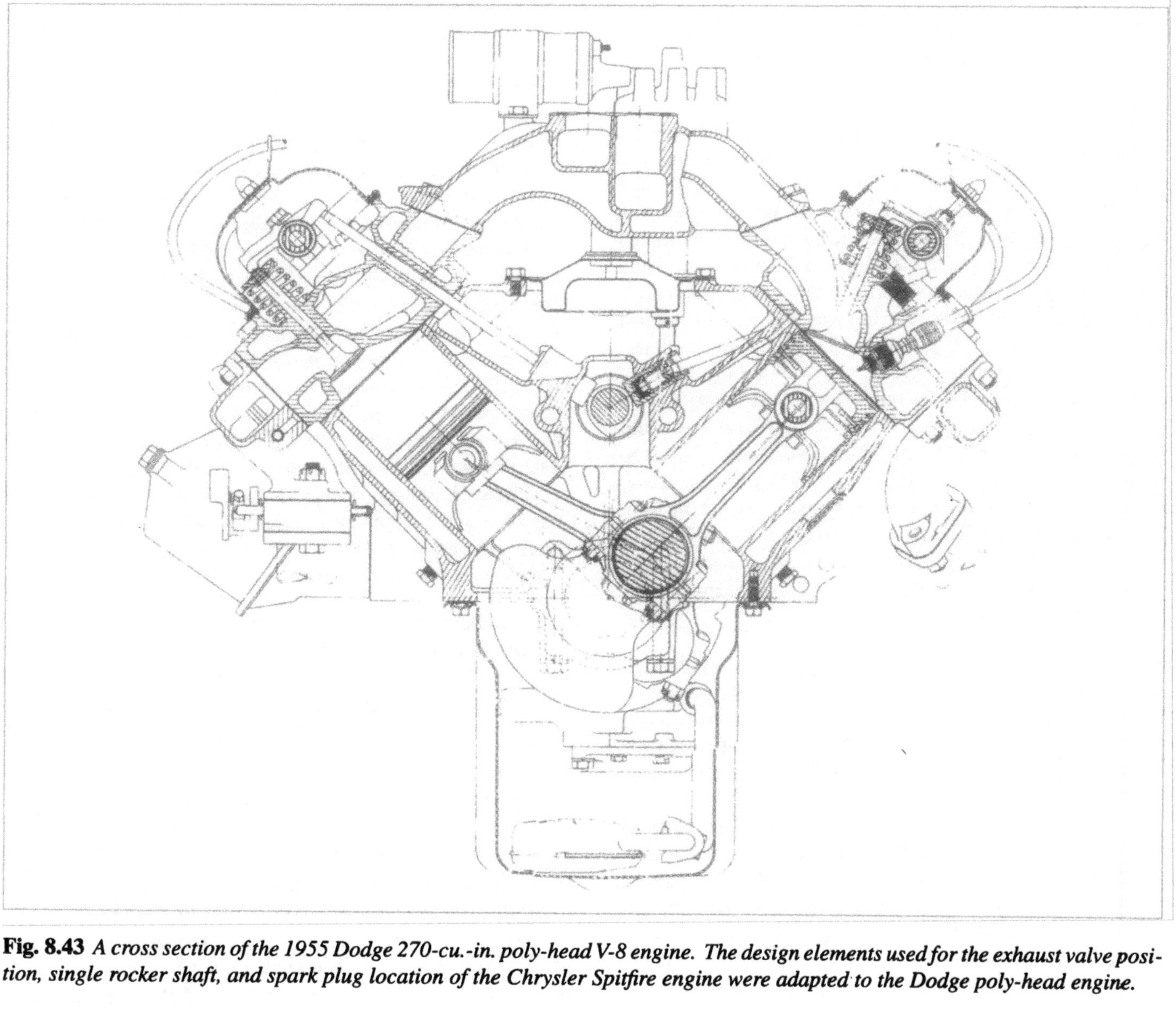

Fig. 8.43 *A cross section of the 1955 Dodge 270-cu.-in. poly-head V-8 engine. The design elements used for the exhaust valve position, single rocker shaft, and spark plug location of the Chrysler Spitfire engine were adapted to the Dodge poly-head engine.*

the engine performance of the two-barrel up to 252 hp and 345 lb-ft torque, and the four-barrel version up to 265 hp and 355 lb-ft torque. Dodge was the only user; DeSoto shut down its own engine production and switched all its engine needs to the new Trenton engine plant-built wedge head big-block B V-8 engine, described in Chapter 9. The 1958 325-cu.-in. single-rocker-shaft, poly-head, 252-hp two-barrel Dodge V-8 became the standard V-8 for the Dodge Coronet LD2-L1 model. The 265-hp four-barrel engine was standard on the Dodge Royal LD2-M model.

In addition to powering passenger cars, this Dodge-built single-rocker-shaft poly-head V-8, in its various displacements over the four years it was manufactured, found extensive use in low- and medium-tonnage Dodge trucks. Chapter 15 presents additional information on truck usage.

Table 8.5 shows the specifications for this Dodge poly-head V-8 engine.

Plymouth Single-Rocker-Shaft Poly-Head V-8—The Mound Road V-8 A-Engine

Following the successive and successful launches of V-8s for Chrysler, DeSoto, and Dodge, Plymouth's hopes and plans for a V-8 engine of its own finally came to fruition with the start of production of V-8 engines at the Mound Road engine plant in Detroit. At the onset of the planning, the manufacturing engineering task force assigned to the project was determined to improve product quality and to reduce labor costs while producing engines at a rate commensurate with Plymouth's high-volume vehicle assembly line speeds. To accomplish the objectives, the plant would be equipped with the latest automated, transfer-line type of machining and assembly equipment available. The fact that Dodge was supplying V-8 engines to Plymouth during 1955 was helpful, but Dodge could not possibly build enough engines for both divisions with the machining lines it had. The corporation already had a building that was somewhat suitable for the new engine plant in the form of a World War II facility where airplane sub-assemblies had been made. Indeed, a few leftover aluminum tail and fuselage sections were still scattered about the plant floor when the plant transformation started in earnest.[19]

After some preliminary designs were made for a new Plymouth V-8, a final design, coded A479, was started that combined the Mound Road engine plant's manufacturing requirements for compatibility with its planned automated equipment with Engine Design's own objectives of simplification and weight reduction. In addition to the Plymouth Mound Road engine plant, corporate plans called for the Windsor engine plant of Chrysler Canada to be tooled for producing the new V-8 engine so that Chrysler Canada could offer Canadian-built V-8 power in Canadian-built vehicles. Except for a larger initial displacement, the engine built at the Windsor engine plant, coded A514, was identical to the engine built at the Mound Road plant. Of interest, in informal conversations within the Engine Design and Development departments, these two engines often were lumped together and referred to as the "A" engine, a shortening of their A program designators, thus paving the way for the use of single alphabetic engine identifiers for some time to come.

The cylinder block bore centerlines of the A-engine were laid down with a spacing of 4.46 in., and its deck height was set at 9.6 in. Both dimensions were greater than the comparable dimensions of either the original Dodge or DeSoto engines. The right and left banks of tappet bores were set at an included angle of 118 deg. The cylinder head generally was similar to the other single rocker shaft, polyspherical-chamber heads but, as a cost reduction and simplification feature, had the rocker shaft pedestals integrated with the cylinder head casting rather than being separate parts. As with the other poly-head engines, the exhaust valve stems were positioned parallel to the cylinder axis, and the intake valve stems were inclined toward the intake manifold. Another simplification feature was a cast iron intake manifold that was designed to cover the entire top of the cylinder block, eliminating the need for separate gasketed stamped-steel covers as used on the prior Chrysler, DeSoto, and Dodge V-8 engines. Die-cast aluminum was used for the chain case cover and water pump body at the front of the engine to reduce cost and weight versus the entirely cast iron front-end designs of the Dodge, DeSoto, and Chrysler engines. Inside the cylinder block, the two oil galleries that were drilled the length of the block to provide oil distribution for the entire engine were designed to intersect the tappet bores, eliminating the need for sixteen small-diameter tappet cross-feed holes. Engine costs were minimized further by the use of solid mechanical tappets, and valve lash adjustment was accomplished by lock-thread screws in the valve rocker arms. The outer steel shell of the oil filter was designed to be attached directly to the side of the cylinder block so that a separate oil filter mounting adapter used on earlier V-8 engines would not be required. Figure 8.44 shows an outside view of the engine. Internal details are shown on the cross section on Fig. 8.45. The parts of the power section of the engine—crankshaft, connecting rods, and pistons—were all similar in design concept to the earlier V-8 engines. As with the other V-8 engines, a vertical shaft at the left-rear of the cylinder block drove the distributor on top of the block and the oil pump attached to the #5 main bearing cap.

The October 1955 issue of the trade magazine *Tooling and Production* was devoted to the launch of the "Plymouth Qualimatic V-8 Engine Plant." The issue contained many detailed articles that gave an excellent description of the manufacturing facilities and planning that went into the plant.

For its introductory year of 1956, the Mound Road engine plant built two- and four-barrel versions of one engine size, 276 cu. in. (often nudged up to 277 cu. in. in advertising literature), with a bore and stroke of 3.75 × 3.125 in. The Windsor engine plant, launching its V-8 engine the same year, built a larger-displacement engine, 303 cu. in., with a bore and stroke of 3.812 × 3.312 in., also in two- and four-barrel versions. The two-barrel Mound Road engine plant-built engine, rated at 187 hp with 265 lb-ft torque and named Hy-Fire, just like

TABLE 8.5
SPECIFICATIONS FOR THE DODGE POLY-HEAD V-8 ENGINE

Model Year	1955	Initial 1955	Later 1955	1956
Car line	Dodge	Plymouth	Plymouth	Dodge
Model	Coronet D55-1	P27	P27	Royal D63-2,3
Displacement (cu. in.)	270	241	259	315
Bore	3.625	3.44	3.563	3.63
Stroke	3.25	3.25	3.25	3.8
Compression ratio	7.6:1	7.6:1	7.6:1	8.0:1
BHP @ rpm	175 @ 4400	157 @ 4400	167 @ 4400	218 @ 4400
Torque (lb-ft) @ rpm	240 @ 2400	217 @ 2400	231 @ 2400	309 @ 2000
Combustion chamber	Polyspherical	Polyspherical	Polyspherical	Polyspherical
Cyl. block mat'l.—skirt type	Cast iron—shallow	Cast iron—shallow	Cast iron—shallow	Cast iron—shallow
Cyl. block deck height	9.29	9.29	9.29	10.38
Cyl. block length	20.344	20.344	20.344	20.344
Cyl. bore spacing	4.1875	4.1875	4.1875	4.1875
Tappet angle (from vertical)	65°	65°	65°	58°
Ck/shaft—no. of main brgs.	5	5	5	5
Ck/shaft main journal dia.	2.375	2.375	2.375	2.5
Crankpin journal dia.	1.9375	1.9375	1.9375	2.25
Piston mat'l.—no. of rings	Aluminum—3	Aluminum—3	Aluminum—3	Aluminum—3
Connecting rod centers	5.9355	5.9355	5.9355	6.618
Valve system	OHV—single rocker shaft	OHV—single rocker shaft	OHV—single rocker shaft	OHV—single rocker shaft
Valve events	244-244-26	244-241-23	244-244-26	244-244-26
Camshaft drive	Chain	Chain	Chain	Chain
Valve head dia.	Intake 1.72; exhaust 1.47	Intake 1.72; exhaust 1.47	Intake 1.72; exhaust 1.47	Intake 1.72; exhaust 1.47
Valve lash adjustment	Hydraulic tappet	Hydraulic tappet	Hydraulic tappet	Hydraulic tappet
Oil pump/distributor drive	Vertical at rear of block	Vertical at rear of block	Vertical at rear of block	Vertical at rear of block
Fuel system	2-barrel carburetor	2-barrel carburetor	2-barrel carburetor	2-barrel carburetor

(Continued on next page)

TABLE 8.5
(Continued)

Model Year	1957	1957	1957
Car line	Dodge	DeSoto	Dodge
Model	D66, D67, D70, D71	Firesweep S27	Royal LD2-M
Displacement (cu. in.)	325	325	325
Bore	3.69	3.69	3.69
Stroke	3.8	3.8	3.8
Compression ratio	8.5:1	8.5:1	9.0:1
BHP @ rpm	245 @ 4400	245 @ 4400	265 @ 4400
Torque (lb-ft) @ rpm	320 @ 2400	320 @ 2400	355 @ 2800
Combustion chamber	Polyspherical	Polyspherical	Polyspherical
Cyl. block mat'l.—skirt type	Cast iron—shallow	Cast iron—shallow	Cast iron—shallow
Cyl. block deck height	10.38	10.38	10.38
Cyl. block length	20.344	20.344	20.344
Cyl. bore spacing	4.1875	4.1875	4.1875
Tappet angle (from vertical)	58°	58°	58°
Ck/shaft—no. of main brgs.	5	5	5
Ck/shaft main journal dia.	2.5	2.5	2.5
Crankpin journal dia.	2.25	2.25	2.25
Piston mat'l.—no. of rings	Aluminum—3	Aluminum—3	Aluminum—3
Connecting rod centers	6.618	6.618	6.618
Valve system	OHV—single rocker shaft	OHV—single rocker shaft	OHV—single rocker shaft
Valve events	248-252-26	248-252-26	248-252-26
Camshaft drive	Chain	Chain	Chain
Valve head dia.	Intake 1.84; exhaust 1.47	Intake 1.84; exhaust 1.47	Intake 1.84; exhaust 1.47
Valve lash adjustment	Hydraulic tappet	Hydraulic tappet	Hydraulic tappet
Oil pump/distributor drive	Vertical at rear of block	Vertical at rear of block	Vertical at rear of block
Fuel system	2-barrel carburetor	2-barrel carburetor	4-barrel carburetor

Fig. 8.44 *A left-front view of the 1956 Plymouth 276-cu.-in. poly-head V-8 engine. Sturdy and built with automated high-volume production tooling, the engine was known simply as the "A" engine.*

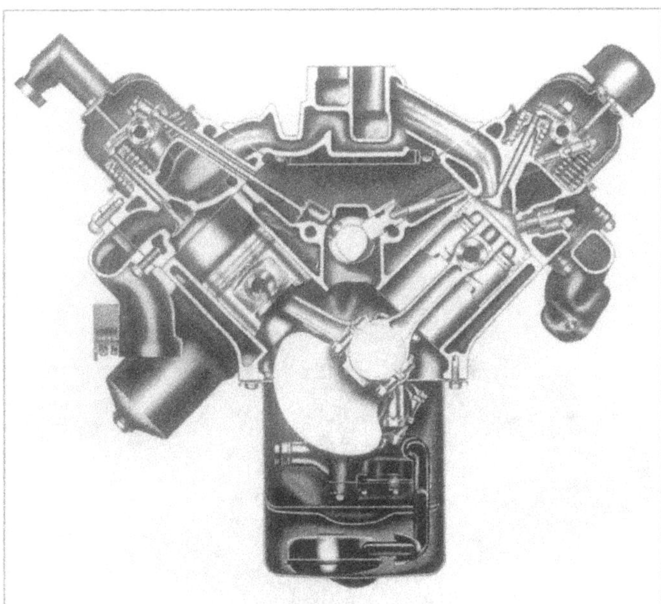

Fig. 8.45 *A cross section of the 1956 Plymouth poly-head V-8 A-engine. The engine repeated the polyspherical chamber of the Chrysler and Dodge engines. Note the intersection of the oil gallery and the tappet bore, which eliminated the need for sixteen small cross drills. The use of mechanical tappets with screw adjust for valve lash setting was for cost savings, not for high-speed engine operation.*

the Dodge-built V-8, was the base V-8 for P29 Belvederes. A high-performance option, available on any P29 Plaza, Savoy, or Belvedere, was the four-barrel Mound Road engine plant-built engine, named Hy-Fire with Power Pack, which produced 200 hp and 272 lb-ft torque. During 1956, a new sporty Plymouth was introduced, the Plymouth Fury that was given a substantial power boost over the other Plymouth models by the installation of a four-barrel 303-cu.-in. V-8 A-engine with a 9.25:1 compression ratio, shipped from the Windsor engine plant to the Plymouth Division Lynch Road assembly plant, putting 240 hp under the hoods of the Furys.

To boost public awareness of the performance capability of this new Plymouth Fury, Plymouth Product Planning arranged with the National Association of Stock Car Auto Racing (NASCAR) for a car to be given speed trials under its jurisdiction. Very ably supported by the talents of crack engine development engineer Bob Cahill, vehicle test engineer Wally Zierer, and engineer Charlie Scales of Product Planning, on January 10, 1956, at Daytona Beach, Florida, the production-stock Fury with its 303-cu.-in. four-barrel-carburetor 240-hp engine, driven by race car pro Phil Walters, broke class records for the flying mile, achieving a speed of 124.01 mph, and the standing start mile hitting 82.54 mph.[20] Following these runs, the stock manifold and carburetor were replaced with a conversion package that had been thoroughly dyno-tested at Chrysler Engineering: an adapter plate topped with a Chrysler 300B intake manifold complete with tandem four-barrel carburetors. Now running in the Factory Experimental class, the car did 148 mph in the flying mile, showing what the engine really was capable of doing. Figure 8.46 shows the victorious group, with the car and NASCAR president Bill France.

The launches of the Mound Road engine plant in calendar year 1955 together with the start of the production of A-engines at both the Mound Road and Windsor engine plants were momentous events, but two additional events occurred before the year ended that deserve special recognition. One event was the formation of the Engine Division under a newly created Engine and Transmission Group. The Engine Division assumed the responsibility of manufacturing all the corporate passenger car and truck engines, transferring that task from the car divisions and paving the way for any future engine plants to be managed by a central corporate manufacturing group rather than by individual car divisions. The other event, a sad one, was the death of Assistant Chief Engineer of Engine Design Mel Carpentier on October 18, 1955. Carpentier's death broke the direct lineage of Engine Design back to the formation of Chrysler because he was one of the small group of designers who stuck with Zeder, Skelton, and Breer when they left Studebaker in 1920 to join Walter P. Chrysler at Willys Corporation in Elizabeth, New Jersey.[21] His design guidance was part of every production engine and many experimental engines built by Chrysler right up to the day he died.

Fig. 8.46 *Gathered around the record-setting 1956 Plymouth Fury at Daytona Beach, Florida, in January 1956 are (from left to right): driver Phil Walters, Chrysler engineers Wally Zierer and Bob Cahill, NASCAR president Bill France, mechanic J. Donaldson, and Chrysler engineer Charlie Scales. Photo from the collection of Robert J. Cahill.*

Subsequent Changes to the A-Engine

Although the Mound Road engine plant was highly automated, its manufacturing engineers showed in 1957 that it also was flexible by producing engines of three different displacements using two bore sizes and two strokes in the process. One engine was a carryover 276-cu.-in. two-barrel having an unchanged bore and stroke of 3.75 × 3.125 in. Used as the base V-8 for the 1957 P31 Plymouth Plaza, its power rating moved up to 197 hp. A second engine displaced 301 cu. in. by the use of the same 3.125-in. stroke but with a larger 3.91-in. bore. Two- and four-barrel versions of this 301-cu.-in. A-engine, both with 8.5:1 compression ratio, provided 215 and 235 hp, respectively. The two-barrel 215-hp engine was a base V-8 on P31 Savoys and Belvederes; the four-barrel 235-hp engine was an option throughout the car line. For the performance-leading 1957 P31 Plymouth Fury, the Mound Road engine plant built a special A-engine that combined the larger 3.91-in. bore with the longer 3.31-in. stroke used on the Canadian A-engines to give the engine a displacement of 318 cu. in. Topped with tandem four-barrel carburetors and offset air cleaners, the engine put out an impressive 290 hp and 325 lb-ft torque, giving the car plenty of acceleration and speed.

The displacement range of engines built at the Mound Road engine plant settled down the next year, 1958, with only the 318-cu.-in. size being produced. Meanwhile, in 1957, the Windsor engine plant introduced a new bore size of 3.875 in., which, combined with its 3.31-in. stroke, provided an A-engine with a displacement of 313 cu. in. The following year and through the 1964 model, this was the only size V-8 built at the Windsor engine plant. The output of the Mound Road engine plant 1958 318-cu.-in. V-8 with a two-barrel carburetor and 9:1 compression ratio was 225 hp and 330 lb-ft torque. The 1958 318-cu.-in. V-8 Power Pack engine with its four-barrel carburetor and the same compression ratio was good for 250 hp and 340 lb-ft torque. The power output of the 1958 Plymouth Fury 318-cu.-in. V-8 engine with a 9.25:1 compression ratio, tandem four-barrel carburetors, and conventional air cleaners as shown on Fig. 8.47 was unchanged at 290 hp, but its torque level of 330 lb-ft was 5 lb-ft higher than in 1957.

In 1957, a Plymouth powered by the new V-8 A-engine took first place in its class in the Mobilgas Economy Run, demonstrating that the engine was not only a good power producer but also frugal on fuel consumption. To underscore that this was no fluke, the next year two 1958 Plymouth Belvedere sedans equipped with the 318-cu.-in. V-8 A-engine were entered in the low-price eight-cylinder class of the 1958 Mobilgas Economy Run. At the completion of the run that had been further lengthened to 1,883 miles, starting in Los Angeles and ending in Galveston, Texas, the two Plymouths placed first and second in their class, so close in total fuel used that the winning car, driven by Pierce Venable, burned only 12 fewer ounces of gasoline for the entire trip than the second-place Plymouth, driven by Mary Davis. Their fuel economies were 20.02 and 19.99 mpg, respectively.

Fig. 8.47 *Top-front view of the 1958 318-cu.-in. A-engine with tandem four-barrel carburetors. Offset air cleaners had been used in 1957, but in 1958, there was sufficient room under the hood for the conventional air cleaners shown.*

A new customer for the engine appeared in 1959—Dodge Division that had built the last V-8 of its own in 1958 and needed the Mound Road V-8 as a replacement. However, Dodge did not like the idea of marketing a new V-8 any smaller than the 325 cu. in. it had been using, so it requested the Mound Road engine plant to furnish a size larger than the 318 cu. in. being built at the highly automated plant. The Mound Road engine plant grudgingly agreed and for 1959 supplied Dodge with 0.04-in.-overbore, 326-cu.-in. A-engines, having a bore and stroke of 3.95 x 3.31 in. that produced 255 hp and 350 lb-ft torque with a two-barrel carburetor and a 9.2:1 compression ratio. Two Dodge Coronets with this 326-cu.-in. engine under the hood were entered in the 1959 Mobilgas Economy Run, routed from Los Angeles to Kansas City, Missouri, a distance of 1899 miles. One of the Coronets, averaging 21.7 mpg, bested all the other cars in its low medium price class, earning the Mobilgas Economy Run Class D first-place trophy.

Plymouth Division continued using two- and four-barrel 318-cu.-in. V-8 A-engines from 1959 through 1962, with minor revisions in output. Dodge dropped its request for a special size engine in 1960, accepting the same 318-cu.-in. two-barrel 230-hp, 340 lb-ft torque A-engine as the Mound Road engine plant shipped to the Plymouth Lynch Road assembly plant. For 1961 and 1962, Dodge added the four-barrel 318-cu.-in. engine to its power offerings, with the Mound Road engine plant obliging by shipping to the Dodge Main plant the same 260-hp, 345 lb-ft torque engine it shipped to the Lynch Road plant. The four-barrel version was dropped by Plymouth and Dodge in 1963, and the two-barrel version of the A-engine continued for both divisions through 1966.

A new engine requirement appeared in 1964, which had a profound effect on V-8 engines built at both the Mound Road and Windsor engine plants. The new requirement was V-8 power for the compact Valiant that had been launched in 1960 with six-cylinder power but would have a much broader appeal with the added performance and smoothness of a V-8 engine. Chapter 9 tells how this new V-8 engine was related to and eventually replaced the poly-head A-engine V-8.

Table 8.6 shows the specifications for the poly-head V-8 A-engine.

Chart 8.1 displays time lines for all the engines described in this chapter.

TABLE 8.6
SPECIFICATIONS FOR THE PLYMOUTH POLY-HEAD V-8 A-ENGINE

Model Year	1956	1956	1957	1957
Car line	Plymouth	Plymouth	Plymouth	Plymouth
Model	P29	Fury P29-3	P31	Fury V-800
Displacement (cu. in.)	276	303	301	318
Bore	3.75	3.8125	3.91	3.91
Stroke	3.125	3.3125	3.125	3.31
Compression ratio	8.0:1	9.25:1	8.5:1	9.25:1
BHP @ rpm	187 @ 4400	240 @ 4800	215 @ 4400	290 @ 5400
Torque (lb-ft) @ rpm	265 @ 2400	310 @ 2800	285 @ 2800	325 @ 4000
Combustion chamber	Polyspherical	Polyspherical	Polyspherical	Polyspherical
Cyl. block mat'l.—skirt type	Cast iron—shallow	Cast iron—shallow	Cast iron—shallow	Cast iron—shallow
Cyl. block deck height	9.6	9.6	9.6	9.6
Cyl. block length	21.4	21.4	21.4	21.4
Cyl. bore spacing	4.46	4.46	4.46	4.46
Tappet angle (from vertical)	59°	59°	59°	59°
Ck/shaft—no. of main brgs.	5	5	5	5
Ck/shaft main journal dia.	2.5	2.5	2.5	2.5
Crankpin journal dia.	2.125	2.125	2.125	2.125
Piston mat'l.—no. of rings	Aluminum—3	Aluminum—3	Aluminum—3	Aluminum—3
Connecting rod centers	6.123	6.123	6.123	6.123
Valve system	OHV—single rocker shaft	OHV—single rocker shaft	OHV—single rocker shaft	OHV—single rocker shaft
Valve events	244-244-26	244-244-26	240-240-16	256-256-38
Camshaft drive	Chain	Chain	Chain	Chain
Valve head dia.	Intake 1.84; exhaust 1.56	Intake 1.84; exhaust 1.56	Intake 1.84; exhaust 1.56	Intake 1.84; exhaust 1.56
Valve lash adjustment	Mech.—rocker arm screw	Mech.—rocker arm screw	Mech.—rocker arm screw	Mech.—rocker arm screw
Oil pump/distributor drive	Vertical at rear of block	Vertical at rear of block	Vertical at rear of block	Vertical at rear of block
Fuel system	2-barrel carburetor	4-barrel carburetor	2-barrel carburetor	Tandem 4-barrel carbs
Other		Engine built at Windsor engine plant		

(Continued on next page)

TABLE 8.6
(Continued)

Model Year	1958	1966
Car line	Plymouth	Plymouth & Dodge
Model	LP2	BR2, BP2, BW2, BD2
Displacement (cu. in.)	318	318
Bore	3.91	3.91
Stroke	3.31	3.31
Compression ratio	9.0:1	9.0:1
BHP @ rpm	225 @ 4400	230 @ 4400
Torque (lb-ft) @ rpm	330 @ 2800	340 @ 2400
Combustion chamber	Polyspherical	Polyspherical
Cyl. block mat'l.—skirt type	Cast iron—shallow	Cast iron—shallow
Cyl. block deck height	9.6	9.6
Cyl. block length	21.4	21.4
Cyl. bore spacing	4.46	4.46
Tappet angle (from vertical)	59°	59°
Ck/shaft—no. of main brgs.	5	5
Ck/shaft main journal dia.	2.5	2.5
Crankpin journal dia.	2.125	2.125
Piston mat'l.—no. of rings	Aluminum—3	Aluminum—3
Connecting rod centers	6.123	6.123
Valve system	OHV—single rocker shaft	OHV—single rocker shaft
Valve events	240-240-16	244-240-20
Camshaft drive	Chain	Chain
Valve head dia.	Intake 1.84; exhaust 1.56	Intake 1.84; exhaust 1.56
Valve lash adjustment	Mech.—rocker arm screw	Mech.—rocker arm screw
Oil pump/distributor drive	Vertical at rear of block	Vertical at rear of block
Fuel system	2-barrel carburetor	2-barrel carburetor

Chart 8.1

Time Lines—Hemi®-Head and Poly-Head V-8s

Engines, Car Lines, and Displacements (cu. in.)

Engine	Car Line	1951	1952	1953	1954	1955	1956	1957	1958
Chrysler Hemi® Head 4.5625 cyl. bore centers	Chrysler	331	331	331	331	331	354	392	392
	Model	C52, C55	C52, C55	C56	C63	C68, 300	C72, 300B	C76, 300C	LC3, 300D
	Imperial	331	331	331	331	331	354	392	392
	Model	C53, C54	C53, C54	C58, C59	C64, C66	C69, C70	C70, C73	IM1	LY1
	Dodge							354	
	Model							D66	
DeSoto Hemi® Head 4.3125 cyl. bore centers	DeSoto		276	276	276	291	330 341	341 345	
	Model		S17	S16	S19	S21, S22	S23, S24, Adv		
Dodge Hemi® Head 4.1825 cyl. bore centers	Dodge			241	241	270	315	325	
	Model			D44, D48	D50, D53	D55	D63	D66, 67, 70, 71	
Chrysler Poly-Head 4.5625 cyl. bore centers	Chrysler					301	331	354	354
	Model					C67	C71	C75	LC1, LC2
Dodge Poly-Head 4.1825 cyl. bore centers	Dodge					270	270 315	325	325
	Model					D55	D63	D66, D67, D70	LD2
	Plymouth					241 259	270		
	Model					P27	P29		
	DeSoto							325	
	Model							S27	
Plymouth Poly-Head A-Engine 4.46 cyl bore centers	Plymouth						276 303	276 301 318	318
	Model						P29	P31	LP2

Chart 8.1
(Continued)

Engines, Car Lines, and Displacements (cu. in.)

Engine	Car Line	1959	1960	1961	1962	1963	1964	1965	1966
Plymouth/ Engine Division A-Engine 4.46 cyl. bore centers	Plymouth	318	318	318	318	318	318	318	318
	Model	MP2	PP2	RP2	SP2	TP2	VP2	AR2	BR2, BP2
	Dodge	326	318	318	318	318	318	318	318
	Model	MD2	PD4	RD4	SD2	TD2	VD2	AW2	BW2

CHAPTER 9

Wedge-Head V-8s—B, RB, and LA Engines

After releasing V-8 engines with hemispherical combustion chambers starting with 1951 model cars and followed by V-8s with polyspherical chambers four years later with 1955 models, Engine Design was faced with a crossroads on what kind of combustion chamber should be placed on top of a planned new large-displacement V-8 engine, the first engine to be tooled from scratch by the newly formed and centralized Engine Division of the manufacturing group of the corporation. Timing was urgent because engines were scheduled to start arriving at the vehicle assembly plants for the 1958 models.

Acknowledging that the hemi chamber was still the best for performance but carried with it the burdens of larger size, greater weight, and higher cost—burdens considerable enough to have caused the creation of the poly-heads—the choice narrowed to the continuation of the poly-head or the possible introduction of the chamber used by virtually all other high-volume manufacturers: the wedge-shaped combustion chamber.

The debate was short lived. Dynamometer tests showed wedge chambers to be the equal of poly-chambers, and design comparisons showed that cylinder heads with wedge chambers and in-line valves were less bulky, lighter, and less costly to manufacture than polyspherical-chambered cylinder heads with their divergent valves. With that, the design of Chrysler engines moved in a new direction, and for decades to come, the in-line valve wedge head would be dominant.

Chrysler Big-Block V-8s—The B and RB Engines

Robert S. (Bob) Rarey, who succeeded Mel Carpentier as Assistant Chief Engineer of Engine Design when Carpentier died in 1955, soon was challenged with the task of putting the corporation into the big engine arena with power suitable for the large C-body cars being prepared for introduction in the late 1950s—power that would match anything the formidable competition could unleash. With his excellent engine background as a test engineer for Powerplant Research during the XI-2220 V-16 aircraft engine program that will be described in Chapter 17, as the resident motor engineer of Plymouth Division, and as manager of the Engine Design department under Mel Carpentier, Rarey rose to the challenge, directing his highly regarded overall engineering acumen to the task, along with his customary personal vigor.

Engineering's requirements for the new V-8 engine, input from the car divisions regarding future engine displacement sizes, and tooling information furnished by Manufacturing's Engine Division merged as guidelines for the new design.

Assigning lead designer Fred Shrimpton to make the initial layout and assisted by Advance Engine Design supervisor Ray Latham, Rarey established the fundamental parameters for the new V-8 family, and the engine took form. The cylinder block was big; its 4.8 in. between cylinder centers were greater than any other production engine since the start of Chrysler Corporation. To eliminate any potential problems with cylinder block flexing or powerplant resonance, the block was designed with a deep skirt extending 3 in. below the crankshaft centerline. For the anticipated displacement range, two cylinder blocks differing in deck heights would be needed: a low deck block that was designed with a 9.98-in. high top deck, and a raised deck block having a deck height of 10.725 in. Clearance for the connecting rod path of a 3.75-in. stroke engine, the longest stroke planned, was a design requirement for the interior of the crankcase. The oil pump and distributor, located at the left-rear of the cylinder block on all the first-generation V-8 engines, were given new locations at the front of the engine. The distributor appeared at the top right-front; the oil pump, mounted on the outside of the cylinder block, appeared at the bottom left-front of the block. Both units were driven by a diagonal shaft whose gear meshed with a gear cut into the camshaft. The impetus to the relocations came from the need to make the distributor less vulnerable to interference with ventilation ductwork at the firewall and to position the oil pump so the oil pan sump could be shaped to give clearance to the steering tie rods under the engine, wherever they might be located. Advantage was taken of the outside-mounted oil pump to design a die-cast aluminum cover that doubled as mounting for the spin-on, full-flow oil filter.

Without question, the design of the cylinder heads for the new engine was the greatest departure from previous Chrysler practice. As noted, the prior hemispherical and polyspherical chamber designs were abandoned, and the new heads were designed with wedge-shaped combustion chambers and in-line valves, which immediately gave benefits for vehicle packaging and vehicle weight. To minimize the iron content and weight of each cylinder head, the water jacket—designed to provide cooling for the combustion chambers, valve seats, and valve stems—had a horizontal roof. On the bottom coolant side of the roof, steam pockets could not form. On the top side of the roof, lubricating oil from the valve rockers was free to flow directly into the tappet chamber of the cylinder block and down to the sump without needing drilled drainback holes in either the cylinder head or the cylinder block.

The extremely short exhaust ports, designed to keep the exhaust manifolds tucked in close to the cylinder heads, reduced heat rejection to the coolant while reducing weight. Side-by-side intake ports further reduced metal content. Five die-cast aluminum pedestals solidly anchored the single rocker shaft of each cylinder head with a bolt through each pedestal. Thanks to the simple geometry of the in-line valves, the two banks of tappet bores in the cylinder block were positioned with a 90-deg included angle.

Because of the large cylinder bore sizes planned for the engine, a decision was made to use a five-bolt head-bolt pattern for each cylinder rather than the four-bolt pattern that had been used on all first-generation V-8s. Arranged in somewhat of a pentagon shape around each bore, the number of bolts per head increased from 10 to 17 for the new V-8s, and the bolt size was reduced from 1/2-13 to 7/16-14. An innovative design feature of the engine was the use of a single embossed sheet-steel gasket that covered the entire top of the block and sealed the joints between the cylinder heads and the intake manifold, providing a savings in weight and cost over the multiple gaskets and covers used in the past.

Little known outside the core engine design and development engineers, the success of the engine hinged on the ability of an outside supplier to make a small but critical engine part—the valve rocker arm, sixteen per engine. The design for the new valve rocker arm was unprecedented, using flat steel stock that was blanked, formed, pierced, heat treated, and coined to exacting tolerances, rather than being machined from steel forgings or malleable iron castings, using expensive machine tools to produce the huge quantities needed. The confidence of the supplier's chief tool engineer was rewarded with the supplier being able to furnish excellent parts in time for launch. This was a fortunate outcome because there was no backup plan.[1]

For the 1958 introduction of the low-deck engine, two displacements were prepared: 350 cu. in. using a 4.06-in. bore, and 361 cu. in. with a 4.125-in. bore, sharing a common 3.375-in. stroke. Coded A561 and A652, respectively, both versions had a 10:1 compression ratio and valve head diameters of 1.95 in. for the intake valves and 1.6 in. for the exhaust valves. Figure 9.1 shows a longitudinal section of the 1958 361-cu.-in. engine. A cross section through the #1 cylinder and a cross section through the front of the engine are shown on Figs. 9.2 and 9.3, respectively. Following the nomenclature lead of the A-engine, the engines quickly became known as the Low B, LB, or simply B engines.

Meanwhile, the central manufacturing group had selected an existing plant, occupied by the Marine and Industrial Division, in Trenton, Michigan, as the facility for producing this new family of V-8 engines. Basic production machine tools were placed on order at the same time as parts for experimental engines were being procured by the Engine Development department, giving a strong sense of urgency to the engineering program. The feeling of urgency never slacked off.

However, when the first experimental engine went on the dynamometer, it abruptly stopped after running under its own power for only a short time. The problem was traced quickly to insufficient running clearance for the oil pump rotors. The rotors had bound up, causing a pin that held together the oil pump and distributor driveshaft and its gear to snap. When that happened, the distributor quit turning, and the engine simply stopped.[2] The problem soon was corrected, and the engine, joined by dozens of other experimental B engines, continued in its role of exposing problems by rigorous testing so improvements could be made before volume production started at the Trenton engine plant. The first pilot engines, built with production tools, came off the engine assembly line of the Trenton engine plant in June 1957. High volume started later, in time for vehicle assembly to start in September 1957. An SAE paper written by Bob Rarey, Assistant Chief Engineer of Engine Design, and Ev Moeller, Manager of Engine Development, presented on March 6, 1958, described in detail the design and development of this initial LB engine with its two displacements.[3]

The 1958 Plymouths, Dodges, and DeSotos shared the Engine Division's Trenton engine plant's first-year production of LB engines, using two- and four-barrel carburetor versions of the 350- and 361-cu.-in. sizes, plus several high-horsepower versions that are described in the section of this chapter titled "Power Programs." The 350-cu.-in. two-barrel engine, rated at 280 hp and 380 lb-ft torque, was installed in only the DeSoto FireSweep model. Dodge named the 350-cu.-in. four-barrel 295-hp 385 lb-ft torque LB engine, shown on Fig. 9.4, the Ram-Fire V-8 and made it standard on its Custom Royal (pictured on Fig. 9.5) and station wagon models. DeSoto offered the same engine (without, of course, the Dodge name) as an option

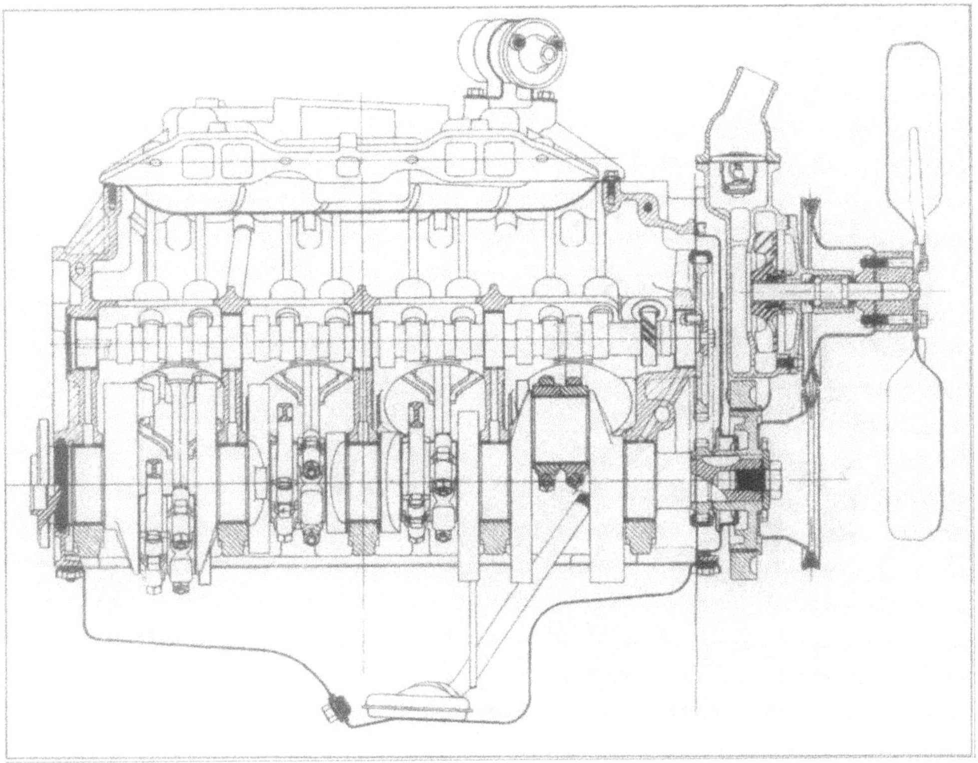

Fig. 9.1 *A longitudinal section of the 1958 361-cu.-in. LB V-8 engine. The large-diameter main journals and crankpins gave the crankshaft the strength to handle future increases in engine displacement and power.*

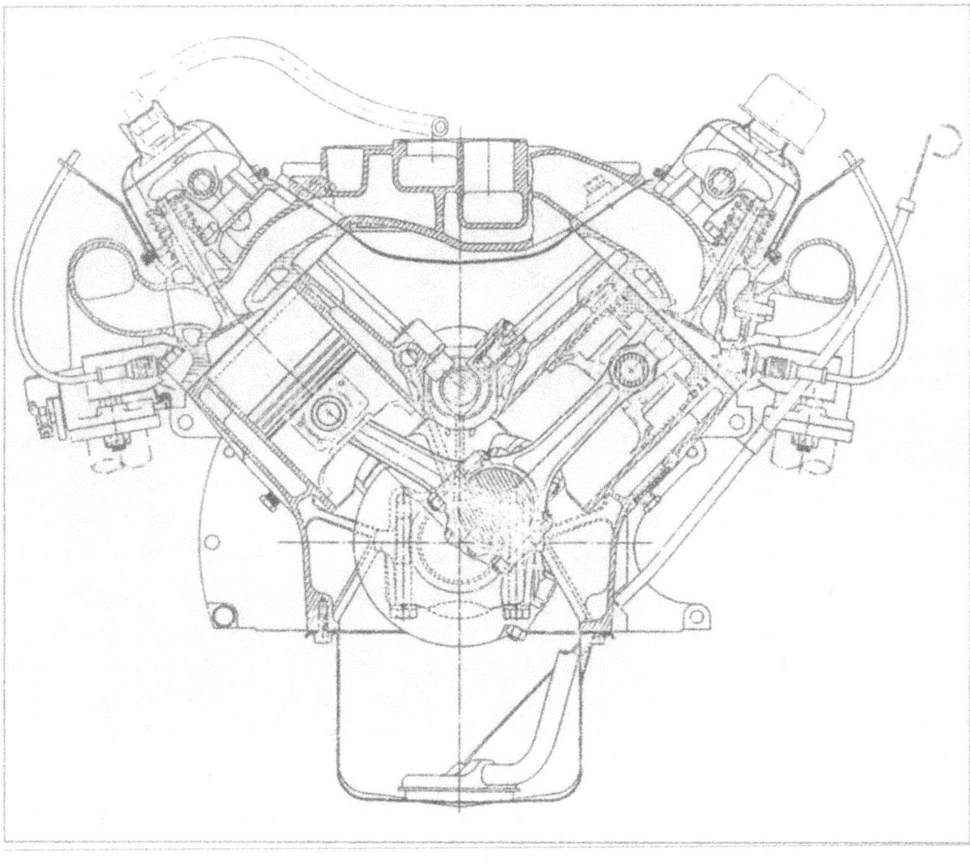

Fig. 9.2 *A cross section of the 1958 LB V-8 engine, showing its wedge-shaped combustion chamber and in-line valves. The short exhaust ports resulted in the spark plugs being located below the exhaust manifold runners.*

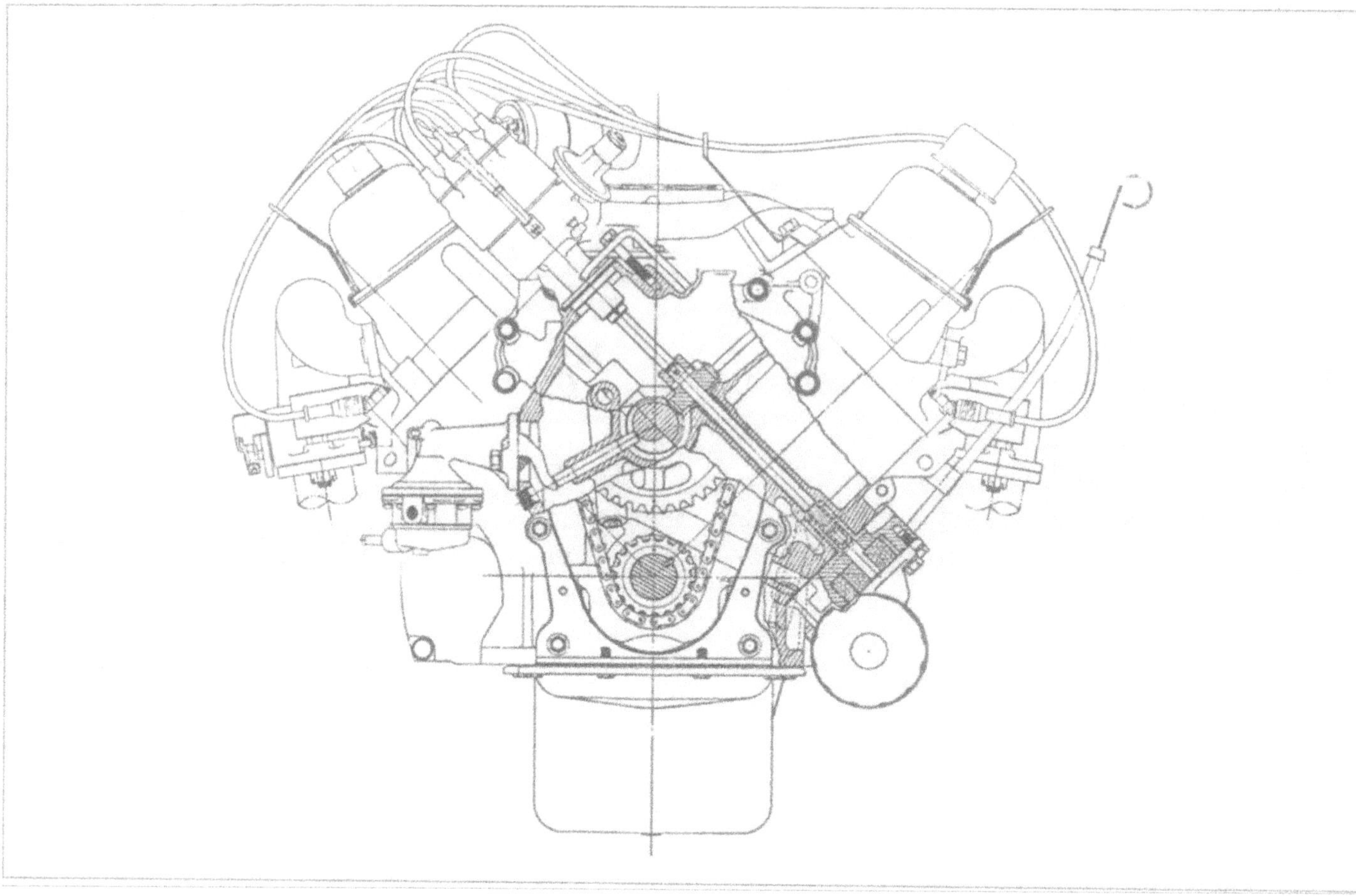

Fig. 9.3 *A cross section through the oil pump and distributor drive of the 1958 LB V-8 engine. Both units were more accessible than in their prior back-of-block location.*

Fig. 9.4 *A left-front view of the 1958 350-cu.-in. LB V-8 engine equipped with a four-barrel carburetor. The narrow cylinder head cover is quite a contrast to the broad cover used on the Hemi® engines. The large fitting in the side of the oil pan indicates the engine shown was a dynamometer test engine.*

Fig. 9.5 *The 1958 Dodge LD3 Custom Royal convertible. Standard power was a 350-cu.-in. LB V-8 engine with a four-barrel carburetor.*

on its FireSweep model. Figure 9.6 shows the 1958 DeSoto Firedome that was the sole customer for the two-barrel version of the 361-cu.-in. LB engine, rated at 295 hp and 390 lb-ft torque. The four-barrel version, with 305 hp and 400 lb-ft torque, was installed as standard in the DeSoto Fireflite. The same four-barrel 361-cu.-in. engine, but with the name D-500 (transferred from the prior year's Dodge-built 325 Hemi® head engine), was an option generally available throughout the Dodge V-8 car line.

Fig. 9.6 *The 1958 DeSoto LS2 Firedome four-door sedan. Power was provided by a 361-cu.-in. LB V-8 engine with a two-barrel carburetor, capable of 295 hp.*

The 350-cu.-in. LB engine lasted only through 1958. The year 1959 saw a continuation of the 361-cu.-in. engine and the emergence of a 383-cu.-in. LB engine. The extra cubic inches were obtained by a larger 4.25-in. bore. With the extra room made available by the bigger bore, intake valves with a 2.08-in. head diameter were released for the 383-cu.-in. engine. A two-barrel version of the 1959 361-cu.-in. engine was shipped by the Trenton engine plant to become the base V-8 power for the 1959 Dodge Royal and Sierra models and the DeSoto FireSweeps. A four-barrel version was supplied to Plymouth as a generally available power option. Titled by Plymouth as the Golden Commando 395, the 361-cu.-in. engine was advertised to produce 305 hp and 395 lb-ft torque. The same engine, shipped to the Dodge Main plant for assembly as the base V-8 for the Dodge Custom Royal and Custom Sierra models, was titled by Dodge as the Super Ram-Fire V-8.

DeSoto powered its 1959 Firedome model with a two-barrel version of the new 383-cu.-in. LB engine that provided the car with an additional 10 hp and 20 lb-ft more torque than it had in 1958, the new levels being 305 hp and 410 lb-ft torque. Four-barrel versions of the new 383-cu.-in. LB engine found use in 1959 Dodge and DeSoto models. Dodge made it a generally available option on all models, calling it the D-500 and giving it a rating of 320 hp and 420 lb-ft torque. DeSoto made the engine standard power on its 1959 Fireflite models and gave it a slight advertised output edge of 325 hp and 425 lb-ft torque. Not wishing to be left behind in the engine naming business, DeSoto bestowed the title of Turboflash V-8 to all three of its 1959 LB V-8 engines.

Big brother to the LB engine, the raised B or RB engine, with its higher decks, longer stroke, longer connecting rods (6.768 versus 6.358 in.), and wider intake manifold, made its roll-out from the Trenton engine plant in time to power the 1959 car models. The longer 3.75-in. stroke did necessitate a crankshaft main bearing diameter increase from the 2.625 in. of the LB to 2.75 in. to ensure infinite fatigue life for the crankshaft. The diameter of the crankpins, which was 2.375 in., was left unchanged. The 1959 engine needs of the Chrysler Division were for two engine sizes: a 413-cu.-in. engine to power New Yorkers, Imperials, and 300Es; and a smaller 383-cu.-in. size to power Windsor and Saratoga models. However, the Chrysler request for 383-cu.-in. engines put the Trenton engine plant in a bind because the plant had two cylinder block machining lines—one for LB blocks, and the other for RB blocks. Although each line could be reset for the other block height, downtime was required to make the changeover, which meant loss of production. Build schedules for the LB 361- and 383-cu.-in. engines for Plymouth, Dodge, and DeSoto were so heavy that the LB block machining line essentially was booked, leaving no excess capacity for the 383-cu.-in. LB blocks for Chrysler. To break this capacity logjam, Manufacturing asked that a 383-cu.-in. Chrysler engine using a raised-deck cylinder block be released instead of an engine with a low-deck block.[4] Engine Design responded with the design and release of a small-bore, long-stroke, 4.03 × 3.75 in. 383-cu.-in. RB engine for 1959 Chryslers in two versions—a two-barrel engine for the Windsor model, and a four-barrel engine for the Saratoga. Outputs were 305 hp and 410 lb-ft torque, and 325 hp and 425 lb-ft torque, respectively. Two years later, capacity for the needs of Chrysler opened up on the LB block machining line, and the Chrysler small-bore RB 383-cu.-in. engine was discontinued.

The 1959 413-cu.-in. RB V-8 engine, coded A694 and looking similar to the 1963 engine shown on Fig. 9.7, was an impressive and brawny powerplant. Its 4.18-in. bore, 3.75-in. stroke, 10.1:1 compression ratio, 2.08-in.-diameter intake valves, 1.6-in. exhaust valves, hydraulic tappets, four-barrel carburetor, 350 hp, and 470 lb-ft torque provided effortless and quiet power for Chrysler New Yorker models, such as the elegant four-door hardtop shown on Fig. 9.8, and for Imperials. A comparison of the longitudinal and cross sections of the 1960 413-cu.-in. RB engine shown on Figs. 9.9 and 9.10 with sections of the 1958 361-cu.-in. LB engine shown on Figs. 9.1 and 9.2 reveals the similarity of the LB and RB engine designs. High-horsepower versions of the engine, including dual four-barrel carburetors, are described in the "Power Programs" section of this chapter. Dodge started using this standard

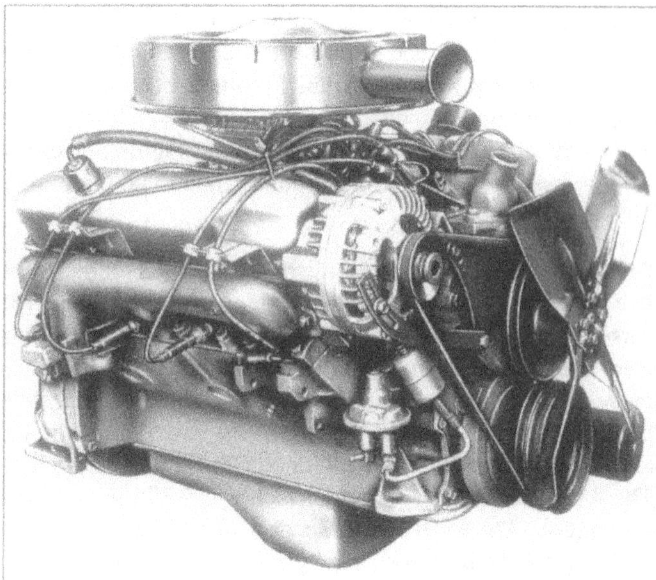

Fig. 9.7 *A right-front view of the 1963 413-cu.-in. RB V-8 engine. The big but compact engine provided smooth and effortless power for the luxury cars of the corporation.*

Fig. 9.8 *The 1959 Chrysler MC3 New Yorker four-door hardtop. Its 413-cu.-in. RB V-8 engine with a four-barrel carburetor was rated at 350 hp.*

413-cu.-in. four-barrel RB V-8 engine in its large Polara, Custom 880, and Monaco models in 1965.

Subsequent Changes—The LB V-8 Engine

The year 1961 was a carryover year for the two-barrel and four-barrel carburetor versions of the 361-cu.-in. engine. However, in 1961, the compression ratio was lowered to 9:1 so both could burn regular fuel, dropping outputs to 265 hp and 380 lb-ft torque, and 305 hp and 395 lb-ft torque, respectively. The four-barrel 361-cu.-in. engine went out of production at the end of 1962, leaving the two-barrel version to continue as the base V-8 for various Plymouth and Dodge models through 1966, which was the final year of the 361-cu.-in. LB V-8 engines.

The two-barrel version of the 383-cu.-in. LB V-8 engine was made a regular fuel burner in 1965 when its compression ratio was dropped to 9.2:1. The ratio was further dropped to 8.7:1 in 1970 and then to 8.5:1 in 1971 as exhaust emission limitations took effect. The advertised output of the 1971 two-barrel 383-cu.-in. engine became 275 hp and 375 lb-ft torque. Another change in 1971 was to the material of the engine crankshaft, which until then had been made of forged steel. The new material was nodular cast iron, also called ductile iron, and was already proven in small-block LA V-8 engines (described later in this chapter). The material change represented a substantial cost reduction for the 383-cu.-in. two-barrel LB V-8 engine.

Starting in 1972, all cylinder blocks machined on the LB line at the Trenton engine plant were bored and honed to a new size—4.34 in.—marking the end of the 383-cu.-in. size and the beginning of the 400-cu.-in. displacement LB V-8 engines. Interestingly, for parts commonality, Engineering initially planned to make the engine 396 cu. in. by using the same 4.32-in. bore already in production for the 440-cu.-in. RB engine. However, the car divisions insisted that the 400 number would have greater sales appeal, and their view was accepted.[5] Concurrent with this 1972 displacement increase to 400 cu. in., the compression ratio of the LB engine was adjusted downward to 8.2:1. The published output of the two-barrel carburetor version of the 1972 engine of 190 hp and 310 lb-ft torque would appear to be a staggering reduction versus the ratings posted for its prior smaller displacements, including the 275 hp and 375 lb-ft level shown for the 1971 two-barrel 383-cu.-in. LB V-8. However, in 1972, the automotive companies started displaying SAE Standard J245 net horsepower and torque outputs of their engines rather than the much larger maximum obtainable gross outputs that had been used, and this resulted in reductions in engine rating numbers starting in 1972. The net rating for the four-barrel version of the 1972 400-cu.-in. LB engine with its high-performance camshaft was 255 hp and 340 lb-ft torque.

Over the span of years from 1961 to the end of production, many changes were made to the LB engine to make it emissions compliant with both California and federal mandates. Among these was the addition of induction hardening to the exhaust valves seats of the cylinder head in 1973, which was done to all Chrysler engines so the engines could operate on unleaded fuel without encountering valve seat sinkage; unleaded fuel operation was mandatory when catalytic converters started arriving in 1975. The Addendum located near the end of this book displays emission control system changes incorporated from 1961 through 1976.

For 1976, its final year, the two-barrel 400-cu.-in. LB engine produced net power and torque of 175 hp and 300 lb-ft, respectively. Exiting at the end of 1978, which marked the end of

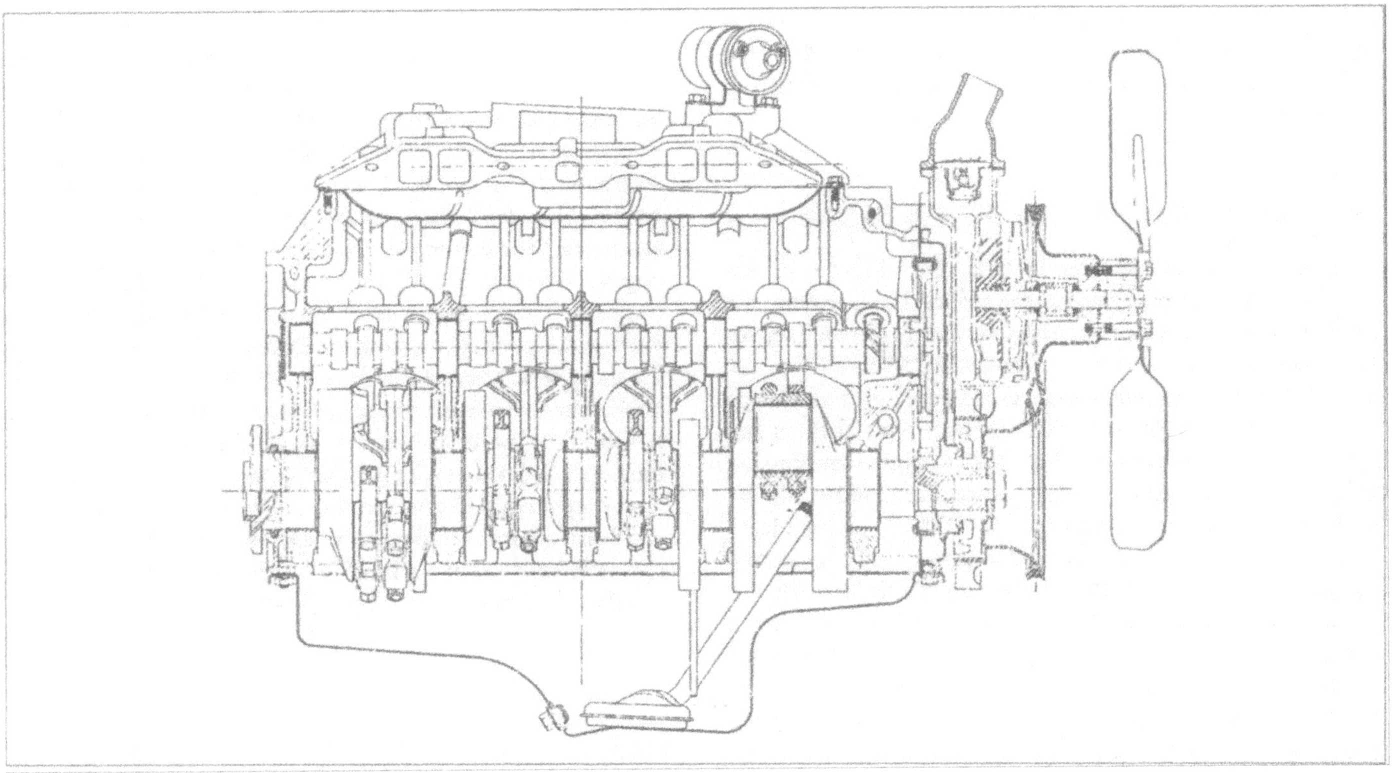

Fig. 9.9 *A longitudinal section of the 1960 413-cu.-in. RB V-8 engine.*

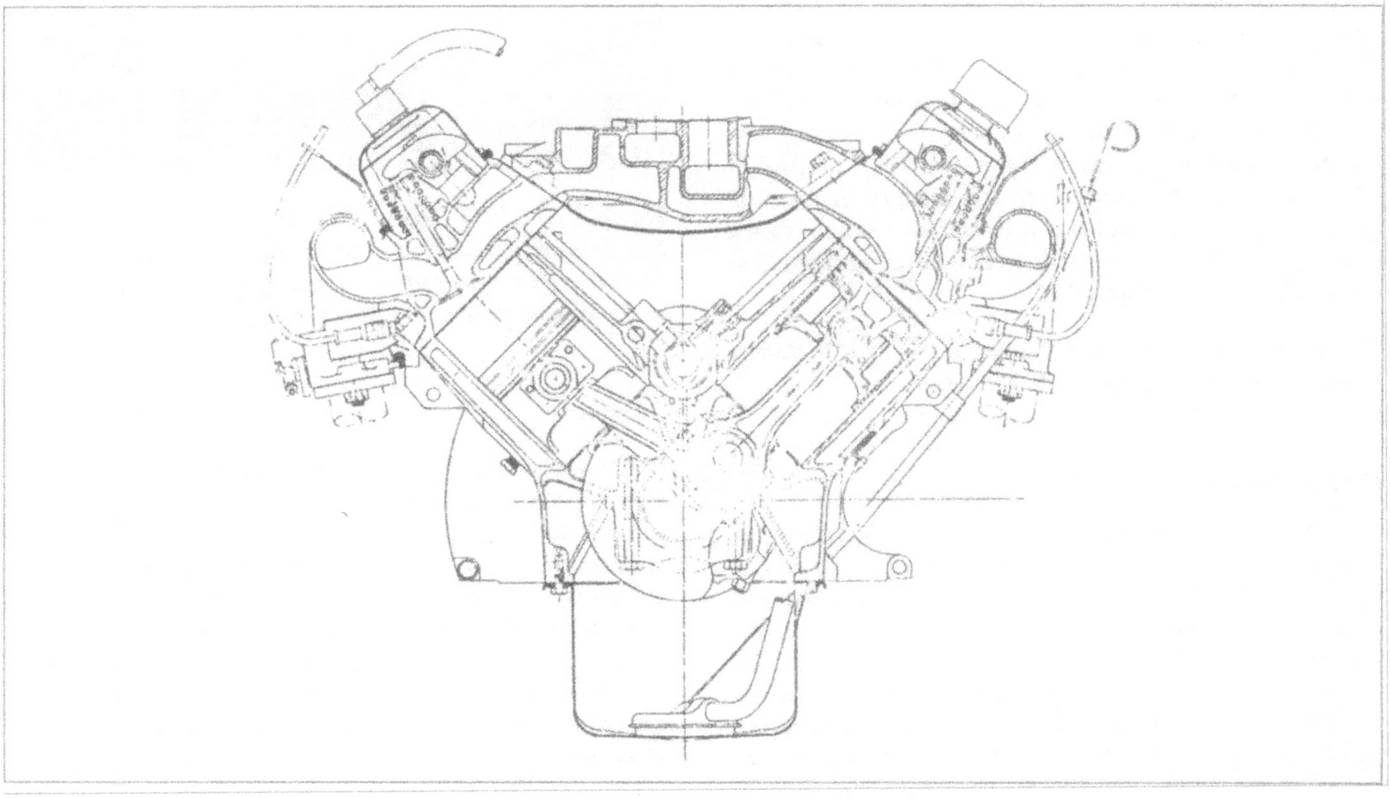

Fig. 9.10 *A cross section of the 1960 413-cu.-in. RB V-8 engine. The higher decks of the RB cylinder block required longer pushrods and a wider intake manifold.*

all big-block B-RB V-8 engine production at the Trenton engine plant, the four-barrel version, now equipped with the same camshaft as the two-barrel engine, produced 190 hp and 305 lb-ft torque.

Subsequent Changes—The RB V-8 Engine

As noted previously, the somewhat oddball 383-cu.-in. RB V-8 engine was discontinued at the end of 1960, leaving the Trenton engine plant building only one size of RB engines for 1961—the 413-cu.-in. engine. The 1959 350-hp 413-cu.-in. base RB V-8 previously described carried over without change as power for the 1960 and 1961 Chrysler New Yorkers and Imperials. In 1962 through 1965, with a slight compression ratio reduction from 10.1:1 to 10:1, power decreased to 340 hp, but the torque remained unchanged at 470 lb-ft, giving plenty of acceleration for New Yorkers and Imperials. An engineering program coded A982 resulted in the size of this base RB V-8 being increased to a final 440 cu. in. for 1966 by a bore change from 4.1875 to 4.32 in., providing 350 hp and 480 lb-ft torque as standard power on Chrysler New Yorkers and Imperials and as an option on the large Dodge models. During the following year, cylinder head port areas were increased, and port shapes were improved as the outcome of a program coded A120; however, the rated output was left unchanged. Naming this standard RB engine the Commando 440, Plymouth joined the list of users in 1967 by making the engine optional on Fury station wagons. A compression ratio reduction from 10:1 to 9.7:1 was made in 1970, followed by another reduction in 1971 to 8.8:1, and a final tumble to 8.2:1 in 1972, as efforts continued to keep the exhaust emission levels of the engine below California and federal mandated maximums. With the 8.2:1 compression ratio, the output of the 1972 engine, now reported as net ratings, was 225 hp and 345 lb-ft torque. In 1973, the crankshaft of the engine was changed from forged steel to nodular cast iron for cost reduction, and the exhaust valve seats in the cylinder heads were induction hardened to resist sinkage with unleaded fuel.

The RB V-8 machining lines stopped at the Trenton engine plant with the build-out of the 1978 engines.

Power Programs

Extra power and performance, followed by product recognition and increased car sales, paved the road to success as mapped out by Bob Roger when he created the sensational Chrysler 300 for the Chrysler Division in 1955. The move triggered enthusiastic similar action by DeSoto, Dodge, and finally Plymouth to field high-powered and high-performance cars using their own engines with multiple carburetors and hot camshafts. When the centrally produced LB V-8 engine was readied to start production in 1958, the momentum was there to come out with big power options for each division—something well beyond the normal four-barrel carburetor in place of a two-barrel carburetor. Plymouth was lined up to use the 350-cu.-in. LB engine and received the extra power version it wanted—tandem four-barrel carburetors and a long-duration camshaft with valve events of 260-260-42. (For an explanation of valve events, see the "Abbreviations and Definitions" section in the back of this book.) This put 305 hp and 370 lb-ft torque of Golden Commando performance under the control of the driver. The leading Dodge engine was the 361-cu.-in. LB engine. Tandem four-barrel carburetors and the 260-260-42 camshaft transformed it into the Super D-500, available as an extra-cost option throughout the Dodge 1958 model offerings for those who wanted 320 hp and 400 lb-ft torque under the hood. The DeSoto Adventurer, a long-time user of engines with tandem four-barrel carburetors, opted for the same engine as the Super D-500 but with a higher compression ratio that gave it an output edge. Thus, the 361-cu.-in. LB V-8 engine in the 1958 Adventurer, similar to the one shown on Fig. 9.11, was specially built to have a 10.25:1 compression ratio that raised its advertised power and torque to 345 hp and 400 lb-ft.

Fig. 9.11 *The 1958 DeSoto LS3 Adventurer two-door hardtop. The performance edge of the car came from a 345-hp 361-cu.-in. LB V-8 engine equipped with tandem four-barrel carburetors and a special 10.25:1 compression ratio.*

Then the opportunity to be a pioneer in the high-tech field of electronically controlled multipoint fuel injection arrived and, after much testing, was approved by Engineering. Marketed by all four car divisions, the so-equipped Plymouth 350-cu.-in. LB engine pictured on Fig. 9.12, titled the Golden Commando with Fuel Injection, produced 315 hp and 370 lb-ft torque. The Dodge 361-cu.-in. D-500 with fuel injection had an output of 333 hp and 400 lb-ft torque, and the DeSoto 361-cu.-in. Adventurer with its 10.25:1 compression ratio and fuel injection was advertised at 345 hp and 400 lb-ft. As related in Chapter 8, Chrysler applied the same type of fuel injection to its 1958 300D 392-cu.-in. Hemi® head engine. Similar to the Chrysler engine, the other engines ran into unexpected problems with

Fig. 9.12 *The engine compartment of the 1958 Plymouth LP2 Golden Commando with Fuel Injection. Its engine was a 350-cu.-in. LB V-8 equipped with Bendix multipoint fuel injection, rated at 315 hp. Four of the eight injectors in the engine can be seen under the fuel rail.*

Fig. 9.13 *The 1959 Chrysler MC3 300E two-door hardtop. Under the hood, a 380-hp 413-cu.-in. RB V-8 engine with tandem four-barrel carburetors gave the car exceptional performance.*

electromagnetic interference, which were severe enough to have the units retrofitted with tandem four-barrel carburetors. The actual number of fuel-injected engines that were built was low—about two dozen total. Although the application honestly must be described as unsuccessful (my personal assessment), it was an important precursor to the almost universal use of electronic multipoint fuel injection by the mid-1990s.

In 1959, DeSoto and Dodge marketed a tandem four-barrel version of their largest 383-cu.-in. engine that produced 345 hp and 420 lb-ft torque for Dodge and 350 hp and 425 lb-ft torque for the DeSoto Adventurer. A 413-cu.-in. RB V-8 engine developed with tandem four-barrel carburetors was released as a suitable powerplant for the handsome 1959 Chrysler 300E shown on Fig. 9.13, succeeding the legendary Hemi® Firepower 392-cu.-in. V-8 engine used in the 1958 300D. Bob Roger, father of the Chrysler 300 models, is appropriately portrayed on Fig. 9.14 with his hand on an air cleaner of a 300E 413-cu.-in. RB V-8 engine. The same engine, putting out the same 380 hp and 450 lb-ft torque three years later under the hood of the 1962 Chrysler 300H, still gained attention. Charles J. Wick, in the March 1962 issue of *Today's Motor Sports*, wrote in an article titled "Chrysler 300H," "The 300-H is an absolute bomb! It goes—it handles—and it stops! I make that statement after a week of all-out testing and regular highway driving in both town and country." Later in the article, "Arriving at the Daytona International Speedway, I obtained permission to take the car around the track and proceeded to do eight to ten laps. Most of these were made holding the rpm to 4000 (110 m.p.h.). Several spurts proved the car's top speed to

Fig. 9.14 *Bob Roger, chief engineer of Chrysler Division and father of the Chrysler 300 series cars, has his hand on one of the air cleaners of a 1959 300E 413-cu.-in. RB V-8 engine.*

be near the 135 m.p.h. mark—a very creditable figure for a strictly 'stock' highway machine."

In 1960, intake pipe tuning experiments of the early 1950s K310 and A311 Hemi® engine programs were applied to the big-block LB and RB engines, with visually startling results. To obtain meaningful resonance lengths for these engines, long pipes were needed—so long that the four-barrel carburetor for each bank was placed outboard of the opposite bank, providing 30 in. of tuned length, and the intake manifold branches crossing at the center of the engine. Figure 9.15 shows an exterior view of this engine with its swooping, eye-catching, long-branch cross-ram intake manifold.

Fig. 9.15 *The 1960 Plymouth long-branch cross-ram LB V-8 engine. The engine came in two versions: a 361-cu.-in. 310-hp 435 lb-ft Sonoramic Commando 435 version, and a 383-cu.-in. 330-hp 460 lb-ft SonoRamic Commando 460 version. Naming the engine D-500 Ram Induction, Dodge used similar versions: the 361-cu.-in. engine on its Dart Phoenix, and the 383-cu.-in. engine on the Matador and Polara. DeSoto used the 383-cu.-in engine, giving it the title of Ram Charge Adventurer.*

Plymouth liked the engine so much that it requested two versions—one for each displacement. Its 361-cu.-in. LB SonoRamic Commando 435 was rated at 310 hp and 435 lb-ft torque, while its 383-cu.-in. SonoRamic Commando 460 engine was rated at 330 hp and 460 lb-ft torque. Dodge marketed the same engines, the 361-cu.-in. engine on its Dart Phoenix model and the 383-cu.-in. engine on the Dodge Matador and Polara models, calling both engines D-500 Ram Induction. DeSoto made the 383-cu.-in. engine optional on its 1960 Adventurer model, with the title of Ram Charge Adventurer. Chrysler requested the release of a 413-cu.-in. RB cross-ram engine, rated at 375 hp and 495 lb-ft torque, as standard on its 1960 300F model, shown on Fig. 19.16.

Fig. 9.16 *The 1960 Chrysler PC3 300F two-door hardtop. Standard power was a 375-hp long-branch cross-ram 413-cu.-in. RB V-8 engine.*

Knowing there were competitive events where an engine tuned for higher horsepower and lower torque would be preferred, Chrysler also asked for the release of a second, optional version of the engine with shorter tuned length branches. This request was easy to accomplish simply by shortening the partitions between the paired branches, reducing the tuned branch length to 15 in. With the shortened partitions, the 413-cu.-in. long-branch cross-ram RB V-8 saw its horsepower rise to 400 hp and its torque drop to a still substantial 465 lb-ft. An intake manifold with the shortened partition is illustrated in the cross section of the engine on Fig. 9.17. Although some exhaust heat was provided for the intake manifold floor underneath the carburetors, as can be seen in the cross section, cold-weather operation was marginal at best. The long- and short-partition versions of the 413-cu.-in. RB cross-ram engine continued as standard and optional power for the 1961 Chrysler 300G.

The following year, the 1962 Chrysler 300H reverted back to a tandem four-barrel 413-cu.-in. RB engine for its standard power, dropping the long-partition cross-ram engine and keeping the short-partition cross-ram engine as an option. The 1963 Chrysler 300J installed the short-partition cross-ram 413-cu.-in. RB engine as standard, providing the car with 390 hp and 485 lb-ft torque. The following year, its last year, the same engine was available as an optional powerplant on the 1964 Chrysler 300K.

The tried and true big-power setup using tandem four-barrel carburetors was not forgotten in 1960 because Dodge sold a

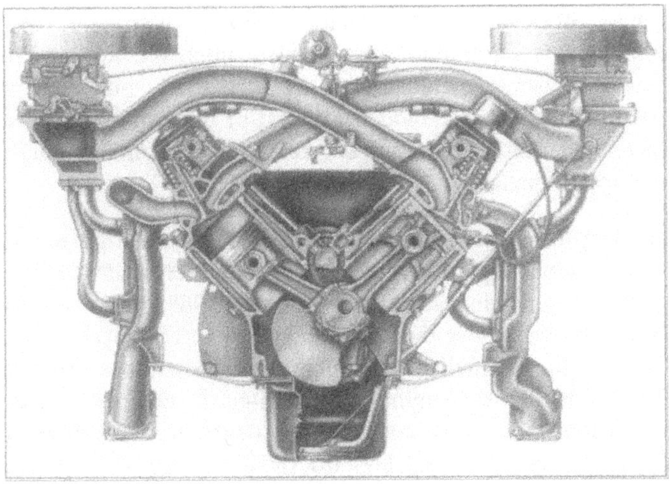

Fig. 9.17 *A cross section of the 413-cu.-in. RB V-8 engine equipped with a long-branch cross-ram intake manifold, having shortened partitions for higher engine speed tuning. The engine was optional on the 1960 Chrysler 300F. Note the mechanical valve lash adjustment.*

so-equipped 383-cu.-in. LB engine as one of its Maximum Performance Police Package engines, and Plymouth made the engine a general option on all V-8 models. Named Golden Commando 425 by Plymouth, the engine had an output of 330 hp and 425 lb-ft torque.

In 1961, the 361-cu.-in. engine dropped out of any big-power versions, but the 383-cu.-in. engine added a long-branch, short-partition, cross-ram version. Both Plymouth and Dodge marketed it as the standard engine on their Maximum Special Police Package #1, listing 340 hp and 440 lb-ft torque as the output. For police departments wanting more power, the larger 413-cu.-in. long-branch, short-partition, cross-ram engine was an option that put 375 hp and 465 lb-ft torque under the hood.

Big power using tandem four-barrel carburetors continued into 1961 in the form of a Plymouth and Dodge 383-cu.-in. LB engine having 330 hp and 425 lb-ft torque, and a 413-cu.-in. RB engine producing 375 hp and 450 lb-ft torque. The two engines were Plymouth and Dodge Maximum Performance Police Package #3 base and optional engines, respectively. Base and optional engines of the Plymouth and Dodge Maximum Performance Police Package #2 were the more mundane 383-cu.-in. LB and 413-cu.-in. RB single four-barrel engines with outputs of 325 hp and 425 lb-ft torque, and 350 hp and 470 lb-ft torque, respectively.

Three displacements of tandem four-barrel B-RB V-8 engines were optional throughout the car lines of Plymouth and Dodge in 1962: 361- and 383-cu.-in. LBs, providing 310 hp and 390 lb-ft torque, and 335 hp and 420 lb-ft torque, respectively, and an RB 413-cu.-in. engine with an output of 385 hp and 455 lb-ft torque. The 1962 Chrysler 300H base engine, a 413-cu.-in. RB engine with tandem four-barrel carburetors, as mentioned previously, produced 380 hp and 450 lb-ft torque. During the middle of the model year, four additional high-performance engine versions appeared in the 1962 Engineering Data Book[6] as optional on the Chrysler 300 and 300H models, all having a 426-cu.-in. displacement obtained by a bore increase from 4.19 to 4.25 in. One version, with a single four-barrel carburetor and 11:1 compression ratio, produced 373 hp and 472-lb-ft torque. A second version, also with a single four-barrel carburetor but with a 12:1 compression ratio, produced 385 hp and 486 lb-ft torque. The third version, equipped with a long-branch, short partition, cross-ram manifold with an 11:1 compression ratio, was rated at 413 hp and 485 lb-ft torque. The fourth version, similar to the third but with a 12:1 compression ratio, produced 421 hp and 498 lb-ft torque. Engineering Data Book[6] text indicated an Automobile Manufacturers Association (AMA) specifications supplement had been issued covering these new engines and that the 426-cu.-in. engines were designed "for owner build-up," inferring that Chrysler would not actually be producing 426-cu.-in. engines but owners of 413-cu.-in. engines had authorization to add 13 cu. in. to their engines' displacements, presumably for use in competitive events.

The following year, 1963, the 426-cu.-in. displacement became a production reality with Plymouth and Dodge dropping their 413-cu.-in. high-performance engine offerings and replacing them with the 426-cu.-in. engines. Two single four-barrel carburetor versions were available: a 370-hp engine with a compression ratio of 11:1, and a 375-hp engine with a 13.5:1 compression ratio. Needless to say, the high compression ratios required premium octane fuel. Dual four-barrel carburetor Max Wedge versions of the 1963 426-cu.-in. engine are described in the following paragraphs. In 1964, a more "streetable" version was released, with a 10.3:1 compression ratio, hydraulic tappets, and a mildly aggressive 268-268-48 camshaft that had been in use for several years. Titled the Commando 426 by Plymouth and described by Dodge as a "426-cu.-in high-performance V-8 engine," the advertised output of the engine was 365 hp @ 4800 rpm and 470 lb-ft @ 3200 rpm. Available as an option throughout both car lines, the engine was a sight to behold under the hood, with its gleaming chrome-plated cylinder head covers and chrome air cleaner. After testing a 1964 Dodge Polara 500 equipped with the engine, Jim Wright, Technical Editor, wrote in the February 1964 issue of *Motor Trend*, " Performance was very satisfying, making the '426' package worth the extra cost...The engine has plenty of pulling power spread over a wide rpm range, which means instant go when you hit the throttle (no matter what speed you're traveling)." Its release was repeated for 1965, and starting in 1966, all wedge-head RB V-8 engines were built with a displacement of 440 cu. in.

For serious competitors, both at the drag strips and on the street, the big-power news for 1962 from Chrysler came late in the model year in the form of a new race-ready 413-cu.-in. RB V-8 engine, released for Plymouth and Dodge cars. This first of a series of engines that would become known as Max Wedge or often as Max Performance. It was the result of a sudden strong desire by Chrysler senior management to have the corporation become a serious supporter of its products in competitive events, in order to replace its existing poor street image among the nation's car-savvy young buyers with a winning, high-performance image. That desire quickly led to an intense dynamometer development program directed by an engineer with plenty of performance experience and a founding member of the Ramchargers drag racing club—Tom Hoover, now coordinator of racing programs for the Engineering Division.

The outcome of the dynamometer development was a big and tough 413-cu.-in. RB V-8 engine with an all-new cast aluminum intake manifold that mounted two four-barrel carburetors in a diagonal arrangement. With each carburetor located above a longitudinal plenum chamber contained inside the manifold, individual intake branches crossed the engine to feed the cylinders of the opposite bank of the cylinder block. Effectively the new intake manifold provided each cylinder with a 15-in. tuned runner fed by a plenum chamber that received its fuel/air mixture from a large-capacity carburetor—all fitting neatly in the space between the two cylinder heads. An internal balance tube connected the two plenums. Supplementing the new intake manifold were streamlined, high-rising exhaust manifolds, which, together with increases in the cylinder head port area, larger (1.88-in.) exhaust valves that required bore notches in the cylinder block, and a long-duration 300-300-75 valve event camshaft, raised the power of this competition engine to 410 hp @ 5200 rpm and 460 lb-ft torque @ 4400 rpm with an 11:1 compression ratio. An optional 13.5:1 compression ratio added 10 hp and 10 lb-ft torque. To handle the higher speed and higher power levels of the engine, a host of heavy-duty parts were used in the engine, including a shot-peened, hardened-journal crankshaft; tri-metal, full-grooved main bearings; Magnafluxed connecting rods; malleable iron valve rocker arms with adjusting screws and lock nuts; large-diameter hollow pushrods; flat-faced mechanical tappets; forged aluminum pistons; dual valve springs; deep-groove pulleys; and a small-diameter water pump impeller.

Plymouth gave the 1962 engine the name Super Stock 413 and Dodge the name 413 Ramcharger. However, Max Wedge was the name commonly used for the engine. There was no pretense about the engine being suitable for ordinary street use; the exhaust crossover passage had been stripped from the cylinder heads, and the intake manifold was totally devoid of any warm-up-assisting heat. A hand choke sufficed to start the engine.

Field results soon came in, and they were great! As reported by Ray Brock in the January 1963 issue of *Hot Rod Magazine*, "The new engine was a whopping success...records in Super Stock classes soon started tumbling at strips all over the country...Plymouth prestige never improved faster. The youth of America, those who are particularly interested in performance, had a sudden reversal of opinion about these cars. Overnight, Plymouth stockers jumped from also-ran to the top of the ladder."

To keep the engine out in front meant no letup in the dynamometer development effort, and upgrades were released whenever a package of changes was ready. In early 1963, the 413-cu.-in. Max Wedge (to use the common name) was replaced with a 426-cu.-in. size, a displacement that had been initiated for the 1962 Chrysler 300 engines. The 426-cu.-in. engines built for Plymouth sported a decal on the cylinder head covers that read "Super Stock 426." With the extra 13 cu. in., output ratings rose modestly to 415 hp @ 5600 rpm for the 11:1 compression ratio engine, and 425 hp @ 5600 rpm for the 13.5:1 version.

Late in the 1963 model year, additional changes were made to the 426-cu.-in. Max Wedge engine; the primary bores of the carbuertors were made the same 1-11/16 in. diameter as their secondary bores, giving more air capacity. Also, a machining cut was added to the combustion chambers of the cylinder heads near the intake valve seats to boost airflow past the valve heads. In addition, a new camshaft with 300-308-74 events lifting the valves 0.52 in. (versus 0.509 in. for the original camshaft) was released. The changes were significant enough to retitle the engines as Super Stock 426-II for Plymouth and as Ramcharger 426-A for Dodge, as shown on Fig. 9.18. Advertised outputs were left unchanged.

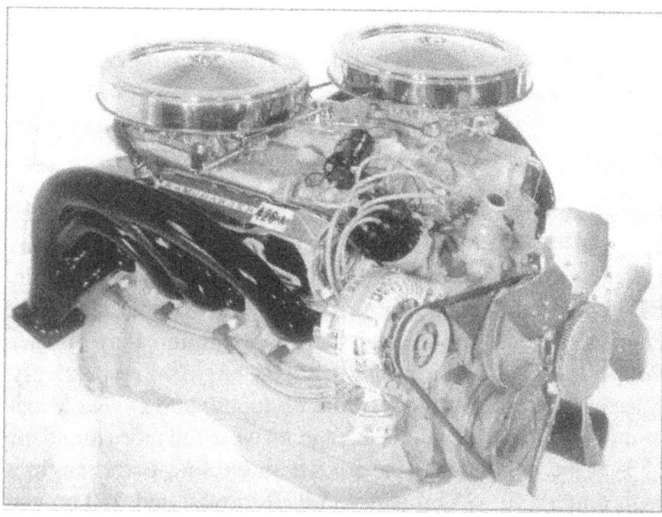

Fig. 9.18 *The 1963 Dodge Ramcharger 426-A Max Wedge RB V-8 engine. Below each carburetor is a plenum chamber that feeds the ram pipes leading the cylinders of the opposite bank.*

At the beginning of 1964, the appearance of the Max Wedge engine took a dramatic visual change with the arrival of the Super Stock 426-III. Gone were the high-rise, gracefully curved exhaust manifolds used since the beginning of the Max Wedges. Instead, a Medusa-like tangle of cast iron was attached to each side of the engine. Displayed on Fig. 9.19, these unusual exhaust manifolds had been introduced on the 1963 closed-track competition engines. Although strange in appearance, the new manifolds were functionally designed to provide all cylinders with individual, tuned, 21-in., equal-length branches. At the downstream end of the 21-in. branches, the passages were "Y'd" together in pairs and shaped to present two circular openings at the outlet flange of each manifold. From the outlet flange of each exhaust manifold, the exhaust gas flowed into a welded collector assembly made with two steel tubes that were side by side at its inlet flange and then "Y'd" into a single 3-1/2-in.-diameter downstream pipe. The three "Ys" were used in the exhaust system on each side of the car; two were inside each cast manifold, and one was in each collector assembly. This gave the system its name of "Tri-Y."

Fig. 9.19 *The 1964 Dodge Ramcharger 426-cu.-in. Max Wedge RB V-8 engine. Last year's exhaust manifolds were replaced with manifolds providing an equal-length branch for each cylinder and progressively joining the branches to a final single outlet on each side.*

With the running start of the Max Wedge engines and continual changes, the competition was kept far behind, and the final Super Stock drag race of many meets would find a 426 Max Wedge engine in each lane.

The year 1964 was the last year of these high-performing dual four-barrel Max Wedge engines. In mid-1964, a new and even mightier RB engine burst onto the competition scene and immediately became the center of high-performance attention and development. Chapter 10 describes this new engine with its all-conquering hemi heads.

However, for street use, the RB V-8 still had a lot to offer in terms of big power. Capitalizing on the extra cubic inches gained when the standard RB V-8 went up to 440 cu. in. in 1966 and on the improved A120 cylinder head developed for the 440 RB for 1967, a high-performance version, coded A134, was readied for 1967. Its larger exhaust valves (1.74 in.), longer-duration camshaft (268-284-46), free-flowing high-rise (somewhat similar to the 1962 Max Wedge) exhaust manifolds, oil pan windage tray on engines with manual transmissions, and larger-capacity four-barrel carburetor brought its rating up to 375 hp and 480 lb-ft torque. The compression ratio was unchanged from the 10.1:1 of the standard 440. Plymouth made the engine the base and only engine on its sporty GTX model shown on Fig. 9.20, naming it the Super Commando 440. Dodge gave it the title of Magnum 440 and made it the sole engine for its Coronet Charger R/T, with availability on other models as an option. Chrysler called it the Chrysler TNT and made it optional on most of its models. Dual exhausts and either a twin-snorkle or an unsilenced air cleaner, depending on the car model, were part of the vehicle installation. Without further significant changes, the engine continued through 1969. Then in 1970, its compression ratio was dropped to 9.7:1, followed by a further drop to 8.2:1 for the years 1971 through 1975. In 1975, its final year, this 440-cu.-in. high-performance engine was offered only as an option on Plymouth and Dodge police cars and was rated at a net 260 hp and 355 lb-ft torque.

Multiple carburetors for big power made another comeback in 1969 when a setup with triple two-barrel carburetors was developed for the 440-cu.-in. RB engine. Using a host of

Fig. 9.20 *The 1967 Plymouth GR2 Belvedere GTX two-door hardtop. Standard power was a high-performance 440-cu.-in. RB V-8 engine with a single four-barrel carburetor that delivered 375 hp.*

performance parts from the A134 engine, including the long-duration 268-284-46 camshaft but with an even higher 10.5:1 compression ratio and named the Magnum Six Pack by Dodge and the Super Commando 440-Six Barrel by Plymouth, the engine, shown on Fig. 9.21, had an output of 390 hp and 490 lb-ft torque. This triple two-barrel carburetor engine was last built in 1972, its apparently precipitous drop in output to 330 hp and 410 lb-ft torque for that year caused partly by its lower 10.3:1 compression ratio but mostly by the switch from gross to net ratings.

Fig. 9.21 *A view of the engine compartment of a 1969 car powered by a high-performance 440-cu.-in. RB V-8 engine with three two-barrel carburetors. Removal of the single oval-shaped air cleaner has allowed a close look at the triple carburetion.*

These big-cubic-inch and multiple-carburetor engine programs certainly deserved the accolades they received for their out-and-out performance capabilities, but the program that really rang up the sales was a modestly priced 1968 car—created by product planners Jack Smith, Paul Sturm, and Gordon Cherry—that could run with the big boys. The car was a sparsely outfitted intermediate Plymouth B-body named "Road Runner," with decals, beep-beep horn, and license approval of the name and cosmetics from Warner Brothers. The engine of the car was a 383-cu.-in. LB V-8 with a single four-barrel carburetor that was given far more punch than the standard-type four-barrel engine by loading it with many of the plus-power parts already developed for the 440-cu.-in. A134 engine previously described, including the aggressive 268-284-46 camshaft. Coded A139, the engine had few new unique parts, making it easy for the Trenton engine plant to build. Street recognition of this affordable, high-performance "muscle car" with its 335 hp and 425 lb-ft torque engine was quick, and the Road Runner's beep-beep horn became a national icon as sales took off. With the launch of the Plymouth Road Runner, Dodge introduced a companion car named Super Bee.

Showing its broad range of engineering excellence, in addition to its high-performance capabilities, the LB-RB V-8 engine had a formidable ability to do well in fuel economy competition, as exemplified by the annual Mobile Economy Run. Powering Chrysler model cars in the runs from 1960 through 1968, the engine, in displacements of 361, 413, or 440 cu. in. depending on year, won its class in eight of the nine runs. Winning models included the Chrysler Newport, Chrysler New Yorker, Chrysler 300, and Imperial.

As noted, 1978 was the last year of production of the raised-deck RB version of the Chrysler big-block V-8. Production of the low-deck LB version ended at the same time, concluding twenty-one years of production of this rugged and versatile big-block engine—and with it, twenty-one years of engineering excitement on the dynamometer, on the tracks, at the drag strips, and on the street.

Table 9.1 shows the specifications of the LB V-8 engine. Table 9.2 shows the specifications of the RB V-8 engine.

Small-Block Wedge-Head—The LA V-8 Engine

The in-line valve, wedge-head LA engine was born of necessity. Despite heroic design efforts, the poly-head A-engine simply could not be revised enough to fit in the engine compartment of the compact Valiant and Lancer A-body car that had been created without any provision for V-8 power but by 1964 needed a V-8 to broaden its market.[7] There was never any question but that the block structure of the existing poly-head A-engine had to be the starting point of the new engine to make the Valiant and Lancer V-8 affordable in terms of tooling cost. Assigned the engineering code A828, the design quickly took form on the drafting table. A wedge-shaped combustion chamber, in-line valves, and single rocker shaft were basics to the new cylinder head that maintained the cylinder spacing and head bolt pattern of the A-engine. Deviating from the low spark-plug location of the big-block LB-RB engines, the A828 spark plugs were raised so they would be closer to the center of the combustion chamber and would be accessible from above the exhaust manifold runner rather than from below it, as was the case with the LB and RB designs. Although fitting the engine into the existing engine compartment was the first priority for the design, an almost equal priority was to have the engine achieve a much lower weight than the A-engine. The importance of this second priority was reflected in the LA designator given to the engine, which stood for "Lightweight A." The in-line valve,

TABLE 9.1
SPECIFICATIONS FOR THE WEDGE-HEAD LB V-8 ENGINE

Model Year	1958	1958	1958	1958
Car line	DeSoto	Plymouth	Dodge, DeSoto	Dodge
Model	Firesweep	Golden Commando	D500, Fireflite	Fuel Injection
Displacement (cu. in.)	350	350	361	361
Bore	4.06	4.06	4.125	4.125
Stroke	3.375	3.375	3.375	3.375
Compression ratio	10.0:1	10.0:1	10.0;1	10.0;1
BHP @ rpm	280 @ 4600	305 @ 5000	305 @ 4600	333 @ 4800
Torque (lb-ft) @ rpm	380 @ 2400	370 @ 3600	400 @ 2800	400 @ 3600
Combustion chamber	Wedge	Wedge	Wedge	Wedge
Cyl. block mat'l.—skirt type	Cast iron—deep	Cast iron—deep	Cast iron—deep	Cast iron—deep
Cyl. block deck height	9.98	9.98	9.98	9.98
Cyl. block length	23.46	23.46	23.46	23.46
Cyl. bore spacing	4.8	4.8	4.8	4.8
Ck/shaft—no. of main brgs.	5	5	5	5
Ck/shaft main journal dia.	2.625	2.625	2.625	2.625
Crankpin journal dia.	2.375	2.375	2.375	2.375
Piston mat'l.—no. of rings	Aluminum—3	Aluminum—3	Aluminum—3	Aluminum—3
Connecting rod centers	6.358	6.358	6.358	6.358
Valve system	OHV—single rocker shaft	OHV—single rocker shaft	OHV—single rocker shaft	OHV—single rocker shaft
Valve events	252-252-30	260-260-42	252-252-30	260-260-42
Camshaft drive	Chain	Chain	Chain	Chain
Valve head dia.	Intake 1.95; exhaust 1.6	Intake 1.95; exhaust 1.6	Intake 1.95; exhaust 1.6	Intake 1.95; exhaust 1.6
Valve lash adjustment	Hydraulic tappet	Hydraulic tappet	Hydraulic tappet	Hydraulic tappet
Oil pump/distributor drive	Diagonal at front of block	Diagonal at front of block	Diagonal at front of block	Diagonal at front of block
Fuel system	2-barrel carburetor	Tandem 4-barrel carbs	2-barrel carburetor	MPI fuel injection

(Continued on next page)

TABLE 9.1
(Continued)

Model Year	1959	1960	1960	1968
Car line	DeSoto	Plymouth, Dodge	Plymouth, Dodge, DeSoto	Plymouth, Dodge
Model	MS2 Firedome	Plymouth: Sonoramic Commando Dodge: D500 Ram Induction (opt. Dart Phoenix)	Plymouth: Sonoramic Golden Commando 460 Dodge: D500 Ram Induction (opt. Matador and Polara) DeSoto: opt. on Adventurer	Plymouth Road Runner Dodge Super Bee
Displacement (cu. in.)	383	361	383	383
Bore	4.25	4.125	4.25	4.25
Stroke	3.375	3.375	3.375	3.375
Compression ratio	10.0:1	10.0:1	10.0;1	10.0;1
BHP @ rpm	305 @ 4600	310 @ 4800	330 @ 4800 (std. ca/s)	335 @ 4800
Torque (lb-ft) @ rpm	410 @ 2400	435 @ 2800	460 @ 2800 (std. ca/s)	425 @ 2800
Combustion chamber	Wedge	Wedge	Wedge	Wedge
Cyl. block mat'l.—skirt type	Cast iron—deep	Cast iron—deep	Cast iron—deep	Cast iron—deep
Cyl. block deck height	9.98	9.98	9.98	9.98
Cyl. block length	23.46	23.46	23.46	23.46
Cyl. bore spacing	4.8	4.8	4.8	4.8
Ck/shaft—no. of main brgs.	5	5	5	5
Ck/shaft main journal dia.	2.625	2.625	2.625	2.625
Crankpin journal dia.	2.375	2.375	2.375	2.375
Piston mat'l.—no. of rings	Aluminum—3	Aluminum—3	Aluminum—3	Aluminum—3
Connecting rod centers	6.358	6.358	6.358	6.358
Valve system	OHV—single rocker shaft	OHV—single rocker shaft	OHV—single rocker shaft	OHV—single rocker shaft
Valve events	252-252-30	268-268-48	Std. hyd.: 268-268-48 Opt. mech.: 284-284-55	268-284-46
Camshaft drive	Chain	Chain	Chain	Chain
Valve head dia.	Intake 2.08; exhaust 1.6	Intake 1.95; exhaust 1.6	Intake 2.08; exhaust 1.6	Intake 2.08; exhaust 1.74
Valve lash adjustment	Hydraulic tappet	Hydraulic tappet	Std.: hyd. tappet Opt.: mech. tappet	Hydraulic tappet
Oil pump/distributor drive	Diagonal at front of block	Diagonal at front of block	Diagonal at front of block	Diagonal at front of block
Fuel system	2-barrel carburetor	2 × 4-barrel Long cross-ram	2 × 4-barrel Long cross-ram	4-barrel carburetor

(Continued on next page)

TABLE 9.1
(Continued)

Model Year	1972	1977–1978
Car line	Plymouth, Dodge, Chrysler	Plymouth, Dodge, Chrysler
Engine code	E63	E64
Displacement (cu. in.)	400	400
Bore	4.34	4.34
Stroke	3.375	3.375
Compression ratio	8.2:1	8.2:1
BHP (SAE net) @ rpm	190 @ 4400	190 @ 3600
Torque lb-ft (SAE net) @ rpm	310 @ 2400	305 @ 3200
Combustion chamber	Wedge	Wedge
Cyl. block mat'l.—skirt type	Cast iron—deep	Cast iron—deep
Cyl. block deck height	9.98	9.98
Cyl. block length	23.46	23.46
Cyl. bore spacing	4.8	4.8
No. of main brgs.—ck/s mat'l.	5—nodular cast iron	5—nodular cast iron
Ck/shaft main journal dia.	2.625	2.625
Crankpin journal dia.	2.375	2.375
Piston mat'l.—no. of rings	Aluminum—3	Aluminum—3
Connecting rod centers	6.358	6.358
Valve system	OHV—single rocker shaft	OHV—single rocker shaft
Valve events	260-268-38	260-268-38
Camshaft drive	Chain	Chain
Valve head dia.	Intake 2.08; exhaust 1.74	Intake 2.08; exhaust 1.74
Valve lash adjustment	Hydraulic tappet	Hydraulic tappet
Oil pump/distributor drive	Diagonal at front of block	Diagonal at front of block
Fuel system	2-barrel carburetor	4-barrel carburetor

TABLE 9.2
SPECIFICATIONS FOR THE WEDGE-HEAD RB V-8 ENGINE

Model Year	1959	1959	1959	1960
Car line	Chrysler	Chrysler	Chrysler	Chrysler
Model	Windsor MC1	New Yorker MC3; Imperial MY1	300E	300F
Displacement (cu. in.)	383	413	413	413
Bore	4.031	4.1875	4.1875	4.1875
Stroke	3.75	3.75	3.75	3.75
Compression ratio	10.0:1	10.0:1	10.0:1	10.1:1
BHP @ rpm	305 @ 4600	350 @ 4600	380 @ 5000	375 @ 5000
Torque (lb-ft) @ rpm	410 @ 2400	470 @ 2800	450 @ 3600	495 @ 2800
Combustion chamber	Wedge	Wedge	Wedge	Wedge
Cyl. block mat'l.—skirt type	Cast iron—deep	Cast iron—deep	Cast iron—deep	Cast iron—deep
Cyl. block deck height	10.725	10.725	10.725	10.725
Cyl. block length	23.46	23.46	23.46	23.46
Cyl. bore spacing	4.8	4.8	4.8	4.8
No. of main brgs.—ck/s mat'l.	5—forged steel	5—forged steel	5—forged steel	5—forged steel
Ck/shaft main journal dia.	2.75	2.75	2.75	2.75
Crankpin journal dia.	2.375	2.375	2.375	2.375
Piston mat'l.—no. of rings	Aluminum—3	Aluminum—3	Aluminum—3	Aluminum—3
Connecting rod centers	6.768	6.768	6.768	6.768
Valve system	OHV—single rocker shaft	OHV—single rocker shaft	OHV—single rocker shaft	OHV—single rocker shaft
Valve events	252-252-30	252-252-30	260-260-42	268-268-48
Camshaft drive	Chain	Chain	Chain	Chain
Valve head dia.	Intake 1.95; exhaust 1.6	Intake 2.08; exhaust 1.74	Intake 2.08; exhaust 1.74	Intake 2.08; exhaust 1.6
Valve lash adjustment	Hydraulic tappet	Hydraulic tappet	Hydraulic tappet	Hydraulic tappet
Oil pump/distributor drive	Diagonal at front of block	Diagonal at front of block	Diagonal at front of block	Diagonal at front of block
Fuel system	2-barrel carburetor	4-barrel carburetor	Tandem 4-barrel carbs	2 × 4-barrel Long cross-ram

(Continued on next page)

TABLE 9.2
(Continued)

Model Year	1962	1962	1963	1964
Car line	Chrysler	Plymouth, Dodge	Plymouth; Dodge; Chrysler	Chrysler
Model	300H	Plymouth: Super Stock 413 Dodge: 413 Ramcharger Max Wedge	Plymouth: Super Stock 426-II Dodge: Ramcharger 426-A Chrysler: 300 (opt.)	300K
Displacement (cu. in.)	413	413	426	413
Bore	4.1875	4.1875	4.25	4.1875
Stroke	3.75	3.75	3.75	3.75
Compression ratio	10.1:1	11.0:1 std. 13.5:1 opt.	11.0:1 std. 13.5:1 opt.	10.0:1
BHP @ rpm	380 @ 5200	410 @ 5200	Std. 415 Opt. 425 @ 5600	360 @ 4800
Torque (lb-ft) @ rpm	450 @ 3600	460 @ 4400	Std. 470 Opt. 480 @ 4400	470 @ 3200
Combustion chamber	Wedge	Wedge	Wedge	Wedge
Cyl. block mat'l.—skirt type	Cast iron—deep	Cast iron—deep	Cast iron—deep	Cast iron—deep
Cyl. block deck height	10.725	10.725	10.725	10.725
Cyl. block length	23.46	23.46	23.46	23.46
Cyl. bore spacing	4.8	4.8	4.8	4.8
No. of main brgs.—ck/s mat'l.	5—forged steel	5—forged steel	5—forged steel	5—forged steel
Ck/shaft main journal dia.	2.75	2.75	2.75	2.75
Crankpin journal dia.	2.375	2.375	2.375	2.375
Piston mat'l.—no. of rings	Aluminum—3	Forged aluminum—3	Forged aluminum—3	Aluminum—3
Connecting rod centers	6.768	6.768	6.768	6.768
Valve system	OHV—single rocker shaft	OHV—single rocker shaft	OHV—single rocker shaft	OHV—single rocker shaft
Valve events	268-268-48	284-284-55	Std. 300-300-82 Opt. 300-308-74	268-268-48
Camshaft drive	Chain	Chain	Chain	Chain
Valve head dia.	Intake 2.08; exhaust 1.6	Intake 2.08; exhaust 1.88	Intake 2.08; exhaust 1.88	Intake 2.08; exhaust 1.6
Valve lash adjustment	Mech.—rocker arm screw	Mech.—rocker arm screw	Mech.—rocker arm screw	Hydraulic tappet
Oil pump/distributor drive	Diagonal at front of block	Diagonal at front of block	Diagonal at front of block	Diagonal at front of block
Fuel system	Tandem 4-barrel carbs	2 × 4-barrel dual-plenum short ram	2 × 4-barrel dual-plenum short ram	4-barrel carburetor

(Continued on next page)

TABLE 9.2
(Continued)

Model Year	1966	1967	1977	1978
Car line	Dodge, Chrysler, Imperial	Plymouth, Dodge, Chrysler	Plymouth, Dodge, Chrysler	Chrysler
Model	Dodge: Monaco 500 Chrysler: New Yorker Imperial: Crown and LeBaron	Plymouth: Commando 440 Dodge: Magnum Chrysler: TNT	Plymouth: Gran Fury (opt.) Dodge: Royal Monaco (opt.) Chrysler: New Yorker, Town & Country	Newport (opt.) New Yorker Brougham (opt.)
Engine code			E85	E85
Displacement (cu. in.)	440	440	440	440
Bore	4.32	4.32	4.32	4.32
Stroke	3.75	3.75	3.75	3.75
Compression ratio	10.0:1	10.0:1	8.2:1	8.2:1
BHP @ rpm	350 @ 4400	375 @ 4600	195 @ 3600*	195 @ 3600
Torque (lb-ft) @ rpm	480 @ 2800	480 @ 3200	320 @ 2000*	320 @ 2000
Combustion chamber	Wedge	Wedge	Wedge	Wedge
Cyl. block mat'l.—skirt type	Cast iron—deep	Cast iron—deep	Cast iron—deep	Cast iron—deep
Cyl. block deck height	10.725	10.725	10.725	10.725
Cyl. block length	23.46	23.46	23.46	23.46
Cyl. bore spacing	4.8	4.8	4.8	4.8
No. of main brgs.—ck/s mat'l.	5—forged steel	5—forged steel	5—nodular cast iron	5—nodular cast iron
Ck/shaft main journal dia.	2.75	2.75	2.75	2.75
Crankpin journal dia.	2.375	2.375	2.375	2.375
Piston mat'l.—no. of rings	Aluminum—3	Forged aluminum—3	Forged aluminum—3	Aluminum—3
Connecting rod centers	6.768	6.768	6.768	6.768
Valve system	OHV—single rocker shaft	OHV—single rocker shaft	OHV—single rocker shaft	OHV—single rocker shaft
Valve events	256-260-32	284-284-55	260-268-38	260-268-38
Camshaft drive	Chain	Chain	Chain	Chain
Valve head dia.	Intake 2.08; exhaust 1.6	Intake 2.08; exhaust 1.74	Intake 2.08; exhaust 1.74	Intake 2.08; exhaust 1.74
Valve lash adjustment	Hydraulic tappet	Hydraulic tappet	Hydraulic tappet	Hydraulic tappet
Oil pump/distributor drive	Diagonal at front of block	Diagonal at front of block	Diagonal at front of block	Diagonal at front of block
Fuel system	4-barrel carburetor	4-barrel carburetor	4-barrel carburetor	4-barrel carburetor

* Starting in 1972, SAE J245 net engine ratings replaced former gross engine ratings.

reduced-cross-section-area cylinder head was a big step in that direction. Another big step was a new coring process installed at the Chrysler Indianapolis Foundry, where the cylinder block castings were made. This new process greatly improved the core dimensional accuracy and permitted thinner walls throughout the block. A third step was a lighter, single-plane intake manifold designed primarily to reduce the height of the engine. The total weight reduction efforts were rewarded with an engine assembly weighing 50 lb less than the A-engine.

Figures 9.22 and 9.23 show a cross section and a right-front view of the engine, respectively. Two items of interest can be seen in the cross section: the substantial angle between the tappet bores and the pushrod centerlines, which was due to the retention of the same tappet bore machining as used on the A-engine; and the long arc-shaped extension of the left exhaust manifold that was needed for clearance to the steering column.

An engine size of 273 cu. in., with a bore of 3.625 in. and a stroke of 3.31 in., was determined to be the best selection as a base V-8 for the 1964 Plymouth Valiant and Dodge Dart compact cars. Built with a two-barrel carburetor, solid tappets, a self-locking screw in each malleable iron rocker arm for adjusting valve lash, a camshaft with valve events of 240-240-16, and a compression ratio of 8.8:1, the engine produced 180 hp and 260 lb-ft torque. During SAE Automobile Week 1964, I presented a paper that gave the design and development details about this engine.[8] Praise for the move to put V-8 power in Valiants and Darts was expressed by Ray Brock in an article titled "Chrysler's New V-8—Tailored for a Compact," published in the December 1963 issue of *Hot Rod* magazine, with the statement, "The most rewarding thing that owners of Dart and Valiant V-8's will notice is the smooth, quiet operation at high speeds. Already better than most compacts when subjected to sustained high speed driving, the 273 V-8 will transform Chrysler's compacts into genuine turnpike champions."

Fig. 9.23 *A right-front view of the 1964 Valiant 273-cu.-in. LA V-8 engine. Good spark plug accessibility was a feature of this lightweight V-8 engine.*

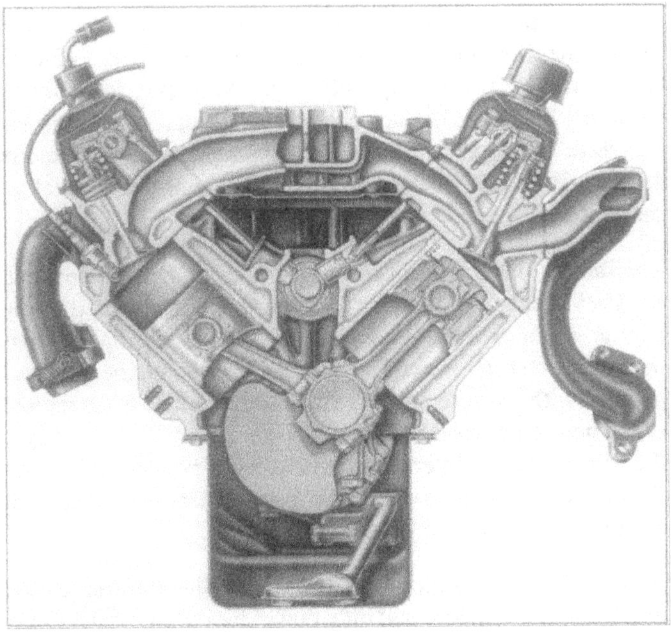

Fig. 9.22 *A cross section of the 1964 Valiant 273-cu.-in. LA V-8 engine. The use of a wedge combustion chamber and in-line valves resulted in a compact cylinder head. The arc-shaped exhaust manifold extension on the right was designed to clear the steering column.*

Although most of the performance engine work at this time was focused on the big-block LB-RB engines, especially the 426-cu.-in. size, this small-block LA V-8 attracted the attention of the compact car planners who wanted to give the car extra zip above its base V-8. A lively four-barrel 235-hp and 280 lb-ft torque version of the 273-cu.-in. engine, coded A861 and shown on Fig. 9.24, with a 10.5:1 compression ratio, an unsilenced air cleaner, a longer-duration 248-248-26 camshaft, and black-crackle-painted cylinder head covers was released as an option for the 1965 Dodge Dart. Also released for the 1965 Plymouth Valiant and Plymouth Barracuda, the engine was named the Commando 273 for those applications.

In 1965, the use of the two-barrel 273-cu.-in. LA engine spread beyond the compacts, becoming the base V-8 in Plymouth Belvederes and Dodge Coronets.

The 318-cu.-in. A-engine that was built alongside the 273-cu.-in. LA engine from 1964 through 1966 was converted to the LA design in 1967, simplifying plant complexity and gaining weight benefits in all applications of the engine. The 230-hp and 340 lb-ft torque rating of the 1967 318-cu.-in. wedge-head LA engine was exactly the same as that of the heavier 1966 poly-head A-engine,

Fig. 9.24 *A three-quarter view of the 1964 Dodge Dart 273-cu.-in. LA V-8 engine with a four-barrel carburetor. The black-crackle cylinder head covers topped with finned, extruded aluminum panels were a visual identifier of this small-block high-performance engine.*

Fig. 9.25 *The 1968 Dodge Dart GTS two-door hardtop. Standard power was a new 275-hp 340-cu.-in. four-barrel LA V-8 engine loaded with performance features.*

showing the relative equality of these two types of combustion chambers. In a cost-reduction move, the crankshaft material of both the 273-cu.-in. LA and the 318-cu.-in. LA engines was changed from forged steel to nodular cast iron in 1967. For quality improvement and more consistent emission control, the tappets of the 318-cu.-in. LA engine were upgraded from mechanical to hydraulic during the same year, with the same change being made to the 273-cu.-in. LA engine a year later. For use with the hydraulic tappets, the valve rocker arms were changed from malleable iron castings to steel stampings, similar in concept to the rocker arms used on the LB and RB big-block engines.

A feisty new version of the LA engine was created in 1968 to make the Plymouth Barracuda Formula "S" model and the Dodge Dart GTS, similar to the two-door hardtop shown on Fig. 9.25, genuinely hot performance cars. Guided by engine performance expert and engineering race coordinator Tom Hoover, the new engine, coded A105, was analyzed for performance and strength improvement leading to a bore size increase to 4.04 in. This was a daring thing to do, considering the cylinders were only 4.46 in. apart, which moved the displacement up to 340 cu. in. using the 3.31-in. stroke of the 318-cu.-in. engine. Then the valves were made as big as possible—2.02 in. for the intake, and 1.60 in. for the exhaust. This change required the spark plug to be moved 0.08 in. outboard. Thicker bulkheads and higher main bearing caps considerably strengthened the bottom end of the cylinder block structure, while the connecting rods were beefed up with added forged steel in the critical shank and big-end areas. A long-duration 268-276-44 camshaft on the automatic transmission engines and even longer 276-284-52 camshaft for the manual transmission engines emphasized top-end power. The mounting flange of the intake manifold for the four-barrel carburetor was raised one inch over the prior single-plane manifold, providing enough room for a two-level manifold with larger runner and branch areas than in the past. New free-flowing, larger-area exhaust manifolds reduced exhaust gas backpressure, while power loss to the cooling fan at high engine speeds was lowered by the use of a viscous fan drive. Four of the main bearing cap attaching screws were drilled and tapped for attachment of a windage tray positioned between the connecting rod paths and the surface of the oil in the sump, reducing friction horsepower at high speeds. Topped off with a circular, non-silenced air cleaner, an outside view of the engine, rated at 275 hp and 340 lb-ft torque, can be seen on Fig. 9.26. An SAE paper, co-authored by James W. Dean, Manager of Engine Design, and Harry M. Casebeer, Manager of Engine Development, was presented in January 1968 and gave details about the design and development of the engine.[9]

The January 1969 issue of *Motor Trend* carried an article about the 1969 'Cuda 340 compact fastback that was powered by the engine. In the article, author Eric Dahlquist wrote, "The 340 is something else, anyway. The nut the Chrysler engine guys had to crack with it was to duplicate or better the explosive potential of the now deceased 350 horse, 327-c.i.d. Chevy—which they did, handsomely. Interacting with a 3400-pound car like the Barracuda, this jells as one of the best accelerating stockers of all time and with good balance, yet."

An even hotter version of the 340 LA engine was introduced in 1970 as the standard and only power for two special A-body cars: the Plymouth AAR 'Cuda, shown on Fig. 9.27, and the Dodge Challenger T/A. Both cars were intended to be contenders in Trans Am racing events. The initials "AAR" stood for international racing star Dan Gurney's All-American Racers team.[10] Of course, the "T/A" stood for Trans Am. This hotter version, coded A340, sported three two-barrel carburetors,

Fig. 9.26 *The 1970 27-hp 340-cu.-in. LA V-8 engine with the air cleaner lifted to show the four-barrel carburetor.*

Fig. 9.27 *The 1970 Plymouth AAR 'Cuda. Under the hood with its air scoop and racing-type hood pins was a 290-hp 340-cu.-in. LA V-8 engine topped with triple two-barrel carburetors.*

as illustrated on Fig. 9.28, in place of the single four-barrel carburetor. Cylinder heads and valve gear were modified for improved air flow and durability. Iron was added to the cylinder block bulkheads to accommodate four-bolt main bearing caps. With these changes, the 340-cu.-in. "Six Pack" put 290 hp and 345 lb-ft torque under the hood. Production of these two special high-performance cars and the six-barrel engine ceased at the end of 1971. The 340-cu.-in. LA engine with its four-barrel

Fig. 9.28 *The three two-barrel carburetors of the 340 Six-Pack take up all the space available on top of the small-block LA V-8 engine.*

carburetor was dropped at the end of 1973. By then, the compression ratio was down to 8.5:1, and the output was down to net 240 hp and 295 lb-ft torque.

In addition to its well-recognized capabilities for high performance, the LA V-8 engine had the ability to do well in terms of fuel economy. Entered in every Mobil Economy Run from 1964 to the last-held run in 1968, the engine brought home a host of top prizes. With 273-cu.-in. LA V-8s providing the power, a Dodge Coronet came in first and a Plymouth Belvedere II placed second in Class D in 1965. In 1966, a Dodge Dart equipped with a 273-cu-in. LA V-8 engine took the Class B top prize, while the Class D top prize went to a Dodge Coronet Deluxe equipped with a similar engine. In 1967, a Plymouth Barracuda and Plymouth Belvedere II were awarded first places in Classes B and D, respectively, while a Dodge Dart earned second place in Class B. The three cars all had 273-cu.-in. LA V-8 engines. The year 1968 gave even better results—two firsts and two seconds for LA V-8s. A Plymouth Barracuda with a 318-cu.-in. LA V-8 engine received first place for Class B, and a 273-cu.-in. LA V-8 powered Dodge Coronet took first place for Class D. Second place in Class D went to a Plymouth Belvedere with a 273-cu.-in. LA V-8, while a 318-cu.-in. LA V-8 provided the power for a Plymouth Fury III to take second place in Class E.

The fourth, last, and largest-displacement version of the LA V-8 wedge-head engine entered production at the start of 1971—the 360-cu.-in. LA engine. Initially built only at the Chrysler Canada Windsor engine plant, the engine was created to fill the displacement gap between the 318-cu.-in. LA engine and the 383-cu.-in. LB engine with an engine that would be much lighter and less expensive to build than any reinstatement of the original big-block 361-cu.-in. LB engine. At the beginning of the design effort, the task appeared daunting because a stroke longer than the 3.31-in. stroke in use since 1956 would be required to obtain 360 cu. in. But raising the block decks to accommodate a longer stroke, as had been done in years past, was ruled out because changing the highly automated block machining line, engine assembly line, and engine test stands to accommodate a second deck height would make the tooling costs far too expensive. After zeroing in on a 4.00-in. bore and a 3.58-in. stroke and giving a careful look at what could be done without raising the decks, a plan emerged to use the same connecting rod as the 318-cu.-in. engine, reduce the height of the piston from the piston pin to the top of the piston, and reduce the radius of the crankshaft counterweights to make clearance for the piston at the bottom of the stroke. The engine had all the necessary internal clearances, but there was a big problem—the engine was out of balance. The solution to that came with the addition of offset weights at both ends of the crankshaft assembly. The torque converter flex plate and flywheel attached to the rear of the crankshaft both were redesigned to have offset weights; at the front of the crankshaft, the vibration damper hub had an offset weight added. This resulted in the engine as an assembly being in complete balance.

The 360-cu.-in. LA engine, seen in the cross section on Fig. 9.29 and from the outside on Fig. 9.30, may appear similar to the 318-cu.-in. LA engine. However, in addition to the bore and stroke differences, it differed in many details, including the use of the same spark plug location as the 340-cu.-in. LA engine previously described, and even in the location of the core holes in the sides of the cylinder block. On the 318-cu.-in. engine, only the two center core holes were located halfway between the cylinder bores. On the 360- and 340-cu.-in. engines, all six core plugs were located halfway between the bores so that the steel shot used for cleaning out water jacket core sand at the foundry could be aimed at the narrow gaps between the bores, a haven for packed-in sand. With nodular cast iron specified as the crankshaft material, the longer 3.58-in. stroke of the 360-cu.-in. engine required an increase in the diameter of the main journals to maintain acceptable torsional and bending strength levels for the crankshaft. The new main journal diameter was 2.81 in., which caused the main bearing caps to be made higher and the distance between the two cap-attaching bolts to be increased by 0.31 in. In its 1971 launch year, the 360-cu.-in. LA engine, equipped with a two-barrel carburetor, 8.7:1 compression

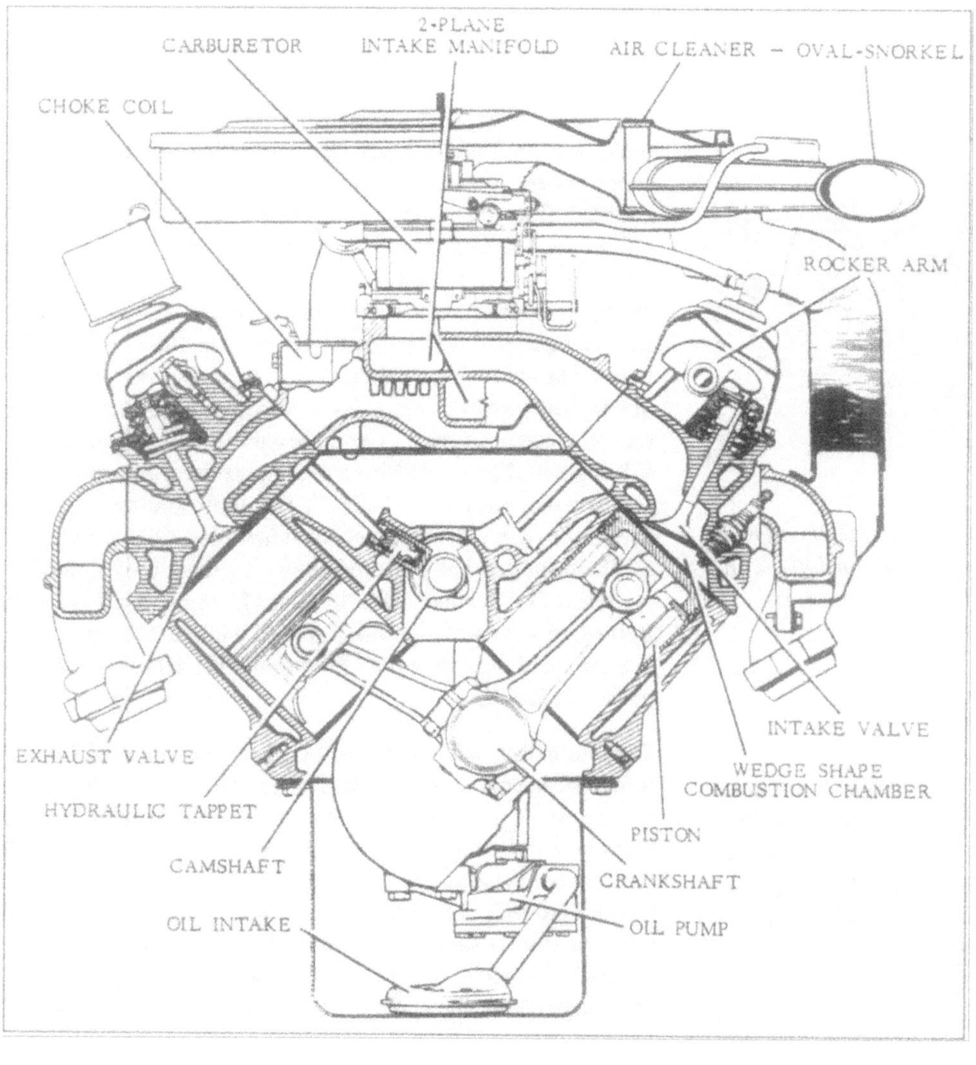

Fig. 9.29 *A cross section of the 1971 360-cu.-in. LA V-8 engine. Although the general design was similar to the 318-cu.-in. LA V-8 engine, there were many differences in the details. The provision for heated inlet air shown was one of its emission control features.*

Fig. 9.30 *A left-front view of the 1971 360-cu.-in. LA V-8 engine.*

ratio, and a 252-256-32 camshaft, had an output of 255 hp and 360 lb-ft torque. It was installed as an optional V-8 on the 1971 Plymouth Fury and Dodge Polara models, such as the four-door hardtop shown on Fig. 9.31, and as a standard engine on the Chrysler Newport Royal, replacing the Newport's customary 383-cu.-in. LB big-block engine in what was a smaller-engine trial balloon for the Chrysler lineup.

Subsequent Changes

The attempt to put a small-block V-8 in the Chrysler Newport did not work; buyers expected the big-block engine. Thus, shortly after the start of 1972, the 360-cu.-in. LA V-8 was pulled, and the 383-cu.-in. LB V-8 engine was reinstated.

Two variations of the 360-cu.-in. LA V-8 were added in 1974: a standard camshaft four-barrel engine for use by Dodge, and a high-performance four-barrel engine with a longer-duration 268-276-44 camshaft, borrowed from the now-departed 340-cu.-in. LA engine. With the compression ratio of all versions down to 8.4:1, the Dodge-only engine with the standard camshaft was rated at a net 200 hp and 290 lb-ft torque; the high-performance engine, used by both Plymouth and Dodge, produced a net 245 hp and 320 lb-ft torque.

In 1975, with growing emphasis on fuel economy triggered by widespread gasoline shortages caused by OPEC oil production restrictions starting in 1973, the LA V-8 engine was reintroduced to the Chrysler line, with a 360-cu.-in. two-barrel LA V-8 released as standard on the popular Cordoba (shown on Fig. 9.32) and Newport models. An even smaller 318-cu.-in. two-barrel LA V-8 was made available as an economy option. On the 1975 Chrysler Newport Custom, the two-barrel 360-cu.-in. LA became an economy option. For the California market with its more severe emission requirements, a standard camshaft, single exhaust, four-barrel 360-cu.-in. LA V-8 having a net output of 190 hp and 270 lb-ft torque was released for various Plymouth, Dodge, and Chrysler models, with full emission controls, including a catalyst and an air pump. The four-barrel carburetor with its primary bores smaller than the bores of a two-barrel carburetor was more amenable than the two-barrel to meeting the California emission levels. The long-duration camshaft, four-barrel, non-California 360-cu.-in. LA continued powering 1975 Plymouth Duster 360s and Dodge Dart 360 Sports with its net 230 hp and 300 lb-ft torque. This lineup stayed more or less intact through 1979, when the 360-cu.-in. LA V-8 engine was no longer sold in California. At the end of the next year, 1980, production of the 360-cu.-in. LA engine for passenger cars ceased altogether, leaving the 318-cu.-in. LA to soldier onward. However, the 360-cu.-in. LA was a popular truck engine, and its use as a truck engine, described in Chapter 15, continued beyond the 1998 purview of this book.

Fig. 9.31 *The 1971 Dodge Polara four-door hardtop. Optional power for the car was a 255-hp, 360-cu.-in. LA V-8 engine with a two-barrel carburetor.*

Fig. 9.32 *The 1975 Chrysler Cordoba two-door sedan. A 180-hp, 360-cu.-in. LA V-8 with a two-barrel carburetor was its standard engine in all states except California, where a four-barrel version was furnished.*

In 1981, a four-barrel, standard camshaft, net 165 hp and 240 lb-ft torque version of the 318-cu.-in. LA V-8 engine entered production for both California and federal cars. The year 1981 also saw the introduction of an electronic fuel injection system, shown on Fig. 9.33, which was designed and developed in-house by the Chrysler Engineering Engine Electrical group for the 318-cu.-in. LA V-8 engine. Using a number of sensors and an engine control computer to establish flow requirements, fuel was injected into the intake air by a spray bar in a two-throat throttle body. With application limited to a reintroduced luxury nameplate, the distinctively styled Chrysler Imperial pictured on Fig. 9.34, the engine remained in production through 1983, when both the car and fuel-injected engine were discontinued. An evaluation of the car by *Motor Trend* writer Jim McCraw, which appeared in the October 1980 issue in an article titled "The Imperial Strikes Back" reported, "The Imperial is no rocket ship—it was never intended to be—but its engine with EFI is easily the quietest, smoothest small-block the corporation has built, with a combination of response, economy and power that's right for this class of car."

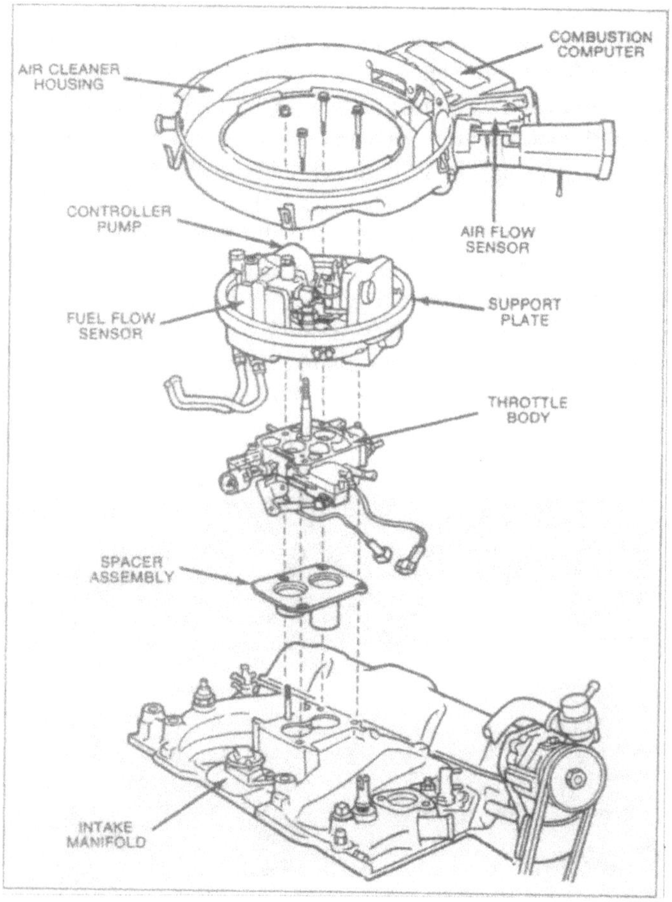

Fig. 9.33 *An exploded view of the 1981 electronic fuel injection system developed by the Engine Electrical department for the 318-cu.-in. LA V-8 engine.*

Fig. 9.34 *The 1981 Imperial two-door hardtop coupe. A 140-hp, 318-cu.-in. LA V-8 engine equipped with Chrysler-developed electronic throttle-body fuel injection was its standard power.*

In 1985, the two-barrel-carburetor 318-cu.-in. LA V-8, standard on the remaining rear-wheel-drive Dodge Diplomat, Plymouth Gran Fury, and Chrysler Fifth Avenue and Newport, received an upgrade that included roller tappets and a slightly higher compression ratio (9.0:1 versus 8.7:1), bringing its power and torque up to 140 hp and 265 lb-ft from the prior year's 130 hp and 235 lb-ft. This was the last significant change to the passenger car engine. Chrysler Corporation finally stopped building conventional driveline rear-wheel-drive cars at the end of 1989. With that move, passenger car use of the 318-cu.-in. (now denoted 5.2-liter) LA V-8 engines ended. Truck use of the 5.2-liter LA V-8 continued, as will be related in Chapter 15.

Table 9.3 shows the specifications of the LA V-8 engines described in the preceding paragraphs.

Chart 9.1 displays time lines for the LB, RB, and LA engines described in this chapter.

The LA Engine Goes Racing

In the late 1960s and early 1970s, with the 426-cu.-in. big-block RB Hemi® V-8 (described in Chapter 10) made increasingly noncompetitive on closed-circuit races by displacement and carburetor restrictor plate handicaps, design and development efforts were directed toward preparing the small-block LA V-8 engine for competition use.

The potential for considerable performance gains had already been explored to some extent with the work that led to the release of the 1968 340-cu.-in. LA high-performance production engine previously described. Preceded by lengthy air-flow tests on intake and exhaust port models, a design for a high-horsepower, competition-use cylinder head incorporating the large, oval-shaped intake port shown on the upper view of Fig. 9.35 and low-restriction exhaust port appearing on the lower view of Fig. 9.35 was finalized in 1969. Initially, the exhaust valve guides were integral, but a foundry problem was encountered due to insufficient venting of the water jacket

TABLE 9.3
SPECIFICATIONS FOR THE WEDGE-HEAD LA V-8 ENGINE

Model Year	1964	1965	1967	1968
Car line	Plymouth, Dodge	Plymouth, Dodge	Plymouth, Dodge	Plymouth, Dodge
Model	Plymouth: Valiant Dodge: Dart	Plymouth: Valiant, Barracuda (opt.) Dodge: Dart (opt.)	Plymouth: Fury Dodge: Charger, Polara	Plymouth: Barracuda S Dodge: Dart GTS
Displacement (cu. in.)	273	273	318	340
Bore	3.625	3.625	3.91	4.04
Stroke	3.312	3.312	3.312	3.312
Compression ratio	8.8:1	10.5:1	9.2:1	10.5:1
BHP @ rpm	180 @ 4200	235 @ 5200	230 @ 4400	275 @ 5000
Torque (lb-ft) @ rpm	260 @ 1600	280 @ 4000	340 @ 2400	340 @ 3200
Combustion chamber	Wedge	Wedge	Wedge	Wedge
Cyl. block mat'l.—skirt type	Cast iron—shallow	Cast iron—shallow	Cast iron—shallow	Cast iron—shallow
Cyl. block deck height	9.6	9.6	9.6	9.6
Cyl. block length	21.4	21.4	21.4	21.4
Cyl. bore spacing	4.46	4.46	4.46	4.46
No. of main brgs.—ck/s mat'l.	5—forged steel	5—forged steel	5—nodular cast iron	5—forged steel
Ck/shaft main journal dia.	2.5	2.5	2.5	2.5
Crankpin journal dia.	2.125	2.125	2.125	2.125
Piston mat'l.—no. of rings	Aluminum—3	Aluminum—3	Aluminum—3	Aluminum—3
Connecting rod centers	6.123	6.123	6.123	6.123
Valve system	OHV—single rocker shaft	OHV—single rocker shaft	OHV—single rocker shaft	OHV—single rocker shaft
Valve events	240-240-16	248-248-26	240-248-20	268-276-44 automatic trans. 276-284-52 manual trans.
Camshaft drive	Chain	Chain	Chain	Chain
Valve head dia.	Intake 1.78; exhaust 1.5	Intake 1.78; exhaust 1.56	Intake 1.78; exhaust 1.5	Intake 2.02; exhaust 1.6
Valve lash adjustment	Mech.—rocker arm screw	Mech.—rocker arm screw	Hydraulic tappet	Hydraulic tappet
Oil pump/distributor drive	Vertical at rear of block	Vertical at rear of block	Vertical at rear of block	Vertical at rear of block
Fuel system	2-barrel carburetor	4-barrel carburetor	2-barrel carburetor	4-barrel carburetor

(Continued on next page)

TABLE 9.3
(Continued)

Model Year	1970	1971	1975	1975
Car line	Plymouth, Dodge	Plymouth, Dodge, Chrysler	Plymouth, Dodge, Chrysler	Plymouth, Dodge, Chrysler
Model	Plymouth: AAR 'Cuda Dodge: Challenger T/A	Plymouth: Fury (opt.) Dodge: Polara (opt.) Chrysler: Newport Royal	Plymouth: Valiant, Duster, Fury Dodge: Dart, Coronet Chrysler: Cordoba (opt.)	Plymouth: Gran Fury Dodge: Monaco, Charger Chrysler: Cordoba
Engine code			E44	E57
Displacement (cu. in.)	340	360	318	360
Bore	4.04	4.0	3.91	4.0
Stroke	3.312	3.58	3.312	3.58
Compression ratio	10.5:1	8.7:1	8.5:1	8.4:1
BHP @ rpm	290 @ 5000	255 @ 4000	150 @ 4000	180 @ 4000
Torque (lb-ft) @ rpm	345 @ 3200	360 @ 2400	255 @ 1600	290 @ 2400
Combustion chamber	Wedge	Wedge	Wedge	Wedge
Cyl. block mat'l.—skirt type	Cast iron—shallow	Cast iron—shallow	Cast iron—shallow	Cast iron—shallow
Cyl. block deck height	9.6	9.6	9.6	9.6
Cyl. block length	21.4	21.4	21.4	21.4
Cyl. bore spacing	4.46	4.46	4.46	4.46
No. of main brgs.—ck/s mat'l.	5—forged steel	5—nodular cast iron	5—nodular cast iron	5—nodular cast iron
Ck/shaft main journal dia.	2.5	2.81	2.5	2.81
Crankpin journal dia.	2.125	2.125	2.125	2.125
Piston mat'l.—no. of rings	Aluminum—3	Aluminum—3	Aluminum—3	Aluminum—3
Connecting rod centers	6.123	6.123	6.123	6.123
Valve system	OHV—single rocker shaft	OHV—single rocker shaft	OHV—single rocker shaft	OHV—single rocker shaft
Valve events	268-276-44	252-256-32	240-248-26	252-256-32
Camshaft drive	Chain	Chain	Chain	Chain
Valve head dia.	Intake 2.02; exhaust 1.6	Intake 1.88; exhaust 1.6	Intake 1.78; exhaust 1.5	Intake 1.88; exhaust 1.6
Valve lash adjustment	Hydraulic tappet	Hydraulic tappet	Hydraulic tappet	Hydraulic tappet
Oil pump/distributor drive	Vertical at rear of block	Vertical at rear of block	Vertical at rear of block	Vertical at rear of block
Fuel system	Triple 2-barrel carbs	2-barrel carburetor	2-barrel carburetor	2-barrel carburetor

(Continued on next page)

TABLE 9.3
(Continued)

Model Year	1975	1979	1981	1989
Car line	Plymouth, Dodge	Plymouth, Dodge, Chrysler	Imperial	Plymouth, Dodge, Chrysler
Model	Plymouth: Duster 360 Dodge: Dart 360	Plymouth: Volare (opt. heavy duty) Dodge: Aspen (opt. heavy duty) Chrysler: Cordoba, Newport, New Yorker (opt. heavy duty)	Imperial	Plymouth: Gran Fury Dodge: Monaco, Charger Chrysler: Cordoba
Engine code	E58	E58	E43	ELA
Displacement (cu. in.)	360	360	318	318
Bore	4.0	4.0	3.91	3.91
Stroke	3.58	3.58	3.312	3.312
Compression ratio	8.4:1	8.0:1	8.5:1	9.0:1
BHP @ rpm	230 @ 4400	195 @ 4000	140 @ 4000	140 @ 3600
Torque (lb-ft) @ rpm	300 @ 3600	280 @ 2400	240 @ 1600	265 @ 1600
Combustion chamber	Wedge	Wedge	Wedge	Wedge
Cyl. block mat'l.—skirt type	Cast iron—shallow	Cast iron—shallow	Cast iron—shallow	Cast iron—shallow
Cyl. block deck height	9.6	9.6	9.6	9.6
Cyl. block length	21.4	21.4	21.4	21.4
Cyl. bore spacing	4.46	4.46	4.46	4.46
No. of main brgs.—ck/s mat'l.	5—nodular cast iron	5—nodular cast iron	5—nodular cast iron	5—nodular cast iron
Ck/shaft main journal dia.	2.81	2.81	2.5	2.5
Crankpin journal dia.	2.125	2.125	2.125	2.125
Piston mat'l.—no. of rings	Aluminum—3	Aluminum—3	Aluminum—3	Aluminum—3
Connecting rod centers	6.123	6.123	6.123	6.123
Valve system	OHV—single rocker shaft	OHV—single rocker shaft	OHV—single rocker shaft	OHV—single rocker shaft
Valve events	268-276-44	252-256-32	240-248-26	240-248-26
Camshaft drive	Chain	Chain	Chain	Chain
Valve head dia.	Intake 1.88; exhaust 1.6	Intake 1.88; exhaust 1.6	Intake 1.78; exhaust 1.5	Intake 1.78; exhaust 1.5
Valve lash adjustment	Hydraulic tappet	Hydraulic tappet	Hydraulic tappet	Hydraulic tappet
Oil pump/distributor drive	Vertical at rear of block	Vertical at rear of block	Vertical at rear of block	Vertical at rear of block
Fuel system	4-barrel carburetor	4-barrel carburetor	Chrysler TBI	2-barrel carburetor
Engine assembly weight*			283.1 kg (624.2 lb)	259.64 kg (572.4 lb)

* For content, see the Abbreviations and Definitions section near the back of this book.

Chart 9.1
Time Lines—Wedge-Head V-8s

Engines, Car Lines, and Displacements (cu. in.)

Engine	Car Line	1958	1959	1960	1961	1962	1963	1964	1965
		- - Model Years - -							
LB V-8 4.8 cyl. bore centers	Plymouth	350	361	361 383	361 383	361 383	361 383	361 383	361 383
	Model	LP2	MP2	PP2	RP2	SP2	TP2	VP2	AR2, AP2
	Dodge	350 - 361	361 - 383	361 - 383	361 383	361 383	361 - 383	361 - 383	361 - 383
	Model	LD2 - LD2,3	MD3 - MD2,3	PD1,4 - PD1,2,4	RD1, RD4	SD2	TA3 - TA3,TD2	VA3 - VA3,VD2	AW2-AW2,AD2
	DeSoto	350 - 361	361 - 383	361 - 383	361				
	Model	LS1 - LS2	MS1 - MS1,2	PS1 - PS1,3	RS1				
	Chrysler				361 - 383	361 - 383	361 - 383	361 383	383
	Model				RC1 - RC2	SC1 - SC2	TC1 - TC2	VC1	AC1
RB V-8 4.8 cyl. bore centers	Plymouth				413	413	426	426	426
	Model				RP2	SP2	TP2	VP2	AP2, AR2
	Dodge				413	413	426	426	413 - 426
	Model				RD4	SD2	TD2	VD2	AD2 - AD2,AW2
	Chrysler		383 - 413	383 - 413	413	413 426	413 - 426	413	413
	Model		MC1,2 - MC3	PC1,2 - PC3	RC1, 3, 4	SC2	TC3 - TC2,3	VC1	AC1
	Imperial		413	413	413	413	413	413	413
	Model		MY1	PY1, 2, 3	RY1	SY1	TY1	VY1	AY1
LA V-8 4.46 cyl. bore centers	Plymouth							273	273
	Model							Valiant , Barra	AV2, AR2
	Dodge							273	273
	Model							Dart	AL2, AW2

Chart 9.1
(Continued)

Engines, Car Lines, and Displacements (cu. in.)

Engine	Car Line	1966	1967	1968	1969	1970	1971	1972	1973	1974
LB V-8 4.8 cyl. bore centers	Plymouth	361 - 383	383	383	383	383	383	400	400	400
	Model	BR2 - BP2,BR2	CP2, CR2	DP, DR	EP, ER	FB, FP, FR	GB, GP, GR	HP, HR	JP, JR	4R, 4P
	Dodge	361 383	383	383	383	383	383	400	400	400
	Model	BD2, BW2	C:D2,L2,W2	D:D,W	E:D,L,W	FD,FJ,FW,FX	GD, GJ, GW	HD, HW	JD, JW	4D, 4W
	Chrysler	383	383	383	383	383	383	400	400	400
	Model	BC1, BC2	CC1	DC	EC	FC	GC	HC	JC	4C
RB V-8 4.8 cyl. bore centers	Plymouth	440	440	440	440	440	440	440	440	440
	Model	BP2	CP2	DP	EP	F: BPR	G: BPR	HP	JP	4P, 4R
	Dodge	440	440	440	440	440	440	440	440	440
	Model	BD2	CD2, CW2	DD, DW	ED, EW	F: DJWX	G: DJW	HD, HW	JD, JW	4D, 4W
	Chrysler	440	440	440	440	440	440	440	440	440
	Model	BC3	C: C1,C2,C3	DC	EC	FC	GC	HC	JC	4C
	Imperial	440	440	440	440	440	440	440	440	440
	Model	BY3	CY1	DY	EY	FY	GY	HY	JY	4Y
LA V-8 4.46 cyl. bore centers	Plymouth	273	273 318	273 318 340	273 318 340	318 340	318 340 360	318 340 360	318 340 360	318 360
	Model	BR1, BV2	C: B2,R2,V2	DR, DV	E: BPRV	F: BPRV	G: BPRV	H: BPRV	J: BPRV	4: BPRV
	Dodge	273	273 318	273 318	273 318	318 340	318 340 360	318 340 360	318 340 360	318 360
	Model	BL2, BW2	C: D2,L2,W2	D: DLW	E: DLW	F: DJLWX	G: DJLW	H: DJLW	J: DJLW	4: DJLW
	Chrysler						360			
	Model						GC			

Chart 9.1
(Continued)

Engines, Car Lines, and Displacements (cu. in.)

Engine	Car Line	1975	1976	1977	1978	1979	1980	1981	1982	1983	1984–1989
LB V-8 4.8 cyl. bore ctrs	Plymouth Model	400 5P, 5R	400 P, R	400 P, R	400 R						
	Dodge Model	400 5D, 5W, 5X	400 D, W, X	400 D, W, X	400 W, X						
	Chrysler Model	400 5C, 5S	400 SS,CL,CM CP, CS	400 SS, CL CP, CS	400 SS, CL,CS						
RB V-8 4.8 cyl. bore ctrs	Plymouth Model	440 5P, 5R	440 P, R	440 P, R							
	Dodge Model	440 5D, 5W	440 D	440 D							
	Chrysler Model	440 5C	440 CL,CM,CP,CS	440 CL,CP,CS	440 CL, CS						
	Imperial Model	440 5Y									
LA V-8 4.46 cyl. bore ctrs	Plymouth Model	318 360 5P, 5R, 5V	318 - 360 318 - H,P,R, V 360 - H, P, R	318 360 H, P, R	318 360 H, R	318 360 H	318 H, J	318 J	318 B	318 B	318 B
	Dodge Model	318 360 5D, 5L, 5W, 5X	318 360 D, L, W, X	318 360 D, G, N, W, X	318 360 G, N, W, X	318 360 E, G, N, X	318 360 318-ENGX 360-EX	318 E, G, X	318 G,X	318 G,X	318 G
	Chrysler Model	318 - 360 5S - 5C,S	318 - 360 318 - SS 360-SS,CL,CM	318 - 360 318 - SS 360-SS,CL,CM	318 - 360 318-F, SS 360-CL,SS	318-ENGX 360-CL,SS	318 360 F, S, T	318 F, S, T	318 F,S	318 F,S	318 F
	Imperial Model							318 Y	318 Y	318 Y	

core.[11] A change to separate wet guides, illustrated in the lower view on Fig. 9.35, overcame this problem by increasing the core thickness and allowing better vents. New offset intake valve rocker arms were designed to accommodate the wider intake ports. To provide an exceptionally solid anchor for the rocker arm shaft, the integral posts with a half-round saddle at the top were replaced with five steel mounting blocks per head. Intended strictly for off-highway competitive events, the cylinder heads and intake manifold were not equipped with exhaust cross-over heat for carburetor warmup. Designated W2, which stood for "Wedge #2," the cylinder head and its related parts are shown on Fig. 9.36.

Actual race use of the W2 cylinder head started in 1971 on a limited basis. A single-winged Dodge Charger Daytona driven by Dick Brooks and powered by a 304-cu.-in. small-block LA engine equipped with a pair of W2 cylinder heads entered the February 14, 1971 Daytona 500 race. (For 1971, NASCAR imposed a severe displacement limit of 305-cu.-in. for any engine powering a winged car, regardless of the valve gear type.) The cylinder heads were especially notable, in that they had been machined on a computer-controlled Cincinnati mill at the Engineering Technical Computer Center, the first time a Chrysler prototype cylinder head had been machined in this manner.[12] The 175.890-mph qualifying speed of the car gave it a back-in-the-pack ninth starting position center; however, during the race, it actually led for 18 laps. Unfortunately, the car was involved in an accident that caused it to drop behind, finishing the race in seventh place.[13] Two years later, for 1973, NASCAR effectively imposed a six-liter (366-cu.-in.) limit on engines by requiring that any engine over this size had to use a small Holley carburetor. The small-block LA V-8, race-developed at 355 cu. in., did reasonably well, especially in 1975 when Dodge entries won 14 of the 30 total NASCAR Grand National races, with Richard Petty entering the winner's circle 13 times in a Dodge.

By mid-1976 calendar year, a complete W2 engine, ready for customizing by individual racers and teams, was part of the competitive Kit Car, engineered by Special Vehicles Manager Larry

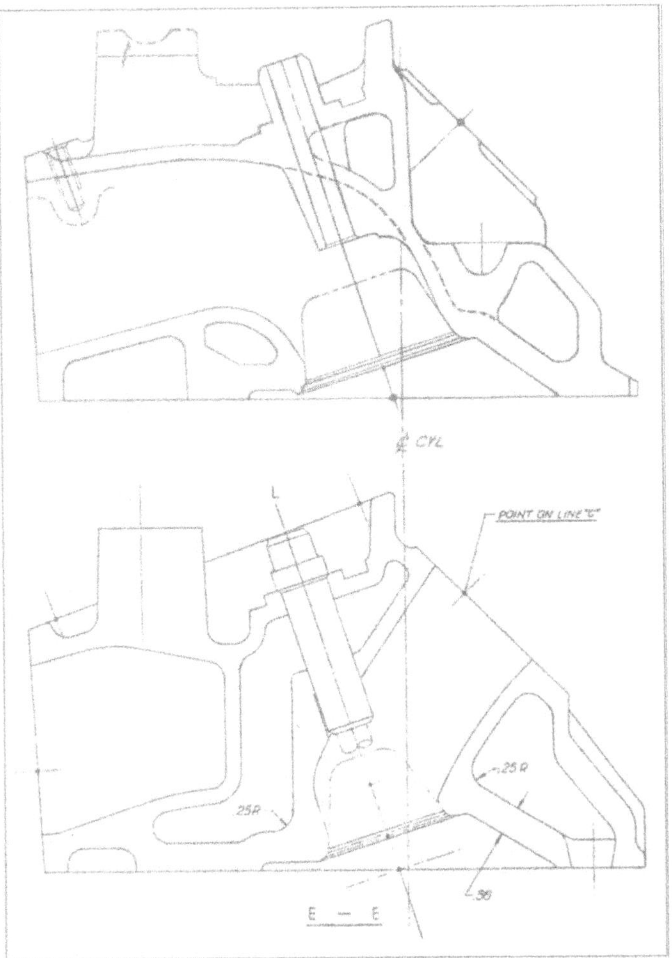

Fig. 9.35 *Sections through the intake port (upper view) and the exhaust port of the LA V-8 engine W2 cylinder head designed for racing. Engine Design Layout 7450-A05-77.*

Rathgeb, which was available through the Mopar sales system. Kit Cars were offered with 340-, 355-, and 360-cu.-in. engines, depending on the model year. The 355-cu.-in. engine had a bore and stroke of 4.05 × 3.46 in. Successful in the short track venues in which it was designed to compete, a typical 1978 Kit Car ad lists wins at Wiscasset Raceway, Wiscasset, Maine; Speedway 99, Stockton, California; West Fargo Fairgrounds, Fargo, North Dakota; Kil-Kare Speedway, Xenia, Ohio; Enid, Oklahoma Speedway; Shady Bowl, Bellefontaine, Ohio; Rice Lake Wisconsin Speedway; Henry's Speedway, Boswell, Indiana; Huron Speedway, Huron, North Dakota; Capital City Raceway; and Crystal Beach Speedway, Woodward, Oklahoma.

The high stresses of competitive operation required strengthening throughout the engine, including the use of forged aluminum pistons and stress-relieved Trans Am-type cylinder blocks with extra-thick bulkheads, designed to accept four-bolt in place of two-bolt main bearing caps. With the entire engine in all its versions developed under the supervision of high-performance engine development manager John Wehrly, the 355-cu.-in. engine achieved gross power output levels somewhat greater than 600 hp using a single four-barrel carburetor.[14] Although Chrysler's passenger car use of the base LA engine ceased at the end of 1989, use of the LA engine continued in Dodge trucks, as noted earlier in this chapter. Spawned by the burgeoning popularity of trucks, race organizers added truck races to their schedules, with NASCAR adding a truck series in 1996. Dodge trucks participated, using an updated version of the 1970s passenger car race LA engine, which included aluminum cylinder heads. Finally in 2001, well beyond the 1998 closing year of this book, the LA V-8 race engine was retired, replaced by a V-8 engine designed specifically for racing.

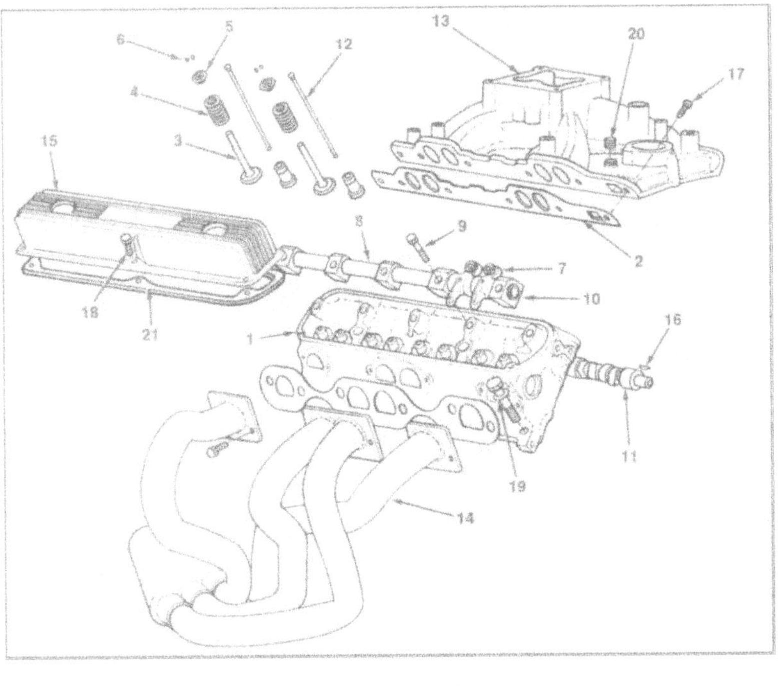

Fig. 9.36 *The W2 cylinder head and related parts for the LA V-8 race engine. From an illustration in the Chrysler Kit Car Catalog.*

CHAPTER 10

Power Personified—The 426 Hemi® Engine

The 426 Hemi® was born with one purpose—to put a Chrysler product in the winner's circle of the longest and fastest stock car race in America at the time, the NASCAR Daytona 500, run on the high-banked, tri-oval track of the Daytona International Speedway in Daytona Beach, Florida. A number of attempts to run the well-developed big-block RB wedge-head V-8 on the super speedways of the NASCAR circuit ended in disappointment. The engine simply did not have enough power to match the winning vehicles fielded by other manufacturers. Challenged with finding a way to change Chrysler's also-ran status, the key high-performance, hemi-engine savvy engineers Bob Roger, Bob Cahill, Don Moore, and Tom Hoover knew exactly what to do—bring out a new Hemi® engine. Because nothing else was available or appropriate as a platform for mounting the hemi heads, the question was: "Could it be done using the big-block RB V-8?"

Preliminary design studies by Engine Design established enough feasibility of turning the wedge head RB V-8 into a Hemi® head RB V-8, to allow Jack Charipar, director of the group responsible for the corporate racing effort, to recommend going ahead. This resulted in program approval together with a firm demand that the engine and car be ready for the premier NASCAR event—the Daytona 500 race on February 23, 1964.

As soon as the decision was relayed to Engine Design, lead designer Frank Bialk, given the assignment of creating the new cylinder head, unrolled a fresh, clean sheet of 36-in. tracing paper on his drafting table, impressed the lower-right corner with a layout title-block ink stamp, named the layout "Hemi Chamber Cyl. Head," wrote "731-342-A" in the layout number box, gave it that day's date March 28, 1963, and started filling the large and empty white space with the precise, carefully thought-out lines for which he was well known. One specific instruction given to Bialk was to use the same 58.5-deg (rounded to 58-deg) included angle between the intake and exhaust valve centerlines as used on the original Chrysler FirePower Hemi® engine[1]—there was no time for a test program to explore other chamber geometries. However, as Bialk found, if the location of the valves relative to the cylinder axis simply was copied from the original Hemi®, the exhaust valve rocker arm would become exorbitantly long. A design resolution came by Bialk's rotating the valves toward the intake manifold until the exhaust rocker arm, sometimes jokingly referred to as a "pump handle," had a length that finally looked reasonable. With a careful selection of 2.42 in. as the spherical radius of the machined chamber, Bialk was able to find room for exceptionally large valves, an intake valve with a head diameter of 2.25 in., and an exhaust valve with a 1.94 in.-head.

An unusual design challenge arose with the five-bolt head pattern of the RB engine because the inboard bolt was exactly where the intake port needed to be for a direct passage to the combustion chamber. Bialk's creative solution to this challenge was to add bosses to the tappet chamber of the cylinder block, which were drilled with clearance holes for screws (later replaced with studs) that threaded into the bottom of the cylinder head, below the intake ports. These and other design details can be seen in the illustration of this initial layout on Fig. 10.1.

As soon as enough lines and dimensions showed on the cylinder head layout to define the shape of the intake and exhaust ports, information was dispatched to air-flow expert Harry Weslake in England so flow models could be made and bench testing could be started. His input resulted in changes giving significant improvements to air flow, which were incorporated in the pattern equipment before the first cylinder head castings were made. Apace with the new cylinder head, the RB cylinder block was redesigned to have cross bolts tying the deep-skirt walls of the cylinder block to the #2, #3, and #4 main bearing caps to greatly strengthen the bottom end of the engine. Other block changes included the addition of a large drilled hole in

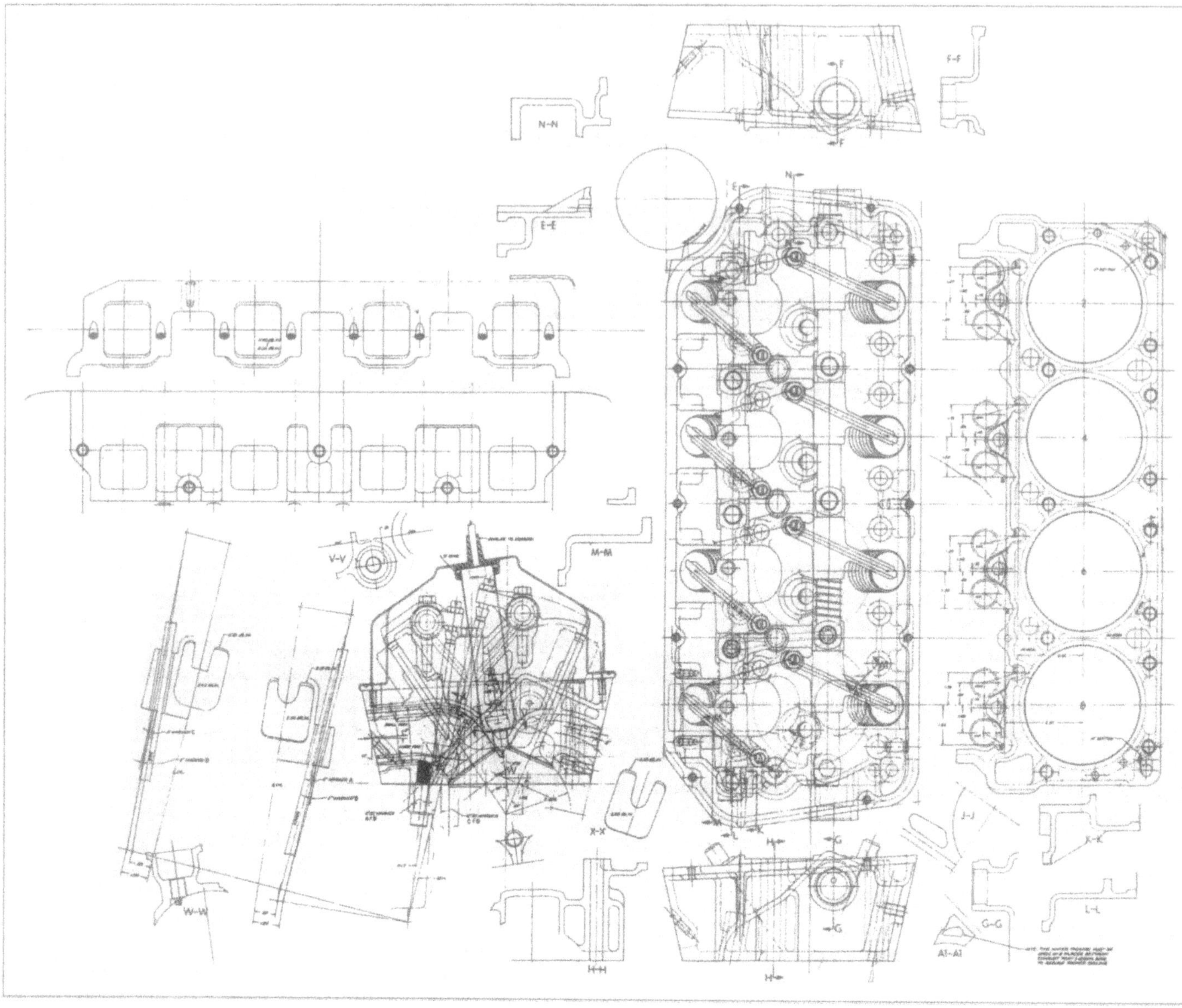

Fig. 10.1 *The design layout of the 426-cu.-in. RB Hemi® V-8 engine started on March 23, 1963 by Frank Bialk.*

each corner from the top deck to the crankcase for draining oil from the cylinder head, and a machined pad with drilled and tapped holes on each side of the block near the front for beefed-up engine mounts. Double valve springs with a damper between the coils for both the intake and exhaust valves pushed the valve bounce speed to beyond the normal operating range of the engine. Sturdy, malleable iron rocker shaft brackets located between the cylinders anchored the two rocker shafts per cylinder head required, similar to the original Chrysler Hemi® head, for the valve geometry. Four forged steel intake rocker arms and four forged steel exhaust rocker arms rotated about the inboard and outboard rocker shafts, respectively.

Valve lash adjustment was accomplished by an adjusting screw with a lock nut at the pushrod end of each rocker arm. Both pushrods and tappets were hollow to minimize valvetrain weight. Greatly strengthened connecting rods went into the 426 Hemi® engine right from the start, the initial connecting rods having an increased I-section area versus the standard RB engines and larger cap bolts—7/16-20 versus 3/8-24. Two years later, in 1966, the race engine connecting rods were beefed up again, this time using 1/2-20 bolts with bolt tightening determined by a micrometer measurement of bolt stretch. Specified stretch was 0.010 in., only one thousandth of an inch below the yield point of the bolt of 0.011 in., getting the maximum load from the alloy-steel bolt.

For the initial race season of the engine, a conventional-appearing, two-level intake manifold mounting a rules-required single four-barrel carburetor was designed. The material of choice for the pistons that would be required to operate failure-free at higher levels of mechanical and thermal stress was impact-extruded aluminum, which had a benefit of providing the complex shape of the domed top of the piston without any machining. Although the forged steel crankshaft dimensionally was similar to passenger car RB engine crankshafts, its shot-peened-all-over plus surface nitriding processing and much smoother journals of 5 versus 15 microinch gave it strength and bearing durability far beyond its everyday counterpart. A double-row roller timing chain was released for the Hemi® engine, with three bolts used to clamp the camshaft sprocket to the nose of the camshaft rather than the single bolt used on the high-volume RB engines. To quickly complete the designs and detail drawings of all the new parts, which included almost every part in the engine, so that prototype procurement could be started as soon as possible, virtually the entire Engine Design department was mobilized. With suppliers making equally vigorous efforts to prepare their parts, the first engine was assembled in the Engine Development Motor Room, placed on a dynamometer stand, and ran under its own power on December 6, 1963.[2] Calibration and development, including selection of a long-duration 312-312-88 race camshaft, went on until late January when a nearly show-stopping event occurred.

Race engine development supervisor Larry Adams took a chair in my office and gave me the bad news. Cracks had been found in the cylinder bores during an endurance test that simulated Daytona 500 race conditions on a dynamometer. Unfortunately, the test engine was well short of completing its endurance run. As Larry put it, "Bill, we're not going to finish the race." That was a call to action! Increasing the thickness of the bore walls in the area of the cracking was quickly agreed upon as the best way of overcoming the problem. That meant starting with new cylinder block castings with revised water jacket cores—usually a long process. Instead, armed with templates, a three-person team consisting of me, Engine Design casting expert Louie Taylor, and Earl Pinches from the Chrysler Central Pattern department flew to the cylinder block casting source, which was the Chrysler Indianapolis foundry. Immediately starting work with the willing staff of the foundry, twelve sets of cores were hand-rubbed to add metal to the bore walls, placed in twelve molds on the production casting line, and 2600-deg molten iron was poured into them. After sand shakeout and cooldown, the castings were given a hopeful close look. Not one was remotely usable! All had large voids instead of solid iron, a discouraging reward for our efforts. However, the Indianapolis Foundry experts were experienced in overcoming casting problems. Thus, by the time the night was over and several more casting runs had been made, good and solid blocks finally emerged from the sand. Additional casting runs produced more and more excellent block castings, enough for all the race teams.

No time was lost in shipping these to the heat treatment furnaces for annealing and then to the Trenton engine plant for top-priority machining, followed by assembly into complete race-ready engines by expert motor room mechanics at Chrysler Engineering and then a quick break-in run on a dynamometer. Trucked to the race teams at the Daytona International Speedway, which had already qualified for starting positions using the more fragile earlier blocks, the fresh and stronger engines finally were bolted in place, ready to race.

As if the events prior to the race weren't exciting enough, soon after the race started on February 23, 1964, Assistant Chief Engineer of Engine Development Ev Moeller and I were sitting together at a good vantage point in the stands across from the pit area. We watched in stunned disbelief as all the Hemi®-powered cars pulled into the pits and the hoods went up—often a sign of terminal mechanical problems. However, after only a few seconds, with huge sighs of relief from both of us, the hoods slammed down and the cars roared back on the tri-oval track. The problem had been debris on the track, which clogged the front of the radiators, causing overheating. Quickly corrected by the race crews, the engines ran perfectly for the remainder of the race. The final outcome was Richard Petty in first place, driving car #43; Jimmy Pardue in second place, driving car #54; and Paul Goldsmith in third place, driving car #25. All of these cars were Hemi®-powered Plymouths, which was a wonderful reward for everybody involved in the program. With his victory, Petty's average speed of 154.334 mph set a new track record. As the local newspaper *Orlando Sentinel* glowingly reported the next day, " He [Petty] drove one of the fantastic 1964 Plymouths whose blazing speed has given other manufacturers something new at which to shoot." Of interest, the cylinder block in Petty's engine was cast on February 10, only 13 days before the race. That July 4, legendary driver A.J. Foyt piloted a Hemi®-powered Dodge to victory in the Firecracker 400 at the same speedway. With these successes of the 426 RB Hemi® V-8 engine, Chrysler Hemi® engines again were the dominant powerplants on stock car race tracks. Figure 10.2 shows a cutaway view of this 1964 track engine that was identified within Engineering by its A864 code.

Hard on the heels of the leadoff Hemi® engine version aimed at closed-circuit track races, a second version of the 426 RB Hemi® engine that was aimed at sanctioned acceleration events, better known as drag races, was prepared. Engine requirements were similar except for the carburetion because the principal drag race sanctioning bodies, the National Hot Rod Association (NHRA) and the American Hot Rod Association (AHRA), permitted the use of two four-barrel carburetors. Assigned to the task of creating an appropriate intake manifold, Engine Design ace manifold designer Forbes Bunting quickly merged the dual-carburetor, dual-plenum, short-ram concept of the Max Wedge intake manifold with the different porting size and arrangement of the new Hemi® heads. The new drag race intake manifold emerged with a tuned ram pipe length of 12.4 in. Sections of this manifold appear on Figs. 10.3 and 10.4, which show a longitudinal view and a cross-section view of the 1964 drag race engine, respectively. Following dynamometer development, the drag engine was released for a small production run by the Chrysler Marine and Industrial Division, a group long accustomed to making small runs of

Fig. 10.2 *A cutaway illustration of the 1964 426-cu.-in. RB Hemi® V-8 track engine. The intake air for the engine was ducted from the cowl area of the car to the rear opening of the air cleaner.*

customized engines; thus, it would be qualified as a stock car engine by the drag race sanctioning bodies. The 426 RB Hemi® V-8 convincingly took over as the top Chrysler drag engine at the NHRA U.S. National meet held on Labor Day weekend in 1964, where the final Super Stock match race pitted Roger Lindamood's Hemi®-engined Plymouth against Jim Thornton's Hemi®-engined Dodge Ramcharger. (Lindamood won.) At the same meet, Jim Thornton drove his Hemi® Dodge to a new NHRA speed record for the quarter-mile—130.06 mph.

Frank Bialk (and others) took amused pride in two nicknames the big Hemi® engine acquired: "Elephant Motor" and "King Kong," reflecting its massive appearance.

As soon as development of the original cast iron cylinder head was completed, work started on an aluminum head for the 426 Hemi® drag engine (coded A990), incorporating thicker walls in high loaded areas, valve seat inserts, valve guides, steel shims under the valve springs to prevent frettage, large-diameter head bolt bosses, and steel washers for the head bolts that were not also used for the rocker shaft brackets. Because of the space needed for the intake valve seat insert, the head diameter of the 1965 intake valve was reduced slightly to 2.23 in. In a later year, a second spark plug was added to each combustion chamber of the aluminum head. For weight savings, the intake manifold of the 1965 drag engine was changed from aluminum to magnesium. Figure 10.5 shows a cutaway view of the 1965

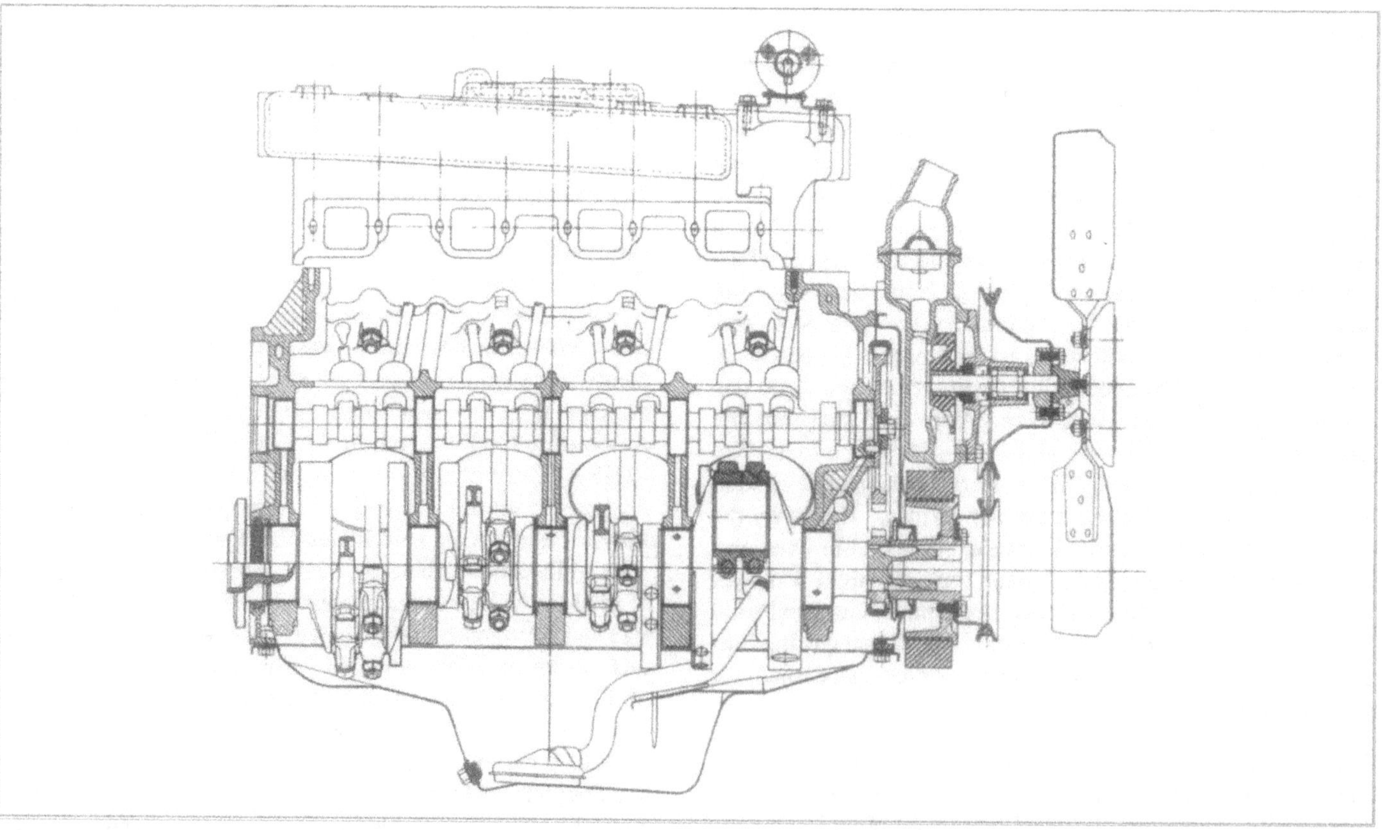

Fig. 10.3 *A longitudinal section of the 1964 426-cu.-in. RB Hemi® V-8 drag engine.*

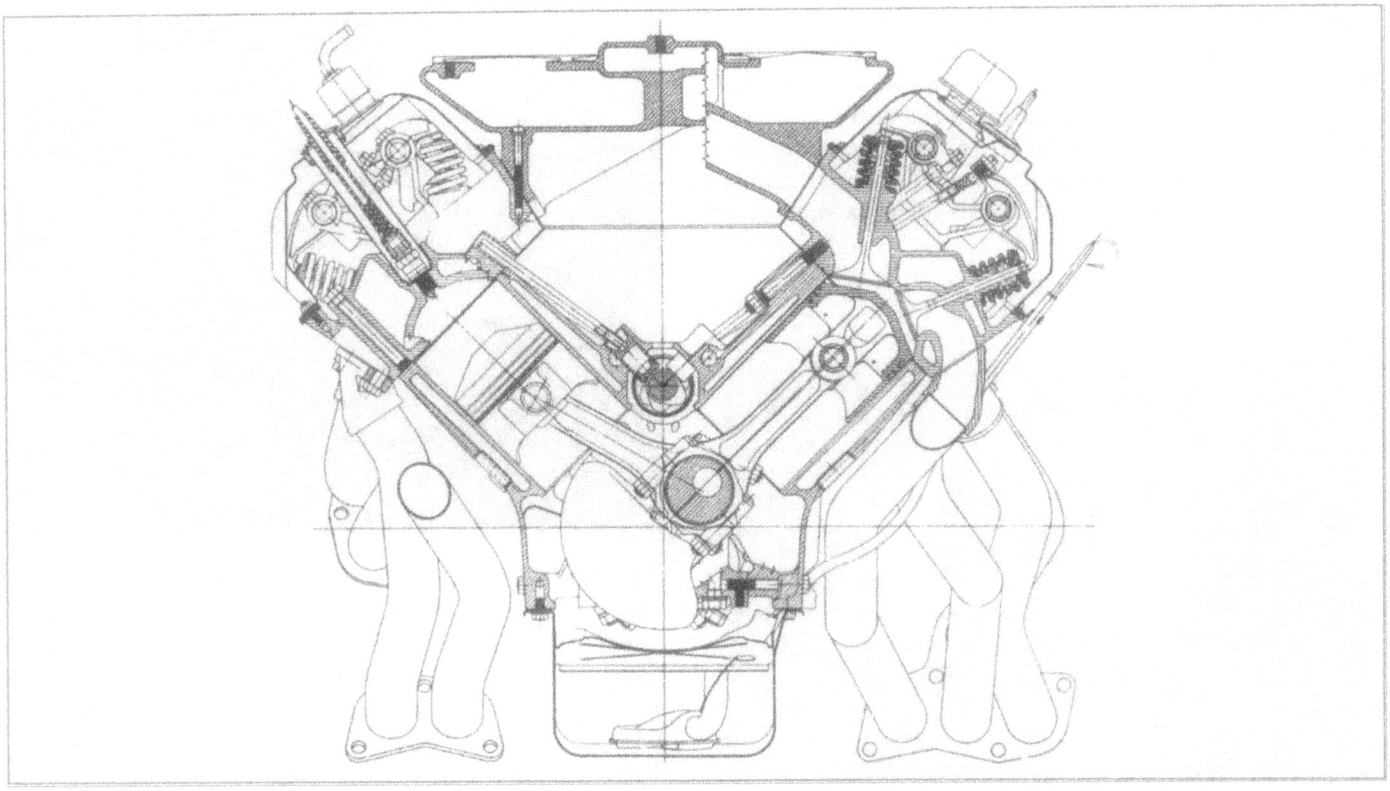

Fig. 10.4 *A cross section of the 1964 426-cu.-in. RB Hemi® V-8 engine. The combustion chamber and the valve positions show its strong heritage with the original 1951 Chrysler FirePower Hemi® V-8 engine.*

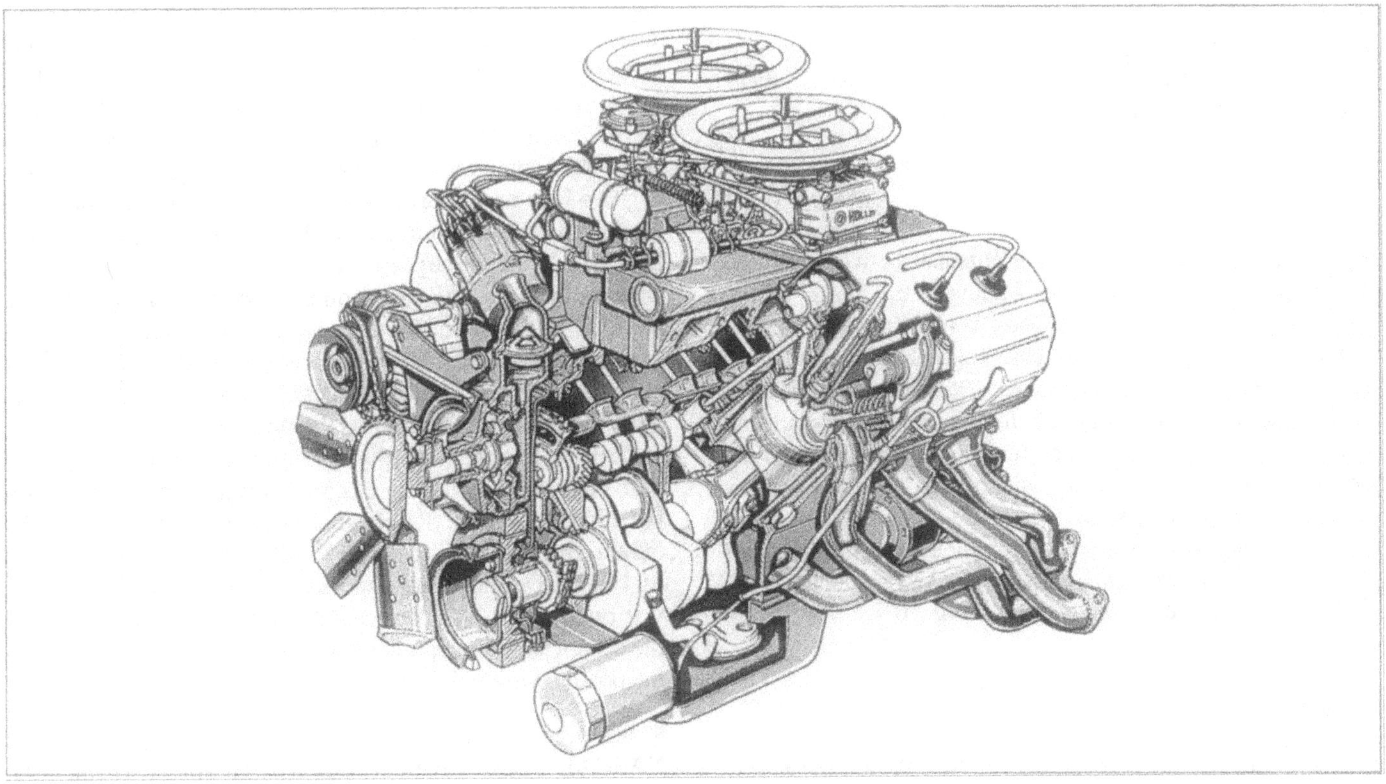

Fig. 10.5 *A cutaway illustration of the 1965 426-cu.-in. RB Hemi® V-8 drag engine with aluminum cylinder heads. The dual-plenum cross-ram Max Wedge type of intake manifold was readily adapted to the Hemi® engine.*

Hemi® drag engine. Two key engineering executives who were involved in the entire 426-cu.-in. RB Hemi® V-8 engine program—Chief Engineer of Power Train Bob Rarey, and Chief Engineer of Engine Development Harold Welch—pose with the engine as shown on Fig. 10.6. Together, these two excellent engineers shouldered the overall responsibilities for the design and development of the RB Hemi® and all other corporate engines.

Fig. 10.6 *Bob Rarey (left), Chief Engineer of Power Train, and Harold Welch (right), Chief Engineer of Engine Development, with the 1965 RB Hemi® V-8 drag engine.*

The concept of placing a plenum chamber between the carburetor and the short ram pipes leading to the intake valves, which worked so well on the Max Wedge intake manifold, was explored on the single four-barrel closed-circuit-track race manifold. The result, released for the 1966 closed-circuit-track racing season, was a new single four-barrel intake manifold with a boxy plenum, having the carburetor depressed into the manifold cover for under-hood clearance. Its basin-like appearance earned it the nickname "bathtub." An engine equipped with one of these four-barrel bathtub intake manifolds is part of the 426 RB Hemi® engine dynamometer room display at the Walter P. Chrysler Museum shown on Fig. 10.7.

Fig. 10.7 *A recreated race engine dyno room at the Walter P. Chrysler Museum, with the 426-cu.-in. RB Hemi® V-8 closed-circuit-track engine equipped with a bathtub intake manifold. The oversized crankcase vent system pictured here was needed for high lateral G-forces encountered during racing. The angle-bracket stand below the engine is for display purposes only; the actual dyno supports were much beefier. Photo by W.L. Weertman.*

Street Hemi®

The winning 1964 year of the 426 Hemi® on the NASCAR tracks was followed by its non-appearance in 1965 because an escalating NASCAR demand for higher production figures to qualify it for race use could not be met. To remove this barrier and to get back on the track, Chrysler laid plans to offer for sale to the general public Plymouth and Dodge cars powered with a street-drivable version of the engine for 1966. The result was the 426 Street Hemi®. Assigned A102 as its Engineering designator, the engine was noticeable in outward appearance by having two four-barrel carburetors arranged in tandem fashion on top of its two-level aluminum intake manifold, as can be seen on Fig. 10.8. A single, large circular unsilenced air cleaner covered both carburetors. All-cast-iron exhaust manifolds were used in place of the fabricated exhaust headers of the competition engines. Because the Hemi® cylinder heads lacked an exhaust gas crossover, steel tubing was used to route exhaust gas from the right exhaust manifold to the floor of the rear carburetor for cold operation. A mildly aggressive 276-276-52 camshaft with ramps appropriate for mechanical tappets was released for the 1966 426-cu.-in. Street Hemi® that was rated at 425 hp @ 5000 rpm and 490 lb-ft torque @ 4000 rpm. A paper, written by Engine Development engineer Robert J. Lechner and me, was presented to SAE on April 11, 1966, giving details about the design and development of the engine.[3] Figure 10.9 shows me posing with a 1966 Street Hemi® engine.

Reflecting the Plymouth and Dodge car models racing on the NASCAR tracks, Plymouth offered the engine on its 1966 intermediate-size Belvedere model and Dodge on its 1966 intermediate-size Coronet and the fastback Dodge Charger

Fig. 10.8 *A top view of the 1966 426-cu.-in. RB Street Hemi® V-8 engine, showing the tandem arrangement of its two four-barrel carburetors.*

Fig. 10.9 *Here, when I was then Assistant Chief Engineer of Engine Design, I posed with a 426-cu.-in. Street Hemi® engine around the time of its release for the 1966 model year.*

shown on Fig. 10.10. The motoring enthusiast magazines quickly spread the word. In the October 1965 issue of *Motor Trend*, technical editor Bob McVay wrote, "It's big news that in 1966 anyone can buy Plymouth's intermediate-sized Belvedere, right off the showroom floor, with a slightly detuned '426-Hemi' and have a darn good chance of winning his class in A/ or AA/Stock right off the bat with only minor modifications... Relatively docile around town, the 'HP^2' offers somewhat terrifying acceleration in all speed ranges and takes a back seat to none at the top end." (The HP^2 chrome letters on the outside of the car stood for Hemi-power-squared, referring to its two four-barrel carburetors.) Awesome as these cars were as street machines, they were eclipsed for performance in 1970 when, by prodigious effort, the 426 RB Hemi® was shoehorned into the Chrysler smaller, lighter, and compact-size cars, as well as the Plymouth 'Cuda and the Dodge Challenger.

Fig. 10.10 *The 1966 Dodge Charger. The 425-hp 426-cu.-in. RB Street Hemi® V-8 engine was an option on this two-door fastback hardtop model.*

Unique among many vehicles powered by the 426 RB Hemi® engine was a long, narrow custom-built car named "Goldenrod," built by the Sommers brothers of Ontario, California, with four 426 Hemi® V-8 engines installed in a single row. Short intake ram pipes, fuel injection, and a dry sump oil system for each engine were needed for the installation and were developed by Bob Lechner. On the salt flats of dry Lake Bonneville, Utah, on November 12, 1965, the car was driven to a new two-way record of 409.277 mph. A marine version of the 426 RB Hemi® engine also was developed and is discussed in Chapter 16.

Subsequent Changes

The NASCAR rules for 1965 effectively banned the 426 Hemi® from its tracks, but the Hemi® came back in 1966. However, it had a displacement handicap requiring it to run at 404 cu. in. against 426-cu.-in. wedge-head engines. The 404-cu.-in. displacement was obtained easily by a destroked (3.558-in.) crankshaft and a longer connecting rod (7.174 in.). Despite the handicap, the Hemi® came through at Daytona International Speedway with a Daytona 500 win on February 27, 1966 for Richard Petty's Plymouth, followed by Sam McQuagg driving his Dodge into the winner's circle at the end of the July 4, 1966 Firecracker 400. The wins were a bit hollow, however, because Ford, unhappy with the return of the Hemi®, had withdrawn its support of NASCAR racing teams. Three years later, in 1969, with the competition heating up, Chrysler unleashed a new way to boost car speed—the pointed-nose winged Dodge Daytonas—opening the way to almost total domination of the NASCAR superspeedways by Dodge Daytonas and similarly winged Plymouth Superbirds until the end of 1970. Then the days of Hemi® engines on NASCAR tracks were over because for 1971, NASCAR limited the engine size for a winged car to a maximum of 305 cu. in. and imposed a graduated restrictor plate rule for the carburetor, which put the Hemi® engine at a considerable disadvantage compared to conventional-valved and chambered engines.

The Street Hemi® went unchanged until 1968, when its camshaft was changed to one with the longer events of 284-284-60. In 1970, to comply with increasing emission stringency, hydraulic tappets replaced the mechanical tappets. After considering that the RB Hemi® had been virtually banned from stock car circuit racing, that emission compliance beyond 1971 would involve far more drastic changes to the engine with the inevitable reduction of engine performance, and that Street Hemi® sales had dropped from an annual volume of about 2,700 in 1966 to a meager 356 in 1971[4] (due in large part to high insurance rates), a consensus decision was made to end production of the 426 RB Hemi® engine at the end of 1971.[5] Thus, the 426-cu.-in. RB Hemi® V-8 engine was retired, its performance image untarnished.

Table 10.1 shows the specifications for the engine.

Chart 10. 1 displays time lines for the RB Hemi® engine.

426 Hemi® Postscript

Although never publicized by Chrysler, the Engine Design and Development departments worked on two engine programs that were related to the 426 Hemi® engine program in interesting ways but never progressed beyond experimental status. One, intended only for racing, was a 426 RB V-8 with four valves per cylinder, coded A925. The other was a high-performance LB V-8 with hemispherical-like combustion chambers, intended for high-volume production and coded A279. Their stories are related here next.

Topping the Hemi®—Thirty-Two Valves of Racing Power—The A925 Engine

When the Hemi® engines emerged in February 1964 and totally outran the competition on the high-speed tracks, the competitors, especially the Ford entries, understandably were upset at this sudden drop in their own potential to win. In a move to field a sure Hemi®-beater, Ford launched a crash program to create a single overhead camshaft hemispherical-chambered version of its well-developed closed-circuit-track 427-cu.-in. big-block overhead valve V-8 engine. Following engine development, a small production run, required by NASCAR rules that had been put in place after the debut of the Chrysler Hemi® engine, was launched. With certainty about the direction of the Ford plans exposed, Jack Charipar's performance engine experts quickly devised a plan on how to more than match the Ford single overhead camshaft threat. The plan was to design and develop all-new cylinder heads for the proven bottom end of the Hemi® engine—cylinder heads with four valves per chamber operated by double overhead camshafts, as had been used on all-out racing engines for years. As a just-in-case backup design, an overhead valve layout with tappets and pushrods operating four valves per chamber would be prepared. This overhead valve design could have a future use if NASCAR banned overhead camshaft engines but did not restrain overhead valve engines. Race engine design supervisor Bob Dent assigned the task of making the definitive layout to lead designer Frank Bialk, who, in his usual masterful style, came up with Engine Design Layout 731-972 shown on Fig. 10.11, which displays a cross section of the engine, with the double overhead camshaft version on the left side of the layout and the overhead valve version on the right.

The engine never received a name. It was identified only by its assigned A-program code number of A925. A close look at the double overhead camshaft version on the layout reveals that the camshafts were direct acting, favoring high-speed operation, with valve lash adjustment accomplished by selection of valve stem caps of the proper thickness. A single flexible toothed timing belt drove the camshafts via gears mounted inside the front walls of the cylinder heads. The angle between the row of intake valves and the row of exhaust valves was much narrower than that of the two-valve Hemi® engine, resulting in the cross section of either type of four-valve cylinder head being smaller than the cylinder head of the Hemi®; thus, vehicle installation would be an easy task. Bialk's elegant solution to the problem of how to provide enough camshaft lobes to drive the thirty-two tappets needed for the valves of the overhead valve design shows on the section view on the right side of the layout. As can be seen, the solution was to mount two side-by-side geared-together camshafts in the cylinder block—one for each bank of the engine, and each with conventionally spaced sixteen lobes.

TABLE 10.1
SPECIFICATIONS FOR THE HEMI® HEAD RB V-8 ENGINE

Model Year	1964	1964	1965	1965
Car line	Plymouth, Dodge	Plymouth, Dodge	Plymouth, Dodge	Plymouth, Dodge
Model	Competition track	Competition drag	Competition track	Competition drag
Displacement (cu. in.)	426	426	426	426
Bore	4.25	4.25	4.25	4.25
Stroke	3.75	3.75	3.75	3.75
Compression ratio	12.5:1	12.5:1	12.5:1	12.5:1
BHP @ rpm	400 @ 5600	425 @ 6000		425 @ 6000
Torque (lb-ft) @ rpm	465 @ 3800	480 @ 4600		480 @ 4600
Combustion chamber	Hemispherical	Hemispherical	Hemispherical	Hemispherical
Cyl. head material	Cast iron	Cast iron	Cast iron	Aluminum
Cyl. block mat'l.—skirt type	Cast iron—deep, cross-bolts	Cast iron—deep, cross-bolts	Cast iron—deep, cross-bolts	Cast iron—deep, cross-bolts
Cyl. block deck height	10.725	10.725	10.725	10.725
Cyl. block length	23.46	23.46	23.46	23.46
Cyl. bore spacing	4.8	4.8	4.8	4.8
No. of main brgs.—ck/s mat'l.	5—forged steel	5—forged steel	5—forged steel	5—forged steel
Ck/shaft main journal dia.	2.75	2.75	2.75	2.75
Crankpin journal dia.	2.375	2.375	2.375	2.375
Piston mat'l.—no. of rings	Forged aluminum—3	Forged aluminum—3	Forged aluminum—3	Forged aluminum—3
Connecting rod centers	6.861	6.861	6.861	6.861
Valve system	OHV—dual rocker shafts	OHV—dual rocker shafts	OHV—dual rocker shafts	OHV—dual rocker shafts
Valve events	312-312-88	300-300-76	312-312-88	312-312-88
Camshaft drive	Double-row roller chain	Double-row roller chain	Double-row roller chain	Double-row roller chain
Valve head dia.	Intake 2.25; exhaust 1.94	Intake 2.25; exhaust 1.94	Intake 2.25; exhaust 1.94	Intake 2.23; exhaust 1.94
Valve lash adjustment	Mech.—rocker arm screw	Mech.—rocker arm screw	Mech.—rocker arm screw	Mech.—rocker arm screw
Oil pump/distributor drive	Diagonal at front of block	Diagonal at front of block	Diagonal at front of block	Diagonal at front of block
Fuel system	4-barrel carburetor	2 × 4-barrel dual-plenum short ram	4-barrel carburetor plenum-ram	2 × 4-barrel dual-plenum short ram

(Continued on next page)

TABLE 10.1
(Continued)

Model Year	1966–1967	1968–1969	1970–1971
Car line	Plymouth, Dodge	Plymouth, Dodge	Plymouth, Dodge
Model	Street Hemi®	Street Hemi®	Street Hemi®
Engine code			E74
Displacement (cu. in.)	426	426	426
Bore	4.25	4.25	4.25
Stroke	3.75	3.75	3.75
Compression ratio	10.25:1	10.25:1	10.25:1
BHP @ rpm	425 @ 5000	425 @ 5000	425 @ 5000
Torque (lb-ft) @ rpm	490 @ 4000	490 @ 4000	490 @ 4000
Combustion chamber	Hemispherical	Hemispherical	Hemispherical
Cyl. head material	Cast iron	Cast iron	Cast iron
Cyl. block mat'l.—skirt type	Cast iron—deep, cross-bolts	Cast iron—deep, cross-bolts	Cast iron—deep, cross-bolts
Cyl. block deck height	10.725	10.725	10.725
Cyl. block length	23.46	23.46	23.46
Cyl. bore spacing	4.8	4.8	4.8
No. of main brgs.—ck/s mat'l.	5—forged steel	5—forged steel	5—forged steel
Ck/shaft main journal dia.	2.75	2.75	2.75
Crankpin journal dia.	2.375	2.375	2.375
Piston mat'l.—no. of rings	Forged aluminum—3	Forged aluminum—3	Forged aluminum—3
Connecting rod centers	6.861	6.861	6.861
Valve system	OHV—dual rocker shafts	OHV—dual rocker shafts	OHV—dual rocker shafts
Valve events	276-276-52	284-284-60	284-284-60
Camshaft drive	Double-row roller chain	Double-row roller chain	Double-row roller chain
Valve head dia.	Intake 2.25; exhaust 1.94	Intake 2.25; exhaust 1.94	Intake 2.25; exhaust 1.94
Valve lash adjustment	Mech.—rocker arm screw	Mech.—rocker arm screw	Hydraulic tappet
Oil pump/distributor drive	Diagonal at front of block	Diagonal at front of block	Diagonal at front of block
Fuel system	Tandem 4-barrel carbs	Tandem 4-barrel carbs	Tandem 4-barrel carbs

Chart 10.1
Time Lines—RB Hemi® Head V-8

Engines, Car Lines, and Displacements (cu. in.)

Engine	Car Line	- - Model Years - - 1964	1965	1966	1967	1968	1969	1970	1971
RB Hemi Head V-8 4.8 cyl. bore centers	Plymouth	426	426	426	426	426	426	426	426
	Model	VP2	AR2	BR2	CR2	DR	ER	FB, FR	GB, GR
	Dodge	426	426	426	426	426	426	426	426
	Model	VD2	AW2	BW2	CW2	DW	EW	FJ, FW, FX	GJ, GW

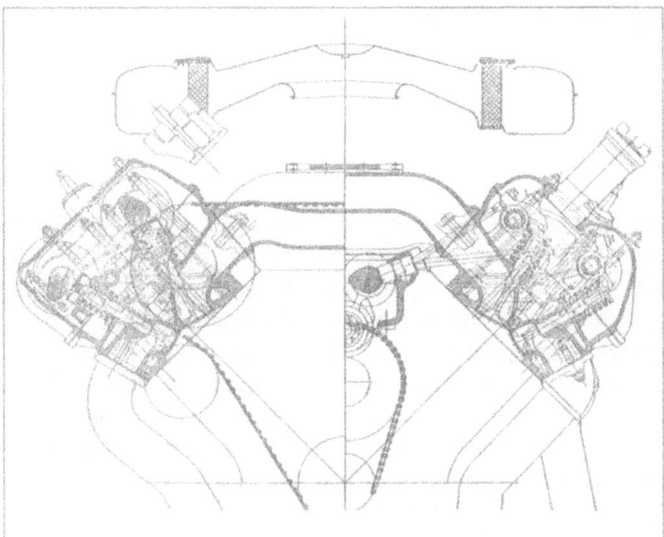

Fig. 10.11 *A cross section of the A925 426-cu.-in. RB V-8 engine with four valves per chamber. The left side of the layout shows the double overhead camshaft version. The right side shows the overhead valve version with dual camshafts and pushrods for operating the valves, 1964.*

Rather than rushing straight into the procurement of multi-cylinder engines for testing the double overhead camshaft concept that was the prime focus, it was decided to first procure enough parts to build a valve gear test engine so the double overhead cam system and its timing belt drive could be checked before procuring and building complete engines. The valve gear test engine did require the procurement of right and left aluminum cylinder heads, machined exactly as intended for the race engine, together with all the valve gear parts. Inside the cylinder block provided for the test engine, connecting rods and pistons were omitted. The crankshaft of the engine was driven by a motoring dynamometer with controls and capacity to drive the rig through the entire anticipated speed range. Problems surfaced shortly after testing started. One problem was slippage of the drive belt system. A more serious problem was cracking of the cast aluminum tappet carriers. Solutions were underway when a message quickly was relayed to Engine Design and Development managers—the A925 program was cancelled! All design and development work was to cease!

This sudden change in plans was the result of a Chrysler victory of sorts at a conference with NASCAR officials. Although work was underway on the A925 engine, Chrysler by far preferred keeping the Hemi® engine as its main race engine, avoiding expensive new programs for both the engine and the cars. Arranging a meeting with NASCAR, Ronnie Householder, the Chrysler circuit race manager and a former champion race car driver, showed a photograph of the A925 valve gear test engine (displayed on Fig. 10.12) that, mounted on a dynamometer with an intake manifold and other parts added, looked similar to an actual operating engine.[6] He explained that development of the engine had started and that if the single overhead camshaft 427 Ford engine was permitted to run, Chrysler would counter as fast as possible with the A925 double overhead camshaft engine-powered cars. NASCAR responded, informing Householder that it was not in favor of these factory-backed escalations of power and would take steps to ensure that neither the Chrysler nor the Ford overhead camshaft engines would be permitted to run on their tracks. Householder lost no time in sending that message to Engineering, triggering the cancellation of the program.

However, NASCAR still wanted to level the playing field and give the other makes a better chance at winning, which led to the progressive handicaps for the Hemi® engine as previously noted. The Ford 427-cu.-in. single overhead camshaft engine did appear on dragstrips and showed itself to be a tough competitor.

Fig. 10.12 *The A925 valve gear test engine mounted on an Engineering dynamometer. Used for testing the double overhead camshaft system, this was not a running engine.*

Some time after program cancellation, with storage space needed for hardware for new programs, the A925 valve gear test engine was sent, as customary, to the By-Products department for scrapping. Sold as-is, it was purchased by an outside enthusiast. Sold and resold several times, it has appeared at various meets. A well-written article about the engine, titled "The Doomsday Machine—Chrysler's 1964 DOHC 426—The Ultimate Hemi," authored by Tom Shaw, was published in the June 1989 issue of *Muscle Car Review*. The article includes a photo of the engine mounted on a dynamometer test stand—the same photo that Ronnie Householder showed to NASCAR.[7]

High Performance for All—The A279 LB V-8 Engine

An affordable engine dedicated to high performance. That's what the vehicle planners wanted to create in the late 1960s. An engine between the high-priced, low-volume, high-power Street Hemi® and the high-volume, standard Low-B and Raised-B V-8s. It would have the same displacement range as the LB and RB engines but, unlike the LB-RB engine, would have only one block height for maximum parts commonality. An engine with a cylinder head much more amenable to high-volume production than the complex double rocker shaft system of the Hemi® but with valve sizes and a combustion chamber shape that would be at least a close match with that of the Hemi®. Could it be done? Engine designers long for that kind of assignment.

Coded A279 when the project was started in 1968, the engine quickly took form on the drafting table. Initially pegged at having two displacements—396 and 440 cu. in., with a common 4.32-in. bore and two strokes, 3.3875 and 3.75-in.—the LB block structure was the starting point. Later during the program, the common bore size was bumped up to 4.34 in., increasing displacements to 400 and 444 cu. in. Revisions to the crankcase gave clearance for swinging the longer 3.75-in. stroke. Designing the cylinder head was a considerably greater challenge. However, recent designs of an overhead valve six-cylinder engine for Chrysler Australia, described in Chapter 11, had incorporated a separate pedestal for each rocker arm, allowing the valves to be positioned so that the combustion chamber achieved a hemispherical-like shape. Similar in principle to the rocker arm systems used for years by Chevrolet and Ford engines, the Chrysler pedestal design used only a single part rather than multiple parts to anchor the stamped-steel rocker arm and to provide a ball-shaped seat about which the arm could rotate. Taking advantage of the resulting freedom in positioning the valves, the valves were splayed or canted to provide free-flowing intake and exhaust ports, large valves with the same head diameters as those of the Hemi® engine, good spark plug cooling, and, as mentioned, to obtain an almost hemispherical shape for the combustion chamber. Of interest, when the first design for the cylinder head was made, a stud was used for anchoring each rocker arm, giving rise to the engine nickname "Ball Stud Hemi," a name always considered temporary—a bit too aggressive as a permanent name for the engine if it were to enter production. The cross section of the engine as shown on Fig. 10.13 illustrates the details of the engine design.

Following procurement of the first of about a dozen 440-cu.-in. prototype engines, dynamometer testing commenced. A development engine soon was producing power where predicted,

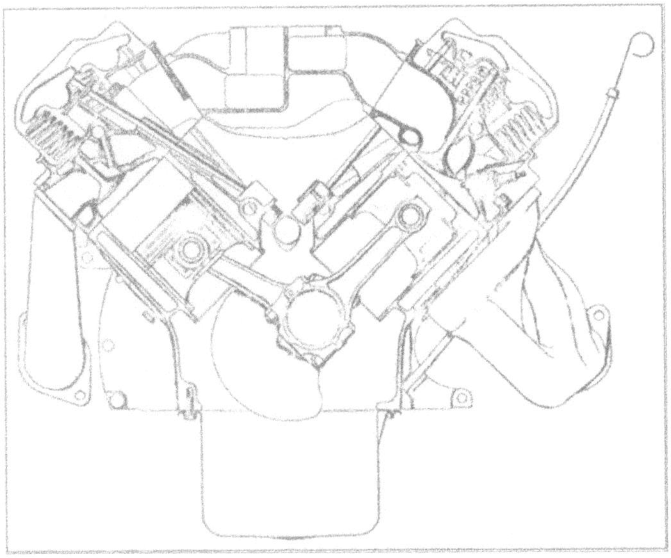

Fig. 10.13 *A cross section of the A279 V-8 engine. Using a modified LB V-8 cylinder block, the high-performance, canted-valve engine with almost hemispherical combustion chambers was designed for high-volume production at displacements of 396 and 440 cu. in., later boosted to 400 and 444 cu. in., 1968.*

better than the wedge-head RB engine but not quite as good as the real Hemi® engine. Then in 1970, about two years after the start of the program, the same outside factors that included the escalating emission requirements and higher insurance premiums as previously discussed and that caused the Hemi® engine to be cancelled likewise caused the A279 engine program to be reappraised. Clearly, investing scarce corporate resources in the continued development and production tooling needed for a new, big-displacement, high-performance engine faced with so many uncertainties would not be a prudent investment. Program termination followed.

Two prototype engine assemblies were known to survive. One engine, after residing in storage for a number of years, was offered by Product Planning to a professional drag racer with strong factory ties, Richard (Dick) Landy. The engine came back home, in a way, when, in 2003, it was loaned by Landy for public display at the Walter P. Chrysler Museum during a special Hemi® engine exhibit. Later, with additional ownership changes, the engine was installed in a suitable contemporary vehicle, a 1969 Plymouth Barracuda hardtop notchback once owned by Tom Hoover, and has made appearances at Chrysler products enthusiasts' meets.[8]

CHAPTER 11

Overhead Valve In-Line Sixes

The Enduring Slant Six Engine

The Slant Six engine produced by Chrysler Corporation from 1960 to 1991 became one of the best recognized and most durable of all engines manufactured by Chrysler. The story of the Slant Six engine is related in this section.

During the late 1950s, many car buyers became attracted to cars with a smaller size and better fuel economy than the standard high-volume Fords, Chevrolets, and Plymouths offered by the "Big Three." A notable import that offered low cost, small size, and excellent fuel economy was the Volkswagen Beetle powered by a horizontally opposed, four-cylinder, air-cooled, overhead valve (OHV) engine. The 1959 Studebaker Lark VI powered with a 169-cu.-in. L-head six-cylinder engine and the AMC Rambler American with its 196-cu.-in. L-head six-cylinder engine were domestic offerings. Substantial sales volumes by these three makes showed that the market was well worth entering.

General Motors, Ford, and Chrysler moved almost simultaneously into this market segment with compacts of their own. Three quite different cars were designed, developed, and produced for the 1960 model year, with each corporation deciding separately on the best approach for capturing the market.

The General Motors entry was the Chevrolet Corvair, powered with a 140-cu.-in. air-cooled, horizontally opposed, aluminum-rich, six-cylinder engine located behind the driver. The Ford entry was the Falcon, a basic car with a conventional driveline and an appropriately basic 144-cu.-in., all-iron but lightweight, in-line six-cylinder engine. The Chrysler entry was the Valiant, which combined a conventional engine-up-front rear-wheel-drive arrangement, with the front torsion bars and the rear leaf springs suspension system introduced on its larger cars in 1957, as can be seen on Fig. 11.1.

A special engineering team, under Robert M. (Bob) Sinclair, the Engineering executive in charge of the Valiant program, was assembled to design the car that was assigned the engineering code A901. Most of the team members were chassis and body engineers. Engine responsibility, however, was retained by the existing Engine Design and Engine Development departments.

Selecting an engine for the Valiant was a special challenge. Engine Design under Assistant Chief Engineer Robert S. (Bob) Rarey and me as Managing Engineer supplied information to the Valiant team on a number of candidate engines, including four-cylinder engines, in-line sixes, and V-6s, with iron and aluminum versions of each type. The need to keep the length of the car to a minimum while providing space for six passengers precluded the use of any in-line, upright, six-cylinder engine having a water pump located in front of the engine—the conventional approach. A true breakthrough came when the lead designer Fred Rose and the Engine Design supervisor Ray Latham came up with the idea of leaning the engine over to one side so the water pump could be located alongside the cylinder block instead of in front of it, thereby considerably shortening the engine.[1]

The Valiant team chassis engineers determined that a lean to the right was best for a number of reasons. Chief among the reasons was that with the engine centerline offset to the right of the car centerline (a common practice to provide room for the driver's foot), the engine fan achieved a desired location near the center of the radiator.

In a meeting on the morning of April 10, 1958, Engine Design requested that the engine type and size be finalized by the Valiant team and senior management by May 1, 1958, so the engine design could be completed and detail drawings started. Bob Rarey and I preferred an in-line, six-cylinder engine with a displacement of 170 cu. in., with a 3.4-in. bore, a 3.125-in. stroke, and cylinders slanted 30 deg to the right.

Later that same day, Harry Chesebrough, vice president of the Plymouth Division, asked Engineering management that, if possible, the new Valiant engine be designed to have a large-displacement version. The large version would be aimed at replacing the obsolete 230-cu.-in. L-head six-cylinder engine

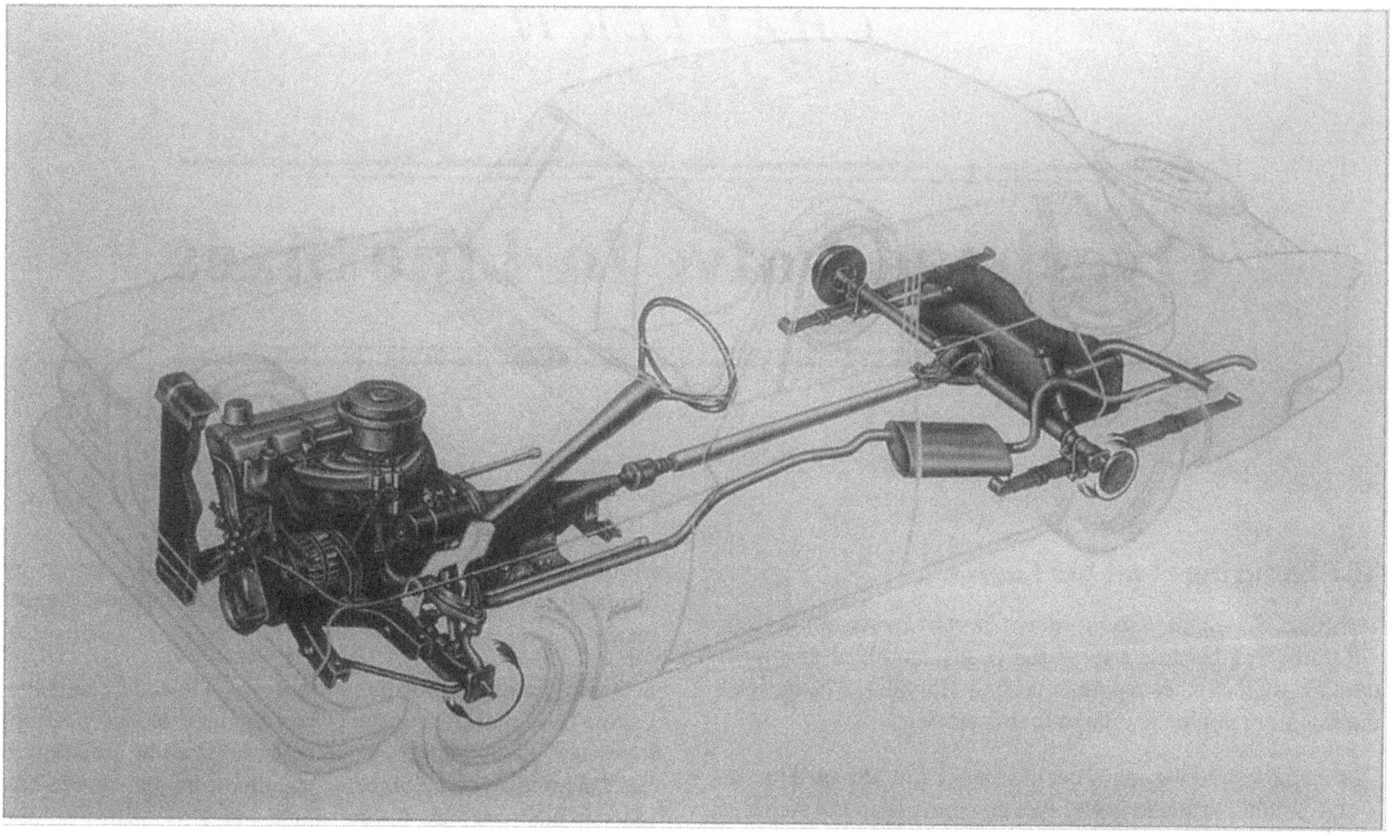

Fig. 11.1 *The 1960 Valiant engine and chassis arrangement. Note the compact engine length, the torsion bars at the front, and the leaf springs at the rear.*

still powering standard-size Plymouth and Dodge cars. This request was quickly relayed to Bob Rarey, and the design investigation started immediately.[2]

An increase in the bore size was ruled out. With the longitudinal tightness of the 170 Valiant engine and engine compartment, no space was available to increase the bores. I remember asking Bob Sinclair, who was in charge of the Valiant program, for a modest one-quarter-inch increase in the engine length. The request was politely but firmly denied; therefore, a stroke increase was the only solution. Within five days, it was determined that there could be a second version of the engine with a 1-in.-longer 4.125-in. stroke, giving a 225-cu.-in. displacement. This second version would require a stroked crankshaft, a cylinder block with a greater deck height, longer connecting rods, and longer pushrods. Other than these four parts, virtually all parts would be identical between the two engines—a real boon to Manufacturing and Service.

On April 16, 1958, a review meeting was held with Paul Ackerman, vice president of Engineering, where three engine proposals—two approaches involving families of four- and six-cylinder engines, and the combined 170/225 Slant Six engine approach—were presented. Later in the month, Ackerman decided Engineering should proceed with only the 170/225 Slant Six engine. With that decision, work on the Slant Six engine moved ahead at wide-open throttle.

With the engine tilted to the right, a great deal of space was available on the left side of the engine for the intake manifold. Engine Design engineer John Hurst took advantage of this space to come up with a manifold having especially long branches opening into a plenum chamber below the carburetor mounting pad. Engine Development engineers John Platner and Don Moore perfected this intake manifold and the complementary exhaust manifold to obtain the best performance and equal fuel/air mixture to all cylinders. The five innovators, Messrs. Rose, Latham, Hurst, Platner, and Moore, were awarded U.S. Patent #3,109,416, its first page illustrated on Fig. 11.2, for the overall engine arrangement.

Further engine details were as follows. Both the intake and exhaust ports were located on the same side of the cylinder head, so that the exhaust gas from all six cylinders could be used to quickly warm the central floor of the intake manifold, minimizing the use of the automatic choke, enhancing cold-start

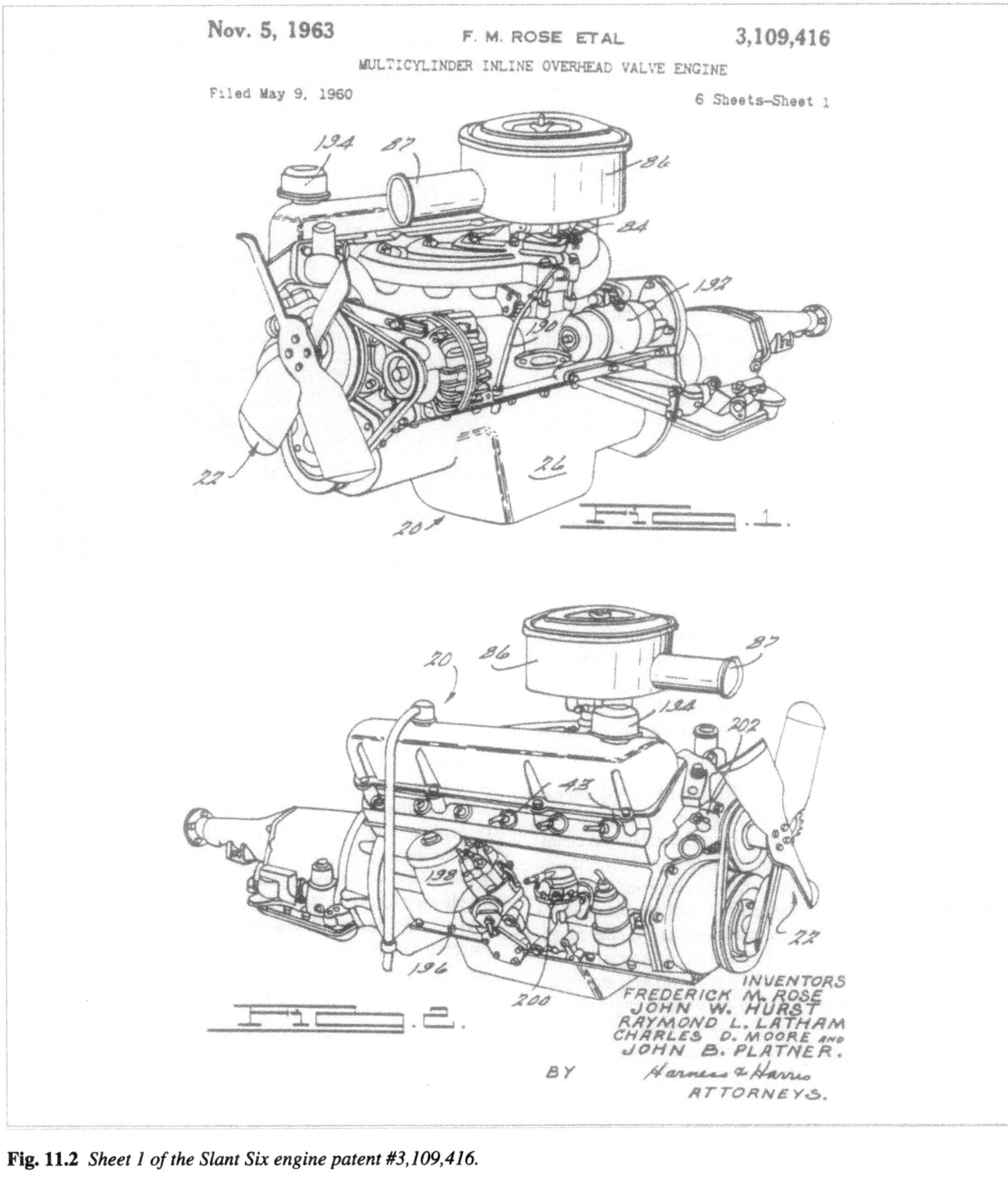

Fig. 11.2 *Sheet 1 of the Slant Six engine patent #3,109,416.*

drivability, and improving fuel economy. No two exhaust ports were located adjacent to each other for optimum valve seat cooling and maximum valve durability.

In terms of the bore and stroke ratio, the 170-cu.-in. engine's 3.4 × 3.125 in. dimensions gave it a desirable oversquare ratio, similar to those of the V-8 engines. The 3.4 × 4.125 in. of the

225-cu.-in. engine made the larger version definitely undersquare, reminiscent of L-head engines. It would have good torque characteristics but, with the bore size restricting the valve sizes, would be limited in power. However, customers wanting extra power certainly could opt for the V-8 engines available on all the larger cars.

A design feature imposed by the engine length restriction was the use of four crankshaft main bearings instead of seven main bearings. It actually was not much of a loss because Advance Engine Design supervisor Ray Latham knew that crankshafts with four bearings generally were less expensive and torsionally stiffer than those with seven bearings. His advice was to steer clear of "those whippy seven main bearing crankshafts."[3] With four bearings and a long 4.125-in. stroke, the main bearing diameter was made a husky 2.75 in., the same diameter as the main bearings in the big-block RB V-8 engine.

Another design challenge caused by the length restriction was the type of tappet to be used. The lead designer, Fred Rose, found there was enough room to use mushroom tappets with large heads and small-diameter stems as were used on the L-head engines, but not enough room for the large-diameter cylindrical tappets that were being used 100% on the V-8 engines. Bob Rarey and I knew that the engine simply had to have the V-8 type of tappet to ensure a low-cost supply and to make the tappets serviceable from the top of the engine. The three of us spent an afternoon and evening at Fred's drafting table, exploring all the possibilities of tappet locations, camshaft lobe locations, camshaft bearing lengths, and everything else along that critical section of the engine until we finally had squeezed the cylindrical tappets in place without needing any change to the engine length.[4]

Following the pattern set by the B/RB big-block V-8 engine, the cylinder head of the Slant Six engine was designed with overhead in-line valves and wedge-shaped combustion chambers. A tappet chamber running the length of the right side of the cylinder block was designed to receive overhead valve lubricating oil from the cylinder head without the need for drilled holes in either the cylinder head or the cylinder block. Rocker arms were precision-fabricated steel stampings. The pushrod end of the rocker arm had a screw with an interference thread for adjusting the valve lash. The stringent engine length restrictions led to the use of thin aluminum tubes with O-rings rather than cast walls for separating the spark plugs from the adjacent pushrods. Although more expensive, the thin tubes did provide a weight savings. Figure 11.3 shows a cutaway view of a portion of the valve arrangement.

For possible weight reduction, an aluminum version of the cylinder head, intended to be made by a semi-permanent mold process, also was designed and experimentally procured. Its design incorporated valve seat inserts, valve guides, and large-diameter head bolt bosses. It later was abandoned in favor of cast iron, as related in the following paragraphs.

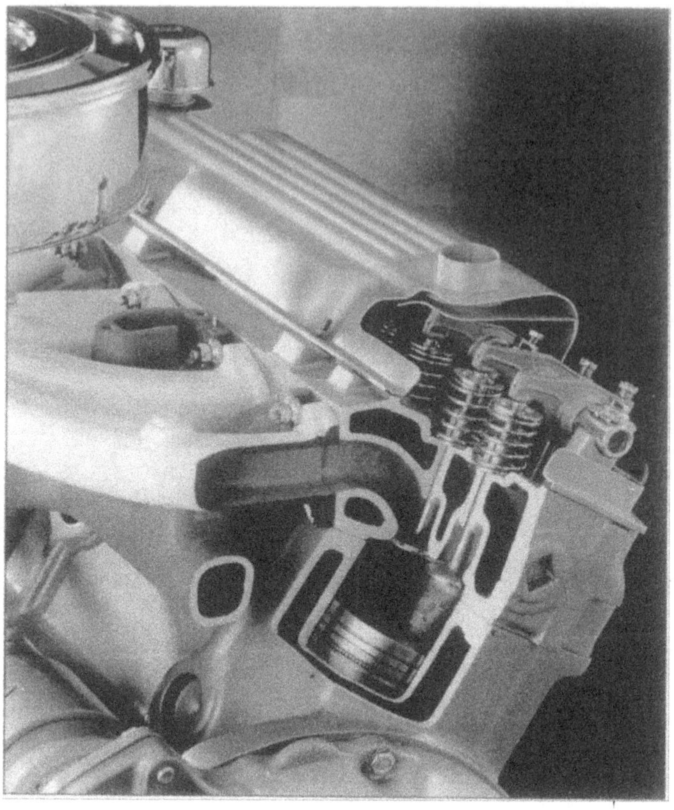

Fig. 11.3 *Cutaway view of the 1960 225-cu.-in. RG Slant Six engine.*

Figures 11.4 and 11.5 show a longitudinal section and a cross section of the 170-cu.-in. engine, respectively. Figures 11.6 and 11.7 the show right-front and left-front outside views, respectively. A longitudinal section and a cross section of the 225-cu.-in. engine appear on Figs. 11.8 and 11.9, respectively. The similarity of the two engines is well illustrated by these hard-to-tell-apart sections. Both engines had iron cylinder blocks and iron cylinder heads. Engineering Program number A907 was assigned to the 170-cu.-in. engine, and A734 was assigned to the 225-cu.-in. engine. They also were given the designators LG (for low G) and RG (for raised G), respectively.

However, Engine Design was far from finished with its tasks. The choice between iron and aluminum for the cylinder block and cylinder head had yet to be made. The use of aluminum was desirable to minimize car weight and possibly to reduce tooling costs because aluminum had much higher machining speeds than cast iron and would require fewer machine tools. A decision was made to proceed with both materials, with cast iron being considered a backup in case unforeseen problems occurred during the development of the aluminum parts.

In an unprecedented move to expedite the aluminum cylinder block design, Bob Kring, a die design engineer with the Chrysler

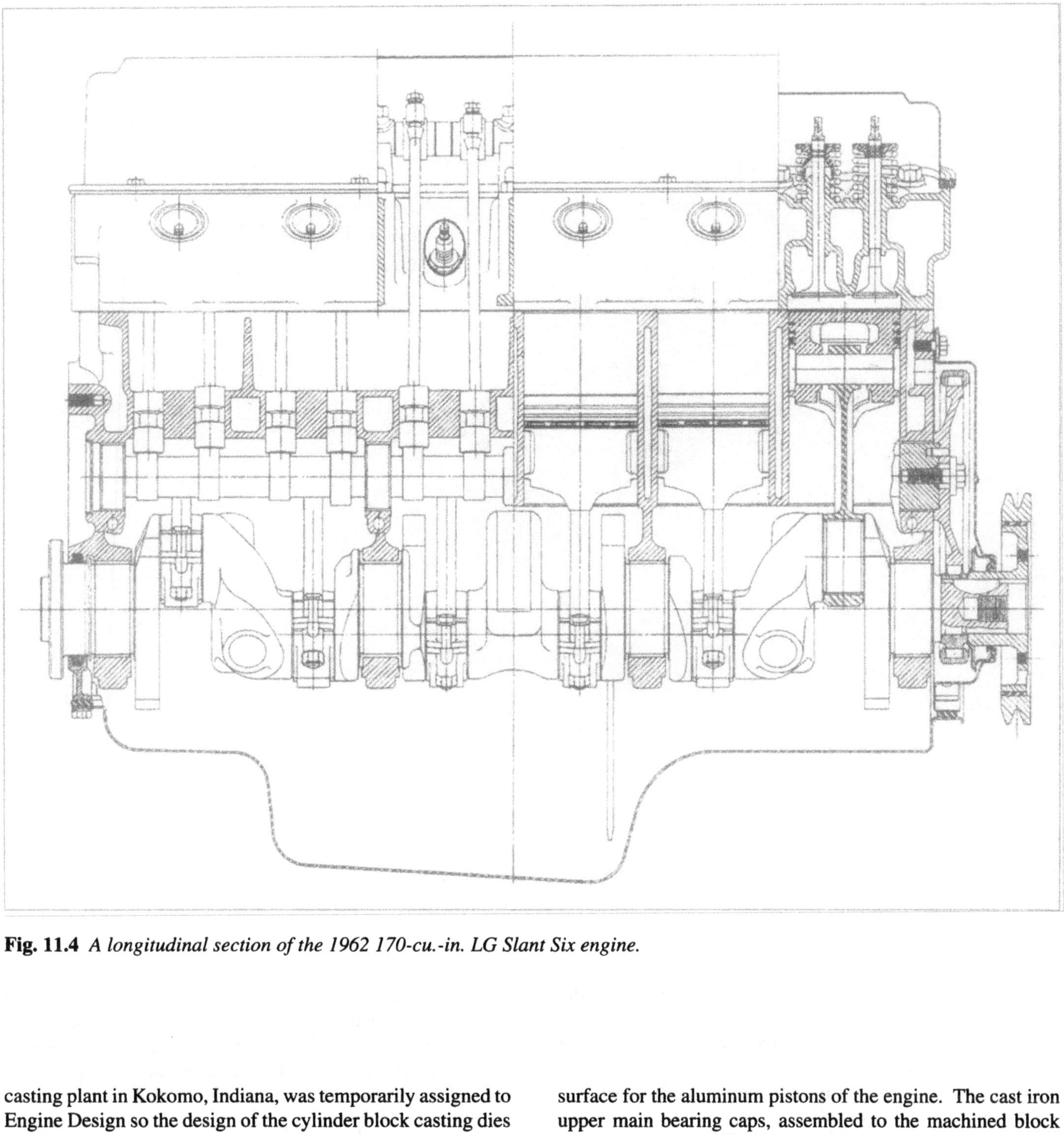

Fig. 11.4 *A longitudinal section of the 1962 170-cu.-in. LG Slant Six engine.*

casting plant in Kokomo, Indiana, was temporarily assigned to Engine Design so the design of the cylinder block casting dies could proceed simultaneously with the design of the cylinder block. The continual information interchange between Kring and the cylinder block designer ensured that the first design would satisfy the requirements of both Engineering and the production casting source of the die-cast blocks.

The die-cast aluminum cylinder block had two features of special interest: cast-in bore liners made of iron, and cast iron upper main bearing caps. The bore liners were preheated and placed in the die before the high-pressure aluminum shot entered the die cavity. These liners gave an excellent wear surface for the aluminum pistons of the engine. The cast iron upper main bearing caps, assembled to the machined block with customary cast iron lower main bearing caps, guaranteed that the main bearing clearances of the forged steel crankshaft would be held constant throughout the operating temperature range of the engine.

Experimental cylinder blocks made of cast iron and sand-cast aluminum for each displacement version of the engine were procured immediately. The sand-cast aluminum blocks would provide some development information prior to arrival of die-cast aluminum blocks, which would require longer procurement times.

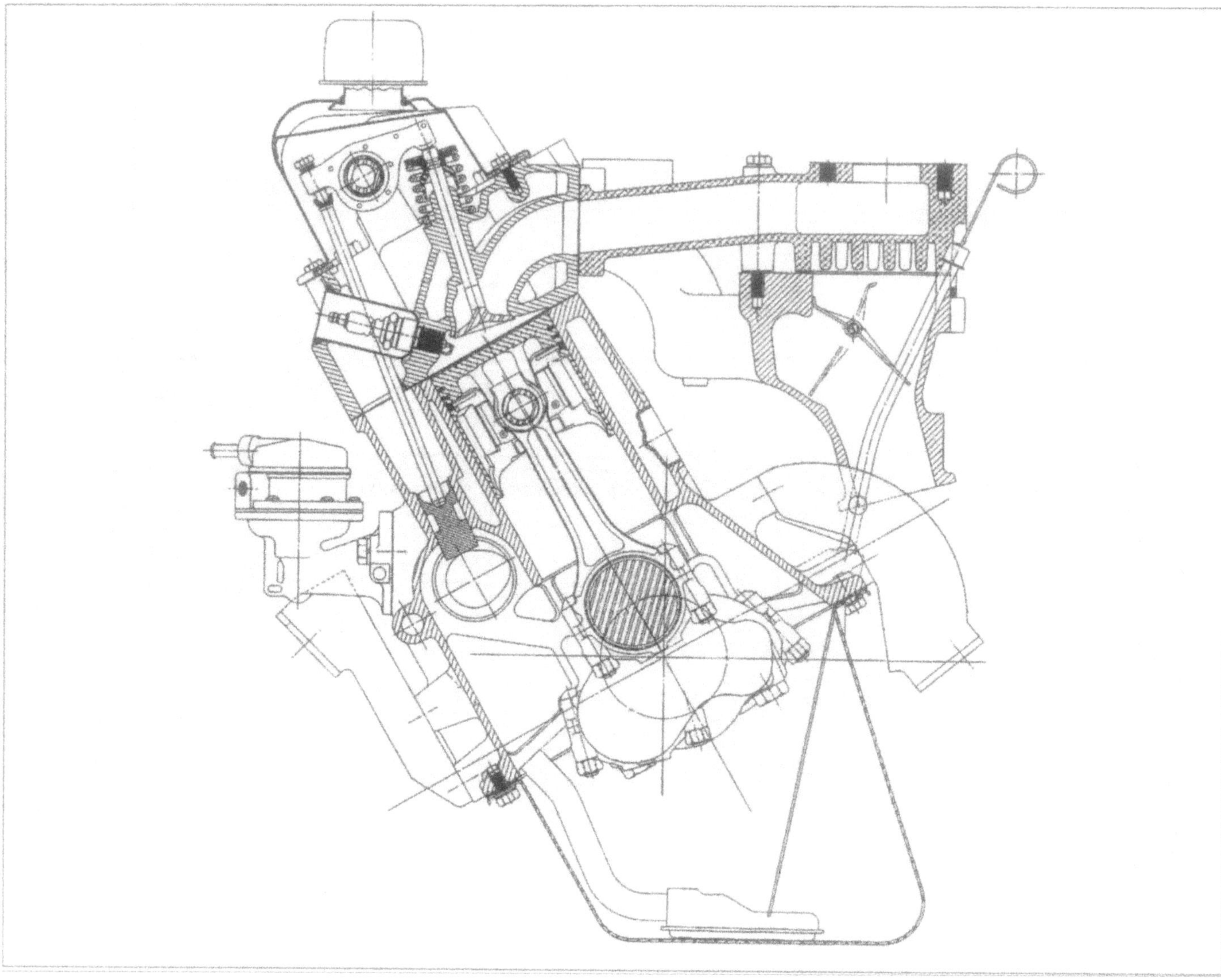

Fig. 11.5 *A cross section of the 1962 170-cu.-in. LG Slant Six engine.*

On November 26, 1958, less than seven months after the May 1958 program received a go-ahead, the first Slant Six engine ran on the dynamometer under the experienced eye of Everett Moeller, managing engineer of Engine Development. It was a 170-cu.-in. engine, built with a cast iron block. Within two weeks, an aluminum version using a sand-cast block was running. Around February 1959, the first 225-cu.-in. RG engine, built with a cast iron block, ran for the first time. In March 1959, the aluminum version of this engine, with a sand-cast block, started running. On March 18, 1959, the first die-cast aluminum LG 170-cu.-in. cylinder block acceptable for machining was delivered to Engineering by the Kokomo casting plant.

However, the entire Valiant program schedule had been moved up, and the Powertrain Manufacturing Division had to know by the end of March 1959 which material would be used for the blocks. Engineering Vice President Paul Ackerman and Director of Engineering Alan Loofbourrow called a special meeting with Engine Design, Engine Development, Valiant Engineering, and Materials Engineering to review the status of the engine. The result was a decision by Ackerman to go with cast iron as the cylinder block material for the initial production engines. Aluminum would be continued as an experimental program for future incorporation when it was ready. The logic of the decision was inescapable. Engines with iron blocks had piled up hundreds of hours of development with excellent results. No test engines had been built yet with a production-level die-cast aluminum block. The launch of the Valiant program could not be jeopardized for the attraction of the weight savings of the aluminum cylinder block.

Fig. 11.6 *A right-front view of the 1960 170-cu.-in. LG Slant Six engine.*

Fig. 11.7 *A left-front view of the 1960 170-cu.-in. LG Slant Six engine.*

Production of 170-cu.-in. engines with cast iron blocks started at the Chrysler Trenton engine plant in August 1959, followed immediately by the start of production of 225-cu.-in. engines, also with cast iron blocks. Slant Six engines being hot-tested after completing assembly are shown on Fig. 11.10. Assembly of 1960 Plymouth and Dodge cars using the 225-cu.-in. RG engine started in September 1959. Assembly of 1960 Valiants with 170-cu.-in. LG engines started in October 1959. Figure 11.11 shows a 1960 Valiant powered with a 170-cu.-in. LG Slant Six engine. Figure 11.12 shows a 1960 Dodge Dart Phoenix powered with a 225-cu.-in. RG Slant Six engine. The 170-cu.-in. engine was rated at 101 hp and 155 lb-ft torque.

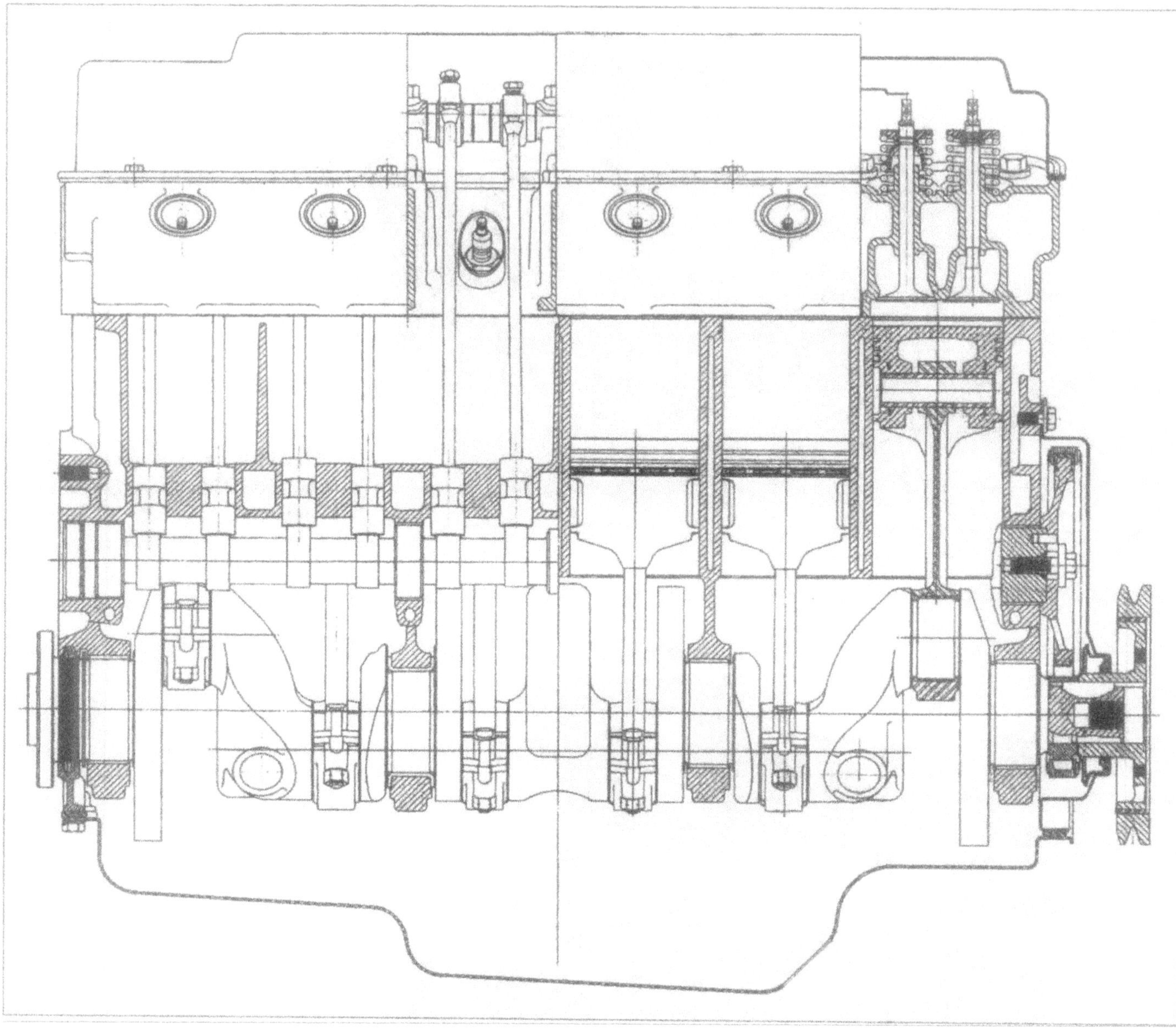

Fig. 11.8 *A longitudinal section of the 1963 225-cu.-in. RG Slant Six engine.*

The rating of the 225-cu.-in. engine was 145 hp and 215 lb-ft torque.

Following the launch of the car, in January 1960 an SAE paper informing SAE members and the public about the Valiant and its Slant Six engine, co-authored by A.G. Loofbourrow, Director of Engineering, V.M. Exner, Vice President and Director of Styling, and R.M. Sinclair, Executive Assistant to the Director of Engineering, was presented during the SAE Annual Meeting in Detroit.[5]

With production of the engine and vehicle well underway, the thought of using racing venues as a method of increasing the car-buying public's awareness of the Valiant car line then grew. This had worked well for the Hemi®-powered Chrysler and Dodge cars of the 1950s and, championed by the Plymouth product planning director Jack Charipar, gave impetus for a stock car racing version of the Valiant. Bill France, president of NASCAR, recognized the public interest in the emerging compact cars and had declared a new compact car category, with two races scheduled for Sunday, January 31, 1960 at the Daytona Beach Speedway in Florida. That race day became the target for the Chrysler factory-backed debut.

Room 37 in Dyno Building 136 was the focus of the engine work. Performance engineer Tom Hoover, with supervision by Don Moore and the able assistance of sheet-metal-fabrication whiz Pat Brady, ran through a series of tests on the 170-cu.-in.

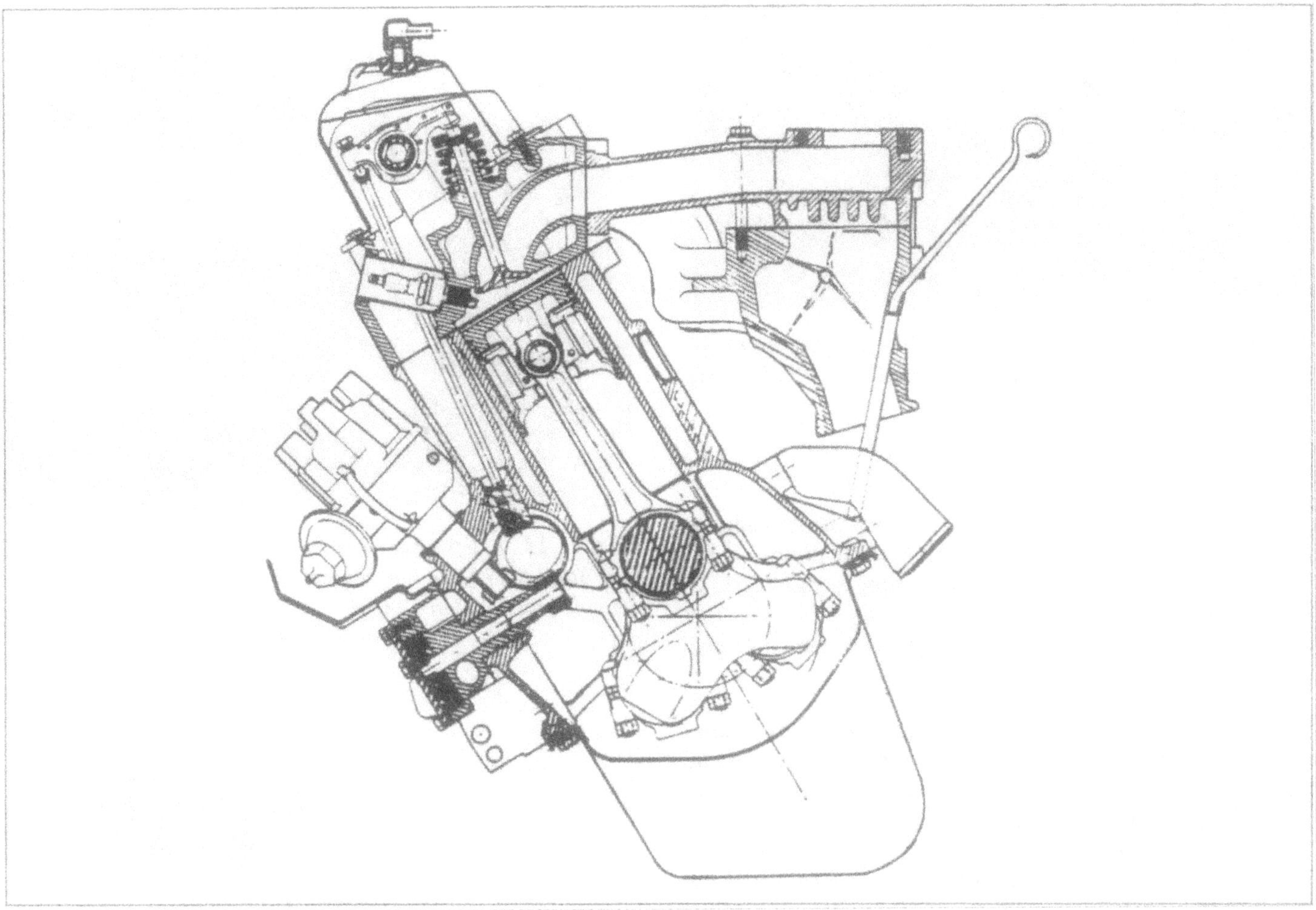

Fig. 11.9 *A cross section of the 1963 225-cu.-in. RG Slant Six engine. The positive crankcase ventilation (PCV) valve at the top of the cylinder head cover was added in 1961 for California cars, replacing the road draft tube.*

Fig. 11.10 *Hot-testing Slant Six production engines at the Trenton engine plant, circa late 1959. A close look reveals L-head engines also being processed, probably for the Marine and Industrial Division.*

Fig. 11.11 *A 1960 Valiant QX1 four-door sedan. The car was powered by a 101-hp 170-cu.-in. LG Slant Six engine.*

Fig. 11.12 *A 1960 Dodge Dart Phoenix two-door hardtop. Standard power was the 145-hp 225-cu.-in. RG Slant Six engine.*

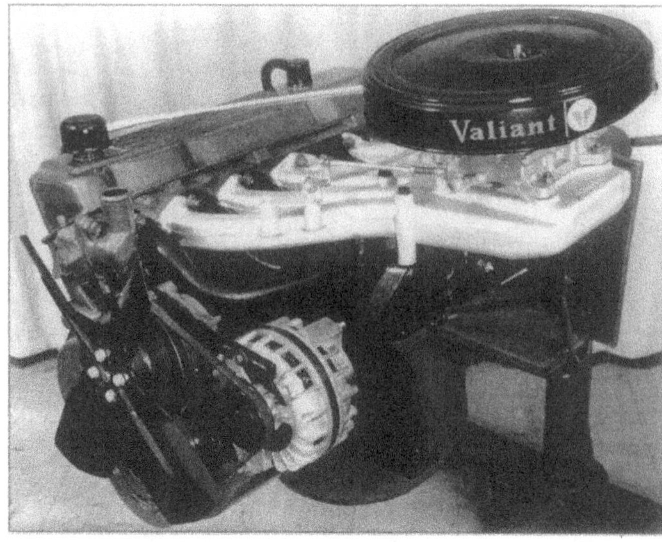

Fig. 11.13 *A left-front view of the 1960 Valiant 170-cu.-in. LG Slant Six Hyper-Pak engine. Sold as a conversion kit, the engine was rated at 148 hp.*

LG Slant Six, using a fabricated intake manifold with adjustable branch lengths. The test engine was equipped with continuous-flow fuel injection, split exhaust manifolds, 11:1 compression ratio pistons, and high-load valve springs.[6] The engine was assigned Engineering program number A785 and was named Hyper-Pak.

Dynamometer runs with a range of high-performance camshafts and various intake manifold branch lengths zeroed in on a 17-in. branch length and a long-duration, high-lift camshaft. Power was in the range of 190 hp. The race package was confirmed using a race-intended Carter four-barrel AFB carburetor. Meanwhile, Engine Design completed the drawing for a cast aluminum production version of the fabricated intake manifold. The first cast manifolds were received and run on the dynamometer test engine barely before the engine conversion kits were sent to the eight Valiant race teams.

The first of the two NASCAR compact car races held on January 31, 1960 at the Daytona Beach Speedway was a road course race, which used a 1-1/2-mile portion of the high-speed, high-banked tri-oval together with an infield road for a lap distance of 3.81 miles. The race length was 10 laps—38.1 miles. The Hyper-Pak-powered Valiants swept the field, taking the first seven places, with Marvin Panch driving the winning car.

The second compact car race of the day used only the tri-oval track. Twenty laps on its full 2-1/2-mile length made the race 50 miles of high-speed competition. Although a multi-car accident knocked out three Valiants during the race, the remaining Valiants easily took the three top places, with Marvin Panch again going into the winner's circle. The lopsidedness of the outcomes was too much for the race sponsor. The compact car division was eliminated from the NASCAR schedule.

Fig. 11.14 *A right-front view of the 1960 Valiant 170-cu.-in. LG Slant Six Hyper-Pak engine.*

The Hyper-Pak engine was not sold as a production car option but rather as an aftermarket conversion kit of parts that could be installed by dealers or individual owners. Figures 11.13 and 11.14 show the engine. The published factory rating of the 170-cu.-in. engine was an understated 148 hp and 153 lb-ft torque.

The 1960 Valiant received good reviews from car enthusiast magazines. The January 1960 issue of *Motor Trend* reported, "In both city and country driving, the car has a definite feeling of ample power and responds very well to sudden depressions of the foot throttle. The over-square, 170-cu.-in. ohv engine is free from valve clatter at all but strained rpm. There is a satisfying sound of spinning wheels when you wish to get away in a hurry, and it's quite possible to break the rear end loose in an accelerating turn." Writer Tom McCahill, relating his impressions about the Valiant versus its competitors in the March 1960 issue of *Mechanix Illustrated*, wrote, "...a handful of Falcons and Corvairs and one lone Valiant (my test car) were taken out on the sports car track (at Daytona Beach International Speedway) and given a squeeze-out. Needless to say, there was very little contest. The Valiant whipped by its lower-calories competitors like Halley's Comet might trickle by Piccard's balloon. This is a superior-performing automobile."

Although Valiant sales did not start until about a month after the sale of other corporate vehicles, 194,000 vehicles were sold by the end of the model year—an excellent beginning for the new Chrysler compact car.

Starting with the 1961 model year, the Dodge Division marketed a companion car to the Valiant, badged as the Dodge Lancer.

During the design phase of the Valiant car, the Valiant Engineering team had declared that the car would never need an engine larger than the 170-cu.-in., short-stroke, low-deck-height engine and accordingly limited the vertical space in the engine compartment available for the engine, as well as the longitudinal space, as noted. However, as soon as Plymouth dealers started selling the car, requests came for an optional engine upgrade. With the extra incentive, the chassis and body engineers somehow managed to find the extra room needed for the raised-deck, long-stroke 225-cu.-in. engine, and it was released as a factory option for both the Plymouth Valiant and the Dodge Lancer, starting with the 1961 model year.

The 225-cu.-in. engine was well received. The February 1961 issue of *Road & Track* reported "...the combination of its 145 bhp and 215 ft/lb [*sic*] of torque along with a 3-speed TorqueFlite transmission seems almost ideal." The November 1960 issue of *Motor Trend* stated, "Equipped with this 145-hp engine, the Lancer should well be serious competition for the new compacts with V-8's." Figure 11.15 shows a 1961 Dodge Lancer that offered the 225-cu.-in. RG engine as an option over the standard 170-cu.-in. LG engine.

Riding on the attention given to the 1960 Valiant 170-cu.-in. Hyper-Pak, the Dodge Division requested for its 1961 Lancer that a similar engine be released together with a Hyper-Pak version of the large 225-cu.-in. engine, which carried a rating of 196 hp and 212 lb-ft torque. Valiant preferred staying with only the 170-cu.-in. Hyper-Pak for 1961. Both the 170- and 225-cu.-in. Hyper-Paks were discontinued at the end of 1961.

With the launch of the engines going well at the Trenton engine plant, Engineering refocused its efforts to complete the work on the aluminum cylinder blocks. By February 1960, the Kokomo casting plant had successfully produced several hundred die-cast aluminum 170-cu.-in. LG and 225-cu.-in. RG blocks, many of which were tested by the Engine Development department.

Fig. 11.15 *The 1961 Dodge Lancer 770 four-door sedan. The optional engine of this car was a 145-hp 225-cu.-in. RG Slant Six.*

More than 100 aluminum 225-cu.-in. prototype engines were built and tested on dynamometers and in vehicles. Engineering program number A826 was assigned to the 225-cu.-in. RG aluminum block engine.

Near the end of the 1960 model year, a small production trial run was made of both low-deck 170-cu.-in. and raised-deck 225-cu.-in. die-cast aluminum cylinder blocks machined on the Trenton engine plant's dual-tooled (for iron and aluminum) high-volume cylinder block line, and the blocks then were assembled into engines. Most were sent to Engineering, but a few were installed in vehicles sold to the public. Following this trial, Senior Design Engineer J.W. (Jim) Dean was temporarily transferred from Engine Design to the Engineering staff at the Trenton engine plant to assist in a larger production run that was the next phase.

This next phase resulted in 10,767 engines being assembled with die-cast aluminum blocks, all 225-cu.-in., during the 1961 model year and installed in vehicles sold to the public. Compared to processing iron blocks, processing aluminum blocks required extra-careful handling to avoid nicks, as well as additional inspection and sealing operations to guard against porosity leakage.

With the experience gained, manufacturing efficiency was improved, and, in 1962, 36,000 additional 225-cu.-in. RG aluminum block engines were built at the Trenton engine plant and were installed in Plymouth and Dodge vehicles sold to the public.

A paper describing the aluminum Slant Six engine, co-authored by E.G. Moeller, me, and H.E. Erickson (plant manager of the Kokomo casting plant) was presented by Ev Moeller during the January 1961 SAE International Congress in Detroit.[7] Figure 11.16 shows a 225-cu.-in. RG Slant Six aluminum engine on display at the Walter P. Chrysler Museum.

Fig. 11.16 *A left-front view of the 225-cu.-in. RG Slant Six engine with an aluminum cylinder block, on display at the Walter P. Chrysler Museum. Photo by W.L. Weertman.*

Then a hard-nosed assessment of the aluminum block program was made by the Trenton engine plant and Engine Division manufacturing staff, with the conclusion that despite substantial improvements, neither the die-casting plant nor the machining plant had been able to achieve the day-in, day-out problem-free processing that was essential for consistent high-quality, high-volume manufacturing. Also, the switching back and forth of the block machining line was costly in terms of plant productivity. Further development runs such as the three that had been made could not be afforded. By the consensus of Engineering and Manufacturing, the aluminum block program was terminated at the end of 1962, and only cast iron blocks were produced from then onward. However, a few aluminum engines were assembled into 1963 cars.

The aluminum cylinder head never entered production. Casting porosity plagued the experimental cylinder heads. Finally, on February 17, 1960, the Chrysler Forge and Foundry Division, supplier of the castings, requested that the program be cancelled.

Slant Six Engines in the Mobilgas Economy Run

Only three months after the 170-cu.-in. LG Slant Six engine made headlines for its Hyper-Pak performance wins in January 1960, as previously described, a car powered by the 225-cu.-in. RG Slant Six engine took home a first-place trophy in another arena: fuel economy. The car, a Plymouth Savoy, was entered in the low-price six-cylinder class of the April 1960 Mobilgas Economy Run from Los Angeles to Minneapolis, Minnesota. The car averaged 24.826 mpg over the 2061-mile distance, giving it the top honors in its class. The victory was repeated in 1961. In 1962, two cars powered by 225-cu.-in. Slant Six engines achieved first-place finishes: a Plymouth Valiant in Class D, and a Plymouth Savoy in Class E. Additional first-place awards were handed to Slant Six engine-powered cars in 1963, 1964, 1966, and 1967. For the 1965 and 1968 runs, respectable second-place finishes were achieved.[8] This overall outstanding fuel economy record was the result of thorough engineering work by Bob Cahill, noted in Chapter 8 for being a member of the group that broke speed records with a 1956 Plymouth powered with the then-new A-engine V-8, assisted by car performance engineering whiz George Wallace. Cahill, starting his Mobilgas Economy Run participation in 1960 as a factory advisor to the Plymouth car dealers' teams that were entered, in 1962 reorganized the program into an engineering-centered team under the auspices of the Chrysler-Plymouth Division. Car testing established techniques to maximize fuel economy, and many schemes were explored to provide road conditions and other information during the run to assist the drivers and navigators or co-drivers. An interesting disclosure of how it was all done, written by Bob Cahill, appeared as a three-part article titled "Mopar and the Mobil Economy Run" in the October 1995, November 1995, and January 1996 issues of *High Performance Mopar* magazine.

This famous fuel economy run that originally was started in 1936 and was renamed the Mobil Economy Run in 1961 was discontinued following completion of the 1968 run.

Subsequent Changes

During its years of usage, the Slant Six received a number of changes. The road draft tube was replaced by a positive crankcase ventilation (PCV) system in California in 1961 and in all markets by 1964 to comply with emission regulations. The exhaust valve seats in the cylinder head were induction hardened starting in 1972 to avoid seat sinkage with non-leaded gasoline. Air injection was added in 1972, and exhaust gas recirculation (EGR) was added in 1973, both for emission control. In 1975, smaller spark plugs were incorporated, which allowed the spark plug tubes to be eliminated from the cylinder head. This, and the internal passages for the air injection system, increased the weight of the head by approximately 20 lb. A major change was made to the crankshaft in 1977 when its material was changed from forged steel to cast nodular iron. This change in material required the lengths of the main bearings and the connecting rod big-end bearings to be reduced to regain the torsional stiffness in the crankshaft. Because of this change, cylinder blocks, crankshafts, connecting rods, and bearings of engines built in 1977 and afterward were not interchangeable with the corresponding parts of engines built prior to 1977.

Intake manifolds started as semi-permanent molded aluminum castings, were changed to iron castings in 1961 due to porosity problems, were changed again to two-piece electron-beam-welded aluminum die-castings in 1978, and then were changed back to cast iron in 1981.

Another significant change occurred in 1981 with the release of hydraulic tappets in place of the mechanical tappets that had been used since the start of production. Although apparently simple to accomplish, this switch posed a real challenge at the time because the original design of the engine did not have direct oil feed to the tappets—the aluminum die design would not permit it. A design for the aluminum block using twelve small-diameter cross holes had been made, and bosses for the holes were included on the casting. However, due to later concerns for drill breakage during machining operations and exposure to external oil leaks, the cross drills were never added to the machining lines. A good solution to the dilemma came from an engineer at the Chrysler Canada Windsor (Ontario) engine plant, who suggested routing oil down hollow pushrods from the cylinder head to feed the hydraulic tappets. The tappet supplier made the experimental tappets, and it worked! The hydraulic tappets were not only nice to have—they were essential to keep the engine compliant with emission regulations.

The 170-cu.-in. low-deck engine was discontinued at the end of 1969, replaced in 1970 with a high-deck, 198-cu.-in. version with a bore of 3.4 in., a stroke of 3.64 in., and power output of 125 hp. This engine used the block and pushrods of the 225-cu.-in. engine, achieving the smaller stroke with a different crankshaft and connecting rods. The change avoided having to reset the cylinder block machining line for low- and high-deck blocks. This 198-cu.-in. RG version was discontinued after 1974.

A two-barrel carburetor version of the 225-cu.-in. engine was released for 1967 Export engines. In 1977, domestic Plymouth and Dodge cars were offered with a similar two-barrel carburetor version of the 225-cu.-in. engine, a cross section of which is shown on Fig. 11.17. The 1978 Chrysler LeBaron pictured on Fig. 11.18 and the 1979 Chrysler Newport also used the engine.

The year 1981 was the beginning of the end of 225-cu.-in. Slant Six engine usage in passenger cars. In 1981, the entire

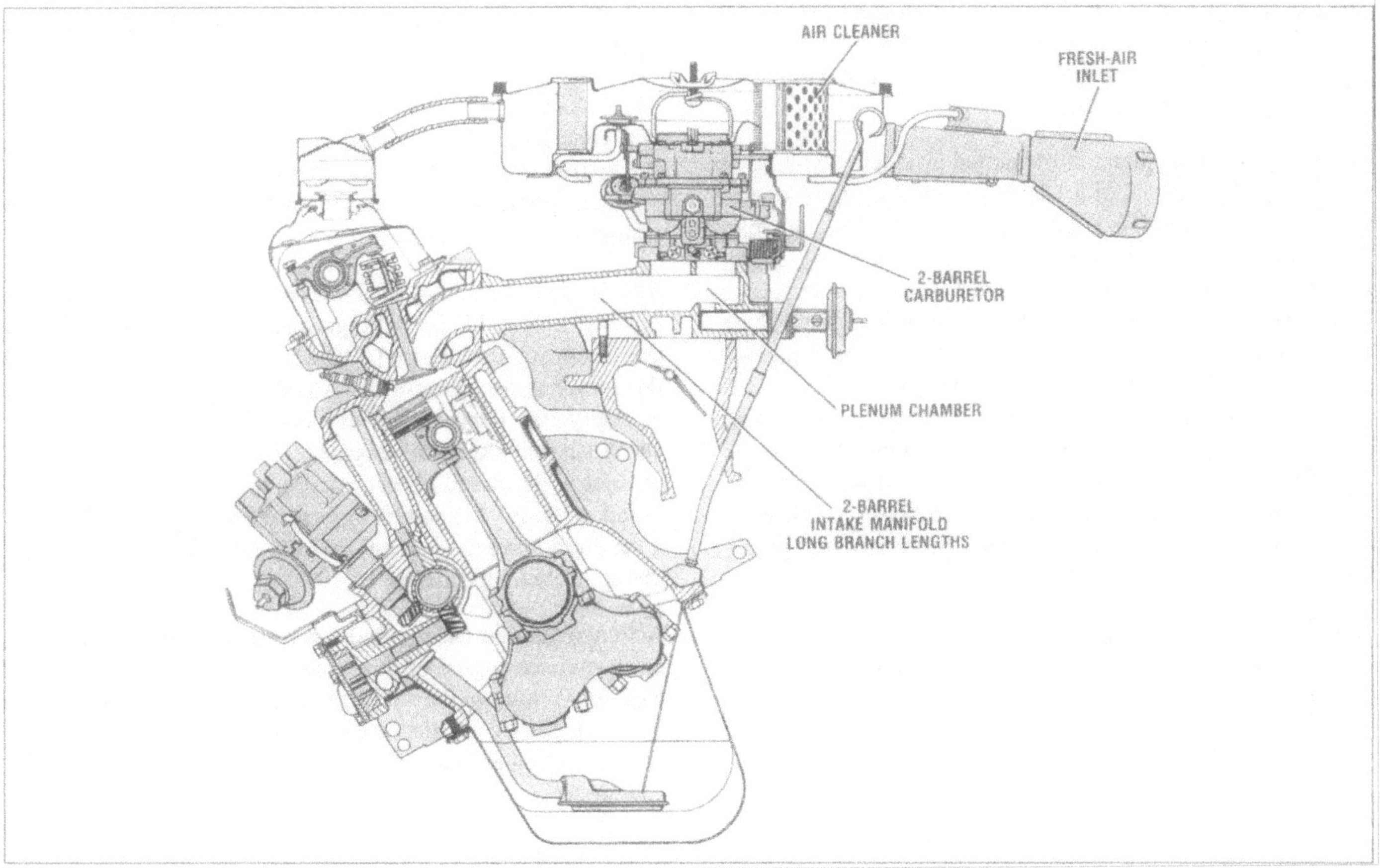

Fig. 11.17 *A cross section of the 1977 225-cu.-in. RG Slant Six engine with a two-barrel carburetor.*

Fig. 11.18 *The 1978 Chrysler LeBaron four-door sedan. A net 110-hp 225-cu.-in. RG Slant Six engine with a two-barrel carburetor was standard power for the car.*

corporate offering of intermediate-size cars was shifted to the new front-wheel-drive K-car, and the Slant Six was limited to standard power for the carryover rear-wheel-drive large Plymouth, Dodge, and Chrysler models. Three years later, starting in 1984, these models became V-8 only, and the days of production of the Slant Six engine for passenger cars were over. Usage in Dodge trucks ended after the 1987 model year was completed; however, Dodge trucks built and sold in Mexico continued using the engine through the 1991 model year. Chapter 15 gives more information about truck use of the engine, and Chapter 16 provides information about marine and industrial uses of the Slant Six engine.

The Trenton engine plant, the principal production source of the engine, built a total of 8,152,768 Slant Six engines, with the last engine completed at the end of the 1983 model year. Additional plants that built Slant Six engines were the Chrysler Canada Windsor engine plant, which was the sole source of the 198-cu.-in. version; the Chrysler de Mexico Toluca engine plant, which continued building 225-cu.-in. engines after August 1984 for Dodge trucks and Mopar service replacements; and the engine plants in Madrid, Spain, in Argentina, and in South Africa. Chrysler Australia manufactured some parts for the Slant Six engines that it assembled at its facilities in Adelaide, South Australia, and imported the rest. Assembly operations using imported parts were performed in Gebez, Turkey, in Peru, and in Columbia.

Table 11.1 shows the specifications for the LG Slant Six engine. Table 11.2 shows the specifications for the RG Slant Six engine.

Chart 11.1 displays time lines for both engines.

An Overhead Valve In-Line Six for Australia

Two years after the Valiant car started production in the United States, a right-hand drive Valiant was marketed in Australia. As originally designed, the Valiant had no provision for right-hand drive, but Chrysler Australia engineers Roy Rainsford and J.M. (Mike) Stacey rose to the challenge and were able to make the conversion in time for the 1962 model year. To meet local content requirements, the vehicle's Slant Six engine, as previously described, was assembled in Australia, using parts largely imported from the United States and Chrysler Canada, with some parts manufactured by Chrysler Australia.

The local content laws had escalating demands for sourcing and manufacturing, which required Australian sourcing of 95% of the content by the early 1970s. This meant that Chrysler Australia would have to produce its own bodies and would have to not only assemble engines but completely manufacture its own engines.

The decision of which engine would be tooled became a paramount issue. With Ed Heins, chief engineer of the Chrysler Export Division, acting as moderator, proposals flew back and forth between Chrysler Center in Highland Park, Michigan, and Chrysler Australia in Adelaide, Australia. The Slant Six was an obvious contender because it was well developed and was already being assembled in Australia with some Chrysler Australia manufactured parts. However, its size limitation of 225 cu. in. was considered a serious disadvantage should future models require larger engines.

The 318-cu.-in. LA V-8, with or without a V-6 derivative, also was considered. This proposal finally was discarded because the engine cost would be inherently higher than that of the competing Ford and General Motors Holden models, with their in-line six-cylinder engines.

This led to the third and final selection, an all-new six-cylinder engine.[9] None of the packaging restrictions that applied to the design of the original Slant Six engine would apply to the new engine because the Australians had the flexibility to change the front-end sheet metal of their Australian-built car to accommodate the engine. The engine could be made upright with spread bores, a large bore size, and the water pump in front. All parties agreed to this proposal, although it required more investment in design and development than the other two approaches.

Actually, Engine Design had been making design studies of a six-cylinder engine of this nature since 1962. Identified as the A941 program, the studies included the use of the LA V-8 bore size of 3.91 in., displacements up to 265 cu. in., and three different cylinder heads. One of the cylinder heads had a single

TABLE 11.1
SPECIFICATIONS FOR THE LG SLANT SIX ENGINES

Model Year	1960	1960	1969
Car line	Valiant	Valiant	Plymouth, Dodge
Model	QX1 V100	QX1 V200 Hyper-Pak	Plymouth: Valiant Dodge: Dart
Displacement (cu. in.)	170	170	170
Bore	3.4	3.4	3.4
Stroke	3.125	3.125	3.125
Compression ratio	8.5:1	10.5:1	8.5:1
BHP @ rpm	101 @ 4400	148 @ 5200	115 @ 4400
Torque (lb-ft) @ rpm	155 @ 2400	153 @ 4200	155 @ 2400
Combustion chamber	Wedge	Wedge	Wedge
Cyl. block mat'l.—skirt type	Cast iron—deep	Cast iron—deep	Cast iron—deep
Cyl. block deck height	9.062	9.062	9.071
Cyl. block length	26.06	26.06	26.06
Cyl. bore spacing	1-2, 3-4, 5-6 = 3.98 2-3, 4-5 = 4.00	1-2, 3-4, 5-6 = 3.98 2-3, 4-5 = 4.00	1-2, 3-4, 5-6 = 3.98 2-3, 4-5 = 4.00
No. of main brgs.—ck/s mat'l.	4—forged steel	4—forged steel	4—forged steel
Ck/shaft main journal dia.	2.75	2.75	2.75
Crankpin journal dia.	2.1875	2.1875	2.1875
Piston mat'l.—no. of rings	Aluminum—3	Aluminum—3	Aluminum—3
Connecting rod centers	5.707	5.707	5.707
Valve system	OHV—single rocker shaft	OHV—single rocker shaft	OHV—single rocker shaft
Valve events	232-228-8	276-268-44	240-236-16
Camshaft drive	Chain	Chain	Chain
Valve head dia.	Intake 1.62; exhaust 1.36	Intake 1.62; exhaust 1.36	Intake 1.62; exhaust 1.36
Valve lash adjustment	Mech.—rocker arm screw	Mech.—rocker arm screw	Mech.—rocker arm screw
Oil pump/distributor drive	Separately geared from camshaft at center of block	Separately geared from camshaft at center of block	Separately geared from camshaft at center of block
Fuel system	1-barrel carburetor	4-barrel carburetor	1-barrel carburetor
Engine angle	30 deg to right	30 deg to right	30 deg to right

TABLE 11.2
SPECIFICATIONS FOR THE RG SLANT SIX ENGINES

Model Year	1960	1961	1961–1962	1966
Car line	Plymouth, Dodge	Dodge	Plymouth, Dodge	Plymouth, Dodge
Model	Plymouth: Savoy, Belvedere, Fury Dodge: Seneca, Pioneer, Phoenix	Lancer 6 Power Pack with Hyper-Pak	Plymouth: Valiant Dodge: Lancer, Dart	Plymouth: Valiant, Barracuda, Belvedere, Fury Dodge: Dart, Coronet
Engine code				E24
Displacement (cu. in.)	225	225	225	225
Bore	3.4	3.4	3.4	3.4
Stroke	4.125	3.125	3.125	3.125
Compression ratio	8.5:1	8.2:1	8.2:1	8.4:1
BHP @ rpm	145 @ 4000	196 @ 5200	145 @ 4000	145 @ 4000
Torque (lb-ft) @ rpm	215 @ 2800	212 @ 4200	215 @ 2800	215 @ 2400
Combustion chamber	Wedge	Wedge	Wedge	Wedge
Cyl. block mat'l.—skirt type	Cast iron—deep	Cast iron—deep	Aluminum (opt.) —deep	Cast iron—deep
Cyl. block deck height	10.691	10.716	10.716	10.70
Cyl. block length	26.06	26.06	26.06	26.06
Cyl. bore spacing	1-2, 3-4, 5-6 = 3.98 2-3, 4-5 = 4.00	1-2, 3-4, 5-6 = 3.98 2-3, 4-5 = 4.00	1-2, 3-4, 5-6 = 3.98 2-3, 4-5 = 4.00	1-2, 3-4, 5-6 = 3.98 2-3, 4-5 = 4.00
No. of main brgs.—ck/s mat'l.	4—forged steel	4—forged steel	4—forged steel	4—forged steel
Ck/shaft main journal dia.	2.75	2.75	2.75	2.75
Crankpin journal dia.	2.1875	2.1875	2.1875	2.1875
Piston mat'l.—no. of rings	Aluminum—3	Aluminum—3	Aluminum—3	Aluminum—3
Connecting rod centers	6.7	6.7	6.7	6.7
Valve system	OHV—single rocker shaft	OHV—single rocker shaft	OHV—single rocker shaft	OHV—single rocker shaft
Valve events	232-228-8	276-268-44	232-228-8	240-236-16
Camshaft drive	Chain	Chain	Chain	Chain
Valve head dia.	Intake 1.62; exhaust 1.36	Intake 1.62; exhaust 1.36	Intake 1.62; exhaust 1.36	Intake 1.62; exhaust 1.36
Valve lash adjustment	Mech.—rocker arm screw	Mech.—rocker arm screw	Mech.—rocker arm screw	Mech.—rocker arm screw
Oil pump/distributor drive	Separately geared from camshaft at center of block	Separately geared from camshaft at center of block	Separately geared from camshaft at center of block	Separately geared from camshaft at center of block
Fuel system	1-barrel carburetor	4-barrel carburetor	1-barrel carburetor	1-barrel carburetor
Engine angle	30 deg to right	30 deg to right	30 deg to right	30 deg to right

(Continued on next page)

TABLE 11.2
(Continued)

Model Year	1970	1977	1977	1983
Car line	Plymouth, Dodge	Plymouth, Dodge	Plymouth, Dodge	Plymouth, Dodge, Chrysler
Model	Plymouth: Valiant, Barracuda Dodge: Dart, Challenger	Plymouth: Volare Dodge: Aspen	Plymouth: Volare Premier, Fury Dodge: Aspen SE and station wagon, Monaco	Plymouth: Gran Fury Dodge: Diplomat, Mirada Chrysler: Cordoba, New Yorker
Engine code		E24	E26	E24
Displacement (cu. in.)	198	225	225	225
Bore	3.4	3.4	3.4	3.4
Stroke	3.64	3.125	3.125	3.125
Compression ratio	8.4:1	8.4:1	8.4:1	8.4:1
BHP @ rpm	125 @ 4400	100 @ 3600	110 @ 3600	90 @ 3600
Torque (lb-ft) @ rpm	180 @ 2000	170 @ 1600	180 @ 2000	165 @ 1600
Combustion chamber	Wedge	Wedge	Wedge	Wedge
Cyl. block mat'l.—skirt type	Cast iron—deep	Cast iron—deep	Cast iron—deep	Cast iron—deep
Cyl. block deck height	10.70	10.70	10.70	10.67
Cyl. block length	26.06	26.06	26.06	26.06
Cyl. bore spacing	1-2, 3-4, 5-6 = 3.98 2-3, 4-5 = 4.00	1-2, 3-4, 5-6 = 3.98 2-3, 4-5 = 4.00	1-2, 3-4, 5-6 = 3.98 2-3, 4-5 = 4.00	1-2, 3-4, 5-6 = 3.98 2-3, 4-5 = 4.00
No. of main brgs.—ck/s mat'l.	4—forged steel	4—nodular cast iron	4—nodular cast iron	4—nodular cast iron
Ck/shaft main journal dia.	2.75	2.75	2.75	2.75
Crankpin journal dia.	2.1875	2.1875	2.1875	2.1875
Piston mat'l.—no. of rings	Aluminum—3	Aluminum—3	Aluminum—3	Aluminum—3
Connecting rod centers	6.7	6.7	6.7	6.7
Valve system	OHV—single rocker shaft	OHV—single rocker shaft	OHV—single rocker shaft	OHV—single rocker shaft
Valve events	240-236-16	244-244-26	244-244-26	240-236-16
Camshaft drive	Chain	Chain	Chain	Chain
Valve head dia.	Intake 1.62; exhaust 1.36	Intake 1.62; exhaust 1.36	Intake 1.62; exhaust 1.36	Intake 1.62; exhaust 1.36
Valve lash adjustment	Mech.—rocker arm screw	Mech.—rocker arm screw	Mech.—rocker arm screw	Hydraulic tappet
Oil pump/distributor drive	Separately geared from camshaft at center of block	Separately geared from camshaft at center of block	Separately geared from camshaft at center of block	Separately geared from camshaft at center of block
Fuel system	1-barrel carburetor	1-barrel carburetor	2-barrel carburetor	1-barrel carburetor
Engine angle	30 deg to right	30 deg to right	30 deg to right	30 deg to right
Engine assembly weight*				227 kg (500 lb)

* For content, see the Abbreviations and Definitions section near the back of this book.

Chart 11.1
Time Lines—OHV Slant Six Engine

Engines, Car Lines, and Displacements (cu. in.)

Engine	Car Line	1960	1961–1969	1970–1974	1975–1977	1978–1983
LG Slant Six	Valiant Model	170 QX1				
	Plymouth Valiant		170			
	Dodge		170			
RG Slant Six	Plymouth Model	225 PP1	225	198 225	225	225
	Dodge Model	225 PD3	225	198 225	225	225
	Chrysler					225

After their 1960 introduction, the G and RG engines had widespread application to many models of the above carlines.

overhead camshaft. Another used conventional in-line overhead valves with a single rocker shaft. The third had canted overhead valves, with the rocker arms mounted on individual pedestals. A truck version of the engine also was designed. With the decision to proceed with an engine for Australia, Engine Design turned the earlier studies into final design reality, using the canted overhead valve type of cylinder head.

The arrangement made with Chrysler Australia was that Engine Design would make the basic design of the engine and that Engine Development would procure a small number of prototype engines for initial development in the United States. Following initial development, the entire engine program would be turned over to Chrysler Australia for completion by its engineering staff. The capable Mike Stacey was assigned the responsibility for the Australian part of the engine program and in June 1966 took station in the Engine Design department at Central Engineering for the critical design and early development of the engine. Engineering program number A167 was assigned to the engine. The engine also was given a "D" identifier.

Regarding the details of the engine, the cylinder block was made of cast iron, and, as noted, the water pump was mounted at the front of the block. The nodular cast iron crankshaft had seven main bearings with a diameter of 2.5 in. each. Intake and exhaust manifolds both were located on the left side of the engine to provide quick warmup, as described in the Slant Six portion of this chapter.

The cylinder head was made of cast iron. The intake and exhaust valves were arranged in-line, as can be seen in the cross section shown on Fig. 11.19. In the fore-and-aft view, the valve stems were not parallel but were canted to the front and rear to improve air flow, to improve cooling between the valves, and to provide an approximate hemispherical shape to the combustion chamber. This canting of the valves, which is illustrated in the longitudinal section of the engine shown on Fig. 11.20, was proposed by R. Dean Engle, who had succeeded Bob Rarey as Chief Engineer of Power Plant Engineering, following the appointment of Rarey as the plant manager of the Mound Road engine plant. The hemispherical-like shape,

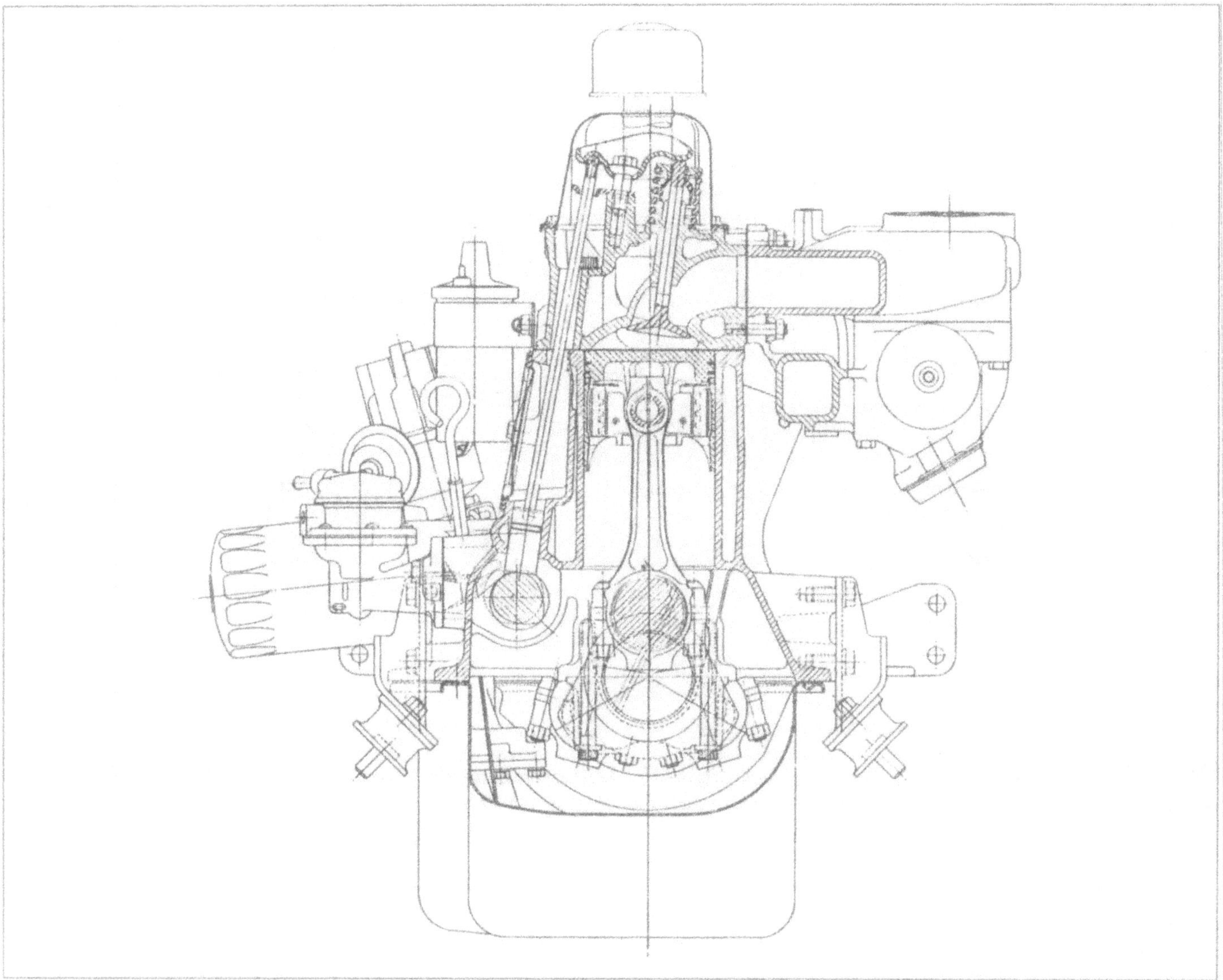

Fig. 11.19 *A cross section of the 1970 Australian 245-cu.-in. overhead valve six-cylinder engine. Note the hemispherical-shaped combustion chamber.*

resulting from the canted valves, was endorsed heartily by Chrysler Australia because it provided a direct association with the legendary Chrysler V-8 Hemi® engines. Officially titled "Hemi®" by Chrysler Australia, the name was used in sales literature and by the motoring press throughout the life of the engine.

Designs were made for two engine sizes—245 cu. in. and 265 cu. in.—sharing the same 3.68-in. stroke and having bores of 3.76 in. and 3.91 in., respectively. The cylinder block deck height was 9.29 in. The crankcase was designed to accommodate a longer 4.16-in. stroke that, with a higher deck block, would result in a 300-cu.-in. engine. If the manufacturing costs were reasonable, an engine of this size had the potential for substantial sales volume in Dodge trucks in the medium-duty six-cylinder market segment dominated by the Ford 300-cu.-in. and Chevrolet 292-cu.-in. six-cylinder truck engines.

A plan to build these large-displacement sixes in Australia and import them to the United States was worked out in some detail with Jim Harbour, head of Manufacturing Facilities Planning; however, the plan was never implemented due to investment costs. Jim Harbour later left the corporation to form his own consulting company, Harbour & Associates, which has gained industry renown for its expertise in assessing car assembly plant productivity efficiencies.

By agreement, the Engine Development department at Chrysler Engineering concentrated on the 245-cu.-in. engine with both one-barrel and two-barrel carburetor versions.

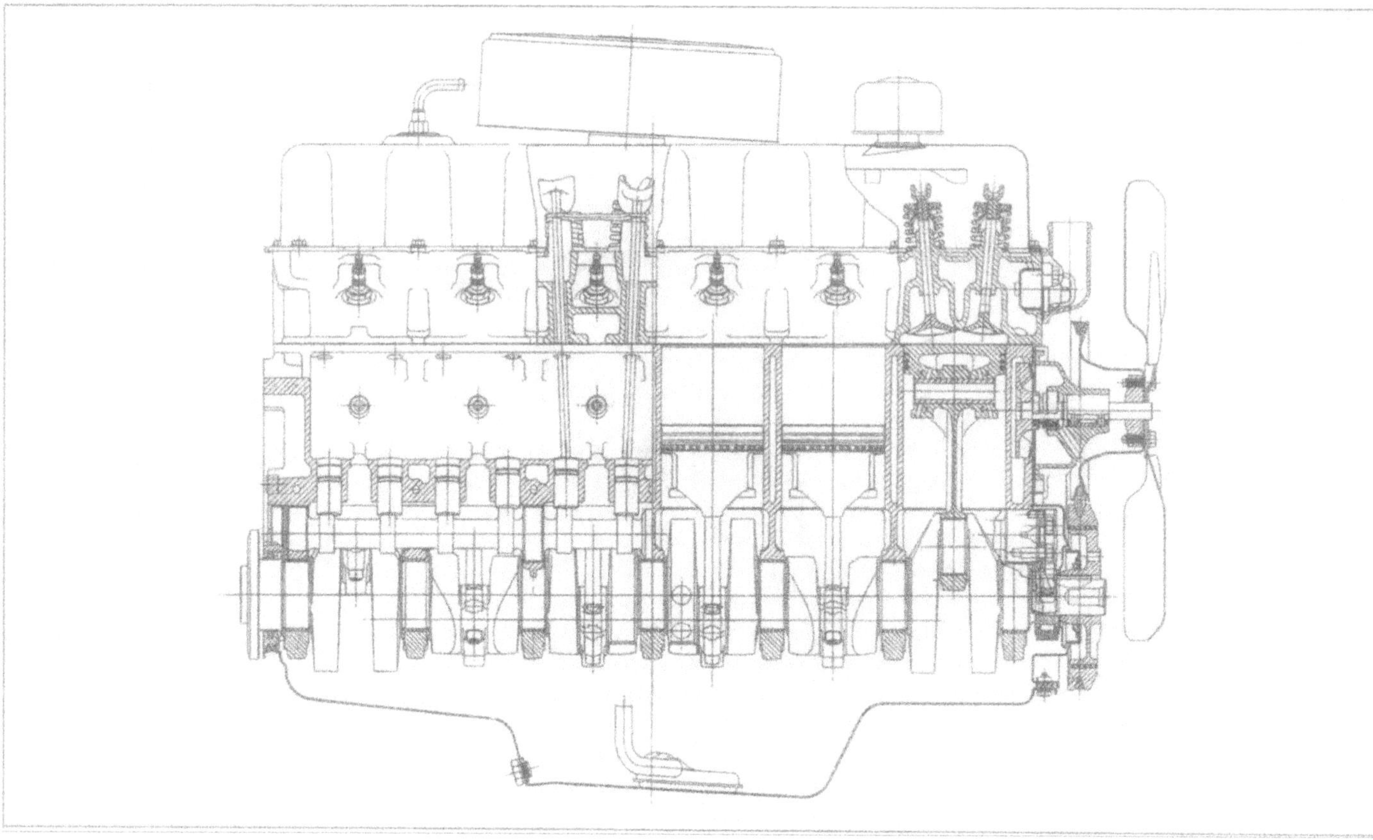

Fig. 11.20 *A longitudinal section of the 1970 Australian 245-cu.-in. overhead valve six-cylinder engine. The section shows the seven-main-bearing crankshaft and canted intake and exhaust valves at the #1 cylinder.*

Prototype procurement and engine test operations generally went smoothly, with one major problem occurring early in the program. The ignition distributor had unexpected rotational gyrations. Instead of rotating steadily throughout the range of engine revolutions per minute, the distributor had bad spark scatter.

The problem quickly was found to be caused by the oil pump. The original design had the distributor and oil pump located at the center of the engine on the right side. The oil pump was gear-driven from the camshaft, and the distributor shaft was driven from the oil pump by a tongue-and-groove connection with the oil pump shaft. Pressure spikes during the pumping action of the oil pump rotors caused the oil pump shaft to have minute rotational variations, which were fed into the distributor shaft, causing spark scatter. The solution was simply to relocate the distributor to a location farther to the rear, away from the oil pump. A second gear was added to the camshaft, and a meshing gear was added to the distributor. In the top-front view of the engine shown on Fig. 11.21, the distributor is visible in this rearward location, and that solved the problem.

E.C. (Zeke) Adams, an engineer with experience in both engine design and engine development, was transferred from Engine Design to Mike Stacey's engineering staff at the new Chrysler Australia engine plant at Lonsdale, near Adelaide in South Australia, to assist in the final development and production launch of the engine.

Production of the engine started at the Lonsdale plant late in the 1969 calendar year, with installations made in the 1970 model year Valiant VG model. A single engine size of 245 cu. in. was manufactured. Equipped with a one-barrel carburetor, its output was 165 hp and 235 lb-ft torque. With a two-barrel carburetor, it produced 185 hp and 240 lb-ft torque. A special high-performance engine for the Valiant Pacer was equipped with a split exhaust manifold in addition to the two-barrel carburetor.

The motoring press was ecstatic about Chrysler Australia having an all-new Hemi® engine. Under an article lead-in headline "Hairy Hemi," the July 1970 *Australian Motor Manual* stated, "Shattering performance, better handling, braking and improved passenger comfort are the main features of this year's Pacer...If ever there was a car with the goods needed to forcibly

Fig. 11.21 *A top-front view of the Australian overhead valve Hemi® six-cylinder engine. Unlike the Slant Six engine, this engine was designed to be straight upright.*

put the case for six cylinder power as opposed to V-8 power, the Valiant Pacer 245 is it...The fact of the matter is that this year's 245 cubic inch, hemi-headed Pacer represents one hell of a lot of performance for very reasonable money. In fact I would say that it has perhaps too much performance for the average driver whose reach it is well within." The October 1971 issue of the magazine *Wheels* had an article titled "Holden v Falcon v Valiant—Which Is Best?" In it, the statement was made, "The Valiant runs away from the others in the six cylinder performance stakes."

Soon after launch, another engine size was added to the lineup. This was a 215-cu.-in. engine with a smaller bore (3.52 versus 3.76 in.) and a lower compression ratio (8.0 versus 9.5:1) than the 245-cu.-in. engine. It was built with only a one-barrel carburetor and had a rating of 140 hp and 200 lb-ft torque.

An all-new Valiant car, coded the VH series, was introduced in June 1971, replacing the Valiant VG. The car and all its components now were almost completely sourced in Australia. With the new car, another engine size entered production—a 265-cu.-in. displacement using the 3.91 bore already noted. In standard form, with a two-barrel carburetor, this engine produced 203 hp and 262 lb-ft torque. A high-performance version with a two-barrel carburetor and split exhaust manifolds produced 218 hp and 273 lb-ft torque. This high-performance engine was available on the Valiant Hemi® Pacer shown on Fig. 11.22 and on a sporty new two-door fastback model, the Valiant Charger pictured on Fig. 11.23. The Charger was both shorter and lighter than the standard Valiant and definitely was aimed at the performance market.

Fig. 11.22 *The 1971 Australian VH Valiant Pacer four-door sedan. The car was powered by a 218-hp 265-cu.-in. Chrysler Australia Hemi® six-cylinder engine equipped with a two-barrel carburetor and split exhaust manifold.*

Fig. 11.23 *The 1972 Chrysler Australia VH Valiant Charger. Three Australian Hemi® six-cylinder engines were available: 215 cu. in. with a one-barrel carburetor, 265 cu. in. with a two-barrel carburetor, and 265 cu. in. with triple two-barrel carburetors. They provided 140, 218, and 248 hp, respectively.*

A Charger model, the R/T, became the performance leader for Chrysler Australia. With an eye toward entering professional car racing to build a reputation for top performance and, of course, to increase sales, Chrysler Australia gave Mike Stacey the assignment of preparing higher-output versions of the 265-cu.-in. engine.

The result was a trio of engines that all were equipped with triple side-draft two-barrel Weber carburetors. Most of the carburetor adaptation and calibration work was done at the Weber facility in Italy, using a car shipped from Australia. One of the engine versions—the Street Six-Pack E37, with a 9.7:1 compression ratio, tubular fabricated exhaust headers, and a single exhaust system—was rated at 248 hp and 306 lb-ft torque. A second version—the Track Six-Pack E38, with a 10.0:1 compression ratio, tubular fabricated headers, and a dual exhaust system—was rated at 280 hp and 318 lb-ft torque. The third version, known by its engine code E49, added a long-duration camshaft and a crankshaft windage tray for an output increase to 302 hp and 320 lb-ft torque. The engine had a number of high-durability parts, including a shot peened crankshaft, larger-section connecting rods, and tri-metal connecting rod big-end bearings. Figure 11.24 shows a Six-Pack engine running at wide-open throttle on a dynamometer.

The performance of the Charger R/T with these Six-Pack engines was impressive. According to road tests conducted by *Wheels* car magazine, the Charger R/T with the E38 engine accelerated from 0 to 60 mph in 6.3 sec. A Charger R/T powered with the E49 engine was quicker at 6.1 sec. The production Charger R/T E49 had the distinction of being the fastest-accelerating automobile ever produced in Australia.[10] A Charger R/T with an E38 engine was victorious in its inaugural September 1971 race at Oran Park, the final round of the Toby Lee series, much to the delight of the Chrysler Australia officials. However, other results were disappointing, with only second-place finishes at the prestigious Hardie-Ferodo 500 at Bathurst in 1971 and 1972.

The Charger R/Ts with their high-performance engines came under rapidly escalating criticism for their emphasis on acceleration and speed at a time when there was widespread concern for the long-term availability of oil supplies. In addition, politicians and others were publicly declaring that such vehicles posed a danger to the general motoring public.[10] The result was that the Charger R/T and the Six-Pack engines were dropped in 1973 at the end of the VH series production run.

The 215-cu.-in. displacement engine was discontinued in 1976 at the end of the VK series. Starting with the CL series, which entered production in October 1976,[11] the 215-cu.-in. engine was replaced with a low-compression 245-cu.-in. engine that was capable of burning low-octane fuel.

The 245- and 265-cu.-in. versions of the engine continued in production until the 1981 calendar year, when production of the engine was discontinued. Throughout its life, the Australian six-cylinder engine gave Chrysler Australia a reputation for having cars with excellent performance, enhanced with the knowledge that the engines were largely designed and developed by Australian engineers.

The end of the engine, as well as the end of production of the Valiant, was occasioned by the complete sale of Chrysler Australia to Mitsubishi Motors Corporation in April 1980.[10] Following the change in ownership, the manufacturing facilities shifted totally to making Japanese-style automobiles in place of the larger American-type cars. Mitsubishi Motors Corporation had become a partial owner prior to this, and production of some of its vehicles already was underway.

Table 11.3 shows the specifications for this Australian-built engine.

Chart 11.2 shows time lines for this engine.

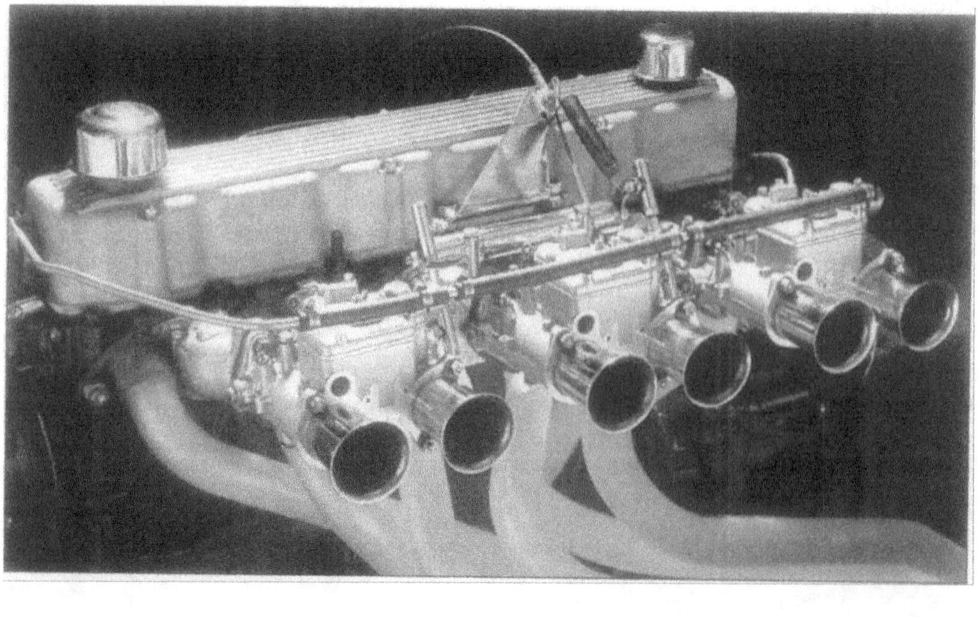

Fig. 11.24 *The 1972 Chrysler Australia Valiant Charger 265-cu.-in. Six-Pack Hemi® engine at full throttle. The engine, running on a dynamometer without air cleaners, is equipped with triple two-barrel side-draft Weber carburetors.*

TABLE 11.3
SPECIFICATIONS FOR THE CHRYSLER AUSTRALIA SIX-CYLINDER ENGINE

Model Year	1970	1971	1971	1971
Car line	Valiant	Valiant	Valiant	Valiant
Model	VG Hemi® 245	VH Hemi® 215	VH Hemi® 245 2-barrel	VH Hemi® 265
Displacement (cu. in.)	245	215	245	265
Bore	3.76	3.52	3.76	3.91
Stroke	3.68	3.68	3.68	3.68
Compression ratio	9.5:1	8.0:1	9.5:1	9.5:1
BHP @ rpm	165 @ 4400	140 @ 4400	185 @ 4600	203 @ 4800
Torque (lb-ft) @ rpm	235 @ 1800	200 @ 1800	240 @ 2000	262 @ 2000
Combustion chamber	Hemispherical	Hemispherical	Hemispherical	Hemispherical
Cyl. block mat'l.—skirt type	Cast iron—shallow	Cast iron—shallow	Cast iron—shallow	Cast iron—shallow
Cyl. block deck height	9.29	9.29	9.29	9.29
Cyl. block length	28.26	28.26	28.26	28.26
Cyl. bore spacing	4.0	4.0	4.0	4.0
No. of main brgs.—ck/s mat'l.	7—nodular cast iron	7—nodular cast iron	7—nodular cast iron	7—nodular cast iron
Ck/shaft main journal dia.	2.5	2.5	2.5	2.5
Crankpin journal dia.	2.0	2.0	2.0	2.0
Piston mat'l.—no. of rings	Aluminum—3	Aluminum—3	Aluminum—3	Aluminum—3
Connecting rod centers	5.75	5.75	5.75	5.75
Valve system	OHV—rocker pedestals	OHV—rocker pedestals	OHV—rocker pedestals	OHV—rocker pedestals
Camshaft drive	Chain	Chain	Chain	Chain
Valve head dia.	Intake 1.84; exhaust 1.5	Intake 1.84; exhaust 1.5	Intake 1.84; exhaust 1.5	Intake 1.96; exhaust 1.5
Valve lash adjustment	Hydraulic tappet	Hydraulic tappet	Hydraulic tappet	Hydraulic tappet
Oil pump/distributor drive	Separately geared to camshaft	Separately geared to camshaft	Separately geared to camshaft	Separately geared to camshaft
Fuel system	1-barrel carburetor	1-barrel carburetor	2-barrel carburetor	2-barrel carburetor

(Continued on next page)

TABLE 11.3 (Continued)

Model Year	1971	1972	1972	1972
Car line	Valiant	Valiant	Valiant	Valiant
Model	VH Pacer Hemi® 265 Hi Perf	VH Charger Hemi® 265 Six-Pack	VH Charger Hemi® 265 Six Pack (opt.)	VH Charger Hemi® 265 Six Pack (opt.)
Engine code		E37	E38	E49
Displacement (cu. in.)	265	265	265	265
Bore	3.91	3.91	3.91	3.91
Stroke	3.68	3.68	3.68	3.68
Compression ratio	9.5:1	9.7:1	10.0:1	10.0:1
BHP @ rpm	218 @ 4800	248 @ 4800	280 @ 5000	302 @
Torque (lb-ft) @ rpm	273 @ 3000	306 @ 3400	318 @ 3700	
Combustion chamber	Hemispherical	Hemispherical	Hemispherical	Hemispherical
Cyl. block mat'l.—skirt type	Cast iron—shallow	Cast iron—shallow	Cast iron—shallow	Cast iron—shallow
Cyl. block deck height	9.29	9.29	9.29	9.29
Cyl. block length	28.26	28.26	28.26	28.26
Cyl. bore spacing	4.0	4.0	4.0	4.0
No. of main brgs.—ck/s mat'l.	7—nodular cast iron	7—nodular cast iron	7—nodular cast iron	7—nodular cast iron
Ck/shaft main journal dia.	2.5	2.5	2.5	2.5
Crankpin journal dia.	2.0	2.0	2.0	2.0
Piston mat'l.—no. of rings	Aluminum—3	Aluminum—3	Aluminum—3	Aluminum—3
Connecting rod centers	5.75	5.75	5.75	5.75
Valve system	OHV—rocker pedestals	OHV—rocker pedestals	OHV—rocker pedestals	OHV—rocker pedestals
Camshaft drive	Chain	Chain	Chain	Chain
Valve head dia.	Intake 1.96; exhaust 1.5	Intake 1.96; exhaust 1.5	Intake 1.96; exhaust 1.5	Intake 1.96; exhaust 1.5
Valve lash adjustment	Hydraulic tappet	Hydraulic tappet	Hydraulic tappet	Hydraulic tappet
Oil pump/distributor drive	Separately geared to camshaft	Separately geared to camshaft	Separately geared to camshaft	Separately geared to camshaft
Fuel system	2-barrel carburetor	3 x 2-barrel Weber	3 x 2-barrel Weber	3 x 2-barrel Weber

Chart 11.2
Time Line—Chrysler Australia OHV Hemi® Head Six

Engines, Car Lines, and Displacements (cu. in.)

Engine	Car Line	1970	1971–1972	1973–1974	1975	1976–1977	1978–1981
Chrysler-Australia Hemi® Head Six	Valiant	245	215 245 265	215 245 265	215 245 265	215 245 265	245 265
	Series	VG	VH	VJ	VK	CL	CM

CHAPTER 12

A New Era—Driving the Front Wheels with Four-Cylinder Engines

We start with an import from Germany—the Volkswagen 1.7-liter single overhead camshaft four-cylinder engine.

The widespread fuel shortages of 1973, noted in Chapter 9, which triggered lines for buying gasoline, had a profound effect on the U.S. car market. Suddenly, the demand for small and fuel-efficient cars, well represented by imports such as the 1974 Volkswagen Rabbit, ballooned. Additional pressure was placed on domestic manufacturers to change course away from the mainstream large-engined, heavy cars toward lighter-weight vehicles with considerably better fuel economy. The pressure came from Congress when, in 1975, reflecting concern over dependence on foreign oil supplies, it enacted Corporate Average Fuel Economy (CAFE) requirements on U.S.-built passenger cars, with escalating levels of miles per gallon fuel economy starting in 1978.

The Chrysler response culminated in the creation of a completely new car that was designated the L-body. This car was powered by a small single overhead camshaft (SOHC) four-cylinder engine, mounted transversely and driving the front wheels. Engine Design quickly prepared a 2-liter-displacement, single overhead camshaft engine with the familiar material of cast iron for both the cylinder block and the cylinder head, coded A405, as a suitable powerplant. In parallel with this engine design effort, negotiations were held with Volkswagen of Germany over the possible purchase of production quantities of its single overhead camshaft four-cylinder engines for use by Chrysler in the planned L-body car. There was an important qualifier, however. For the car to achieve its targeted performance levels, the engine had to have a displacement of at least 1.7-liters, more than 200 cc greater than the 1471-cc Rabbit engine. Design studies by both Engine Design and Volkswagen concluded that the 1.7-liter displacement was feasible without any change to the basic engine structure, including the height of the cylinder block.[1] Some minor changes to the block would be needed to swing the longer stroke and for the attachment of the Chrysler A404 automatic transmission. The engine compartment environment of the L-body was so different from that of the Rabbit that many new parts would be needed, such as intake and exhaust manifolds, engine mounts, accessory mounting, and the fuel system. With Chrysler assuming engineering, procurement, and assembly responsibility for these parts, the engine purchase agreement called for Volkswagen to ship a "short" engine without these parts and devoid of the timing belt and sprockets to the Chrysler Trenton engine plant, where the buildup would be completed and the engine would be hot-tested prior to shipment to the Chrysler L-body assembly plant in Belvidere, Illinois.

With the agreement approved and the Engineering program code A409 assigned, all the engineering work of prototype parts procurement, engine testing, fuel system calibration, and vehicle testing and production releases took place in time for Horizon and Omni (the names for the Plymouth and Dodge editions of the L-body, respectively) vehicles to start rolling off the end of the assembly line in January 1978.

More details of the engine—illustrated in the cutaway view shown on Fig. 12.1, the cross section shown on Fig. 12.2, and the longitudinal section shown on Fig. 12.3—are as follows. The bore and stroke were 80 × 86 mm, giving a displacement of 1716 cc. Engine output was 55 kW (75 hp) and 120 N-m (90 lb-ft) torque for all states except California, with California certified engines being slightly lower. Materials for the cylinder block and the cylinder head were cast iron and aluminum, respectively. The four cylinders were spaced 88 mm apart. The valve gear was composed of a single overhead camshaft operating inverted bucket tappets that had replaceable shims for valve lash adjustment. The in-line valves were 90 deg to the cylinder head joint face. A loop flow pattern was used for the fuel/air into and the exhaust gas out of the cylinder head, with both the intake and exhaust manifolds located at the rear of the head (referring to the in-car position of the transversely mounted engine).

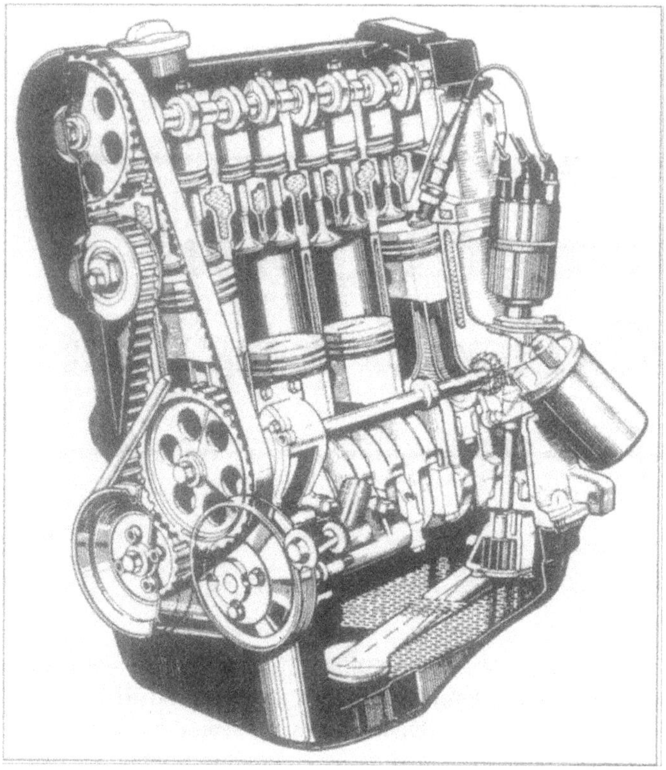

Fig. 12.1 *A cutaway illustration of the 1978 75-hp Volkswagen 1.7-liter single overhead camshaft four-cylinder engine.*

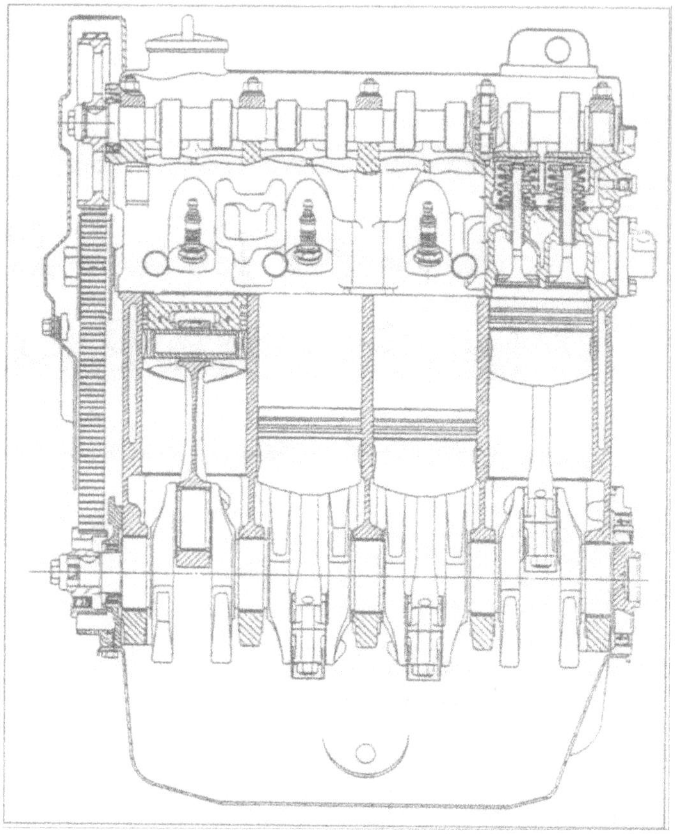

Fig. 12.3 *A longitudinal section of the 1978 Volkswagen 1.7-liter single overhead camshaft four-cylinder engine as used in the Dodge Omni and Plymouth Horizon cars.*

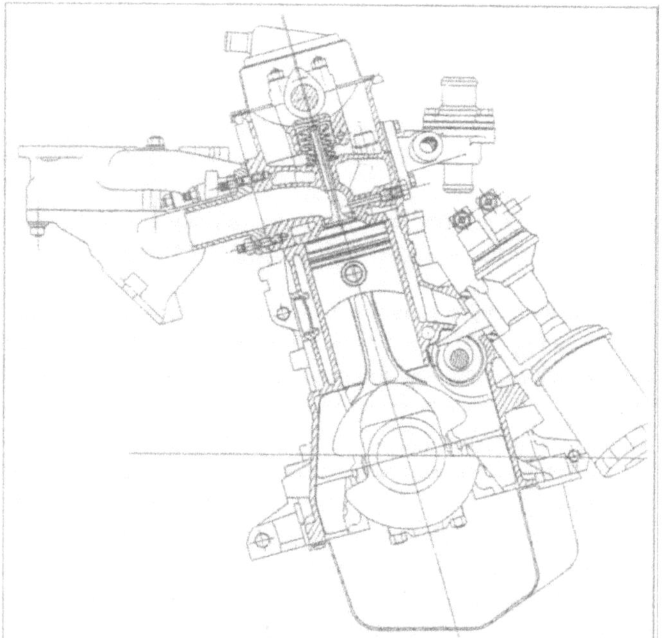

Fig. 12.2 *A cross section of the 1978 Volkswagen 1.7-liter single overhead camshaft four-cylinder engine as used in the Dodge Omni and Plymouth Horizon cars.*

The five main journals of the forged steel crankshaft had a diameter of 54 mm; the crankpins had a diameter of 46 mm. A flexible, toothed timing belt from the crankshaft drove both the camshaft and an auxiliary shaft, located parallel to the crankshaft, which in turn drove the oil pump and the distributor. A two-barrel carburetor was the fuel delivery system of the engine. Engine cooling was accomplished by a thermostatically controlled, electric-motor-driven fan mounted behind the radiator. For exhaust emission control, a V-belt-driven air pump was used to inject air into the four exhaust ports of the cylinder head.

The lead-off 1978 Plymouth Horizon, shown on Fig. 12.4, and the Dodge Omni four-door hatchbacks were immediate sales successes, leading to the addition during the following year of sporty fastback models: the Plymouth Horizon TC3, and the Dodge Omni O24 that appears on Fig. 12.5. Both of these new models were equipped with a carryover Volkswagen-based 1.7-liter engine, the only engine available. The 1978 Horizon and Omni models were recognized by the motoring public as excellent cars with peppy and fuel-efficient emission-compliant engines, roomy bodies, and fine road handling. All these characteristics

Fig. 12.4 *A 1978 Plymouth Horizon four-door hatchback. Similar to its Dodge Omni counterpart, the Plymouth Horizon was front-wheel drive, with the 75-hp 1.7-L Volkswagen single overhead camshaft four-cylinder engine mounted transversely. The crankshaft nose is to the right.*

Fig. 12.5 *This sporty-looking 1979 Dodge Omni 024 two-door fastback was powered by the 1.7-L Volkswagen single overhead camshaft four-cylinder engine.*

contributed to the Omni and Horizon collectively being selected as *Motor Trend's* 1978 Car of the Year.[2]

The strategy of purchasing rather than manufacturing the engine needed for the initial L-body cars paid off in several ways. Prototype vehicles with production-intent engines were running sooner, capital investment for the L-body car program was much lower, and Engineering had more time to devote to the design and development of its own home-grown engine. The home-grown engine, described later in this chapter, arrived in 1981 with a displacement of 2.2 liters, immediately assuming the role of an optional power upgrade for L-body cars and leaving the 1.7-liter engine in place as the standard powerplant. With concern about the supply of Volkswagen engines beyond 1982 (that was the end of the five-year purchase agreement), arrangements were made to qualify a 1.6-liter Peugeot Societe Anonyme (PSA) engine (described in the next section of this chapter) for use as the entry-level L-body powerplant for 1983. As events turned out, Volkswagen agreed to continue its supply of engines for another year, and commitments were made to purchase 1983 engines from both Volkswagen and PSA, with the PSA engine being limited to manual transmission, non-air conditioned, non-California cars. It was the sixth and last year of Chrysler's use of the Volkswagen 1.7-liter engine that served the corporation well in its major shift to space-efficient, fuel-efficient front-wheel-drive cars. In 1981, when Volkswagen assembled its one-millionth 1.7-liter engine for Chrysler, the event was commemorated in Germany with recognition, as shown on Fig. 12.6.

Table 12.1 shows the specifications for this Volkswagen-built 1.7-L engine.

Fig. 12.6 *The completion of the one-millionth 1.7-liter four-cylinder engine for Chrysler is celebrated at the Volkswagen engine plant in Germany.*

TABLE 12.1
SPECIFICATIONS FOR THE 1.7-L VOLKSWAGEN FRONT-WHEEL-DRIVE FOUR-CYLINDER ENGINE
Note: All dimensions are in metric units (millimeters) rather than customary units (inches).

Model Year	1978	1983
Car line	Plymouth, Dodge	Plymouth, Dodge
Model	Plymouth: Horizon Dodge: Omni	Plymouth: Horizon, Turismo, Custom Dodge: Omni, Charger, Custom
Engine code	E12	E12
Displacement (liters)	1.7	1.7
Bore	79.5	79.5
Stroke	86.4	86.4
Compression ratio	8.2:1	8.2:1
Power @ rpm	55 kW (75 hp) @ 5600	47 kW (63 hp) @ 4800
Torque @ rpm	120 N-m (90 lb-ft) @ 3200	113 N-m (83 lb-ft) @ 2400
Cyl. block mat'l.—skirt type	Cast iron—deep	Cast iron—deep
Cyl. block deck height	220	220
Cyl. bore spacing	88	88
No. of main brgs.—ck/s mat'l.	5—forged steel	5—forged steel
Ck/shaft main journal dia.	54	54
Crankpin journal dia.	46	46
Connecting rod centers	136	136
Valve system	SOHC—8 valve	SOHC—8 valve
Valve events	266-260-42	250-244-27
Camshaft drive	Timing belt	Timing belt
Valve head dia.	Intake 37; exhaust 31	Intake 37; exhaust 31
Valve lash adjustment	Mech.—tappet shim	Mech.—tappet shim
Oil pump/distributor drive	Geared to auxiliary shaft	Geared to auxiliary shaft
Fuel system	2-barrel carburetor	2-barrel carburetor
Engine angle	16 deg to rear	16 deg to rear
Engine assembly weight*		128.1 kg (282.4 lb)

* For content, see the Abbreviations and Definitions section near the back of this book.

A 1.6-Liter Overhead Valve Engine from France for the L-Body Car

Under the guidance of engine program manager Mike Royce, an existing PSA engine that had been in production in various displacements for a number of years and was powering Simca cars was modified to fit the L-body engine compartment. It was developed to be in compliance with its intended 1983 launch-year federal emission requirements. The program accomplishing this work was coded A487. The 1.6-liter PSA engine produced at Poissy, France, near Paris, for use by Chrysler was a more complete assembly than the 1.7-liter Volkswagen engine previously noted and was hot-tested at Poissy before being shipped directly to the Belvidere, Illinois L-body assembly plant for its final dress-up and vehicle installation. The dress-up consisted of attachment of the alternator, starter, two-barrel electronic feedback Holley carburetor, exhaust gas recirculation (EGR) controls, and other add-on parts. Figure 12.7 shows a powerplant assembly complete with the power steering pump, air pump, alternator, carburetor, air cleaner, and transmission.

Technical details of the engine are illustrated in the engine longitudinal section shown on Fig. 12.8 and in the cross section shown on Fig. 12.9. The engine had a bore and stroke of 80.6 × 78 mm and provided a displacement of 1592 cc. The

Fig. 12.7 *A right-front (as installed) view of the 1.6-L PSA overhead valve four-cylinder engine assembly.*

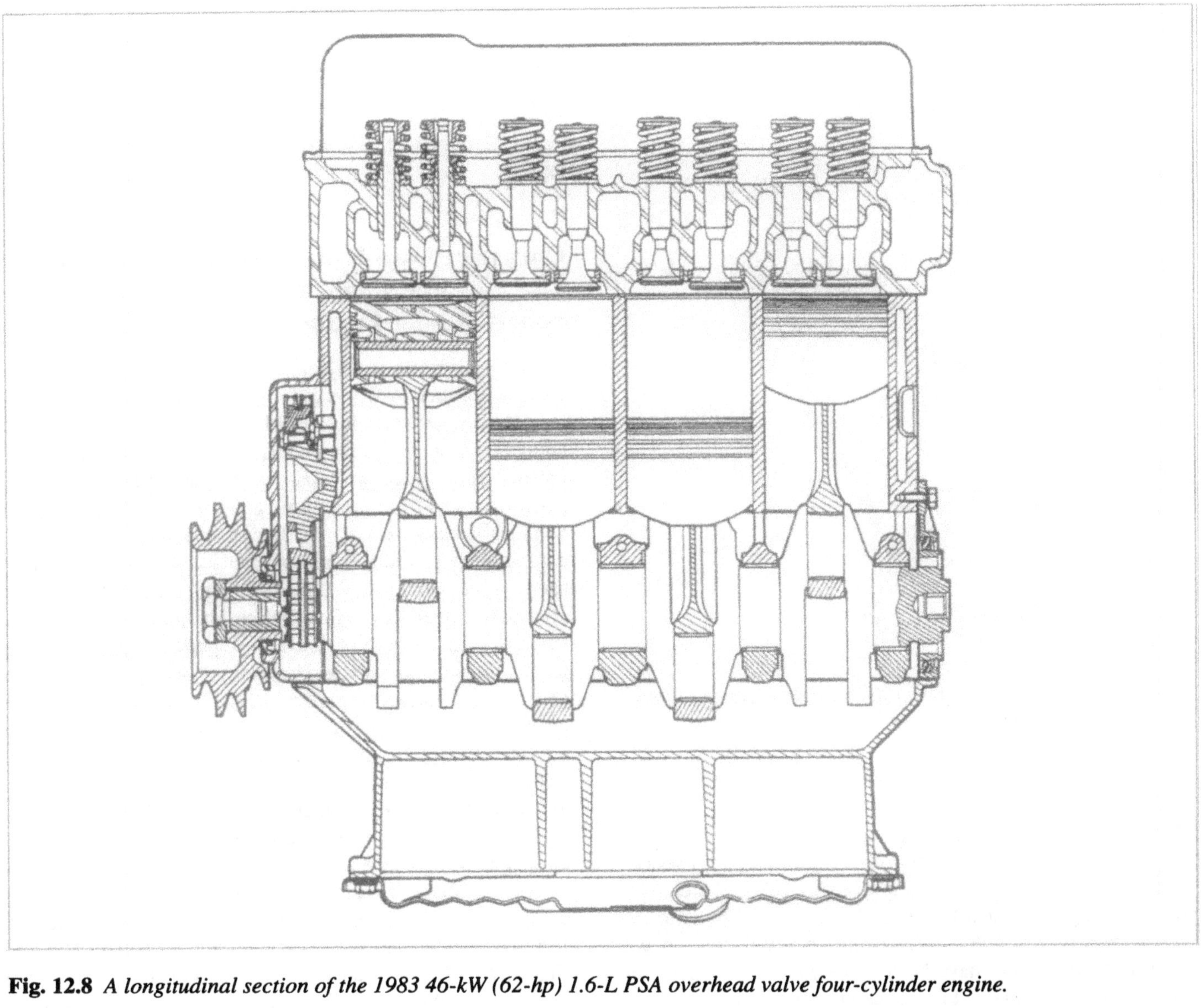

Fig. 12.8 *A longitudinal section of the 1983 46-kW (62-hp) 1.6-L PSA overhead valve four-cylinder engine.*

four cylinders of the cast iron cylinder block were spaced 88 mm apart, too tight for coolant passages between the bores. The aluminum cross-flow cylinder head had a wedge-shaped combustion chamber with in-line overhead valves (OHV) operated by rocker arms, pushrods, and tappets from the block-mounted camshaft. A screw in each rocker arm, with a lock nut, provided valve lash adjustment. The aluminum intake manifold of the engine was located at the front of the cylinder head; the exhaust manifold was located at the rear. The distributor (mounted on the front side of the cylinder block) and the oil pump (attached to the rear side) both were driven by a cross shaft located between cylinders #1 and #2. The five main journals of the forged steel crankshaft had a diameter of 52 mm; the four crankpins had a diameter of 41 mm. The belt-driven water pump was housed in the cast aluminum upper oil pan. For the L-body cars, the engine was installed with its cylinder centers leaning 38 deg to the rear.

Fig. 12.10 *The 1983 Dodge Omni four-door hatchback. Standard power was a 1.6-L PSA overhead valve four-cylinder engine but was available with only a four-speed manual transmission.*

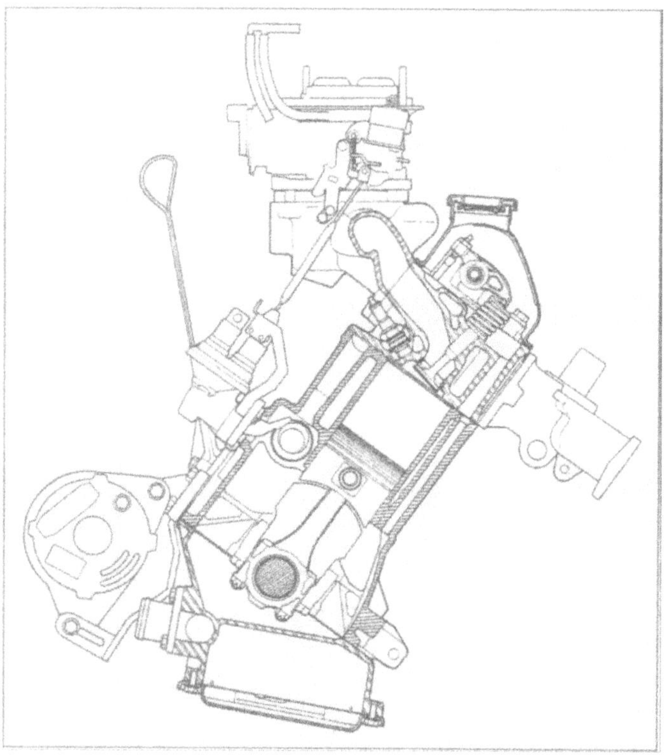

Fig. 12.9 *A cross section of the 1983 1.6-L PSA overhead valve four-cylinder engine.*

Although its 1592-cc displacement was smaller than the 1729 cc of the Volkswagen engine, the 1983 PSA engine output of 46 kW (62 hp) and 117 N-m (86 lb-ft) torque was a close match with the Volkswagen engine's 47 kW (63 hp) and 113 N-m (83 lb-ft) of the same year. A 1983 Dodge Omni, such as the one shown on Fig. 12.10, would have been powered with this 1.6-liter engine, if it had a manual four-speed transmission.

After four years of L-body use, the engine was discontinued at the end of 1986.

Table 12.2 shows the specifications for this PSA-built 1.6-L engine.

Chrysler's Own Front-Wheel-Drive Engine— The 2.2-Liter Single Overhead Camshaft Four-Cylinder Engine

Engineering work never stopped on a four-cylinder engine to be manufactured by Chrysler for its front-wheel-drive vehicles. By the time of its production launch for the 1981 cars, its application had broadened considerably with the addition of the first front-wheel-drive K-cars to the corporate lineup, a platform that was to prove both adaptable to a wide range of vehicles and slow to age. The K-cars, heavier than the L-body cars, influenced the increase of displacement of the Chrysler four-cylinder engine to 2.2 liters from the earlier 2.0-liter size previously mentioned. In weighing the merits of cross flow versus loop flow for the cylinder head porting, importance was given to making exhaust gas heat closely available to the floor of the intake manifold underneath the carburetor should the need arise for faster warmup of the fuel/air mixture during cold start, to positioning the carburetor behind the engine for the best protection during a front-end impact, and, to some extent, to making the engine installation in the L-body cars similar to that of the Volkswagen engine for assembly-line ease.[3] This arrangement also gave excellent access to the spark plugs and oil filter, probably the most frequently serviced feature of any engine. After production began, *Motor Trend* in its September 1980 issue published an article about the car titled "The New Chrysler," which included an observation that "Aside from its light weight (282 pounds) and impressive performance

TABLE 12.2
SPECIFICATIONS FOR THE 1.6-L PSA FRONT-WHEEL-DRIVE FOUR-CYLINDER ENGINE
Note: All dimension are in metric units (millimeters) rather than customary units (inches).

Model Year	1983	1986
Car line	Plymouth, Dodge	Plymouth, Dodge
Model	Plymouth: Horizon, Turismo, Custom Dodge: Omni, Charger, Custom	Plymouth: Horizon, Turismo, SE Dodge: Omni, Charger, SE
Engine code	E82	ECA
Displacement (liters)	1.6	1.6
Bore	80.6	80.6
Stroke	78	78
Compression ratio	8.8	8.8
Power @ rpm	46 kW (62 hp) @ 4800	48 kW (64 hp) @ 4800
Torque @ rpm	117 N-m (86 lb-ft) @ 3200	118 N-m (87 lb-ft) @ 2800
Cyl. block mat'l.—skirt type	Cast iron—deep	Cast iron—deep
Cyl. block deck height	202	202
Cyl. bore spacing	88	88
No. of main brgs.—ck/s mat'l.	5—forged steel	5—forged steel
Ck/shaft main journal dia.	52	52
Crankpin journal dia.	41	41
Connecting rod centers	124	124
Valve system	OHV—single rocker shaft	OHV—single rocker shaft
Valve events	248-238-26	238-248-33
Camshaft drive	Chain	Chain
Valve head dia.	Intake 36; exhaust 29	Intake 36; exhaust 29
Valve lash adjustment	Mech.—rocker arm screw	Mech.—rocker arm screw
Oil pump/distributor drive	Cross shaft through block	Cross shaft through block
Fuel system	2-barrel carburetor	2-barrel carburetor
Engine angle	38 deg to rear	38 deg to rear
Engine assembly weight*	116.3 kg (256.3 lb)	120.5 kg (265.5 lb)

* For content, see the Abbreviations and Definitions section near the back of this book.

(especially in the torque department), the 2.2-liter is a study in componentry and serviceablility. It has been designed with most all auxiliary components to the front of the car; oil filter, distributor, belt-drive adjustment and power steering pump reservoir are all located near or at the top of the engine."

A flexible, cog-toothed timing belt was chosen for driving the overhead camshaft instead of a timing chain to minimize weight and cost while simplifying oil sealing requirements of the engine. Figure 12.11 shows the front of the engine, and Fig. 12.12 shows a three-quarter right-rear view. Figures 12.13 and 12.14 show a cross section and a longitudinal section, respectively. This 1981 engine, coded A452, had a bore and stroke of 87.5 × 92 mm, providing a displacement of 2213 cc. The cast iron cylinder block, with 96-mm cylinder centers, had a deck height of 237.8 mm and a shallow skirt (selected instead of a deep skirt) to keep the weight of the engine to a minimum. The cylinder head, made of cast aluminum, had oval-shaped combustion chambers with front and rear quench areas and a centrally located spark plug. The valves were operated by cast iron finger followers placed below the camshaft, with valve lash adjustment being accomplished by a hydraulically movable pivot at the end of the follower opposite the valve. Camshaft lobe contours for the intake and exhaust valves were designed so that, in the unlikely event of a broken timing belt, the engine had freewheeling clearance and the valves would never hit the pistons. The five-main-bearing crankshaft, made of nodular

Fig. 12.11 *A front view of the 1981 2.2-L single overhead camshaft four-cylinder engine, rated at net 63 kW (84 hp). Designed and developed by Chrysler, the engine was built at the Trenton engine plant of the Power Train Division.*

Fig. 12.12 *A right-rear view of the 1981 2.2-L single overhead camshaft four-cylinder engine.*

cast iron, had 60-mm main journals and 50-mm crankpin journals with all journal fillets deep-rolled for fatigue strength improvement. A block-mounted auxiliary shaft carried a gear for driving the oil pump and distributor near the front-center of the engine, as well as an eccentric near the left side of the engine for the fuel pump lever. Faced with having to add an air pump for exhaust emission control to an already overcrowded group of accessories at the right side of the engine, a solution came with the idea of extending the left end of the camshaft so

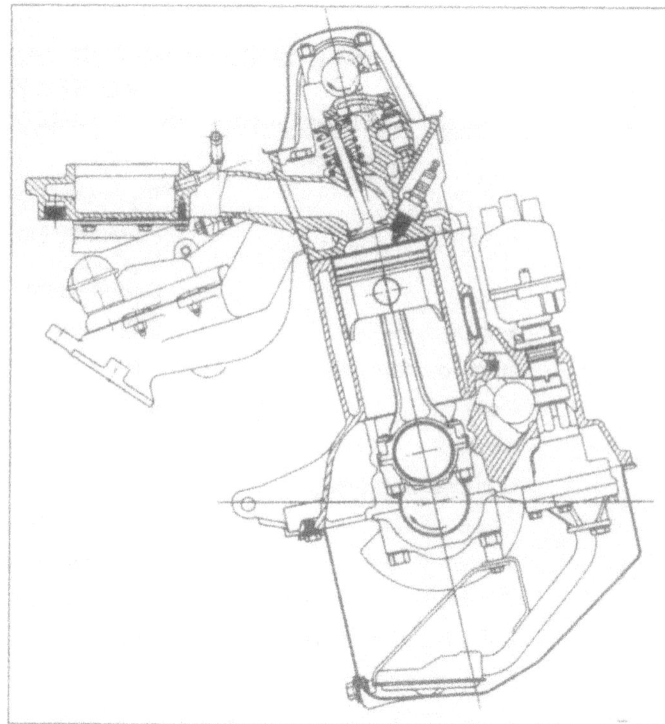

Fig. 12.13 *A cross section of the 1981 2.2-L single overhead camshaft four-cylinder engine. As installed in the car, the manifolds were oriented toward the firewall.*

it could be used to drive a V-belt for the air pump. It worked fine. An SAE paper presented in February 1981 and titled "Chrysler Corporation's New 2.2 Liter 4-Cylinder Engine," authored by me and J.W. Dean, manager of Engine Design, gives additional technical details about the engine.[4]

The Trenton engine plant of the Power Train Division welcomed the task of building the 1981 2.2-liter engine, especially because the big-block B and RB engines that the plant had been building had ceased production at the end of 1978. The most up-to-date production equipment was installed for the 2.2-L engine, including two sets of computer-controlled engine assembly test stands, the first "cold-test" stands for checking engine parameters with the engine being motored by the stand, and the second "hot-test" stands for checking parameters with the engine running under its own power. As part of the ramp-up schedule leading to volume production, the Trenton engine plant built a few engines six months ahead of the start of volume production and a larger group three months ahead of volume production. All of these engines were tested extensively to ensure that production engines would hit the quality targets right from the beginning.

The year 1981, which was the first year of the 2.2-L engine, also was the first year of the new Chrysler front-wheel-drive K-car platform. The Plymouth Reliant four-door sedan, shown

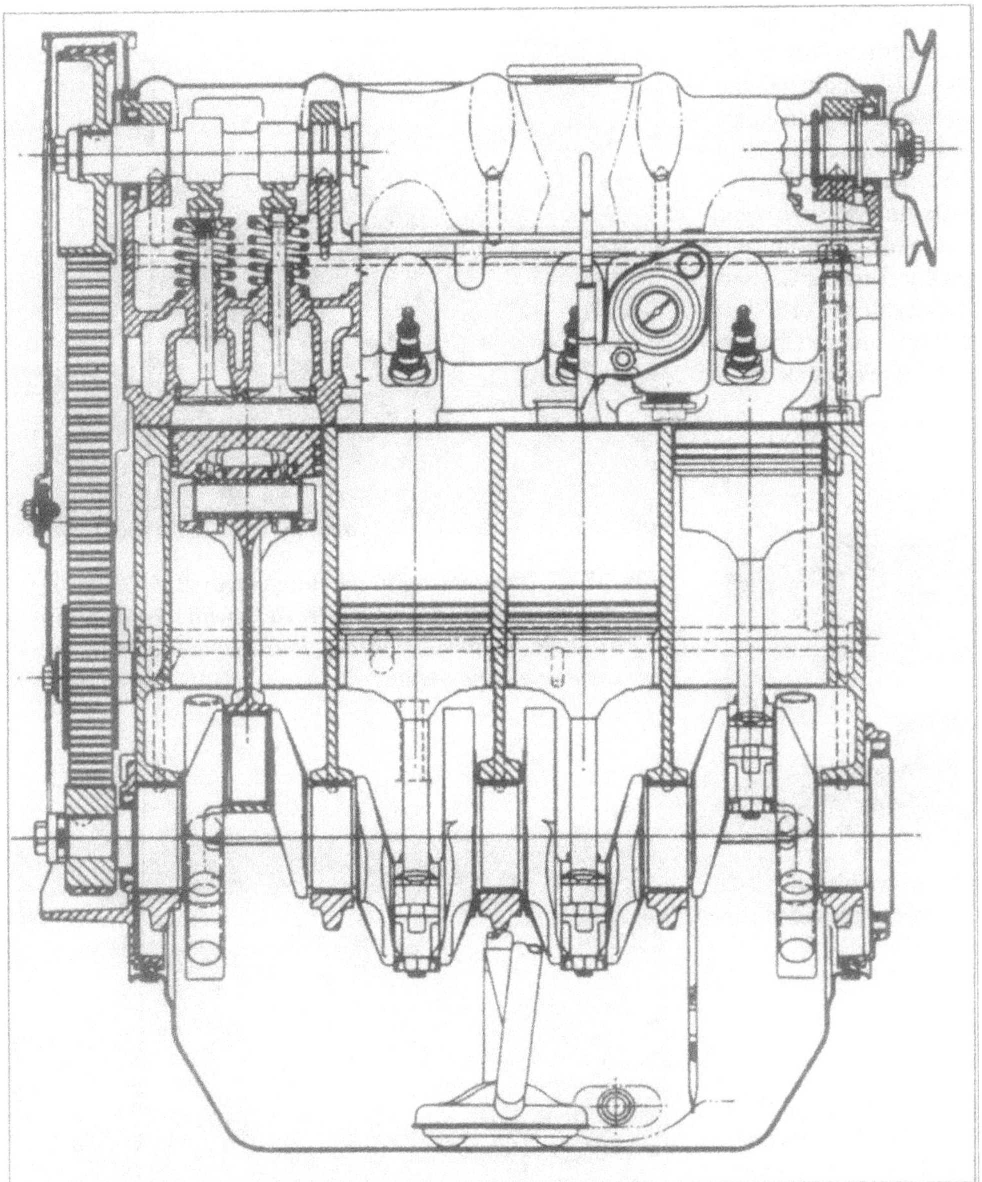

Fig. 12.14 *(Left) A longitudinal section of the 1981 2.2-L single overhead camshaft four-cylinder engine. The V-belt pulley at the right end (as shown) of the camshaft was used to drive the air injection pump, which was needed for exhaust emission control.*

on Fig. 12.15, was among the several 1981 Plymouth Reliant and Dodge Aries K-car models powered by the 2.2-liter single overhead camshaft four-cylinder engine. At the time of introduction, it was well known that the future of the corporation hinged on the sales success of these cars. To the great relief of everyone involved in the program, by the end of 1981, sales had increased substantially, the financial crisis had passed, and the way was paved for the entire passenger car fleet to be converted to front-wheel drive—a conversion that finally was completed ten years later with the 1990 models. A nice recognition for the effort of getting the 1981 Reliant and Aries car into production was its being named *Motor Trend's* 1981 Car of the Year.

The 1981 engine, with a compression ratio of 8.5:1 and equipped with a two-barrel carburetor, had an output of 63 kW (84 hp) and 150 N-m (111 lb-ft) torque. Chrysler joined the

Fig. 12.15 *The 1981 Plymouth Reliant four-door sedan. A 2.2-L single overhead camshaft four-cylinder engine provided power for the front-wheel-drive K-car platform vehicle.*

front-wheel-drive K-car bandwagon with its 1982 LeBaron series, including the trim-looking convertible shown on Fig. 12.16, which was powered with the 2.2-L engine as the base engine. In 1983, with higher-flow intake and exhaust port shapes, improved combustion control, and a compression ratio increase to 9.0:1, output of the 2.2-L engine rose to 70 kW (94 hp) and 158 N-m (117 lb-ft) torque. Also in 1983, two high-performance versions were introduced: a 75-kW (100-hp) and 165 N-m (122 lb-ft) version for the Plymouth Turismo 2.2 and Dodge Charger 2.2 models, and a high-power 80-kW (107-hp) and 172 N-m (127 lb-ft) version for the Dodge Shelby Charger cars, accomplished by a compression ratio increase to 9.6:1.

Fig. 12.16 *The 1982 Chrysler LeBaron convertible. A 2.2-L single overhead camshaft four-cylinder engine provided the power to its front wheels.*

Three years after the first 2.2-liter engine rolled off the final line, on April 22, 1983, the Trenton engine plant built its one-millionth 2.2-liter engine, shown on Fig. 12.17 with factory manager Bob Garlo and engine assembler Leslie Tyler. It was a fine accomplishment for the sole source of this engine.

The growing power of computers for engine control and the growing capability of suppliers brought fuel injection into high-volume use for 1984. A system using a single-point injector located in a throttle body, replacing the two-barrel carburetor, was released as standard for both Dodge and Chrysler front-wheel-drive models, including the Chrysler New Yorker and the new fastback Dodge Daytona. Signals from a host of sensors were sent to an engine control computer that regulated the throttle body injection (TBI) system and the ignition system for engine operation with much greater precision than the standard two-barrel feedback carburetor could provide. The output of the 1984 2.2-L throttle body injection engine was slightly improved at 74 kW and 164 N-m (99 hp and 121 lb-ft). Volume use of the 2.2-L two-barrel carbureted engine took a significant jump in 1984 with the launch of the instantly successful Dodge Caravan and Plymouth Voyager, portrayed on Fig. 12.18. These minivans used the engine as standard power.

Fig. 12.17 *Trenton engine plant manager Bob Garlo (right) and engine assembler Leslie Tyler (left) with the one-millionth 2.2-L single overhead camshaft four-cylinder engine built by the Trenton engine plant.*

Fig. 12.18 *The 1984 Plymouth Voyager. Standard power for the new front-wheel-drive minivan came from a 2.2-L single overhead camshaft four-cylinder engine. This vehicle is on display at the Walter P. Chrysler Museum. Photo by W.L. Weertman.*

The year 1985 saw the introduction of a hot Dodge Omni GLH, whose standard power, a 2.2-L two-barrel carbureted engine with a higher compression ratio (9.6:1 versus 9.0:1) and producing 110 hp, also was standard on the 1985 Dodge Charger 2.2. Carried over to 1986, both the Omni GLH and the Charger 2.2 were discontinued at the end of that year.

Improvements were made to the combustion chamber in 1986 for emission control, giving it the name of Fast Burn, without any output change.

The year 1987 was the last year of production for the two-barrel carburetor version of the 2.2-L engine, leaving the throttle body injection version to carry on. Chrysler dropped its use of the throttle body injection engine at the end of 1989; Plymouth and Dodge continued the 2.2-L throttle body injection engine through the end of 1994, the engine being last used in the almost-twin Plymouth Sundance and Dodge Shadow. The front-wheel-drive Sundance and Shadow used the 2.2-L throttle body injection engine continuously as standard power from their introduction in 1987 to their end of production in 1994.

A significant change for 1988 that was incorporated in all corporate four-cylinder engines was a new roller-type camshaft follower for the intake and exhaust valves, giving a reduction in friction versus the prior sliding-type followers. Smoother idle quality, increased fuel economy at idle and low speeds, and a quieter idle were benefits from these roller followers.

Shortly after the Trenton engine plant started building 2.2-L engines for 1981 cars, a second plant was tooled—the new Chrysler de Mexico engine plant in Saltillo, Mexico.[5] which was equipped with one cylinder block machining line as compared to the three lines at the Trenton engine plant. Of interest, by the late 1980s, with V-6s entering production and four-cylinder volume requirements dropping, the Trenton engine plant had a surplus of 2.2-L engine machining capacity. An opportunity to profitably dispose of surplus machining lines for blocks, cylinder heads, and other parts came in the form of a request from First Automotive Works of the Peoples Republic of China for production tools for the 2.2-L engine, together with a license to build and technical assistance. Despite problems of substantial damage to the machine tools when the freighter carrying the equipment encountered a heavy storm while underway to China, the program was successful. Production of the engine started in the First Automotive Works Changchun factory in northern China in the early 1990s.

Table 12.3 shows the specifications for this 2.2-L single overhead camshaft four-cylinder engine.

A Larger Version—The 2.5-L Engine

With a strong legacy of finding extra displacement for almost every engine ever designed by the Engine Design department, the 2.2-L engine, following its introduction in 1981, was given a careful look to see what could be done. The 2.2-L size was well suited for its assigned car models, but a larger version of the engine possibly could reduce dependency on an overseas supply of the 2.6-L four-cylinder single overhead camshaft engine (described later) in use since 1981 on a number of models while reducing engine compartment variables, a benefit for the vehicle assembly division. Any displacement increase would require a bore increase, a longer stroke, or a combination of both. A bore increase was ruled out quickly because the bores were already packed together so tightly that the cylinder walls were Siamesed between bores without any space for coolant. However, similar to the G-RG engine family before it, a way appeared open—to raise the deck and to increase the stroke. This was a rather straightforward design exercise that resulted in a cylinder block that was 12 mm higher (249.8 versus 237.8 mm) with strengthened bulkheads, a connecting rod that was 6 mm longer (157 versus 151 mm), a crankshaft (still nodular cast iron) with a 104-mm stroke (versus 92 mm) giving a displacement of 2.5 liters, and a dish-top piston with valve notches to provide a 9.0:1 compression ratio and freewheeling clearance for the valves. However, the displacement change was the easy part for Engine Design. The real challenge was in solving the problem of how to handle the increased secondary vibrations generated by the longer stroke. Secondary vibrations that make the engine jump up and down along its bore axis at twice the engine speed are inherent in an in-line four-cylinder engine. With a stroke increase, the vibration grows worse. As the engine speed increases, the vibration becomes really bad because its force varies with the square of the speed. Getting rid of the vibration requires the addition of two counter-rotating balance shafts rotating at twice the engine speed, which was the approach taken for the 2.5-L engine. It also was the approach used by the 2.6-L Mitsubishi Motors Corporation single overhead camshaft four-cylinder engine already in production. However, the cylinder block design of the 2.6-L engine had provision for balance shafts from the beginning, a provision lacking on the 2.2-L design. An add-on system was needed for the 2.5-L engine. A design emerged for the 2.5-L using a cast aluminum housing, bolted to the bottom of the cylinder block, which contained two geared-together counter-rotating balance shafts. These shafts were driven at twice the crankshaft speed by a single-row roller chain from the nose of the crankshaft. Because the shafts were below the oil level in the oil pan, the housing was enclosed and had provision for pumping out any oil that collected inside the housing. I was awarded a U.S. patent for the design.[6] This balance shaft system can be seen on the cross section shown on Fig. 12.19 and on the longitudinal section shown on Fig. 12.20 of the 1986 2.5-L engine, coded A512. An ancillary change was an increase in the capacity of the engine oil pump to handle the extra demand of the four bearings of the balance shafts, an oil pump change that was applied to the 2.2-L engine for commonality.

With its single-point throttle-body fuel injection system, a compression ratio of 9.0:1, and the latest fast-burn combustion chamber technology, the engine produced 75 kW (100 hp) and 184 N-m (136 lb-ft). Its initial application was that of standard power for the 1986 Chrysler LeBaron Town & Country convertible pictured on Fig. 12.21, the New Yorker four-door sedan, and the sporty two-door hatchback Laser XE models, and as optional power on many Dodge, Plymouth, and other Chrysler models.

In 1989, advancing piston technology allowed the piston height to be reduced enough that the extra deck height of the

TABLE 12.3
SPECIFICATIONS FOR THE 2.2-L SOHC FOUR-CYLINDER ENGINE
Note: All dimensions are in metric units (millimeters) rather than customary units (inches).

Model Year	1981	1983	1983
Car line	Plymouth, Dodge	Plymouth, Dodge	Dodge
Model	Plymouth: Reliant, Horizon, TC3 Dodge: Aries, Omni, O24	Plymouth: Turismo 2.2 Dodge: Charger 2.2	Shelby Charger
Engine code	E62	E62	E68
Displacement (liters)	2.2	2.2	2.2
Bore	87.5	87.5	87.5
Stroke	92	92	92
Compression ratio	8.5	9.0:1	9.6
Power @ rpm	63 kW (84 hp) @ 4800	75 kW (100 hp) @ 5200	80 kW (107 hp) @ 5600
Torque @ rpm	150 N-m (111 lb-ft) @ 2800	165 N-m (122 lb-ft) @ 3200	172 N-m (127 lb-ft) @ 3600
Cyl. block mat'l.—skirt type	Cast iron—shallow	Cast iron—shallow	Cast iron—shallow
Cyl. block deck height	237.8	237.8	237.8
Cyl. block length	418	418	418
Cyl. bore spacing	96	96	96
No. of main brgs.—ck/s mat'l.	5—nodular cast iron	5—nodular cast iron	5—nodular cast iron
Ck/shaft main journal dia.	60	60	60
Crankpin journal dia.	50	50	50
Connecting rod centers	151	151	151
Valve system	SOHC—8-valve	SOHC—8-valve	SOHC—8-valve
Valve events	244-244-28	244-244-28	
Camshaft drive	Timing belt	Timing belt	Timing belt
Valve head dia.	Intake 40.6; exhaust 35.4	Intake 40.6; exhaust 35.4	Intake 40.6; exhaust 35.4
Valve lash adjustment	Hyd. rocker arm pivot	Hyd. rocker arm pivot	Hyd. rocker arm pivot
Oil pump/distributor drive	Geared to auxiliary shaft	Geared to auxiliary shaft	Geared to auxiliary shaft
Fuel system	2-barrel carburetor	2-barrel carburetor	2-barrel carburetor
Engine angle	10 deg 16 min. to rear	10 deg 16 min. to rear	10 deg 16 min. to rear
Engine assembly weight*	135.2 kg (298.1 lb)	133.3 kg (293.8 lb)	133.3 kg (293.8 lb)

* For content, see the Abbreviations and Definitions section near the back of this book.

(Continued on next page)

TABLE 12.3
(Continued)

Model Year	1984	1994
Car line	Dodge, Chrysler	Plymouth, Dodge
Model	Dodge: Daytona, 600 Chrysler: Laser, LeBaron, New Yorker	Plymouth: Sundance Dodge: Shadow
Engine code	EDF	EDF
Displacement (liters)	2.2	2.2
Bore	87.5	87.5
Stroke	92	92
Compression ratio	9.0:1	9.5
Power @ rpm	74 kW (99 hp) @ 5600	69 kW (93 hp) @ 4800
Torque @ rpm	164 N-m (121 ft-lb) @ 3200	165 N-m (122 lb-ft) @ 3200
Cyl. block mat'l.—skirt type	Cast iron—shallow	Cast iron—shallow
Cyl. block deck height	237.8	237.8
Cyl. block length	418	418
Cyl. bore spacing	96	96
No. of main brgs.—ck/s mat'l.	5—nodular cast iron	5—nodular cast iron
Ck/shaft main journal dia.	60	60
Crankpin journal dia.	50	50
Connecting rod centers	151	151
Valve system	SOHC—8-valve	SOHC—8-valve
Valve events	244-244-28	236-232-8
Camshaft drive	Timing belt	Timing belt
Valve head dia.	Intake 40.6; exhaust 35.4	Intake 40.6; exhaust 35.4
Valve lash adjustment	Hyd. rocker arm pivot	Hyd. rocker arm pivot
Oil pump/distributor drive	Geared to auxiliary shaft	Geared to auxiliary shaft
Fuel system	TBI	TBI
Engine angle	10 deg 16 min. to rear	10 deg 16 min. to rear
Engine assembly weight*	134.4 kg (295.7 lb)	142.26 kg (313.5 lb)

* For content, see the Abbreviations and Definitions section near the back of this book.

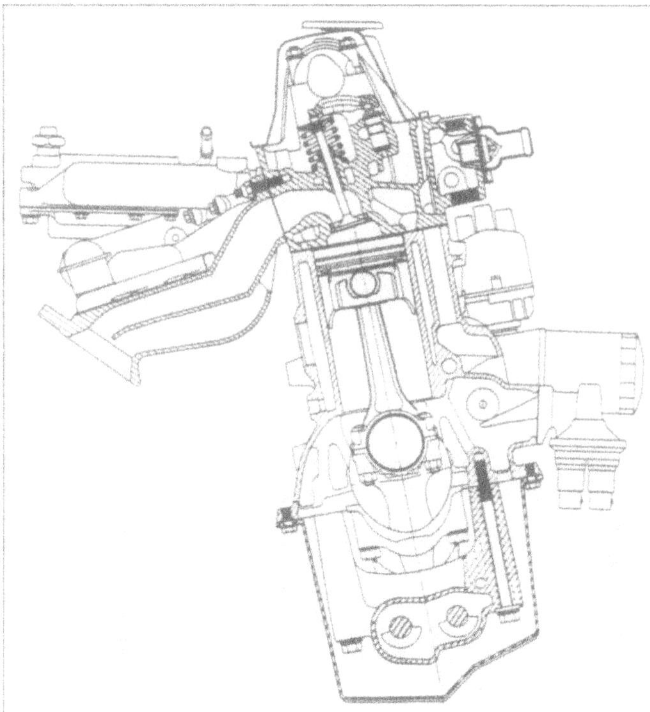

Fig. 12.19 *A cross section of the 1987 2.5-L single overhead camshaft four-cylinder engine equipped with two counter-rotating balance shafts located below the crankshaft.*

2.5-L cylinder block was no longer needed. With the shorter piston and a reversion to the shorter 2.2-L connecting rod, the 237.8-mm deck height of the 2.2-L block was able to accommodate the 104-mm stroke of the 2.5-L engine. Capitalizing on this, a universal block casting was designed, which could be machined into 2.2-L and 2.5-L cylinder blocks in both the naturally aspirated and turbocharged versions.

In 1993, small-diameter cross-drilled holes were added to the cylinder block and the cylinder head to improve the cooling of the narrow cylinder head gasket areas between bores. During 1995, its last year of production, the 2.5-L single overhead camshaft engine was specified as standard power for the Plymouth Voyager and Dodge Caravan minivans, and the Plymouth Acclaim and Dodge Spirit models.

Table 12.4 shows the specifications for this 2.5-L single overhead camshaft balanced four-cylinder engine.

Turbocharging the 2.2-L Engine

Prior to 1984, the Plymouth, Dodge, and Chrysler K-car platform models included two-door sedans, four-door sedans, and station wagons, with Dodge and Chrysler also having a convertible. The available 2.2-L Trenton engine plant-built four-cylinder and 2.6-L Mitsubishi Motors Corporation four-cylinder engines were appropriate power for these traditionally high-volume basic models. However, a sporty-looking, K-platformed car that was being prepared for Dodge and Chrysler—to be named Daytona and Laser, respectively—needed much more power under the hood to give substance to the visual performance image.

A power boost for the 2.2-L engine came in the form of an exhaust gas driven turbocharger. Turbochargers had been around for years, especially during World War II when they were used by the thousands for aircraft engines, with the turbocharger (called turbosupercharger at the time) having the task of maintaining sea-level charge density regardless of the altitude of the airplane. During the early 1980s, a small turbocharger, developed for automotive use by Garrett AiResearch, became available and was applied to the 2.2-L engine. The program, coded A472, combined Garrett's T3 turbocharger with engine control computers, multipoint fuel injection, a pneumatic wastegate for limiting boost pressure to 7.2 psi, and a knock sensor to retard spark timing should detonation occur. However, the existing turbocharger as designed at the time had a chronic oil coking problem caused by a sudden temperature rise in the turbocharger bearings and lubricating oil passages following engine shutdown. Only a few weeks before the scheduled start of production of the Chrysler turbo-equipped engines, Garrett engineers came up with two possible solutions. One solution involved additional cooling fins cast onto the turbocharger housing. The other solution, a more dramatic change, added a water coolant jacket to the turbo center housing so engine coolant would absorb the heat spike that followed shutdown. Test data showed that the coolant jacket approach was by far the most effective solution but had a degree of risk because of an almost total lack of hours of operation. In a joint Garrett/Chrysler meeting chaired by Robert M. Sinclair, director of Chassis Engineering, a unanimous decision was made to abandon the air-cooled center housing and to proceed only with the water-cooled housing.[7] The program proved a success, and with the start of 1984 production, Chrysler became the first manufacturer offering a front-wheel-drive car powered by a turbocharged engine having a water-cooled turbocharger housing.[8]

In addition to the turbocharger itself, other technical highlights of the turbo engine included a unique aluminum intake manifold and nodular iron exhaust manifold that were required; the four fuel injectors were mounted on the intake manifold. A dish-top piston with an 8.1:1 compression ratio was provided. Upgrades were made to the valves, piston rings, and cylinder head gasket. The added lubricating oil flow for the turbocharger bearings normally would have required a larger-capacity oil pump for the engine. However, by selective fitting of the bearing shells of the crankshaft main bearings and connecting rod lower bearings to use only the tighter half of the normal clearance range, excess oil flow to the crankshaft was reduced, which compensated for the extra oil required for the turbocharger bearings, thus eliminating the need for a new oil pump.

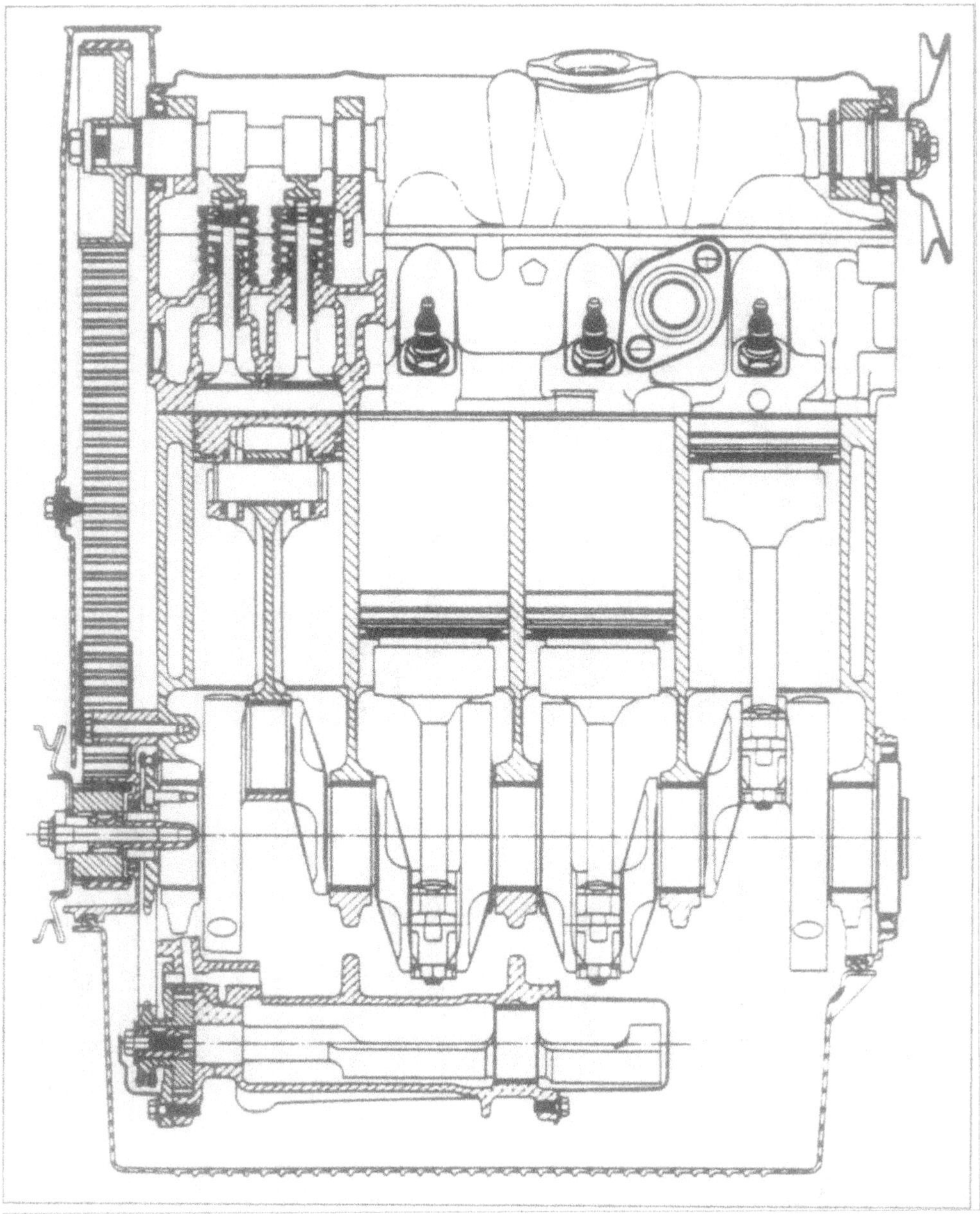

Fig. 12.20 *A longitudinal section of the 1987 2.5-L single overhead camshaft four-cylinder engine. The housing below the crankshaft encloses the two balance shafts.*

Fig. 12.21 *The 1986 Chrysler Town & Country convertible. Standard power was a 2.5-L single overhead camshaft four-cylinder engine with balance shafts and throttle-body fuel injection.*

TABLE 12.4
SPECIFICATIONS FOR THE 2.5-L SOHC FOUR-CYLINDER
Note: All dimensions are in metric units (millimeters) rather than customary units (inches).

Model Year	1986	1988	1989	1995
Car line	Plymouth, Dodge, Chrysler	Plymouth, Dodge, Chrysler	Plymouth, Dodge, Chrysler	Plymouth, Dodge
Model	Plymouth: Reliant, Caravelle (opt.) Dodge: general option Chrysler: New Yorker (std.); Laser, LeBaron (opt.)	Plymouth: Voyager, Caravel std. Dodge: std. and opt. Chrysler: Town & Country, LeBaron (std.)	Plymouth: Voyager, Caravelle (std.) Dodge: std. and opt. Chrysler: Town & Country, LeBaron (std.)	Plymouth: Voyager, Acclaim Dodge: Caravan, Spirit
Engine code	EDM	EDM	EDM	EDM
Displacement (liters)	2.5	2.5	2.5	2.5
Bore	87.5	87.5	87.5	87.5
Stroke	104	104	104	104
Compression ratio	9.0:1	8.9:1	8.9:1	9.6
Power @ rpm	75-kW (100-hp) @ 4800	72 kW (96 hp) @ 4800	75 kW (100 hp) @ 4800	75 kW (100 hp) @ 4800
Torque @ rpm	184 N-m (136 lb-ft) @ 2800	180 N-m (133 lb-ft) @ 2800	183 N-m (135 lb-ft) @ 2800	183 N-m (135 lb-ft) @ 2800
Cyl. block mat'l.—skirt type	Cast iron—shallow	Cast iron—shallow	Cast iron—shallow	Cast iron—shallow
Cyl. block deck height	249.8	249.8	237.8	237.8
Cyl. block length	418	418	418	418
Cyl. bore spacing	96	96	96	96
No. of main brgs.—ck/s mat'l.	5—nodular cast iron	5—nodular cast iron	5—nodular cast iron	5—nodular cast iron
Ck/shaft main journal dia.	60	60	60	60
Crankpin journal dia.	50	50	50	50
Connecting rod centers	157	157	151	151
Balance system	Twin balance shafts	Twin balance shafts	Twin balance shafts	Twin balance shafts
Valve system	SOHC—8-valve	SOHC—8-valve	SOHC—8-valve	SOHC—8-valve
Valve events	240-240-28	236-232-8	236-232-8	236-232-8
Camshaft drive	Timing belt	Timing belt	Timing belt	Timing belt
Valve head dia.	Intake 40.6; exhaust 35.4	Intake 40.6; exhaust 35.4	Intake 40.6; exhaust 35.4	Intake 40.6; exhaust 35.4
Valve lash adjustment	Hyd. rocker arm pivot	Hyd. rocker arm pivot	Hyd. rocker arm pivot	Hyd. rocker arm pivot
Oil pump/distributor drive	Geared to auxiliary shaft	Geared to auxiliary shaft	Geared to auxiliary shaft	Geared to auxiliary shaft
Fuel system	TBI	TBI	TBI	TBI
Engine angle	10 deg 16 min. to rear	10 deg 16 min. to rear	10 deg 16 min. to rear	10 deg 16 min. to rear
Engine assembly weight*		142.38 kg (313.9 lb)		157.3 kg (346.8 lb)

* For content, see the Abbreviations and Definitions section near the back of this book.

High-pressure oil for the turbocharger bearings was obtained from the drilled and tapped hole in the front side of the block normally used only for the oil pressure switch of the engine. Oil from the turbocharger was returned to the oil pan via a new boss added to the rear of the block above the oil pan rail, which was drilled and tapped for the oil return line fitting. Existing drilled and tapped holes in the cylinder block and cylinder head were used to route coolant in and out of the center housing of the turbocharger.

A special camshaft was released for the turbo engine with, perhaps surprisingly, shorter events. The intake duration, exhaust duration, and overlap degrees of the turbo engine were 240-240-20 versus the 244-244-28 of the naturally aspirated 2.2-L engine. This was done to pep up the low- and mid-speed torque levels of the engine and to make the inherent turbo lag less noticeable. A clear-coated, die-cast aluminum cylinder head cover was released for the turbo engine to provide identity and underhood appearance enhancement of the engine.

The key to success in developing the engine to produce power levels impressively beyond the naturally aspirated engine while keeping it within the realm of acceptable temperature and pressure limits and, very importantly, not exceeding allowable exhaust emission ceilings was in the computer-controlled electronic engine management system. The individual cylinder fuel injector system, worked on for years and now perfected and furnished by Robert Bosch, GmbH, together with the Chrysler-built (at Huntsville, Alabama) engine management computers, were combined to accomplish the complex task.

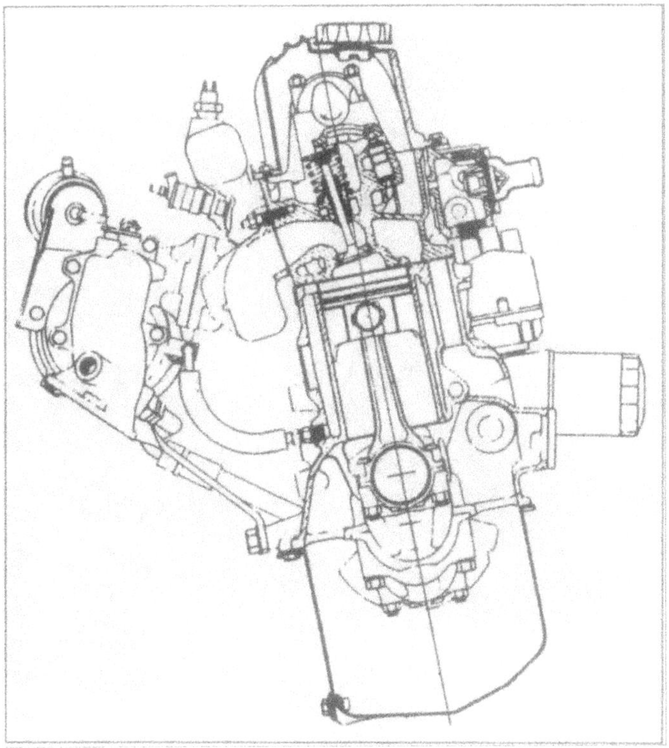

Fig. 12.22 *A cross section of the 1984 2.2-L turbocharged 106-kW (142-hp) four-cylinder engine.*

Figure 12.22 shows a cross section of the 1984 engine; Fig. 12.23 shows a rear view. An SAE paper titled "Turbocharging the Chrysler 2.2 Liter Engine," written by Floyd E. Allen and Gordon L. Rinschler,[9] provides additional information on the design and development of this engine that had an output of 106 kW (142 hp) and 217 N-m (160 lb-ft) torque. Released as optional power on the 1984 Dodge Daytona (shown on Fig. 12.24), Dodge 600, Chrysler Laser, Chrysler LeBaron, Chrysler E-Class, and Chrysler New Yorker, it was installed as standard on a new 1984 Dodge Daytona model badged as the Dodge Daytona Turbo Z.

For 1985, the performance of the 2.2-L Turbo engine was increased slightly to 146 hp @ 5200 rpm and 168 lb-ft @ 3600 rpm torque, and its application was broadened by being made optional on the new Plymouth K-platform-based four-door Caravelle and the hot performing Dodge Omni GLH. Also for 1985, the 2.2-L Turbo went under the hood of another hot performing car—the fastback L-body Dodge Shelby Turbo Charger.

Carried over to 1986, the Dodge Daytona Turbo Z was featured in an article in the April 1986 issue of *Motor Trend*, titled "Pontiac TransAm vs. Dodge Daytona Turbo Z," written by John Hanson, which declared, "At 146 hp, the Turbo Z is one of the most powerful front-drive cars sold in America. Predictably, its high-rpm powerplant performs best at partial-throttle turbo boost at 3500 rpm and above. It delivers excellent power during freeway-passing maneuvers and is an unqualified blast during foot-to-the-floor quarter-mile acceleration runs at full boost..."

In 1986, the 2.2-L turbocharged engine became standard power for another model, the Chrysler K-platform-based limousine, replacing the prior year's Mitsubishi-built 2.6-L four-cylinder engine. In 1987, with the release of the Turbo II engine, described in the next section, this engine was designated the Turbo I.

Continued dynamometer development resulted in a new dual-tuned intake system being released for 1988. The original log-type intake manifold was replaced with a single-piece cast aluminum manifold having a plenum chamber and 246-mm-long individual runners feeding the intake ports of the cylinder head. Concurrently, a smaller and lighter Mitsubishi Heavy Industries (MHI) TE04H turbocharger replaced the original Garrett AiResearch turbocharger.

On January 27, 1988, a 2.2-L Turbo I engine had the distinction of being the twenty-two millionth engine produced by the Trenton engine plant, the momentous occasion recorded as shown on Fig. 12.25. At the end of the 1988 model year, the engine was discontinued, replaced the following year by a 2.5-L Turbo I engine, described here next.

Fig. 12.23 *A rear view of the 1984 2.2-L turbocharged four-cylinder engine.*

Fig. 12.24 *The 1984 Dodge Daytona Turbo. The 106-kW (142-hp) 2.2-L turbocharged engine under its hood gave the car extra performance and the last part of its model name.*

Fig. 12.25 *The twenty-two millionth engine built by the Trenton engine plant on January 27, 1988. The engine was a 2.2-L Turbo.*

Table 12.5 shows the specifications for this 2.2-L single overhead camshaft turbocharged four-cylinder engine (2.2-L Turbo I).

2.5-L Turbo I Engine

With an objective of substantially improving the low- and mid-speed torque of the 2.2-L Turbo I engine, a program coded A519 was initiated to apply the components already developed for the 2.2-L engine to the longer-stroke, higher-torque 2.5-L engine. An extra incentive was that with the 1989 2.5-L engine using the same cylinder block as the 2.2-L engine, the need for new and unique parts would be minimized. By retaining the balance shaft system already in use on the 2.5-L engines, this new 2.5-L Turbo I engine would be free of the secondary vibrations of the 2.2-Turbo I engine, a noticeable quality upgrade. A cross section and a longitudinal section of the engine are shown on Figs. 12.26 and 12.27, respectively.

The new universal cylinder block, already described, was given the strength to handle the extra power loading from the turbocharging. Also, the main bearing caps were increased in height to provide the necessary additional strength.

TABLE 12.5
SPECIFICATIONS FOR THE 2.2-L SOHC FOUR-CYLINDER TURBO I
Note: All dimensions are in metric units (millimeters) rather than customary units (inches).

Model Year	1984	1988
Car line	Dodge, Chrysler	Plymouth, Dodge, Chrysler
Model	Std.: Dodge Daytona Turbo, Opt.: Daytona, 600, Chrysler Laser, LeBaron, New Yorker	Std.: Dodge Daytona Pacifica and Chrysler New Yorker Turbo. Opt. on many other models.
Engine code	EDG	EDG
Displacement (liters)	2.2	2.2
Bore	87.5	87.5
Stroke	92	92
Compression ratio	8.1:1	8.1:1
Power @ rpm	106 kW (142 hp) @ 4800	109 kW (146 hp) @ 4800
Torque @ rpm	217 N-m (160 lb-ft) @ 2000	230 N-m (170 lb-ft) @ 2000
Cyl. block mat'l.—skirt type	Cast iron—shallow	Cast iron—shallow
Cyl. block deck height	237.8	237.8
Cyl. block length	418	418
Cyl. bore spacing	96	96
No. of main brgs.—ck/s mat'l.	5—nodular cast iron	5—nodular cast iron
Ck/shaft main journal dia.	60	60
Crankpin journal dia.	50	50
Connecting rod centers	151	151
Valve system	SOHC—8-valve	SOHC—8-valve
Valve events	240-240-20	236-232-8
Camshaft drive	Timing belt	Timing belt
Valve head dia.	Intake 40.6; exhaust 35.4	Intake 40.6; exhaust 35.4
Valve lash adjustment	Hyd. rocker arm pivot	Hyd. rocker arm pivot
Oil pump/distributor drive	Geared to auxiliary shaft	Geared to auxiliary shaft
Fuel system	MPI	MPI
Turbocharger	Garrett T3	MHI TE04H
Engine assembly weight*	145.2 kg (319.5 lb)	136.71 kg (301.4 lb)

* For content, see the Abbreviations and Definitions section near the back of this book.

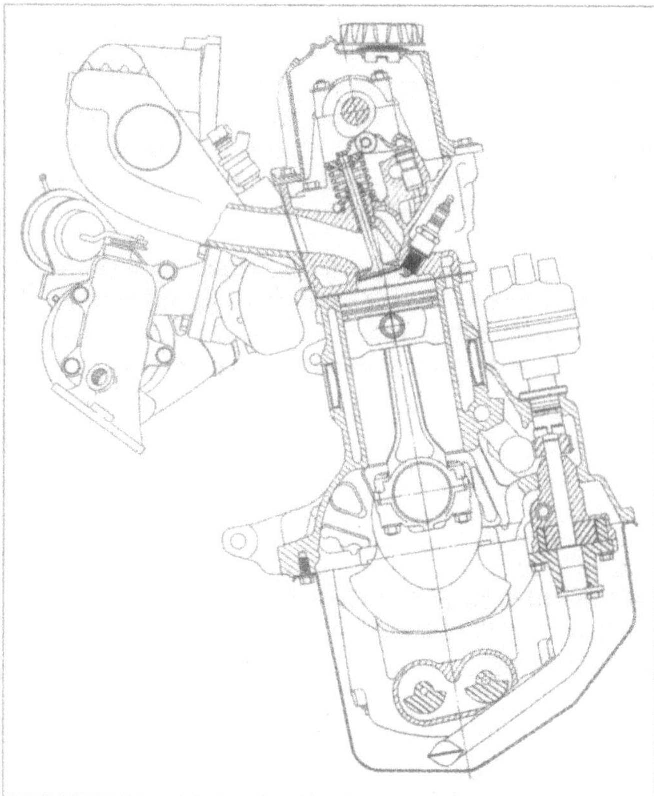

Fig. 12.26 *A cross section of the 1989 2.5-L Turbo I single overhead camshaft four-cylinder engine. Note the lengthened intake passage, the roller follower for valve actuation, and the balance shafts below the crankshaft.*

Testing showed that the standard nodular iron crankshaft of the high-volume 2.5-L naturally aspirated engine would not be strong enough to withstand the turbocharged power levels. Upgrading to forged steel or austempered ductile iron was considered seriously but finally was ruled out in favor of simply selecting production nodular iron crankshafts at the high end of the hardness range for use in the 2.5-L turbocharged engines.

A lighter and shorter cast aluminum piston, developed by Chrysler and the piston supplier Mahle, Inc., was provided with lock ring grooves for a full-floating piston pin, a feature needed to withstand the loads and speeds of the turbo engine. The piston top was dished to reduce the compression ratio to 7.85:1, a requirement for the turbo application.

Roller camshaft followers and the deep-tooth high-torque drive timing belt, both introduced in 1988 on Chrysler-built four-cylinder engines, were carried over to this 2.5-L Turbo I engine. Camshaft events were new and unique. Intake duration, exhaust duration, and overlap, which were 236-232-12 on the naturally aspirated 2.5-L engine, were reduced to 228-228-0 on the 2.5-L Turbo I engine; the events favored low- and mid-speed torque and idle quality while relying on the turbocharger to provide boost for plenty of power at the high range of revolutions per minute.

For adequate cylinder head gasket life, small diagonal coolant holes in the cylinder head and top of the block between the bores, originally developed for the 1987 2.2-L intercooled Turbo II engine (described in the next section) were incorporated in this 2.5-L Turbo I engine.

With the lubrication needs of the four balance shaft bearings added to the already taxed engine oil pump, a problem of adequate oil supply again arose, triggering a careful look at all areas where oil flow possibly might be reduced. Changes that included a reduction of the turbocharger bearing clearances and the addition of a restrictor to the drilled hole leading to the oil gallery in the cylinder head dropped the engine oil demand to a level that could be handled by the existing oil pump.

The single-piece cast aluminum intake manifold with a plenum chamber and 246-mm long branches, developed for the 1988 2.2-L Turbo I engine, carried over to this 1989 2.5-L Turbo I engine. The turbocharger was a Mitsubishi Heavy Industries TE04H unit and was carried over, but internal changes were made to increase the air flow commensurate with the larger engine displacement. A stepped boost curve, dependent on engine speed, was used and topped out at 11 psi. Engine output was 112 kW (150 hp) and 244 N-m (180 lb-ft). Additional technical information about the engine can be found in the SAE paper titled "Turbocharging the Chrysler 2.5 Liter Engine," written by Floyd E. Allen and Thomas W. Witte, presented in 1990.[10]

Assigned to generally replace the 2.2-L Turbo I engine throughout the Dodge, Plymouth, and Chrysler car lines, the 2.5-L Turbo I engine was optional on many 1989 models and standard on the 1989 Dodge Daytona ES Turbo, Dodge Lancer ES, and the new Dodge Spirit ES shown on Fig. 12.28. Among the 1989 models hosting the 2.5-L Turbo I engine as an option were the Dodge Caravan and Plymouth Caravan—the first time a turbocharged engine was available in Chrysler minivans. In 1991, turbo boost calibration changes added 2 hp and a hefty 30 lb-ft torque to the output of the engine, the new ratings being 113 kW (152 hp) and 285 N-m (210 lb-ft).

The 2.5-L Turbo I engine was discontinued at the end of 1992, leaving the 16-valve 2.2-L Turbo III engine (described later) as the sole turbocharged four-cylinder engine in use. It would last for only one additional year.

Table 12.6 shows the specifications of the 2.5-L Turbo I engine.

Turbo II—Intercooling the 2.2-L Turbo I Engine

To provide exceptional performance for a new-for-1987 Dodge Daytona Shelby Z car, a second turbo version was created that

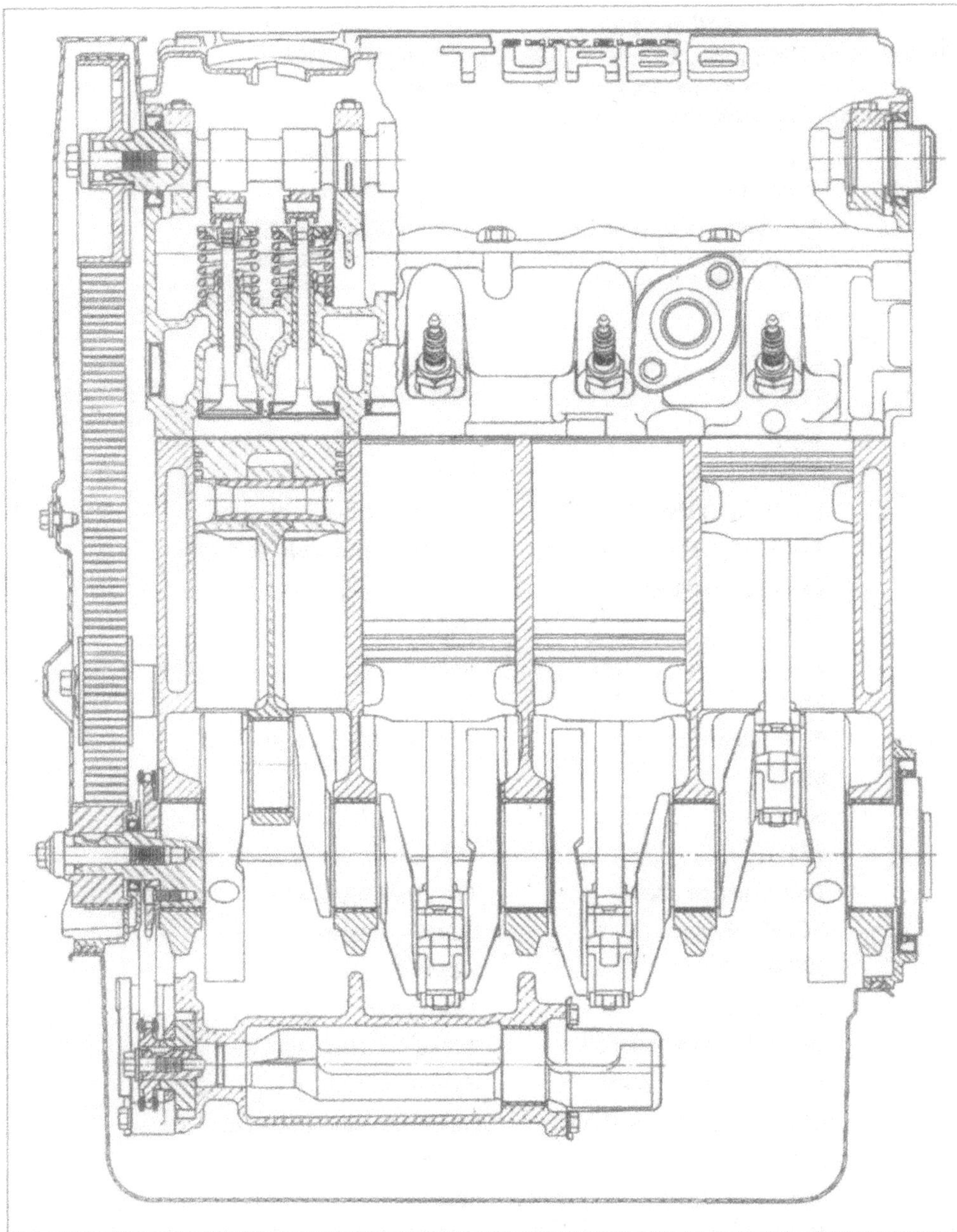

Fig. 12.27 *A longitudinal section of the 1989 2.5-L Turbo I single overhead camshaft four-cylinder engine.*

Fig. 12.28 *The 1989 Dodge Spirit ES four-door sedan. This new Dodge model had a 2.5-L Turbo I engine as its standard powerplant.*

used an air-to-air intercooler for lowering the temperature of the pressurized air leaving the turbocharger prior to it entering the throttle body. The aluminum finned intercooler, which was mounted next to the radiator, and its connecting hoses were ruggedly constructed to withstand the 12-psi boost pressure of the AiResearch water-cooled turbocharger. The intercooler was effective, providing a temperature drop of up to 120 deg F, permitting the 12-psi boost pressure versus the 7.2 psi of the Turbo I. Operating at the same 8.1:1 compression ratio, the output of the 2.2-L Turbo II engine was an impressive 130 kW (174 hp) and 271 N-m (200 lb-ft) torque.

This Turbo II engine had a dual-tuned intake system with a two-piece aluminum manifold that combined a plenum chamber with primary 14.5-in.-long individual runners to the intake

TABLE 12.6
SPECIFICATIONS FOR THE 2.5-L SOHC FOUR-CYLINDER TURBO I
Note: All dimensions are in metric units (millimeters) rather than customary units (inches).

Model Year	1989	1992
Car line	Plymouth, Dodge, Chrysler	Dodge, Chrysler
Model	Std.: Dodge Daytona ES Turbo, Spirit ES, Lancer ES Opt. on other models	Std.: Dodge Spirit ES Opt. on other models
Engine code	EDT	EDT
Displacement (liters)	2.5	2.5
Bore	87.5	87.5
Stroke	104	104
Compression ratio	7.85:1	7.85:1
Power @ rpm	112 kW (150 hp) @ 4800	113 kW (152 hp) @ 4800
Torque @ rpm	244 N-m (180 lb-ft) @ 2000	286 N-m (211 lb-ft) @ 2800
Cyl. block mat'l.—skirt type	Cast iron—shallow	Cast iron—shallow
Cyl. block deck height	237.8	237.8
Cyl. block length	418	418
Cyl. bore spacing	96	96
No. of main brgs—ck/s mat'l.	5—nodular cast iron	5—nodular cast iron
Ck/shaft main journal dia.	60	60
Crankpin journal dia.	50	50
Connecting rod centers	151	151
Balance system	Twin balance shafts	Twin balance shafts
Valve system	SOHC—8-valve	SOHC—8-valve
Valve events	228-228-0	228-228-0
Camshaft drive	Timing belt	Timing belt
Valve head dia.	Intake 40.6; exhaust 35.4	Intake 40.6; exhaust 35.4
Valve lash adjustment	Hyd. rocker arm pivot	Hyd. rocker arm pivot
Oil pump/distributor drive	Geared to auxiliary shaft	Geared to auxiliary shaft
Fuel system	MPI	SMPI
Turbocharger	MHI	MHI
Engine assembly weight*	156.35 kg (344.7 lb)	161.36 kg (355.0 lb)

* For content, see the Abbreviations and Definitions section near the back of this book.

valves. A 12-in. secondary tuned length fed air into the plenum from the throttle body. A number of changes were made to the engine to give it durability commensurate with the increased power output. The crankshaft was changed from nodular cast iron to forged steel, and heavy-duty trimetal connecting rod bearings were used. Full-floating piston pins that were better able to resist scuffing and pistons made of an aluminum alloy with greater hot strength were specified. Importantly, to increase cylinder head gasket life, small diagonal holes were drilled between cylinder bores at the top deck and in the cylinder head to allow engine coolant to cool these hot bridge areas of the gasket. A cross section and a rear view of the 2.2-L Turbo II engine are shown on Figs. 12.29 and 12.30, respectively.

The intake manifold, which was much larger than the manifold of the Turbo I, and the cylinder head cover were painted in a black crackle finish to give the engine a distinctive appearance. In its initial 1987 year, the 2.2-L Turbo II engine was exclusive to the Dodge Daytona Shelby Z model pictured on Fig. 12.31. In 1988, it became an option on the Dodge Lancer ES and was specified as the only engine available with automatic transmission on the Italian-built "Chrysler's TC by Maserati," further described later in this chapter. Winding up its three-year run

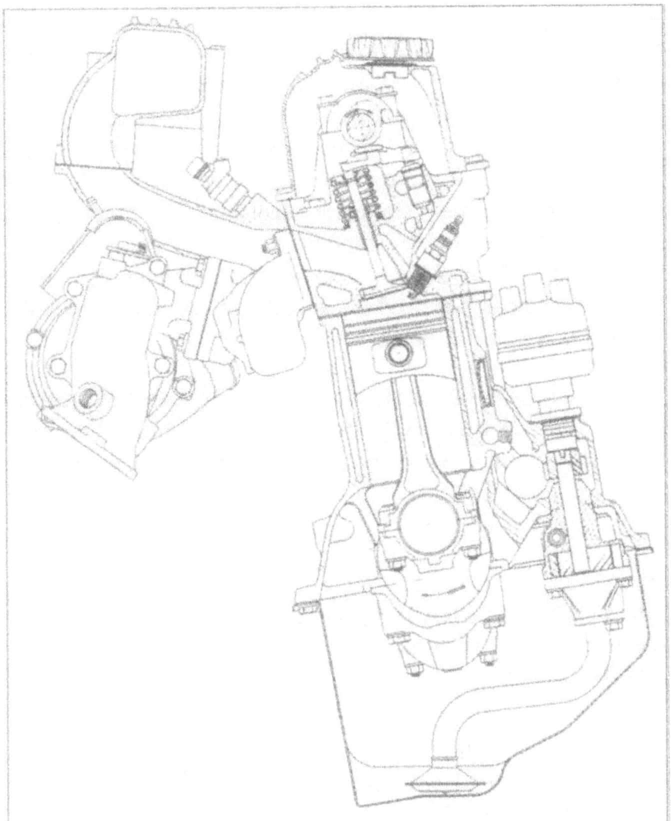

Fig. 12.29 *A cross section of the 1987 2.2-L Turbo II single overhead camshaft four-cylinder engine. Compressed, cooled intake air entered the plenum, shown in the section at the top of the drawing.*

at the end of 1989, during its last year it went under the hoods of all Dodge Daytona Shelbys and Dodge Lancer Shelbys as standard equipment, and under the hoods of the Dodge Daytona CS cars whose owners opted for the turbocharged, intercooled Turbo II engine. Chrysler made it an option on LeBaron high-line coupes and convertibles, and it was standard on the LeBaron GTS. Its power gave the cars plenty of get-up-and-go. In the August 1989 issue of *Motor Trend*, Daniel C. Ross wrote in an article titled "Dodge Daytona CS," "This is as much sports car as you'll ever use...when those ponies taste the whip, they still want to wake up all at once. The explosion of acceleration that follows can be ferocious for a driver not ready for it. If you are ready for it (and it only takes a few runs to get dialed in), the frenzy is at once exciting and challenging. In the straight-ahead, the Daytona CS accelerates passionately."

Table 12.7 shows the specifications of this 2.2-L Turbo II engine.

Turbo IV—2.2-L Variable Nozzle Turbocharged Engine

(Note: The apparently missing Turbo III engine will be described later in this chapter.)

Development efforts to reduce the trait of turbo lag, inherent with exhaust gas turbochargers, led to the release of a new Garrett AiResearch turbocharger for 1990 that had a variable nozzle system as illustrated on Fig. 12.32. This variable nozzle turbocharger (VNT) reduced lag, accelerating to full boost as much as 50% faster than Turbo II, and had a broader torque curve.

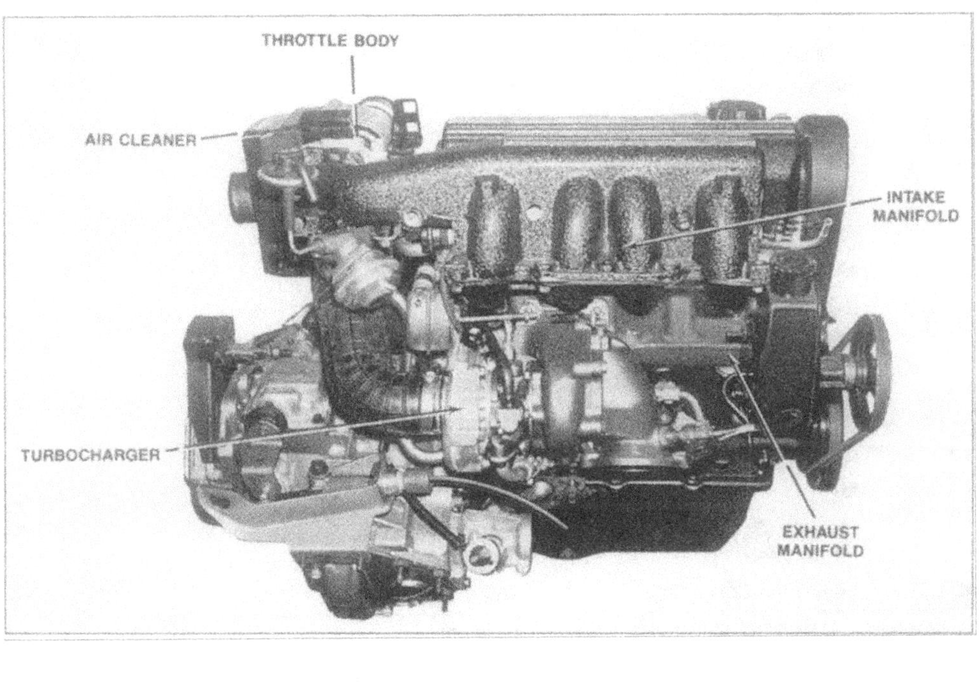

Fig. 12.30 *A rear view of the 1987 2.2-L Turbo II single overhead camshaft four-cylinder engine. The large-diameter ribbed elbow shown here transported air from the air cleaner to the turbocharger inlet. High-pressure, flexible ducting (not visible) connected the turbocharger outlet to the intercooler. From the intercooler, the compressed and cooled air was routed to the throttle body.*

Fig. 12.31 *The 1987 Dodge Daytona Shelby Z. For the 1987 model year, this car had exclusive use of the turbocharged, intercooled 2.2-L Turbo II engine.*

The variable nozzle system used twelve movable vanes mounted to the turbine wheel housing. As the vanes were moved simultaneously by a "unison" ring, the angle that exhaust gas impinged on the turbine blades changed, as did the area through which the gas had to flow. On throttle opening, the engine control computer (ECC) moved the vanes to reduce the area to speed up the exhaust gas and to direct the gas at the best angle for increasing the turbine revolutions per minute. By moving the vanes, the ECC also limited the maximum boost to a preset level, eliminating the need for a wastegate as was used on the Turbo I and Turbo II engines.

The engine was designated the Turbo IV, and Engineering code A538 was assigned to the design and development program, incorporating this variable nozzle turbocharger. Balance

TABLE 12.7
SPECIFICATIONS FOR THE 2.2-L SOHC FOUR-CYLINDER TURBO II
Note: All dimensions are in metric units (millimeters) rather than customary units (inches).

Model Year	1987	1989
Car line	Dodge	Dodge, Chrysler
Model	Dodge Shelby Z	Std.: Dodge Daytona Shelby, Lancer Shelby, Chrysler LeBaron GTS Opt. on others
Engine code	EDR	EDR
Displacement (liters)	2.2	2.2
Bore	87.5	87.5
Stroke	92	92
Compression ratio	8.1:1	8.1:1
Power @ rpm	130 kW (174 hp) @ 4800	130 kW (174 hp) @ 5200
Torque @ rpm	271 N-m (200 lb-ft) @ 3200	271 N-m (200 lb-ft) @ 2400
Cyl. block mat'l.—skirt type	Cast iron—shallow	Cast iron—shallow
Cyl. block deck height	237.8	237.8
Cyl. block length	418	418
Cyl. bore spacing	96	96
No. of main brgs.—ck/s mat'l.	5—forged steel	5—forged steel
Ck/shaft main journal dia.	60	60
Crankpin journal dia.	50	50
Connecting rod centers	151	151
Valve system	SOHC—8-valve	SOHC—8-valve
Valve events	240-240-20	228-228-0
Camshaft drive	Timing belt	Timing belt
Valve head dia.	Intake 40.6; exhaust 35.4	Intake 40.6; exhaust 35.4
Valve lash adjustment	Hyd. rocker arm pivot	Hyd. rocker arm pivot
Oil pump/distributor drive	Geared to auxiliary shaft	Geared to auxiliary shaft
Fuel system	MPI	MPI
Turbocharger	Garrett T3 Intercooled	Garrett T3 Intercooled
Engine assembly weight*	141.75 kg (312.9 lb)	152.7 kg (336.7 lb)

* For content, see the Abbreviations and Definitions section near the back of this book.

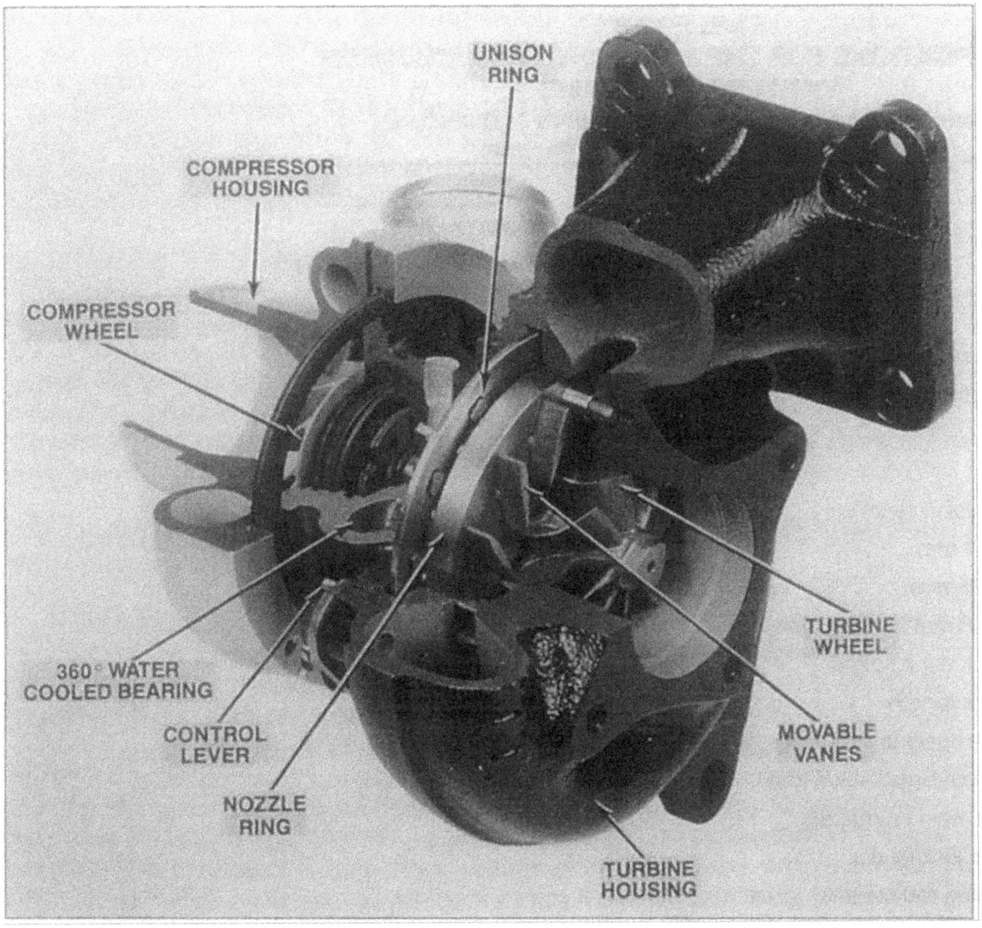

Fig. 12.32 *A cutaway illustration of the variable nozzle turbocharger used on the 1990 2.2-L Turbo IV four-cylinder engines.*

shafts, similar to the balance shaft system used on the 2.5-L single overhead camshaft engine, were employed to enhance vibration smoothness of the engine. The recently developed sequential multipoint fuel injection system was used to improve engine response, improve idle quality, and reduce emissions.

The power of the 2.2-L Turbo IV variable nozzle turbocharger engine remained unchanged from the 1989 Turbo II's 130 kW (174 hp) @ 5200 rpm, but its torque moved up from 271 N-m (200 lb-ft) @ 2400 rpm to 285 N-m (210 lb-ft) @ 3200 rpm.

Standard on the 1990 Dodge Daytona Shelby and Chrysler LeBaron GTC, and available with only manual transmission, the Turbo IV variable nozzle turbocharger engine was offered as optional on the Dodge Daytona, Dodge Shadow, and Chrysler LeBaron models. In writing about the LeBaron GTC in the December 1989 issue of *Motor Trend*, Daniel C. Ross noted, "The Turbo IV VNT engine (replacing the Turbo II) now has twin counter-rotating balance shafts to cancel most of the annoying engine vibration and noise about which we have long complained...The startling difference in perceived performance, however, is derived from the new turbocharging system in place on the Turbo IV motor. It makes turbo lag a thing that, like social diseases, happens to other people...It enables the turbocharger to achieve full-boost as much as 50% faster than the Turbo II engine it makes obsolete."

The year 1990 was the one and only year of production of the 2.2-L Turbo IV variable nozzle turbocharger engine. In 1991, the crown of lead high-performance engine was passed to quite a different version of the 2.2-L engine, as described next.

Table 12.8 shows the specifications of this 2.2-L Turbo IV variable nozzle turbocharger engine.

Big Power for the 2.2-L: Double Overhead Camshaft, 16-Valve Turbo III

With an aim of bringing out a version of the 2.2-L that would benefit from the specialized experience of the one of the world's premier high-performance car and engine developers—Lotus Engineering of England—a project was initiated to have Lotus create a double overhead camshaft, four-valves-per-cylinder conversion of the intercooled turbocharged 2.2-L engine. The 16-valve Engineering program, coded A522, under the leadership of the program manager Michael Royce, actually was started before the variable nozzle turbocharger Turbo IV work

TABLE 12.8
SPECIFICATIONS FOR THE 2.2-L SOHC FOUR-CYLINDER VNT TURBO IV ENGINE
Note: All dimensions are in metric units (millimeters) rather than customary units (inches).

Model Year	1990
Car line	Dodge, Chrysler
Model	Std.: Dodge Daytona Shelby Opt.: Dodge Daytona, Shadow; Chrysler LeBaron
Engine code	EDR
Displacement (liters)	2.2
Bore	87.5
Stroke	92
Compression ratio	8.1:1
Power @ rpm	130 kW (174 hp) @ 5200
Torque @ rpm	285 N-m (210 lb-ft) @ 3200
Cyl. block mat'l.—skirt type	Cast iron—shallow
Cyl. block deck height	237.8
Cyl. block length	418
Cyl. bore spacing	96
No. of main brgs.—ck/s mat'l.	5—forged steel
Ck/shaft main journal dia.	60
Crankpin journal dia.	50
Connecting rod centers	151
Balance system	Twin balance shafts
Valve system	SOHC—8-valve
Valve events	228-228-0
Camshaft drive	Timing belt
Valve head dia.	Intake 40.6; exhaust 35.4
Valve lash adjustment	Hyd. rocker arm pivot
Oil pump/distributor drive	Geared to auxiliary shaft
Fuel system	SMPI
Turbocharger	Garrett Variable Nozzle Intercooled
Engine assembly weight*	168.18 kg (370 lb)

* For content, see the Abbreviations and Definitions section near the back of this book.

began. Following the Turbo I and Turbo II sequence, the engine was given the designator Turbo III at the time.

Instead of applying the most common double overhead camshaft approach using camshafts mounted directly above the valves and inverted bucket tappets with shims for valve lash adjustment, quite a different design was created. To keep the height of the engine assembly to a minimum, the two camshafts were located alongside rather than on top of their banks of valves. Each bank of valves had a rocker shaft and eight rocker arms; each rocker arm had a roller follower at the camshaft end and a mini-hydraulic adjuster at the valve tip end. The included angle of the two banks of valves in the aluminum cylinder head was a fairly narrow 29.5 deg. However, it was wide enough to place the spark plug at the center of the combustion chamber. Valve sizes were 35.6 mm for the eight intake valves and 32.5 mm for the eight exhaust valves. The four branches of the aluminum intake manifold fed the eight intake ports in the cylinder head. The transversely located engine positioned the intake manifold to the front of the car.

To handle the extra power, the piston material was upgraded to forged aluminum. The flat top of each piston had four scallops for freewheeling valve clearance. A computerized engine control system controlled both the sequential multipoint fuel injectors and the newly released distributorless direct ignition system. Charge air boost to 11 psi was accomplished by a Garrett AiResearch TB03 turbocharger. The variable nozzle turbocharger feature of the Turbo IV engine was not repeated. Pressurized air from the turbocharger outlet was routed to an air-to-air intercooler as already described for the Turbo II engine. Similar to the 2.5-L Turbo I engine and the preceding 2.2-L variable nozzle turbocharger Turbo IV engine, this 2.2-L 16-valve Turbo III engine was fitted with balance shafts to reduce engine vibrations.

The mechanical arrangement of the engine can be seen in the cross section and longitudinal section shown on Figs. 12.33 and 12.34, respectively. Figure 12.35 shows a close-up of the engine compartment. With an 8.5:1 compression ratio, the 2.2-L double overhead camshaft 16-valve engine achieved an output level of 167 kW (224 hp) and 286 N-m (217 lb-ft) torque, marking a new high of 76 kW (102 hp) per liter or 1.66 hp per cubic inch specific power for a Chrysler production engine.

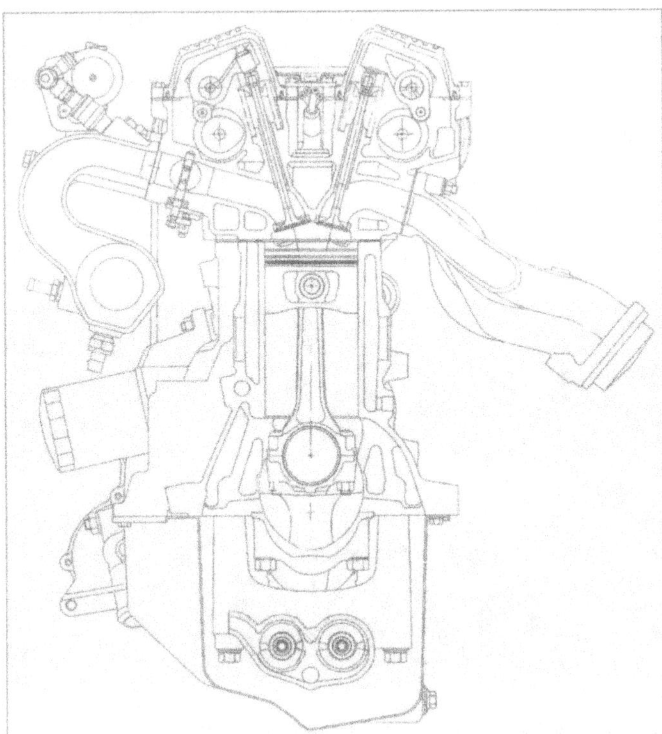

Fig. 12.33 *A cross section of the 1991 2.2-L double overhead camshaft 16-valve Turbo III four-cylinder engine. The valve system was designed to minimize the height of the engine assembly. Valve lash was controlled by mini-hydraulic adjusters in the rocker arms. As installed in the car, the exhaust manifold, shown on the right of the drawing, was oriented toward the firewall.*

In addition to the design and development tasks performed by Lotus Engineering, Lotus produced kits of the new unique parts—cylinder head assemblies, manifolds, and so forth—which were shipped to the Chrysler de Mexico Saltillo engine plant for complete engine buildup and testing.

Available with only a five-speed manual transmission, the few models that received this engine were the 1991 and 1992 Dodge Spirit R/T, and the 1992 and 1993 Dodge Daytona IROC R/T that appears on Fig. 12.36. In the February 1993 issue of *Motor Trend* that contained an article about the Dodge Daytona IROC R/T, Daniel C. Ross wrote, "The engine is without question mighty powerful, replete with an intake charge cooler and twin balance shafts for demonstrably improved engine smoothness. The turbomotor produces a peaky but frenetic 224 horsepower at 6000 rpm (the maximum rev threshold is 6500 rpm) and just has wads of torque for a four-banger, 217 pound-feet peaking at 2800 rpm. Look at this vehicle cross-eyed, and it spins its tires like a cartoon car."

The year 1993 was the last hurrah for four-cylinder turbocharged engines of any sort. Starting in 1994, V-6 engines took over as the performance leaders.

Table 12.9 shows the specifications for this 2.2-L 16-valve Turbo III engine.

2.2-L 16-Valve Turbocharged Power from Maserati

In 1984, Chrysler acquired 15.6% ownership of Officine Alfieri Maserati SpA of Italy, a marque synonymous with fast cars and auto racing victories. This relationship led to the creation of a Maserati-built high-performance car with the formal model name of "Chrysler's TC by Maserati" and an appropriate high-output engine. The engine, designed by Maserati and developed jointly by Chrysler (with the assistance of an outside automotive engineering contractor) and Maserati, used 2.2-L cylinder block assemblies and other 2.2-L engine parts manufactured at the Trenton engine plant, which were shipped to the Maserati plant in Modena, Italy for final engine assembly.

The 16-valve aluminum cylinder head, designed by Maserati, had classic direct-acting camshafts above the valves and inverted bucket tappets with replaceable shims for valve lash adjustment. Maserati also designed the cog-belt camshaft drive, the intake and exhaust manifolds, the accessory drive system, beefed-up connecting rods, the forged steel crankshaft with increased counterweighting, and the installation of the IHI RHB52 turbocharger. Dish-topped pistons were furnished by Mahle, and the induction system included a remotely mounted intercooler. Figures 12.37 and 12.38 show a cross section and a longitudinal section of the engine, respectively. Figure 12.39 shows a photo of an engine assembly.

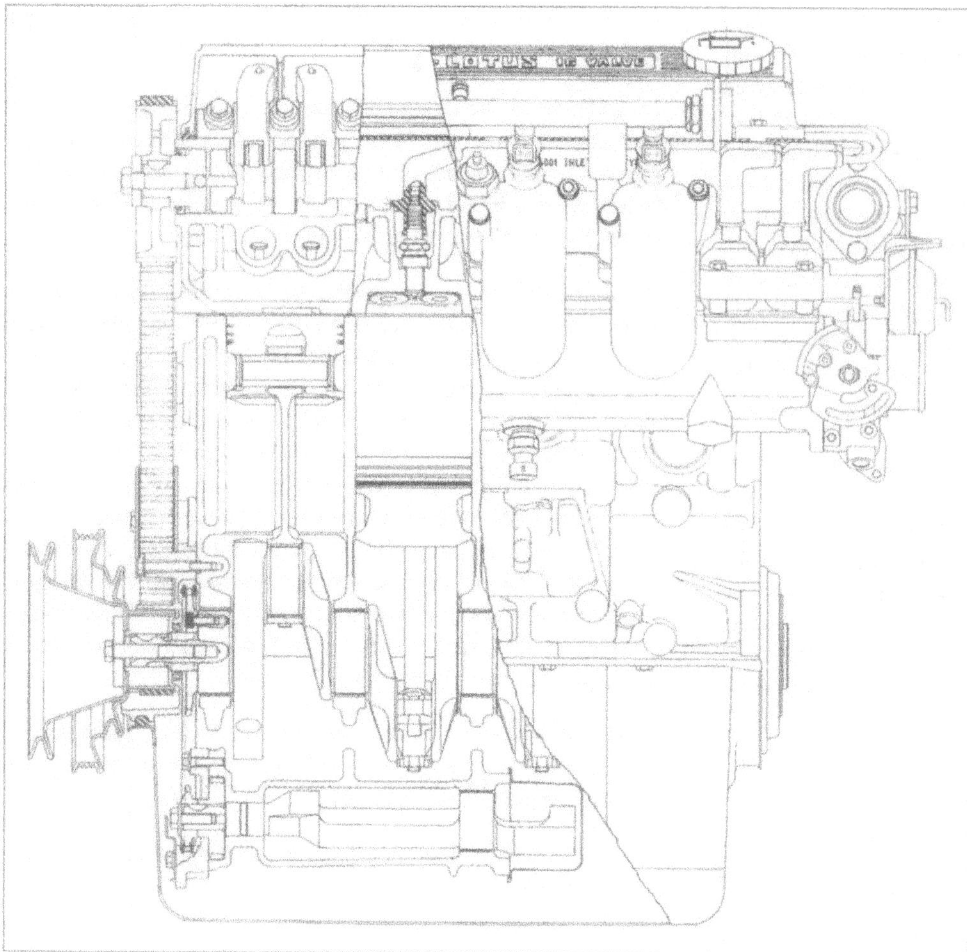

Fig. 12.34 *A longitudinal section of the 1991 2.2-L double overhead camshaft 16-valve Turbo III four-cylinder engine. For improved smoothness, the engine was equipped with balance shafts.*

Fig. 12.35 *A top-front view of the 1991 2.2-L double overhead camshaft 16-valve Turbo III four-cylinder engine. The tubular fuel rail fed the four fuel injectors mounted on the intake manifold branches.*

Fig. 12.36 *The 1992 Dodge Daytona IROC R/T. This car and the Dodge Spirit R/T were the two models in the 1992 lineup powered by the 2.2-L double overhead camshaft 16-valve Turbo III engine.*

TABLE 12.9
SPECIFICATIONS FOR THE 2.2-L DOHC TURBO III FOUR-CYLINDER ENGINE
Note: All dimensions are in metric units (millimeters) rather than customary units (inches).

Model Year	1991	1992	1993
Car line	Dodge	Dodge	Dodge
Model	Spirit R/T	Spirit R/T Daytona IROC R/T	Daytona IROC R/T
Displacement (liters)	2.2	2.2	2.2
Bore	87.5	87.5	87.5
Stroke	92	92	92
Compression ratio	8.1:1	8.1:1	8.1:1
Power @ rpm	130 kW (224 hp) @ 5800	130 kW (224 hp) @ 5800	130 kW (224 hp) @ 5800
Torque @ rpm	285 N-m (217 lb-ft) @ 2800	285 N-m (217 lb-ft) @ 2800	285 N-m (217 lb-ft) @ 2800
Cyl. block mat'l.—skirt type	Cast iron—shallow	Cast iron—shallow	Cast iron—shallow
Cyl. block deck height	237.8	237.8	237.8
Cyl. block length	418	418	418
Cyl. bore spacing	96	96	96
No. of main brgs.—ck/s mat'l.	5—forged steel	5—forged steel	5—forged steel
Ck/shaft main journal dia.	60	60	60
Crankpin journal dia.	50	50	50
Connecting rod centers	151	151	151
Balance system	Twin balance shafts	Twin balance shafts	Twin balance shafts
Valve system	DOHC—16-valve	DOHC—16-valve	DOHC—16-valve
Valve events	242-234-14	242-234-14	242-234-14
Camshaft drive	Timing belt	Timing belt	Timing belt
Valve head dia.	Intake 33.9; exhaust 29.3	Intake 33.9; exhaust 29.3	Intake 33.9; exhaust 29.3
Valve lash adjustment	Hyd. adjuster in rocker arm	Hyd. adjuster in rocker arm	Hyd. adjuster in rocker arm
Oil pump drive	Geared to auxiliary shaft	Geared to auxiliary shaft	Geared to auxiliary shaft
Ignition	Distributorless ignition system	Distributorless ignition system	Distributorless ignition system
Fuel system	SMPI	SMPI	SMPI
Turbocharger	Garrett TB03 Intercooled	Garrett TB03 Intercooled	Garrett TB03 Intercooled
Engine assembly weight*	167.38 kg (369.0 lb)	167.38 kg (369.0 lb)	167.38 kg (369.0 lb)

* For content, see the Abbreviations and Definitions section near the back of this book.

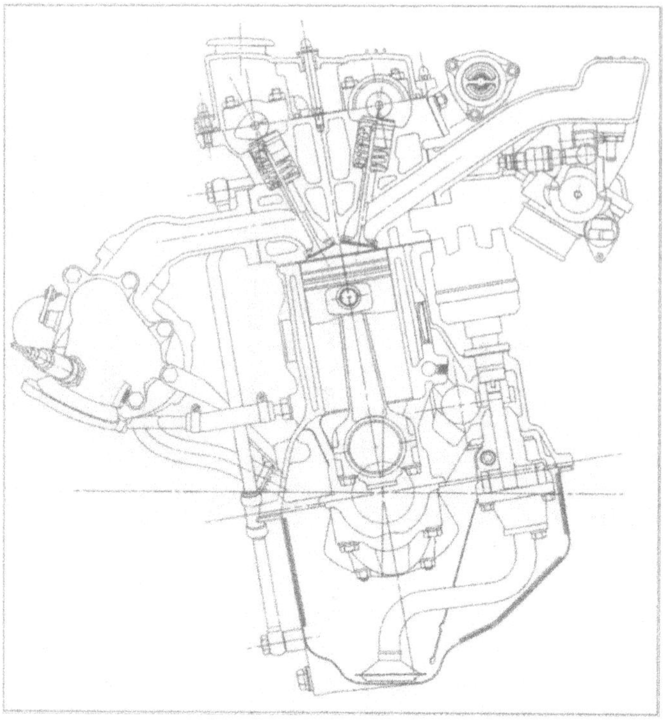

Fig. 12.37 *A cross section of the 1989 Maserati 2.2-L double overhead camshaft 16-valve turbocharged, intercooled four-cylinder engine. Note the direct-acting camshafts and the long inlet passage.*

With a compression ratio of 7.3:1 and multipoint fuel injection, the output of the 2.2-L double overhead camshaft 16-valve turbocharged, intercooled engine was 150 kW (200 hp) and 298 N-m (220 lb-ft). The engine was used only with a five-speed manual transmission; automatic transmission vehicles were equipped with the 2.2-L Turbo II engine during 1989, the first year of production, and a 3.0-L V-6 engine (described in the next chapter) in 1990.

The first production "Chrysler's TC by Maserati" car powered by the 2.2-L double overhead camshaft Maserati engine was built at the Maserati vehicle assembly plant in Milan, Italy on January 17, 1989,[11] with engine production preceding that date by about two weeks. Figure 12.40 portrays one of the stylish 1989 vehicles. At the end of 1990, manufacture of the 2.2-L double overhead camshaft 16-valve Maserati engine ceased. However, the car continued in production for another year, through 1991, powered solely by the 3.0-L Mitsubishi Motors Corporation V-6 engine with an automatic transmission.

Table 12.10 shows the specifications for this Maserati double overhead camshaft 16-valve turbocharged engine.

The Mitsubishi 2.6-L Silent-Shaft Single Overhead Camshaft Four-Cylinder Engine

The new front-wheel-drive Dodge Aries and Plymouth Reliant K-cars, planned for 1981, were expected to enjoy substantial sales with the Chrysler-built 2.2-L single overhead camshaft four-cylinder engine. However, their market range in the upward direction would benefit if a powerplant with more displacement than the Chrysler-built 2.2-L engine could be obtained, and if it could be fitted into the K-car engine compartment. Mitsubishi Motors Corporation (MMC) of Japan had such an engine already in production, powering a car of its own with a conventional driveline. As a quality plus, the engine was equipped with dual cylinder-block-mounted balance shafts that cancelled out secondary shaking forces inherent with in-line, four-cylinder engines. With a displacement of 2.6 liters, the engine would provide both extra performance and smoothness. A cross section and a longitudinal section of the engine are shown on Figs. 12.41 and 12.42, respectively. Successful negotiations between Chrysler and Mitsubishi Motors Corporation resulted in an engineering and purchase agreement for engines to be built at the Mitsubishi Kyoto Works in Japan for use in Chrysler vehicles.

The deep-skirt cast iron cylinder block of the 2.6-L engine had a bore spacing of 101 mm, which was considerably larger than the 96 mm of the 2.2-L engine. The two block-mounted balance shafts were driven at twice the engine revolutions per minute by a single chain from the crankshaft. Aluminum was used for the cross-flow cylinder head that mounted a single overhead camshaft. The combustion chambers were hemispherical, with the four intake valves of the engine located at the front and the four exhaust valves located at the rear of the cylinder head (car position). All valves were operated by rocker arms equipped with adjusting screws and lock nuts for valve lash adjustment. Valve head diameters were 46 and 38.1 mm for the intake and exhaust valves, respectively. For emission control, each combustion chamber was fed with a second, small-diameter intake valve designed to impart a strong swirling action to the fuel/air mixture at low speeds. This feature was titled the MCA (Mitsubishi Clean Air) Jet Engine System. The crankshaft of the engine, with five 60-mm-diameter main journals and four 53-mm crankpin journals, was made of forged steel. With a compression ratio of 8.2:1 and fitted with a two-barrel Mikuni carburetor, the output of the 2.6-L Mitsubishi Motors Corporation single overhead camshaft engine was 69 kW (92 hp) and 178 N-m (131 lb-ft) torque. Hot-tested production engine assemblies such as the one pictured in Fig. 12.43 were shipped directly to the Chrysler U.S. vehicle assembly plants.

For its 1981 introductory year (along with the K-car and the 2.2-L engine), the 2.6-L Mitsubishi Motors Corporation engine was specified as optional on all Plymouth Reliant and Dodge Aries models. The next year, Chrysler LeBaron models and the Town & Country station wagon joined the carryover

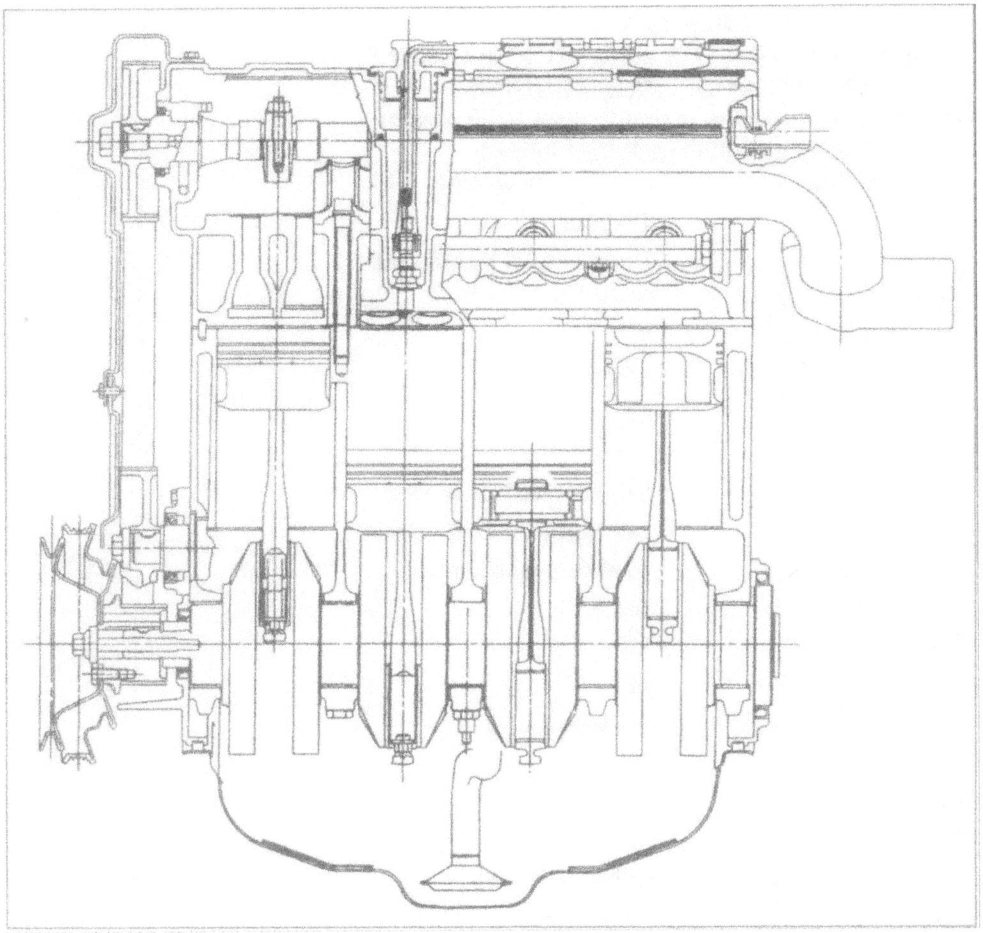

Fig. 12.38 *A longitudinal section of the 1989 Maserati 2.2-L double overhead camshaft 16-valve turbocharged, intercooled four-cylinder engine.*

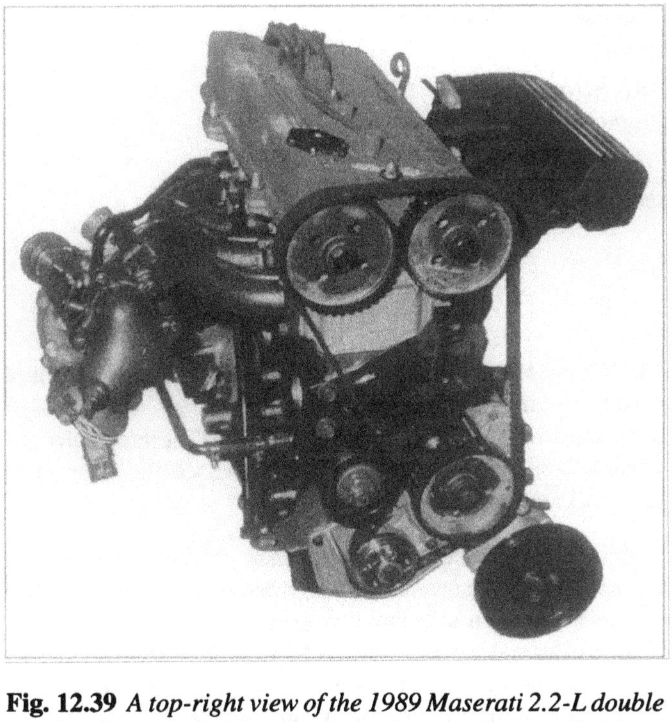

Fig. 12.39 *A top-right view of the 1989 Maserati 2.2-L double overhead camshaft 16-valve turbocharged, intercooled four-cylinder engine assembly. The covers have been removed from this end of the engine to show the timing belt system.*

Fig. 12.40 *The 1989 "Chrysler's TC by Maserati" convertible. A five-speed manual transmission was the only transmission available with the Maserati-built 2.2-L double overhead camshaft 16-valve turbocharged, intercooled engine.*

TABLE 12.10
SPECIFICATIONS FOR THE 2.2-L DOHC MASERATI TURBO FOUR-CYLINDER ENGINE
Note: All dimensions are in metric units (millimeters) rather than customary units (inches).

Model Year	1989–1990
Car line	Chrysler
Model	Chrysler's TC by Maserati
Engine code	EDD
Displacement (liters)	2.2
Bore	87.5
Stroke	92
Compression ratio	7.3:1
Power @ rpm	149 kW (200 hp) @ 5500
Torque @ rpm	298 N-m (220 lb-ft) @ 3400
Cyl. block mat'l.—skirt type	Cast iron—shallow
Cyl. block deck height	237.8
Cyl. block length	418
Cyl. bore spacing	96
No. of main brgs.—ck/s mat'l.	5—forged steel
Ck/shaft main journal dia.	60
Crankpin journal dia.	50
Connecting rod centers	151
Valve system	DOHC—16-valve
Camshaft drive	Timing belt
Valve head dia.	Intake 31.5; exh 29
Valve lash adjustment	Mech.—tappet shims
Oil pump/distributor drive	Geared to auxiliary shaft
Fuel system	MPI
Turbocharger	IHI RHB52 Intercooled
Engine assembly weight*	163.86 kg (360.5 lb)

* For content, see the Abbreviations and Definitions section near the back of this book.

vehicles built on the K-car platform. The 2.6-L Mitsubishi Motors Corporation engine became standard power for the LeBaron convertible and the Town & Country station wagon and optional on the other 1982 K-platformed Chrysler models. In 1983, another Chrysler model was added—the luxurious, long-wheelbase Executive Sedan shown on Fig. 12.44, powered solely by the smooth-running 2.6-L engine.

The year 1984 saw the introduction of the wonderfully successful minivans—Plymouth Voyager and Dodge Caravan, both of which offered the 2.6-L Mitsubishi Motors Corporation engine as an option—available with only an automatic transmission. Because the minivans were classified as trucks (from its beginning, the vehicle carried a T-115 truck designator, and the model line included a commercial version with solid panels replacing the side glass windows), they were subject to the less stringent truck emission limits. Therefore, the jet valve previously described was not used in the 2.6-L engines installed in the minivans.

Starting in 1986, the 2.6-L Mitsubishi Motors Corporation engines were no longer used in any passenger cars. The use of this engine in minivans continued until the end of 1987, when it essentially was replaced by the Chrysler-built 2.5-L single overhead camshaft four-cylinder engine and a Mitsubishi Motors Corporation-built 3.0-L single overhead camshaft V-6 engine, both of which had already entered production.

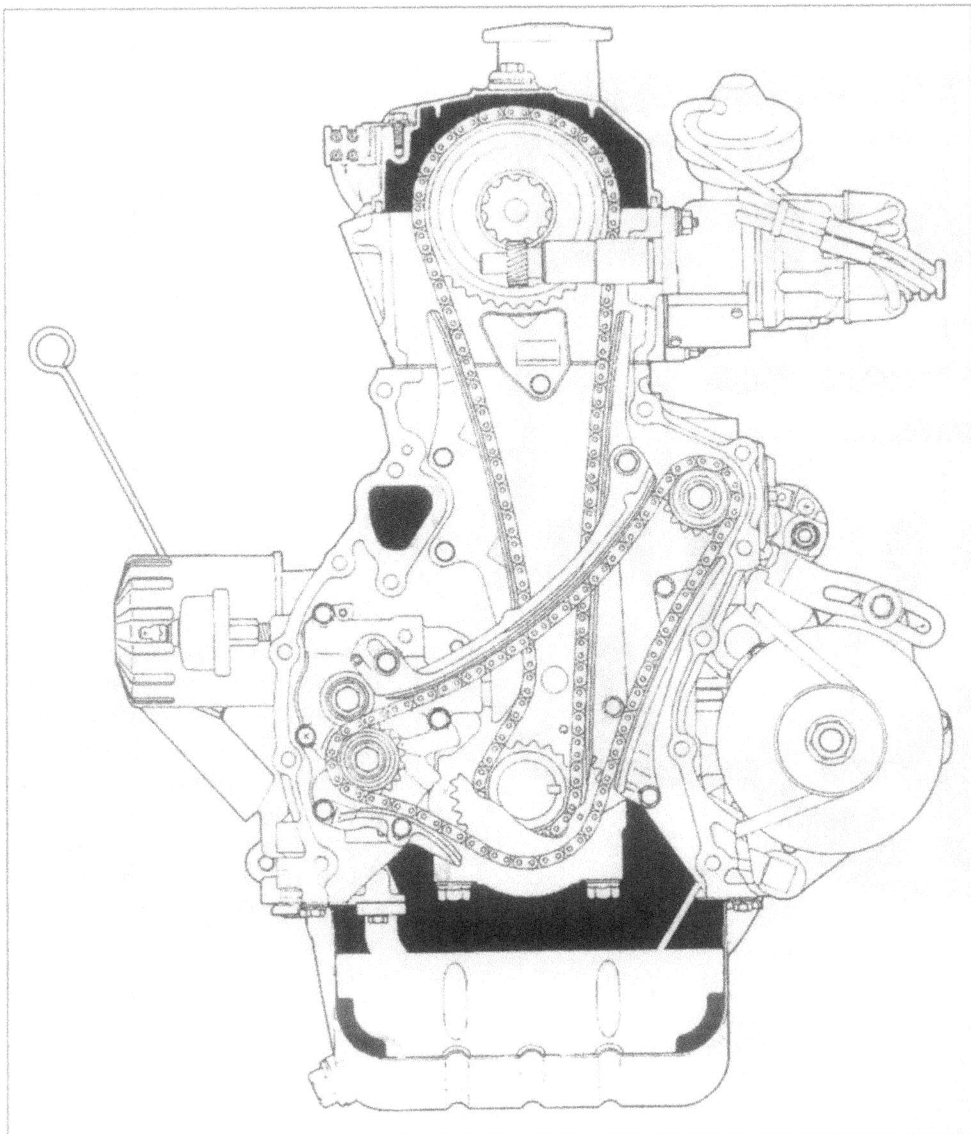

Fig. 12.41 *A cross section through the chain case of the 1981 2.6-L Mitsubishi Motors Corporation single overhead camshaft four-cylinder engine. The chain nearest the cylinder block drives the camshaft. The other chain drives the oil pump and the two counter-rotating balance shafts.*

Table 12.11 shows the specifications of the Mitsubishi Motors Corporation 2.6-L single overhead camshaft balanced four-cylinder engine.

2.0-L Single Overhead Camshaft 16-Valve Four-Cylinder Engine

Chrysler reentered the subcompact market in 1995 with an all new car, the Neon, to be sold as both a Dodge and a Plymouth model. The prior Chrysler offering in this market segment was the L-body Dodge Omni and Plymouth Horizon that ended production at the end of 1990. The Neon was powered with a pair of new engines designed to provide spirited performance with excellent fuel economy—a 2.0-L single overhead camshaft 16-valve four-cylinder engine described here, and a closely related 2.0 double overhead camshaft 16-valve four-cylinder engine described in the next section. The Small Car Platform Engine Group, headed by Executive Engineer Floyd Allen and Engine Engineering Manager Doug Livermore, was assigned the responsibility for the design and development of these two engines.

Faced, as the L-body was, with space constraints of a tightly designed front-wheel-drive car with a transverse engine and transmission, the same 96-mm cylinder centers used on the 2.2-L engine were selected for the new engine, along with the same 87.5 mm bore. A shorter 83-mm (versus 92-mm) stroke provided the planned 2.0-L displacement. Engineering code A597 was assigned to the engine.

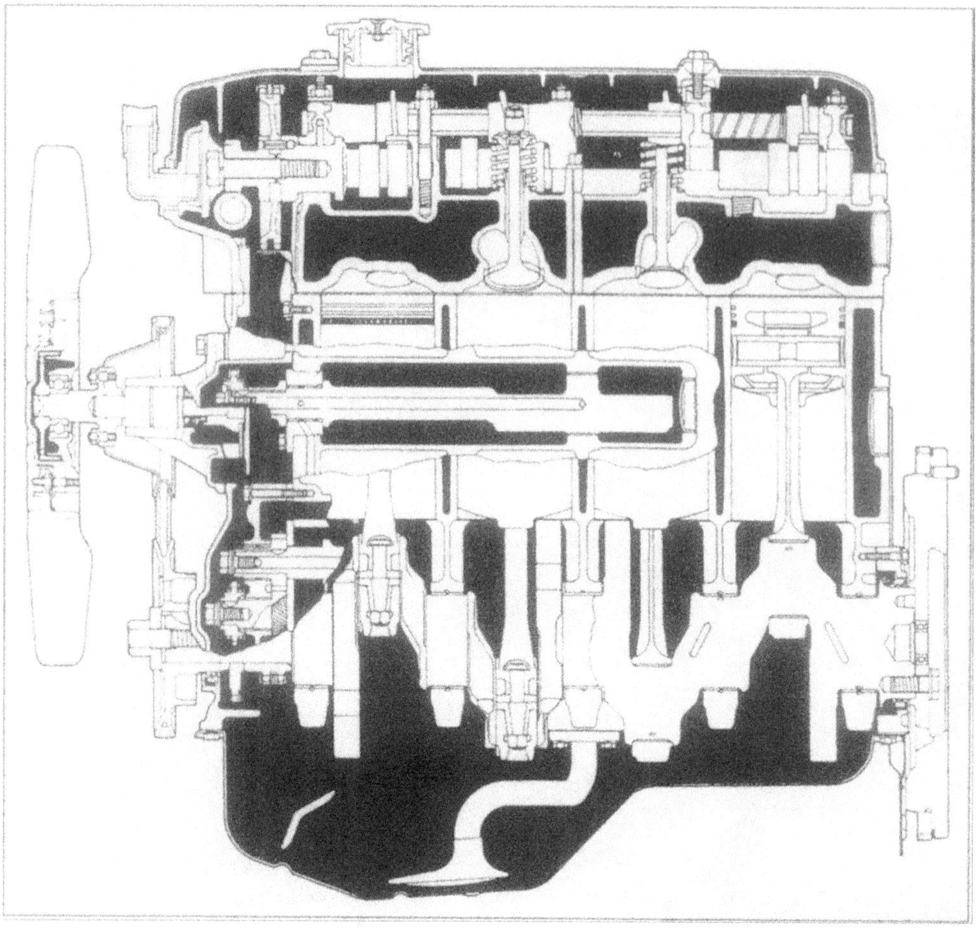

Fig. 12.42 *A longitudinal section of the 1981 2.6-L Mitsubishi Motors Corporation single overhead camshaft four-cylinder engine. The Chrysler front-wheel-drive applications for the engine used an electric motor to drive the cooling fan and did not use the engine-mounted fan shown in the section.*

Fig. 12.43 *A top-front view of the 1984 2.6-L Mitsubishi Motors Corporation single overhead camshaft four-cylinder engine assembly.*

Fig. 12.44 *The 1983 Chrysler Executive Sedan. The 2.6-L Mitsubishi Motors Corporation single overhead camshaft four-cylinder engine with its balance shafts supplied the power for this extended-wheelbase luxury vehicle.*

The cast iron cylinder block had a shallow skirt but was given ample strength by the addition of a bed plate, also made of cast iron, that attached to the bottom of the block and replaced the separate bearing caps normally used. With a low 212-mm deck height and a low-profile cylinder head, the vertical height of the engine was compatible with the new low hood line of the Neon. Figure 12.45 shows a cutaway view of the engine.

The aluminum cylinder head had four valves per cylinder with a pent-roof-shaped combustion chamber. The engine cross section shown on Fig. 12.46 and the longitudinal section shown on Fig. 12.47 illustrate the compact design of the valves, rocker arms, rocker arm shafts, spark plugs, spark plug tubes, and camshaft that was necessary to operate the 16 valves of the engine with a single camshaft. The two banks of valves, the intake and exhaust, were positioned with an included angle of 42 deg; the intake valves were located on the front (car

TABLE 12.11
SPECIFICATIONS FOR THE 2.6-L MMC SOHC FOUR-CYLINDER ENGINE
Note: All dimensions are in metric units (millimeters) rather than customary units (inches).

Model Year	1981	1987
Car line	Plymouth, Dodge	Plymouth, Dodge
Model	Opt.: Plymouth Reliant, Dodge Aries	Opt.: Plymouth Voyager, Dodge Caravan
Engine code	E72	EEA
Displacement (liters)	2.6	2.6
Bore	91.1	91.1
Stroke	98	98
Compression ratio	8.2	8.2
Power @ rpm	69 kW (92 hp) @ 4500	69 kW (92 hp) @ 4500
Torque @ rpm	178 N-m (131 lb-ft) @ 2500	178 N-m (131 lb-ft) @ 2500
Cyl. block mat'l.—skirt type	Cast iron—deep	Cast iron—deep
Cyl. block deck height	251	251
Cyl. bore spacing	101	101
No. of main brgs.—ck/s mat'l.	5—forged steel	5—forged steel
Ck/shaft main journal dia.	60	60
Crankpin journal dia.	53	53
Connecting rod centers	166	166
Balance system	Twin block-mtd. bal. shafts	Twin block-mtd. bal. shafts
Valve system	SOHC—8-valve	SOHC—8-valve
Valve events	264-264-45	264-264-45
Camshaft drive	Chain	Chain
Valve head dia.	Intake 46; exhaust 38.1	Intake 46; exhaust 38.1
Valve lash adjustment	Mech.—rocker arm screw	Mech.—rocker arm screw
Distributor drive	Geared to camshaft	Geared to camshaft
Fuel system	2-barrel Mikuni carburetor	2-barrel Mikuni carburetor
Engine angle	6 deg 16 min. to rear	6 deg 16 min. to rear
Engine assembly weight*	161.6 kg (356.3 lb)	156.0 kg (343.9 lb)

* For content, see the Abbreviations and Definitions section near the back of this book.

Fig. 12.45 *A cutaway illustration of the 1995 2.0-L single overhead camshaft 16-valve four-cylinder engine.*

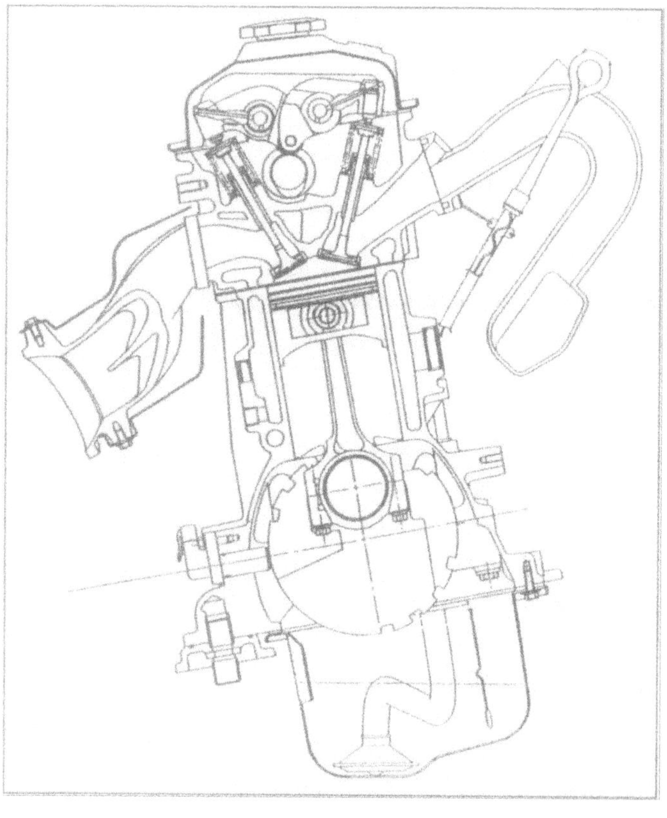

Fig. 12.46 *A cross section of the 2.0-L single overhead camshaft 16-valve four-cylinder engine. Note the long intake passage, the mini-hydraulic adjusters in the rocker arms, and the notches in the crankshaft disk used by the crankshaft position sensor.*

position) side of the cylinder head. In the longitudinal view, the exhaust valves were parallel, but the intake valves were not, being located at a slight 3.6-deg angle to each other. This slight splaying of the intake valves was necessary to make room for the centrally located spark plug.

The valve gear for each cylinder included three aluminum rocker arms for opening the valves, two for the two intake valves, and a single rocker arm for the two exhaust valves. Each intake rocker arm had a mini-hydraulic adjuster in its valve tip end, while the single exhaust rocker arm had two mini-hydraulic adjusters for its pair of exhaust valves. Every rocker arm was furnished with a roller follower for contacting the camshaft lobe. Head diameters were 33 and 28 mm for the intake and exhaust valves, respectively. The camshaft was driven from the crankshaft by a cog-tooth flexible timing belt with a spring-loaded hydraulically dampened automatic tensioner.

Connecting rods were made not of forged steel as had been done since the beginning of the auto industry but of forged powdered metal that combined the strength of steel with precise dimensional control, permitting a considerable weight reduction. The connecting rod and cap were made as one piece, forged together. At the end of the machining process, the cap end was fractured from the rod, resulting in perfect alignment between the cap and rod when reassembled and, with fewer steps in the process, lower machining costs.

Because the engine was equipped with a direct distributorless ignition system and an oil pump driven by the nose of the crankshaft, an auxiliary shaft as used on the 2.2-L and 2.5-L engines was not required and was eliminated. Taking advantage of the space vacated by the distributor, the intake manifold, made of injection-molded glass-filled nylon, was designed with a longitudinal plenum chamber and four long curved branches for intake tuning. This was the first application of this construction to any Chrysler engine.

Both the cylinder head cover and oil pan were made from bonded metal-plastic-metal stock to reduce engine noise, vibration, and harshness levels. A powertrain control module provided information to the sequential multipoint fuel injectors and ignition coil packs for engine operation using many input sensors, including new precise crankshaft and camshaft position sensors.

At its introduction as standard power for the 1995 Dodge Neon shown on Fig. 12.48 and the Plymouth Neon models, the 2.0-L single overhead camshaft 16-valve, four-cylinder engine, with a compression ratio of 9.8:1, was rated at 98 kW (132 hp) and 174 N-m (129 lb-ft). Recognizing the excellence of the Plymouth and Dodge Neon, *Automobile Magazine* pronounced Neon its "Automobile of the Year."

When the 1995 Dodge Stratus was introduced later in the model year, this engine also was specified as its standard power. These net power ratings gave the engine the distinction of having the highest specific power (66 hp/liter or 1.08 hp/cu.in.) of any

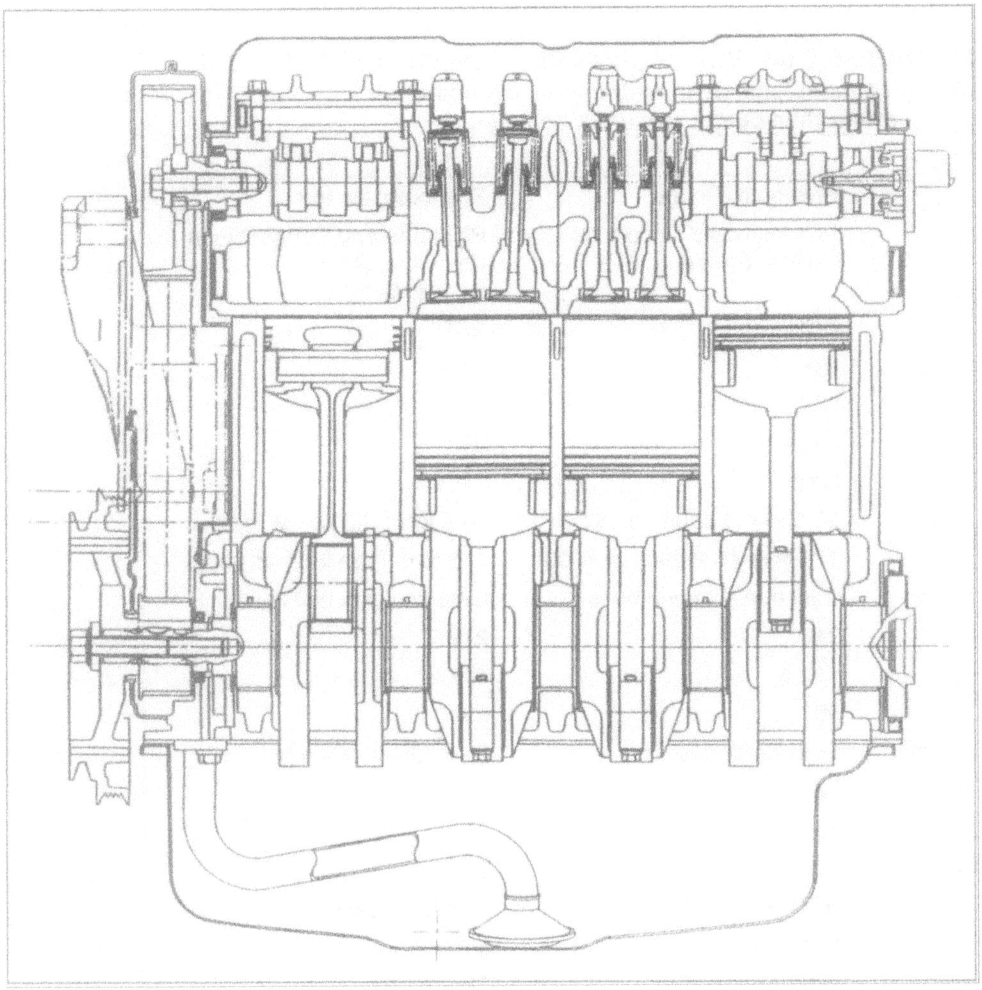

Fig. 12.47 *A longitudinal section of the 2.0-L single overhead camshaft 16-valve four-cylinder engine. Intake valves shown above cylinder #2 are slightly canted. Dual exhaust valves share a single rocker arm with two adjusters, shown above cylinder #3, and a single roller follower, shown above cylinder #4. Thin coolant slots between the bores are visible just below the top deck.*

Fig. 12.48 *The 1995 Dodge Neon two-door sedan. For its introductory year, the Neon was powered by the new 98-kW (132-hp) 2.0-L single overhead camshaft 16-valve four-cylinder engine.*

naturally aspirated Chrysler production engine built to date. The distinction did not last long, however. About a month later, the double overhead camshaft version of the 2.0-L 16-valve four-cylinder engine (described in the next section) started production and, with its identical displacement and higher ratings, took away the honors.

In a vehicle comparison article appearing in the July 1996 issue of *Motor Trend*, written by Jeff Karr and Rik Paul, the observation was made that "Turn the key and the fun continues as the standard 132-horsepower single overhead camshaft inline-four snaps to life. It has a lively character, abetted by the solid shift feel of the five-speed manual transmission. Although it's adequately peppy, a mid-pack 0–60-mph time of 9.1 seconds confirms that it isn't downright fast. Neon drivers looking for more excitement will want to invest in the optional 150-horse twin-cam four."

Table 12.12 shows the specifications for this 2.0-L single overhead camshaft 16-valve engine.

TABLE 12.12
SPECIFICATIONS FOR THE 2.0-L SOHC 16-VALVE FOUR-CYLINDER ENGINE
Note: All dimensions are in metric units (millimeters) rather than customary units (inches).

Model Year	1995	1998
Car line	Plymouth, Dodge	Plymouth, Dodge
Model	Plymouth and Dodge: Neon Dodge: Stratus	Plymouth and Dodge: Neon Plymouth: Breeze Dodge: Stratus
Engine code	ECB	ECB
Displacement (liters)	2.0	2.0
Bore	87.5	87.5
Stroke	83	83
Compression ratio	9.8:1	9.8:1
Power @ rpm	98.5 kW (132 hp) @ 6000	98.5 kW (132 hp) @ 6000
Torque @ rpm	175 N-m (129 lb-ft) @ 5000	175 N-m (129 lb-ft) @ 5000
Cyl. block mat'l.—skirt type	Cast iron with bed plate	Cast iron with bed plate
Cyl. block deck height	212	212
Cyl. block length	417.3	417.3
Cyl. bore spacing	96	96
No. of main brgs.—ck/s mat'l.	5—nodular cast iron	5—nodular cast iron
Ck/shaft main journal dia.	52	52
Crankpin journal dia.	48	48
Connecting rod centers	139	139
Valve system	SOHC—16-valve	SOHC—16-valve
Valve events	217-229-0.5	207-229-0
Camshaft drive	Timing belt	Timing belt
Valve head dia.	Intake 33.3; exhaust 28.7	Intake 33.3; exhaust 28.7
Valve lash adjustment	Hyd. adjuster in rocker arm	Hyd. adjuster in rocker arm
Oil pump drive	Nose of crankshaft	Nose of crankshaft
Ignition	Distributorless ignition system	Distributorless ignition system
Fuel system	SMPI	SMPI
Engine angle	10.5 deg to rear	10.5 deg to rear
Engine assembly weight*	129.3 kg (285.05 lb) ATX	

* For content, see the Abbreviations and Definitions section near the back of this book.

2.0-L Double Overhead Camshaft 16-Valve Four-Cylinder Engine

An increased performance version of the new 2.0-L 16-valve four-cylinder engine using dual overhead camshafts was part of the engine design parameters right from the beginning of the engine program. From the cylinder block top deck downward, the two engines were virtually identical with the same cylinder block, bedplate, nodular iron crankshaft, powder metal connecting rods, and so forth. However, the pistons were different, with the pistons of the double overhead camshaft version having valve relief cuts that were not used on the single overhead camshaft pistons. From the top deck upward, the parts were unique to each engine.

The aluminum cylinder head of the double overhead camshaft engine had four valves per cylinder with an included angle of 48 deg between the row of intake valves, located on the front side of the cylinder head, and the exhaust valves, located on the rear side. This 48-deg angle was somewhat greater than the 42 deg of the companion single overhead camshaft engine, as previously noted, and allowed larger valves to be used. The intake and exhaust valve head diameters were 34.8 and 30.5 mm, respectively, versus 33.3 and 28.7 mm for the single overhead camshaft version. For valve actuation and valve lash adjustment, rocker arms similar in principle to the stamped-steel, roller follower, stationary hydraulic pivot type used on the 2.2-L and 2.5-L engines since 1988 were used. The bores for the 16 hydraulic pivots were located in the cylinder head between the two rows of valves. The spark plug was positioned at the center of the pent-roof-shaped combustion chamber. A single cog-tooth timing belt, equipped with a spring-loaded hydraulically damped automatic tension adjuster, drove the two camshafts and the water pump of the engine. Unlike the single-piece, glass-filled nylon intake manifold of the single overhead camshaft engine, the two-piece intake manifold of the double overhead camshaft engine was made of aluminum. A longitudinal plenum and four branches feeding the eight intake ports of the cylinder head were incorporated in the double overhead camshaft intake manifold, with each of the branches mounting a fuel injector close to the cylinder head joint face. As with the single overhead camshaft engine, a computerized engine control system managed both the distributorless ignition system and the sequential multipoint fuel injection for optimum engine operation. Figure 12.49 shows a cutaway view of the engine. Figs. 12.50 and 12.51 show a cross section and a longitudinal section, respectively.

Fig. 12.49 *A cutaway illustration of the 2.0-L double overhead camshaft 16-valve four-cylinder engine. A fixed pulley located below the intake camshaft sprocket ensured that the intake camshaft sprocket had plenty of belt wrap.*

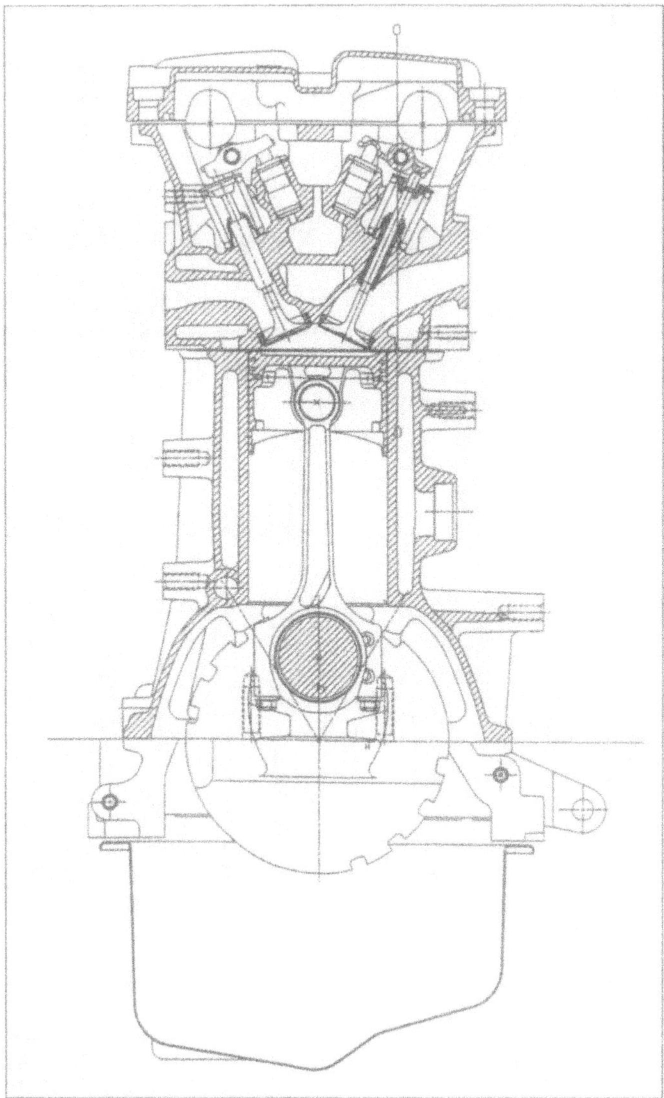

Fig. 12.50 *A cross section of the 2.0-L double overhead camshaft 16-valve four-cylinder engine. The arrangement of the valves, camshafts, rocker arms, and rocker arm pivots can be seen.*

With an excellent output for its size—112 kW (150 hp) and 180 N-m (133 lb-ft)—this 1995 2.0-L double overhead camshaft 16-valve four-cylinder engine was specified as the standard engine on the 1995 Chrysler Sebring two-door coupes and on the Plymouth and Dodge Neon Sport two-door coupes. It also was available as optional power on the other Plymouth and Dodge Neon models. The specific power output of 75 hp/liter or 1.23 hp/cu. in. easily made it the champion of any naturally aspirated Chrysler production engine built to date.

Table 12.13 shows the specifications for this 2.0-L double overhead camshaft 16-valve engine.

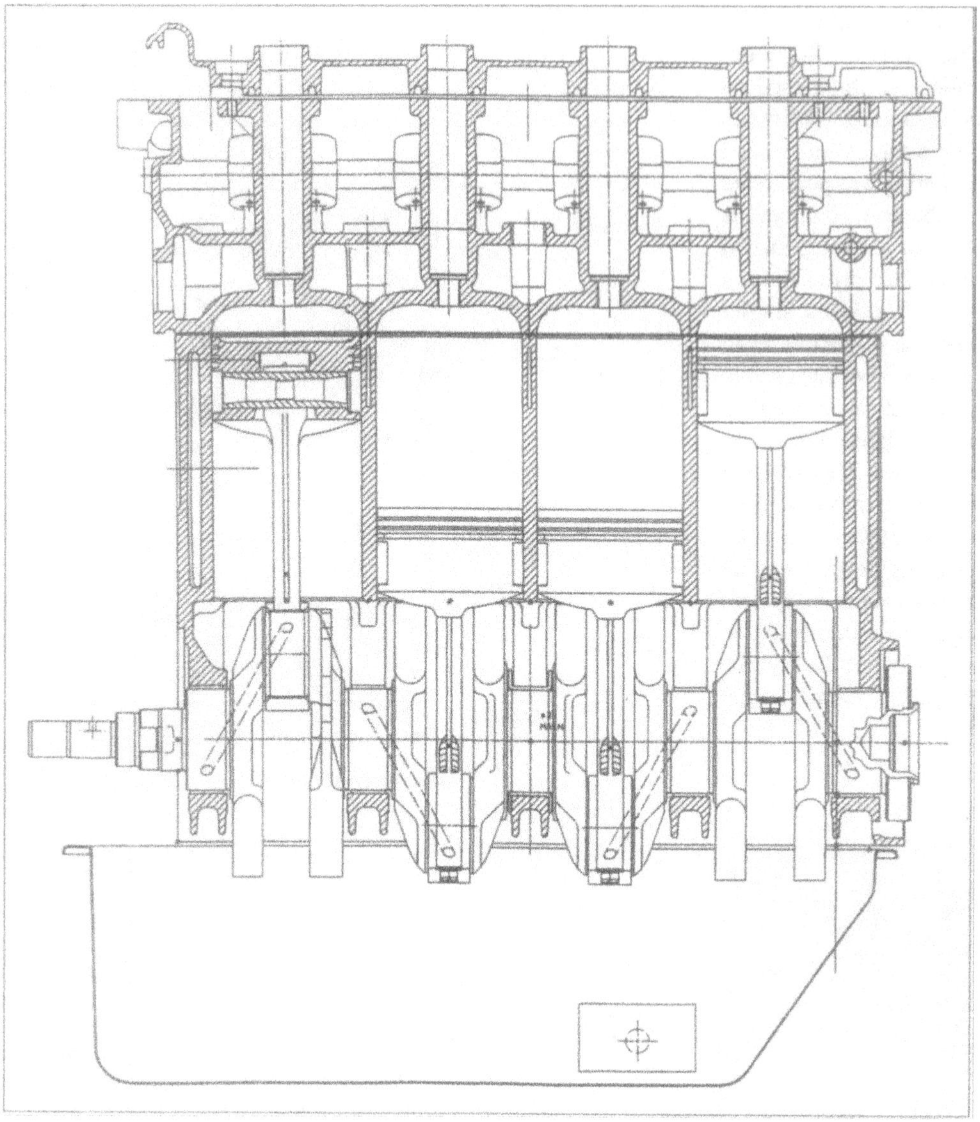

Fig. 12.51 *A longitudinal section of the 2.0-L double overhead camshaft 16-valve four-cylinder engine. The spark plug wells located the spark plugs at the center of the combustion chambers without the need for separate access tubes.*

2.4-L Double Overhead Camshaft 16-Valve Four-Cylinder Engine

As part of the strategy of replacing the first-generation overhead camshaft, eight-valve four-cylinder engines—the 2.2-L and 2.5-L engines—with 16-valve overhead camshaft engines, a longer-stroke engine with torque higher than the 2.0 L was needed to replace the first-generation 2.5-L engine, especially for powering the heavier Chrysler Cirrus compact car and the minivans. This need was filled by a version of the new 16-valve four-cylinder engines having the same 87.5-mm bore as the 2.0-L engine but a longer 101-mm stroke (1 mm less than the replaced 2.5-L), resulting in a displacement of 2.4 L. The cylinder block had to be raised from 212 to 237.8 mm (exactly the same as the old 2.2-L engine) to accommodate this stroke. With the longer stroke, the nodular cast iron crankshaft had to be beefed up, which was done by increasing its main and crankpin journal diameters from 52 and 48 mm, respectively, to 60 and 50 mm. These were the same diameters used on the replaced 2.2-L and 2.5-L engines.

The double overhead camshaft cylinder head of the 2.0-L engine was shared with the 2.4-L engine. A single overhead camshaft version of the 2.4-L engine was never built in production. The balance shaft system of the 2.5-L engine carried over to the 2.4-L 16-valve engine, with modifications for the reciprocating weights of the new engine.

As released for 1995 model year passenger cars, the 2.4-L engine was rated at the same 112 kW (150 hp) power level as the 2.0-L 16-valve double overhead camshaft engine but with a significantly higher torque level—226 N-m (167 lb-ft) versus

TABLE 12.13
SPECIFICATIONS FOR THE 2.0-L DOHC 16-VALVE FOUR-CYLINDER ENGINE
Note: All dimensions are in metric units (millimeters) rather than customary units (inches).

Model Year	1995–1998
Car line	Plymouth, Dodge, Chrysler
Model	Plymouth and Dodge: Neon Chrysler: Sebring
Engine code	ECC
Displacement (liters)	2.0
Bore	87.5
Stroke	83
Compression ratio	9.6:1
Power @ rpm	112 kW (150 hp) @ 6500
Torque @ rpm	180 N-m (133 lb-ft) @ 5500
Cyl. block mat'l.—skirt type	Cast iron with bed plate
Cyl. block deck height	212
Cyl. block length	417.3
Cyl. bore spacing	96
No. of main brgs.—ck/s mat'l.	5—nodular cast iron
Ck/shaft main journal dia.	52
Crankpin journal dia.	48
Connecting rod centers	139
Valve system	DOHC—16-valve
Valve events	1998: 219-219-0
Camshaft drive	Timing belt
Valve head dia.	Intake 34.8; exhaust 30.5
Valve lash adjustment	Hyd. rocker arm pivot
Oil pump drive	Nose of crankshaft
Ignition	Distributorless ignition system
Fuel system	SMPI
Engine angle	10.5 deg to rear
Engine assembly weight*	129.3 kg (285.05 lb) ATX

* For content, see the Abbreviations and Definitions section near the back of this book.

180 N-m (133 lb-ft). Peak power and torque occurred at 5200 and 4000 rpm, respectively, for the larger engine versus the much higher 6500 and 5500 rpm for the 2.0-L engine.

The first assignment for the 2.4-L engine was to power the new JA body, transverse engine, front-wheel-drive Chrysler Cirrus and Dodge Stratus, including the two-door Stratus shown on Fig. 12.52. This engine was standard on the Cirrus and optional on the Stratus. Its use during the following year was expanded to include the new 1996 JX body Chrysler Sebring convertible and the JA body Plymouth Breeze. Importantly, its use also was expanded to include the assignment as standard power on the continuing popular 1996 Plymouth Voyager and Dodge Caravan minivans.

Table 12.14 shows the specifications for this 2.4-L double overhead camshaft 16-valve four-cylinder engine.

Chart 12.1 shows time lines for all the naturally aspirated front-wheel-drive four-cylinder engines described in this chapter. Chart 12.2 shows time lines for the turbocharged front-wheel-drive four-cylinder engines.

Fig. 12.52 *The 1995 Dodge Stratus four-door sedan. Optional power for the car was the 2.4-L double overhead camshaft 16-valve four-cylinder engine.*

TABLE 12.14
SPECIFICATIONS FOR THE 2.4-L DOHC 16-VALVE FOUR-CYLINDER ENGINE
Note: All dimensions are in metric units (millimeters) rather than customary units (inches).

Model Year	1995	1998
Car line	Dodge	Plymouth, Dodge, Chrysler
Model	Stratus	Plymouth: Breeze, Voyager Dodge: Stratus, Caravan Chrysler: Sebring convertible
Engine code	EDZ	EDZ
Displacement (liters)	2.4	2.4
Bore	87.5	87.5
Stroke	101	101
Compression ratio	9.4:1	9.4:1
Power @ rpm	110 kW (148 hp) @ 5200	112 kW (150 hp) @ 5200
Torque @ rpm	218 N-m (161 lb-ft) @ 4400	226 N-m (167 lb-ft) @ 4000
Cyl. block mat'l.—skirt type	Cast iron with bed plate	Cast iron with bed plate
Cyl. block deck height	237.8	237.7
Cyl. block length	417.3	417.3
Cyl. bore spacing	96	96
No. of main brgs.—ck/s mat'l.	5—nodular cast iron	5—nodular cast iron
Ck/shaft main journal dia.	52	52
Crankpin journal dia.	48	48
Connecting rod centers	151	151
Balance system	Twin balance shafts	Twin balance shafts
Valve system	DOHC—16-valve	DOHC—16-valve
Valve events	236-240-15	232-240-9
Camshaft drive	Timing belt	Timing belt
Valve head dia.	Intake 34.8; exhaust 30.5	Intake 34.8; exhaust 30.5
Valve lash adjustment	Hyd. rocker arm pivot	Hyd. rocker arm pivot
Oil pump drive	Nose of crankshaft	Nose of crankshaft
Ignition	Distributorless ignition system	Distributorless ignition system
Fuel system	SMPI	SMPI
Engine angle	10.5 deg to rear	10.5 deg to rear
Engine assembly weight*	129.4 kg (285.3 lb) ATX	

* For content, see the Abbreviations and Definitions section near the back of this book.

Chart 12.1
Time Lines—Front-Wheel-Drive Naturally Aspirated Four-Cylinder Engines

Engines, Car Lines, and Displacements (Liters)

Engine	Car Line	1978	1979	1980	1981	1982	1983	1984	1985
1.7-L VW SOHC	Plymouth	1.7	1.7	1.7	1.7	1.7	1.7		
	Code	M Horizon	M	M	M	M	M		
	Dodge	1.7	1.7	1.7	1.7	1.7	1.7		
	Code	Z Omni	Z	Z	Z	Z	Z		
1.6-L PSA OHV	Plymouth						1.6	1.6	1.6
	Code						M	M	M
	Dodge						1.6	1.6	1.6
	Code						Z	Z	Z
2.2-L SOHC	Plymouth				2.2	2.2	2.2	2.2	2.2
	Code				M, P	M, P	M, P	M, P, H	M, P, H
	Dodge				2.2	2.2	2.2	2.2	2.2
	Code				D, Z	D, V, Z	D, E, V, Z	D, E, V, Z, K	D, E, V, Z, K
	Chrysler							2.2	2.2
	Code					C	C, T	C, T	C
2.6-L MMC SOHC	Plymouth				2.6	2.6	2.6	2.6	2.6
	Code				P	P	P	P, H	P, H,
	Dodge				2.6	2.6	2.6	2.6	2.6
	Code				D	D, V	D, E, V	D, E, K, V	D, E, K, V
	Chrysler					2.6	2.6	2.6	2.6
	Code					C	C, T	C, T	C, T

Chart 12.1
(Continued)

Engines, Car Lines, and Displacements (Liters)

Engine	Car Line	1986	1987	1988	1989	1990	1991	1992	1993
		- - Model Years - -							
1.6-L PSA OHV	Plymouth	1.6							
	Code	M							
	Dodge	1.6							
	Code	Z							
2.2-L SOHC	Plymouth	2.2	2.2	2.2	2.2	2.2	2.2	2.2	2.2
	Code	H, J, M, P	H, M, P	J, P, M	P, Z	ALM, APP			
	Dodge	2.2	2.2	2.2	2.2	2.2	2.2	2.2	2.2
	Code	D, E, K, V, Z	D, E, K, Z	D, Z	D, Z	ALZ, APD			
	Chrysler	2.2	2.2	2.2	2.2				
	Code	C	C	C	C				
2.5-L SOHC	Plymouth	2.5	2.5	2.5	2.5	2.5	2.5	2.5	2.5
	Code	J, P	H, J, P	H, J, P	H, P	AAP, APP, ASH			
	Dodge	2.5	2.5	2.5	2.5	2.5	2.5	2.5	2.5
	Code	D, E, V	D, E, K, V	D, E, K, V	D, K, V	AGV, APD, AAD ACD, ASK			
	Chrysler	2.5	2.5	2.5	2.5	2.5	2.5	2.5	2.5
	Code	C	C, T	C	C	AJC			
2.6-L MMC SOHC	Plymouth	2.6	2.6						
	Code	H	H						
	Dodge	2.6	2.6						
	Code	K	K						

Chart 12.1
(Continued)

Engines, Car Lines, and Displacements (Liters)

Engine	Car Line	1994	1995	1996	1997	1998
2.2-L SOHC	Plymouth	2.2				
	Model	Sundance				
	Dodge	2.2				
	Model	Shadow				
2.5-L SOHC	Plymouth	2.5	2.5			
	Model	Sndnc, Acclm, Voy	Acclaim, Voyager			
	Dodge	2.5	2.5			
	Model	Shdw, Spirit, Cara	Spirit, Caravan			
	Chrysler	2.5				
	Model	LeBaron				
2.0-L SOHC	Plymouth		2.0	2.0	2.0	2.0
	Model		Neon	Neon, Breeze	Neon, Breeze	Neon, Breeze
	Dodge		2.0	2.0	2.0	2.0
	Model		Neon, Stratus	Neon, Stratus	Neon, Stratus	Neon, Stratus
2.0/2.4-L DOHC	Plymouth		2.0	2.0 Neon	2.0 Neon	2.0 Neon
	Model		Neon	2.4 Voyager	2.4 Voyager	2.4 Voyager
	Dodge		2.0 Neon	2.0 Neon	2.0 Neon	2.0 Neon
	Model		2.4 Stratus	2.4 Strtus Crvan	2.4 Strtus Crvan	2.4 Strtus Crvan
	Chrysler		2.0 Sebring	2.0 Sebring	2.0 Sebring	2.0 2.4
	Model		2.4 Cirrus	2.4 Cirrus Sebring	2.4 Cirrus Sebring	Sebring n
	Eagle		2.0	2.0	2.0	2.0
	Model		Talon	Talon	Talon	Talon

Chart 12.2
Time Lines—Front-Wheel-Drive Turbocharged Four-Cylinder Engines

Engines, Car Lines, and Displacements (Liters)

Engine	Car Line	1984	1985	1986	1987	1988	1989	1990
2.2-L SOHC Turbo	Plymouth Code		2.2 J	2.2 J	2.2 J, P	2.2 J, P		
	Dodge Code	2.2 E, V	2.2 D, E, V, Z	2.2 D, E, V, Z	2.2 D, E, V, Z	2.2 D, E, V, Z		
	Chrysler Code	2.2 C, T	2.2 C, T	2.2 C, T	2.2 C, T	2.2 C, T		
2.2-L SOHC Turbo II (Intercooled)	Dodge Model/Code				2.2 Shelby Z	2.2 Shelby Z, Lancer	2.2 D,V	
	Chrysler Model						2.2 LeB, TC by Mas	
2.5-L SOHC Turbo I	Plymouth Code						2.5 H, P	2.5 AAP, APP, ASH
	Dodge Code						2.5 D, K, V	2.5 AAD, AGV, APD
	Chrysler Code/Model						2.5 C	2.5 LeBaron
2.2-L SOHC Turbo IV (VNT)	Dodge Code							2.2 AGV, APD
	Chrysler Model							2.2 LeBaron
2.2-L Maserati DOHC 16-v Turbo	Chrysler Model						2.2 Chry TC by Mas	2.2 Chry TC by Mas

Chart 12.2
(Continued)

Engines, Car Lines, and Displacements (Liters)

Engine	Car Line	1991	1992	1993
2.5-L SOHC Turbo I	Plymouth Model	2.5 Sundance		
	Dodge Model	2.5 Day, Shdw, Spirit	2.5 Daytona, Spirit	
	Chrysler Model	2.5 LeBaron	2.5 LeBaron	
2.2-L DOHC 16-valve Turbo III	Dodge Model	2.2 Spirit R/T	2.2 Spirit R/T Daytona IROC R/T	2.2 Daytona IROC R/T

CHAPTER 13

V-6 Engines—More Power for Front-Wheel-Drive Cars

During the mid-1980s, plans were laid by Product Planning to expand the model offering of the hugely popular Plymouth Voyager and Dodge Caravan minivans by adding an extended wheelbase version. More power than available from the largest four-cylinder engine on the market, the 2.6-L Mitsubishi Motors Corporation engine would be needed to give the longer and heavier vehicle a good performance feel. Also, the midsize front-wheel-drive passenger cars such as the Chrysler New Yorker, already in production since 1983, certainly could use the smoothness and extra power of a V-6 engine. The need for a V-6 engine would become especially pressing with the end of production of the luxury-level rear-wheel-drive Chrysler, Dodge, and Plymouth models powered with the 5.2-L V-8 engine (which happened at the end of 1989) due to emission and fuel economy constraints.

With these incentives, Chrysler launched into an engine type it had never used in production passenger cars—the V-6.

3.0-L Single Overhead Camshaft V-6 from Mitsubishi Motors Corporation

At the time that the Chrysler need for a V-6 engine became apparent, Mitsubishi Motors Corporation (MMC) had a 3.0-L single overhead camshaft V-6 under development for its own vehicles. This opened an opportunity for Chrysler to obtain a V-6 much sooner than by starting with a new design and, importantly at the time, to avoid the vast tooling investment required for a new engine that would be built in-house.

Paralleling the procedure used with the 2.6-L Mitsubishi Motors Corporation four-cylinder engine described in Chapter 12, negotiations took place and agreement was reached on a joint program for Chrysler to purchase and use the engine. The program included modifications required to install the engine in the intended Chrysler vehicles and for the development needed for meeting all mandated safety and emission standards. Chrysler provided Mitsubishi Motors Corporation with information on its engine and vehicle standard tests used to establish satisfactory operation and durability for any engine used in Chrysler vehicles so that Mitsubishi Motors Corporation could ensure compliance with these rigorous tests.[1]

A description of the initial 1987 engine, the details of which are shown on the cross section on Fig. 13.1 and the longitudinal section on Fig. 13.2, is that the engine had a bore and stroke of 91.1 × 76 mm, giving it a displacement of 2972 cc. The two banks of the shallow-skirt cast iron cylinder block had an included angle of 60 deg. Together with the six-throw crankshaft, this provided equal 120-deg firing intervals for the engine. The cylinder block deck height was 210.5 mm, and the cylinder centers of each bank were located 108 mm apart. The four main bearing caps were connected together in a single, ladder-like casting that made the cylinder block assembly exceptionally strong and resistant to vibrations.

The cast aluminum cylinder head had cross-flow intake and exhaust ports with the valves angled away from each other, giving the combustion chamber a hemispherical shape. Each cylinder head had a single overhead camshaft and dual valve rocker arm shafts. The die-cast aluminum rocker arms had hardened powdered-metal wear surfaces for contacting the camshaft lobes and mini-hydraulic lash adjusters in the ends of the arms that contacted the valve tips. Intake and exhaust valve head diameters were 43 and 35 mm, respectively. The camshafts were driven from the crankshaft by a single timing belt that had a slack-side tensioner for belt tension control. The backside of the timing belt was used to drive the water pump, which was mounted in the vee of the engine.

The sturdy nodular cast iron crankshaft had four 60-mm-diameter main journals and six 50-mm-diameter crankpins spaced at 60-deg intervals for equal firing. The forged steel connecting

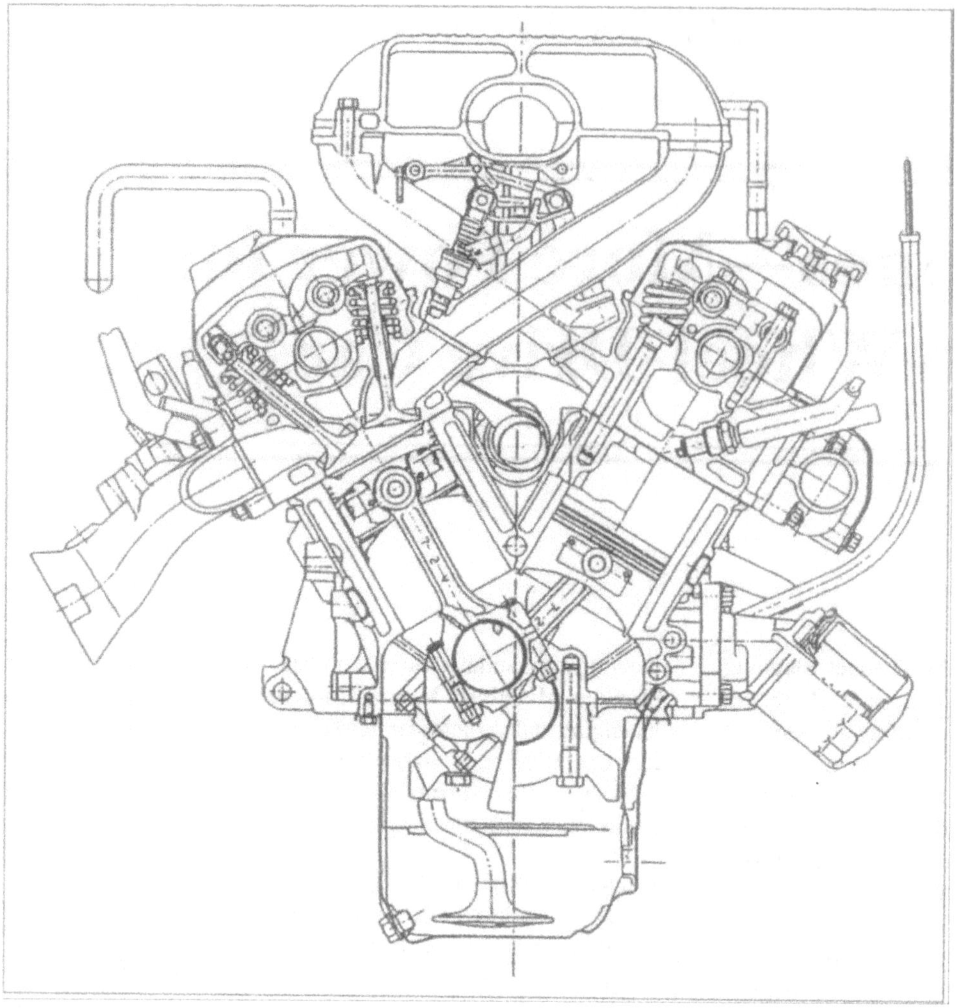

Fig. 13.1 *A cross section of the 1987 3.0-L Mitsubishi Motors Corporation single overhead camshaft V-6 engine. The profile of the engine was minimized by mounting the camshafts low in the cylinder heads. Power and torque were boosted by the cross-ram intake manifold system.*

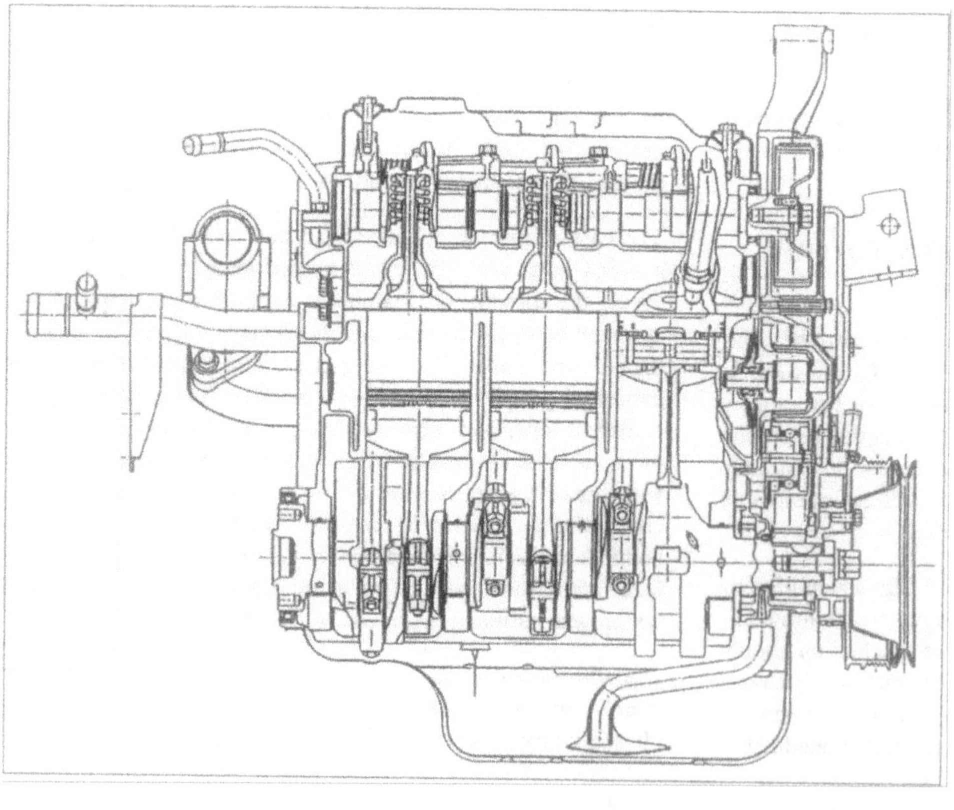

Fig. 13.2 *A longitudinal section of the 1987 3.0-L Mitsubishi Motors Corporation single overhead camshaft V-6 engine. The thick webs between adjacent connecting rods of the crankshaft permitted the crankpins to be splayed 60 deg apart.*

rods were designed with 141-mm centers. The distributor, driven by a camshaft gear, was mounted on the right end of the forward cylinder head, while the oil pump, integrated into the aluminum front cover, had its inner rotor driven by the nose of the crankshaft.

The three-piece aluminum intake manifold incorporated three areas of tuning to enhance maximum power and to maintain close-to-peak torque from 2000 to 4000 rpm for excellent driving performance feel. The lower part of the intake manifold provided the first area of tuning, a long ram pipe for each cylinder. The two upper parts of the intake manifold, bonded and bolted together, formed the second area of tuning, a large plenum chamber that fed the six ram pipes and also formed the third area of tuning: a tuned passage for intake air flowing from the throttle body into the plenum chamber. Fuel injectors, one for each cylinder, were mounted to the lower part of the intake manifold near the cylinder head joint faces. Figure 13.3 shows an exterior view of the front side of the 1987 engine. Figure 13.4 shows the engine installed in a 1990 "Chrysler's TC by Maserati" owned by Douglas Keyes, a retired Chrysler engine designer.

Fig. 13.4 *The engine compartment of the 1990 "Chrysler's TC by Maserati," with a 3.0-L Mitsubishi Motors Corporation single overhead camshaft V-6 engine installed. Vehicle in the collection of Douglas Keyes. Photo by W.L. Weertman.*

Fig. 13.3 *A front view of the 1987 3.0-L Mitsubishi Motors Corporation single overhead camshaft V-6 engine. The distributor, shown at the top-left of the view, was gear-driven from the left bank camshaft. A sheet metal heat shield surrounds the front exhaust manifold.*

Fig. 13.5 *The 1988 Chrysler New Yorker four-door sedan. The 3.0-L Mitsubishi Motors Corporation single overhead camshaft V-6 engine was its standard power.*

For its introductory mid-1987 year, the 3.0-L engine, rated at 104 kW (140 hp) and 230 N-m (170 lb-ft), was offered as optional power on both the standard and long-wheelbase Plymouth Voyager and Dodge Caravan minivans. Use of the engine during the following year was greatly expanded, with the engine specified as standard power on the 1988 Chrysler New Yorker shown on Fig. 13.5, the Chrysler Landau, the premium-level Plymouth Grand Voyager LE, and the Dodge Grand Caravan LE. It was an option on other minivan models and the Dodge Dynasty four-door sedan. In a November 1990 article about the recently released front-wheel-drive Chrysler LeBaron sedan, *Motor Trend* reported, "...the [Mitsubishi Motors Corporation] 60-degree V-6 is a willing partner in the LeBaron's quest for comfort and serenity. It's fed fuel through a multipoint injection system, and its single-overhead cams operate two valves per cylinder. While perfectly conventional in technology, the Mitsubishi-built V-6 has performed admirably in a number of Chrysler products and delivers more smoothness than a big four-banger...Armed with ample low-end torque and quick-flexed electronically controlled four-speed transmission, the motor scoots the car along with something resembling alacrity."

As larger Chrysler-built V-6 engines became available, the use of the 3.0-L Mitsubishi Motors Corporation engine was reduced until, for 1996, similar to its introductory year, it was specified only as optional power on minivans. Its use continued at least through 1998, the last year covered by this book.

During its 1987 to 1998 span, few significant changes were made to the 3.0-L engine. Among them was a change to the valve rocker arms in 1990 to use roller followers for contacting the camshaft lobes. In 1996, changes also were made to the air induction system and exhaust system. With the latter changes, power rose to 112 kW (150 hp) and 239 N-m (176 lb-ft).

From 1987 through 1989, the 3.0-L Mitsubishi Motors Corporation V-6 was the only V-6 powering Chrysler front-wheel-drive cars. Then, for 1990, the first of two Chrysler-built closely related V-6s arrived, as related next.

Table 13.1 shows the specifications for this 3.0-L Mitsubishi Motors Corporation single overhead camshaft V-6 engine.

A Two-Engine Family of V-6s—The 3.3-L and 3.8-L Overhead Valve Engines

Faced with the rising demand for V-6 engines in its minivans and cars and with limitations to the number of engines that could be furnished by Mitsubishi Motors Corporation, Chrysler put into action a V-6 engine program of its own. After considering the desired power and torque characteristics for the intended applications, a conclusion was reached that the focus should be on mid-speed torque rather than high-speed performance. The concept of an up-to-date overhead valve engine with roller tappets, aluminum cylinder heads, a tuned intake system, multi-point fuel injection, and distributorless ignition emerged as a strong contender.[2] Multi-camshaft, multi-valve arrangements were seen to add complexity, bulk, weight, and cost—both piece cost and tooling cost—as compared to pushrod valve actuation, without appreciably adding to customer benefit. With program approval to proceed with the overhead valve engine, a stipulation was made that the basic cylinder block design should be made so that the tooling would be adaptable to a future conversion to an overhead camshaft engine, if needed.

The 1990 lead-off engine, with a displacement of 3301 cc, had a bore and stroke of 93 × 81 mm. Figure 13.6 shows a cutaway view of the engine. Its deep-skirt cast iron cylinder block had a cylinder spacing of 113.28 mm (4.46 in., the same as that of the LA V-8 engine) and a deck height of 229 mm. The included angle between the two three-cylinder banks was 60 deg. The main journals of the nodular iron crankshaft had a diameter of 64 mm, giving the crankshaft a fundamental strength that also would suffice for the large-displacement version. To provide equal firing intervals for the engine, the two 58-mm-diameter crankpins located between the crankshaft main journals were angled 60 deg apart. The fillets of both the main journals and the crankpins were rolled to increase the fatigue strength of the crankshaft. A center distance of 157 mm was used for the forged steel connecting rods.

As illustrated on the cross section of the engine shown on Fig. 13.7 and on the longitudinal section shown on Fig. 13.8, the aluminum cylinder heads had in-line valves operated by hydraulic roller tappets, short pushrods, and stamped-steel rocker arms. Double-quench high-turbulence combustion chambers permitted the 8.9:1 compression ratio engine to operate on regular unleaded fuel. The valve rocker arms for each cylinder head rotated about a single rocker arm shaft. The vee of the cylinder block was the location of the chain-driven camshaft. Head diameters for the intake and exhaust valves were 45.5 and 37.5 mm, respectively.

An intake air plenum with a volume of 2.6 liters and six primary runners, each 470 mm long, were provided by the two-piece aluminum intake manifold. Manifold tuning was designed to boost torque in the mid-speed 3600-rpm range and to broaden the torque curve for enhanced engine response and usable power output. The six fuel injectors were mounted on the lower section of the intake manifold, which attached to the cylinder heads.

An aluminum chain case cover module on the right side of the engine served as the water pump housing and the oil pump housing. Similar to the Mitsubishi Motors Corporation 3.0-L V-6 engine, the inner rotor of the oil pump was driven by the nose of the crankshaft.

The 3.3-L V-6 engine was the first Chrysler engine to be equipped with a direct (i.e., distributorless) ignition system (DIS) instead of the conventional distributor with its rotating shaft that had been used virtually since the beginning of the auto industry.[3] Using signals from camshaft and crankshaft position sensors, the engine control computer gave commands to an engine-mounted coil pack for sending high-voltage current to the proper spark plug at the right time for optimum ignition. Not having to mount and drive a distributor simplified the mechanical design of the engine and freed up space that could be used by other components needing room on the compact front-wheel-drive engine assembly.

For its introductory 1990 year, the 3.3-L overhead valve engine was given the somewhat awesome assignment of being standard power on the flagship Chrysler Imperial together with the luxurious Chrysler New Yorker and the largest of the minivans: Chrysler Town & Country (pictured on Fig. 13.9), Dodge Grand Caravan, and Plymouth Grand Voyager. It was optional on the large front-wheel-drive Dodge Dynasty LE. In the minivans, the output of the engine was 112 kW (150 hp) and 251 N-m (185 lb-ft). In the sedans, the output was slightly less at 110 kW (147 hp) and 248 N-m (183 lb-ft) torque.

As planned, during the following year, 1991, a large-displacement version of the engine went into production alongside the 3.3-L V-6 engine at the Trenton engine plant. The displacement of

TABLE 13.1
SPECIFICATIONS FOR THE 3.0-L MMC SOHC V-6 ENGINE
Note: All dimensions are in metric units (millimeters) rather than customary units (inches).

Model Year	1987	1993	1998
Car line	Plymouth, Dodge	Plymouth, Dodge, Chrysler	Plymouth, Dodge
Model	Plymouth: Voyager Dodge: Caravan	Plymouth: Sundance, Acclaim, Voyager Dodge: Daytona, Shadow, Spirit, Dynasty, Caravan Chrysler: LeBaron	Plymouth : Voyager Dodge: Caravan
Engine code	EFA	EFA	EFA
Displacement (liters)	3.0	3.0	3.0
Bore	91.1	91.1	91.1
Stroke	76	76	76
Compression ratio	8.9:1	8.9:1	8.9:1
Power @ rpm	104 kW (140-hp) @ 4800	105 kW (141 hp) @ 5200	112 kW (150 hp) @ 5200
Torque @ rpm	230 N-m (170 lb-ft) @ 2800	232 N-m (171 lb-ft) @ 2400	239 N-m (176 lb-ft) @ 4000
Cyl. block mat'l.—skirt type	Cast iron—shallow "Ladder" main bearing cap	Cast iron—shallow "Ladder" main bearing cap	Cast iron—shallow "Ladder" main bearing cap
Cyl. block deck height	210.5	210.5	210.5
Cyl. block length	384	384	384
Cyl. bore spacing	108	108	108
No. of main brgs.—ck/s mat'l.	4—nodular cast iron	4—nodular cast iron	4—nodular cast iron
Ck/shaft main journal dia.	60	60	60
Crankpin journal dia.	50	50	50
Connecting rod centers	141	141	141
Valve system	SOHC—12-valve	SOHC—12-valve	SOHC—12-valve
Valve events	258-258-38	258-258-38	258-258-38
Camshaft drive	Timing belt	Timing belt	Timing belt
Valve head dia.	Intake 43; exhaust 35	Intake 43; exhaust 35	Intake 43; exhaust 35
Valve lash adjustment	Hyd. adjuster in rocker arm	Hyd. adjuster in rocker arm	Hyd. adjuster in rocker arm
Oil pump drive	Nose of crankshaft	Nose of crankshaft	Nose of crankshaft
Distributor drive	Geared to left camshaft	Geared to left camshaft	Geared to left camshaft
Fuel system	MPI	SMPI	SMPI
Engine assembly weight*	158.94 kg (350.4 lb)	168.18 kg (370.0 lb)	

* For content, see the Abbreviations and Definitions section near the back of this book.

Fig. 13.6 *A cutaway illustration of the 1990 3.3-L overhead valve V-6 engine. The engine was light, compact, and easy to manufacture.*

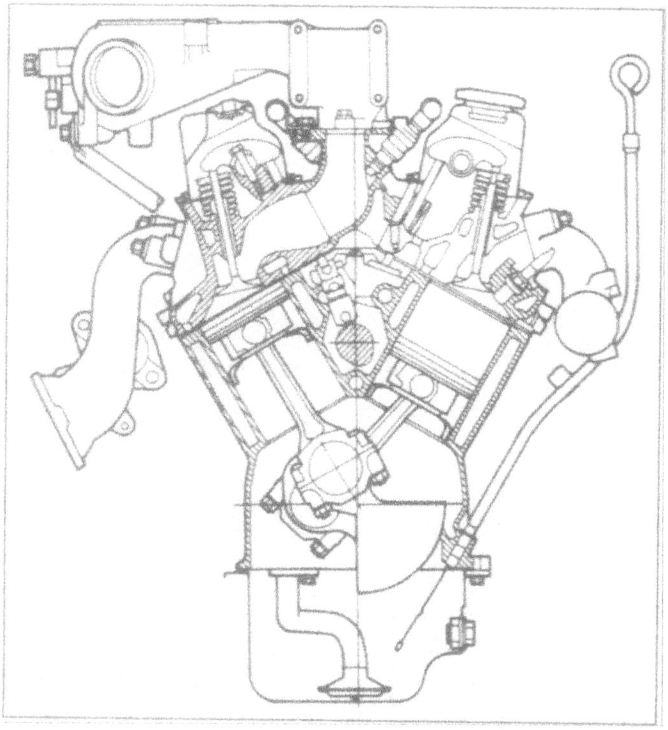

Fig. 13.7 *A cross section of the 1990 3.3-L overhead valve V-6 engine. The latest techniques of multipoint fuel injection and intake tuning were combined with a well-developed block-mounted camshaft, tappets, pushrods, and rocker arms.*

the larger version was 3.8 liters (3778 cc) and was obtained by a slight increase in bore to 96 mm (versus 93 mm) and stroke to 87 mm (versus 81 mm). Externally, the two engines were identical. Despite the longer stroke of the 3.8-L engine, the same cylinder block deck height was used, the cylinder block of the 3.8-L V-6 being different only because of its larger bore. Pistons, rings, and crankshafts were unique to each engine, but the connecting rods with their 157-mm center distance were common. Cylinder head assemblies and manifolds were common. The engine was an easy drop-in replacement for any engine compartment already using the 3.3-L engine and, for 1991, was assigned to be the sole power for the Chrysler Imperial shown on Fig. 13.10 and as optional power on the Chrysler New Yorker Fifth Avenue four-door sedan. The 3.8-L engine, with its output of 112 kW (150 hp) and 275 N-m (203 lb-ft), was well suited for these top-of-the-line luxury cars.

In 1993, a completely new car arrived in the Chrysler lineup. The car had an engine orientation that was different from the east-west position used in all Chrysler front-wheel-drive vehicles until that time. The new car was the LH, with its engine mounted longitudinally (also called north-south) ahead of the front-wheel-drive transmission instead of the transverse east-west orientation used on Chrysler front-wheel-drive vehicles ever since the 1978 L-body Plymouth Horizon and Dodge Omni. The 3.3-L engine, swung around 90 deg and fitted with the mounts, accessories, and manifolds suitable for the north-south position, became standard power for the three models sharing this new body and chassis: the 1993 Chrysler Concorde shown on Fig. 13.11, the Dodge Intrepid, and the Eagle Vision ESi. The new intake manifold of the engine, configured to fit under the low sloping hood of the car, resulted in the north-south-installed 3.3-L V-6 engine having a higher power rating of 114 kW (153 hp) but a lower torque rating of 240 N-m (177 lb-ft) versus the ratings of 110 kW (147 hp) and 248 N-m (183 lb-ft) when installed east-west in the 1993 Chrysler New Yorker and Dodge Dynasty. In 1993, the fuel injection system on both the 3.3-L and 3.8-L engines was changed to sequential operation.

The use of the 3.8-L V-6 engine in Chrysler sedans ended in 1993 with the termination of the front-wheel-drive transverse-engined Chrysler New Yorker and Imperial models at the end of that year. However, during the following year, 1994, the 3.8-L V-6 engine replaced the 3.3-L V-6 as the standard and only power for the Chrysler long-wheelbase Town & Country minivan. Two years later, in 1996, the 3.8-L V-6 added Plymouth and Dodge minivans, both standard-wheelbase and long-wheelbase models, to its applications.

The 3.3-L V-6 engine provided power for sedans until the end of 1997, when its use in the front-wheel-drive, north-south engined Dodge Intrepid was terminated. The use of the 3.3-L and 3.8-L V-6 engines in minivans continued beyond the 1998 ending year covered in this book.

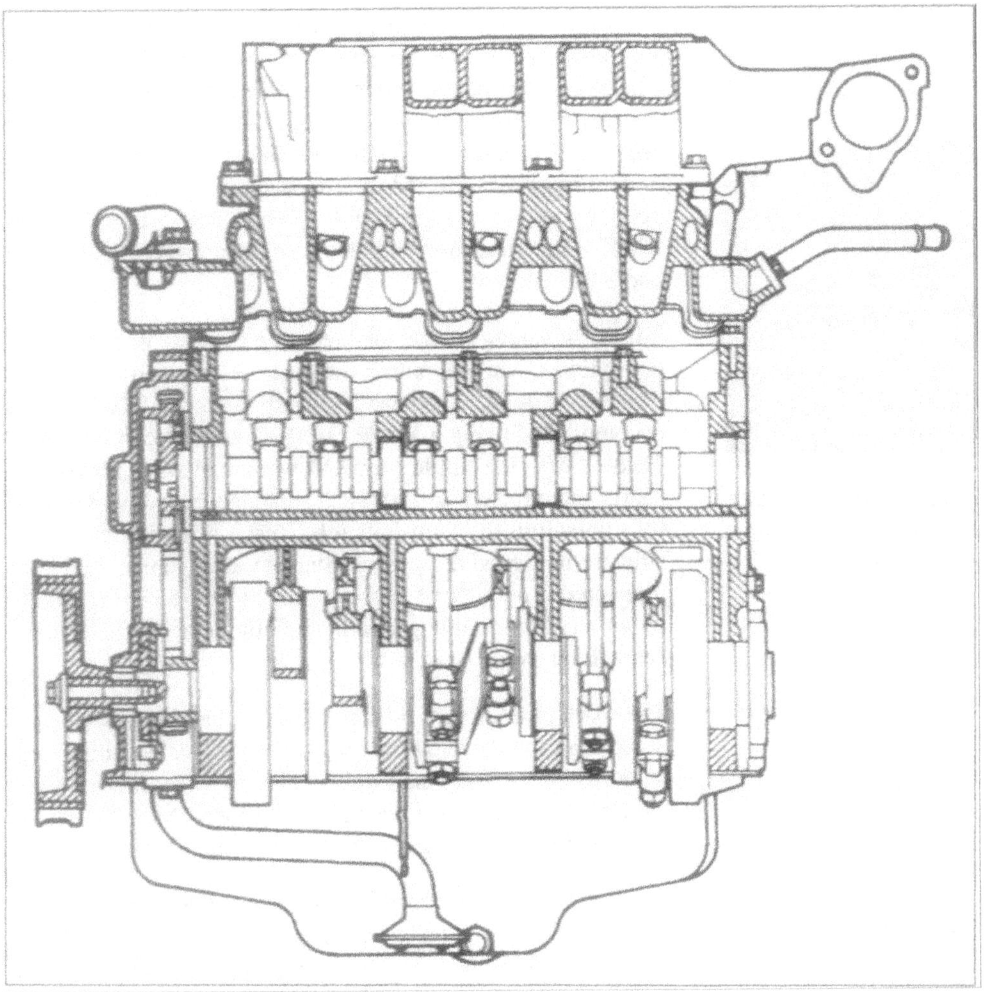

Fig. 13.8 *A longitudinal section of the 1990 3.3-L overhead valve V-6 engine. The thick webs of the crankshaft between adjacent crankpins were designed to allow 60-deg splay, a requisite for equal 120-deg firing intervals.*

Fig. 13.9 *The 1990 Chrysler Town & Country minivan. The 112 kW (150 hp) 3.3-L overhead valve V-6 was its standard engine.*

Fig. 13.10 *The 1991 Chrysler Imperial four-door sedan. The overhead valve V-6 engine with a new displacement of 3.8 L was its powerplant.*

Fig. 13.11 *The 1993 Chrysler Concorde four-door sedan. Still front-wheel drive, the 3.3-L overhead valve V-6 engine powering this all-new car was positioned longitudinally.*

Table 13.2 shows the specifications for the 3.3-L and 3.8-L overhead valve V-6 engines.

Chart 13.1 displays time lines for the 3.3-L and 3.8-L overhead valve V-6 engines.

The First Chrysler Engine Family with 24 Valves—The 3.5 and 3.2-L Single Overhead Camshaft V-6

The 3.5-L Single Overhead Camshaft 24-Valve V-6 Engine

As noted, for 1993 Chrysler introduced a new LH large-car platform that was the basis for the Chrysler Concorde, Dodge Intrepid, and Eagle Vision models of that year. The 3.3-L overhead valve V-6 engine was judged to be satisfactory as the base engine. However, to give the car world-class stature, it needed an engine with overhead camshafts and multiple valves. The Large Car Platform Engine Group under executive engineer Gordon Rinschler responded to the need by creating a new engine with a bore and stroke of 96 × 81 mm, giving a displacement of 3518 cc (3.5 L). The identical cylinder bore spacing of 113.28 mm as was used on the 3.3-L and 3.8-L V-6 engine family was applied to the deep-skirt cast iron cylinder block of this new engine. The cylinder block deck height of 229 mm and an included angle between the banks of 60 deg also were identical to the cylinder block of the 3.3-L and 3.8-L engines. Advantage was taken of the deep skirt of the 3.5-L V-6 cylinder block to add cross bolts to the #2 and #3 main bearing caps for added crankcase strength and quietness. The mechanical arrangement of the engine is shown on the cutaway view on Fig. 13.12 and on the cross section on Fig. 13.13.

Anticipating high output levels for the engine, forged steel was specified as the material for the six-throw crankshaft. The main journal and crankpin diameters of 64 and 58 mm, respectively, were carryovers from the 3.3-L and 3.8-L V-6 engines. A combined windage tray and oil pan gasket, located under the crankshaft, reduced parasitic power loss. The connecting rod was designed with a center distance of 151 mm, which was 6 mm shorter than the connecting rod of the 3.3-L and 3.8-L engines.

To take advantage of the air-breathing capacity of four valves per cylinder while avoiding the cost, weight, and complexity penalties of double overhead camshafts, a successful design was accomplished using a single camshaft in each cylinder head. Teamed with the single camshaft were dual valve rocker arm shafts and a rocker arm for each valve, equipped with a roller for contacting the camshaft lobe and a mini-hydraulic lash adjuster in the valve tip end. Such an arrangement had been pioneered a number of years earlier in future engine studies by an advance engine group headed by department manager John Hurst.

The spark plugs, inside oil-tight access tubes, were positioned at the center of the combustion chambers. The head diameters of the twelve intake and twelve exhaust valves were 35 and 29 mm, respectively. Freewheeling clearance for the valves was provided by notches in the heads of the pistons.

A single timing belt, equipped with a self-contained hydraulic belt tensioner, drove the two camshafts of the engine. Camshaft events were 242-252-12. The water pump, mounted in the vee of the cylinder block, was driven by the backside of the timing belt. The oil pump, integrated in the aluminum front cover, was driven, similar to that of the 3.3-L and 3.8-L engines, directly by the nose of the crankshaft.

Individual fuel injectors were mounted on the cylinder heads, close to the lower intake manifold joint faces. The aluminum intake manifold was composed of two sections: a low-profile upper section containing two widely separated plenum chambers, each mounting a throttle body and feeding three intake branches; and a lower section with six passages to distribute intake air to the six cylinders. The design provided a torque-boosting intake branch length of 315 mm for each cylinder. The upper section also contained a large cast air passage connecting the two plenums. A normally closed quick-acting vacuum-operated tuning valve, located in the connecting passage, was programmed to open during 3200- to 4000-rpm wide-open-throttle accelerations for additional engine torque by combining the volumes of both plenums over this range. An SAE paper titled "Chrysler 3.5 Liter V-6 Engine" written by Thomas W. Witte and Keith G. Wright provides further information on the design and development of this engine.[4]

Assigned as optional power in the 1993 Chrysler Concorde and Dodge Intrepid and as standard power on the Eagle Vision Tsi shown on Fig. 13.14, the 3.5-L single overhead camshaft

TABLE 13.2
SPECIFICATIONS FOR THE 3.3-L AND 3.8-L OHV V-6 ENGINES
Note: All dimensions are in metric units (millimeters) rather than customary units (inches).

Model Year	1990	1991	1998	1998
Car line	Plymouth, Dodge, Chrysler	Chrysler	Plymouth, Dodge, Chrysler	Dodge, Chrysler
Model	Plymouth: Voyager Dodge: Dynasty, Caravan Chrysler: New Yorker, Imperial, Town & Country	Chrysler: New Yorker, Imperial	Plymouth: Voyager Dodge: Caravan Chrysler: Town & Country	Dodge: Caravan Chrysler: Town & Country
Engine code	EGA	EGH	EGA	EGH
Displacement (liters)	3.3	3.8	3.3	3.8
Bore	93	96	93	96
Stroke	81	87	81	87
Compression ratio	8.9:1	8.9	8.9:1	8.9
Power @ rpm	112 kW (150 hp) @ 4800	112 kW (150 hp) @ 4400	115 kW (154 hp) @ 4850	134 kW (180 hp) @ 4400
Torque @ rpm	251 N-m (185 lb-ft) @ 3600	275 N-m (203 lb-ft) @ 3200	275 N-m (203 lb-ft) @ 3250	325 N-m (240 lb-ft) @ 3200
Cyl. block mat'l.—skirt type	Cast iron—shallow	Cast iron—shallow	Cast iron—shallow	Cast iron—shallow
Cyl. block deck height	229	229	229	229
Cyl. block length	409.6	409.6	409.6	409.6
Cyl. bore spacing	113.28	113.28	113.28	113.28
No. of main brgs.—ck/s mat'l.	4—nodular cast iron	4—nodular cast iron	4—nodular cast iron	4—nodular cast iron
Ck/shaft main journal dia.	64	64	64	64
Crankpin journal dia.	58	58	58	58
Connecting rod centers	157	157	157	157
Valve system	OHV—12-valve	OHV—12-valve	OHV—12-valve	OHV—12-valve
Valve events	240-240-12	240-240-12	240-240-14	240-240-14
Camshaft drive	Chain	Chain	Chain	Chain
Valve head dia.	Intake 45.5; exhaust 37.5	Intake 45.5; exhaust 37.5	Intake 45.5; exhaust 37.5	Intake 45.5; exhaust 37.5
Valve lash adjustment	Roller hyd. tappets	Roller hyd. tappets	Roller hyd. tappets	Roller hyd. tappets
Oil pump drive	Nose of crankshaft	Nose of crankshaft	Nose of crankshaft	Nose of crankshaft
Ignition	Distributorless ignition system	Distributorless ignition system	Distributorless ignition system	Distributorless ignition system
Fuel system	MPI	SMPI	SMPI	SMPI
Engine assembly weight*	193.18 kg (425.0 lb)	195.05 kg (430.0 lb)		

* For content, see the Abbreviations and Definitions section near the back of this book.

Chart 13.1
Time Lines—OHV Front-Wheel-Drive V-6 Engines

Engines, Car Lines, and Displacements (Liters)

Engine	Car Line	Model Years: 1990–1992	1993	1994–1996	1997	1998
3.3/3.8-L OHV 8-valve V-6	Plymouth	3.3	3.3	3.3, 3.8	3.3	3.3
	Model	Voyager	Voyager	Voyager	Voyager	Voyager
	Dodge	3.3	3.3	3.3, 3.8	3.3, 3.8	3.3, 3.8
	Model	Dynasty, Caravan	Dynasty, Intrepid Caravan	Intrepid 3.3 Caravan 3.3, 3.8	Intrepid 3.3 Caravan 3.3, 3.8	Caravan
	Chrysler	3.3	3.3, 3.8	3.3, 3.8	3.3, 3.8	3.3, 3.8
	Model	NY, Imperial Twn & Cntry	NY, Imperial Twn & Cntry	Concorde 3.3 T & C: 3.3 & 3.8	Twn & Cntry	Twn & Cntry
	Eagle		3.3	3.3	3.3	
	Model		Vision	Vision	Vision	

Fig. 13.12 *A cutaway illustration of the 1993 3.5-L single overhead camshaft 24-valve V-6 engine. Sequential fuel injection, tuned intake system, distributorless ignition, and cross-bolted main bearing caps were part of the engine design right from the beginning.*

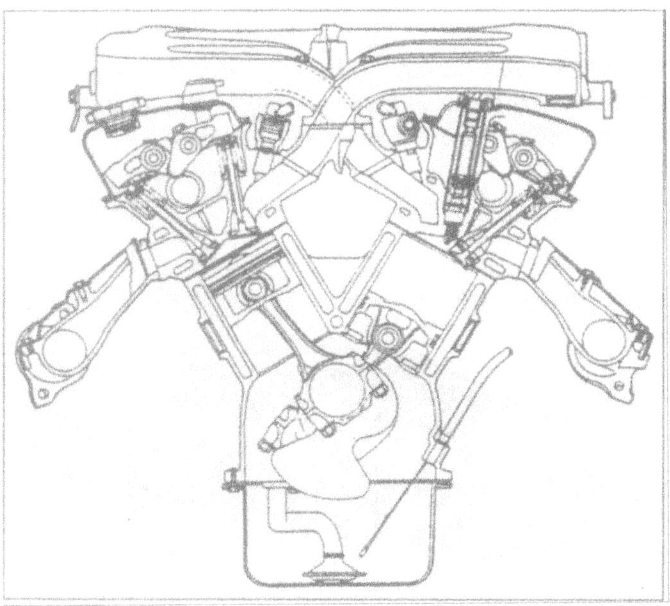

Fig. 13.13 *A cross section of the 1993 3.5-L single overhead camshaft 24-valve V-6 engine. The north-south installation of the engine in the car allowed the exhaust manifold runners to be spaced away from the cylinder head for improved gas flow.*

Fig. 13.14 *The 1993 Eagle Vision Tsi four-door sedan. Standard power for the front-wheel-drive car was a 160-kW (214-hp) 3.5-L single overhead camshaft 24-valve V-6 engine positioned north-south.*

V-6 engine, with its 10.44:1 compression ratio, was rated at 160 kW (214 hp) and 301 N-m (222 lb-ft) torque. In 1994, another Chrysler model was added, the Chrysler LHS. In the June 1993 issue of *Motor Trend* that included a road test of the 1994 Chrysler LHS, author Daniel C. Ross wrote, "The powertrain is borrowed from the LH clan. Chrysler supplies the LHS only with its 214-horsepower single overhead camshaft 24-valve 3.5-liter V-6. This is a lusty all-purpose engine scarcely handicapped by the longer and heavier LHS." Although the compression ratio was dropped to 9.6:1 in 1995, peak power and torque remained unchanged.

For fresh-air driving enthusiasts, a new, sporty, and eye-catching vehicle debuted in late 1997—the Plymouth Prowler, which is shown on Fig. 13.15. Its powerplant, which was positioned north-south, was the same 3.5-L single overhead camshaft 24-valve 9.6:1 compression ratio, 160-kW (214-hp) and 301 N-m (222 lb-ft) torque V-6 engine previously described. This was somewhat of an expedient because the 1997 program time requirements precluded the use of the desired more powerful and lighter aluminum block upgrade that was under development for the 3.5-L engine. The 1997 cast iron block Prowler engine carried over to the start of 1999, when it was replaced with the all-aluminum 3.5-L V-6 engine that, equipped with a new-concept active intake system described in the 3.2-L V-6 engine section that follows, substantially boosted its output to 189 kW (253 hp) and 346 N-m (255 lb-ft).

Fig. 13.15 *The 1997 Plymouth Prowler. This sporty rear-wheel-drive two-seat roadster was powered by a 160-kW (214-hp) 3.5-L single overhead camshaft 24-valve V-6 engine.*

The 3.2-L Single Overhead Camshaft 24-Valve V-6 Engine

A new and smaller-displacement (3.2-L) version of the 3.5-L engine was released for use in the 1998 cars, which was the first of the 3.2-L and 3.5-L single overhead camshaft 24-valve related pair of engines to receive an aluminum cylinder block. The new displacement, 3231 cc (3.2 L), was accomplished by a bore reduction to 92 mm from the 96-mm bore of the 3.5-L engine. Prominent among the new features of the 3.2-L engine was its lightweight aluminum cylinder block with cast-in-place iron liners. The cross bolts of the 3.5-L engine were retained to give exceptional strength to the crankcase. Although the basic design of the cylinder heads was unchanged, modifications were made to improve engine output, including larger intake valves (36.5 versus 35 mm) and revised porting for higher flow. With the larger intake valves, the engine lost its capability for freewheeling should the crankshaft and camshaft get out of phase. A completely new and actively controlled intake system replaced the former dual plenum system. The new system added a central plenum with short branches leading to the intake valves of the engine. The dual throttle bodies of the 3.5-L engine were replaced with a single throttle body upstream of the central plenum. Air valves throughout the system were orchestrated by the engine control computer to direct intake airflow through either the two outboard plenums with their long branches or the central plenum with its short branches to optimize the engine for best power or best torque, depending on the output demands. The system was complex but highly effective in increasing both peak power and peak torque to the levels of 168 kW (225 hp) and 305 N-m (225 lb-ft). Co-inventors Michael Patyi, William Hughs, Mary Chappell (Tymrak), Frederick Hummer, and Donald Kerska were awarded two patents[5,6] for their collective creative efforts. As noted, the 3.5-L engine was upgraded in 1999 with these same new features.

This 3.2-L single overhead camshaft V-6 all-aluminum 24-valve 225-hp engine, shown on Fig. 13.16, was quickly put to work providing the pulling power for the Chrysler 1998 Concorde LXi and the 1998 Dodge Intrepid ES shown on Fig. 13.17.

Table 13.3 shows the specifications for the 3.5-L and 3.2-L V-6 engines.

Fig. 13.16 *A front-top view of the 1998 168-kW (225-hp) 3.2-L single overhead camshaft 24-valve all-aluminum V-6 engine with the timing belt cover removed.*

Fig. 13.17 *The 1998 Dodge Intrepid four-door sedan. Power for the car was a 168-kW (225-hp) 3.2-L single overhead camshaft 24-valve all-aluminum V-6 engine.*

TABLE 13.3
SPECIFICATIONS FOR THE 3.5-L AND 3.2-L SOHC 24-VALVE V-6 ENGINES
Note: All dimensions are in metric units (millimeters) rather than customary units (inches).

Model Year	1990	1997	1998
Car line	Dodge, Chrysler, Eagle	Plymouth	Dodge, Chrysler
Model	Dodge: Intrepid Chrysler: Concorde Eagle: Vision	Prowler	Dodge: Intrepid Chrysler: Concorde
Engine code	EGG	EGG	EGW
Displacement (liters)	3.5	3.5	3.2
Bore	96	96	96
Stroke	81	81	81
Compression ratio	10.44:1	9.6:1	9.5:1
Power @ rpm	160 kW (214 hp) @ 5800	160 kW (214 hp) @ 5850	168 kW (225 hp) @ 6300
Torque @ rpm	301 N-m (221 lb-ft) @ 2800	301 N-m (221 lb-ft) @ 3100	305 N-m (225 lb-ft) @ 3800
Cyl. block mat'l.—skirt type	Cast iron—deep, cross-bolts	Cast iron—deep, cross-bolts	Aluminum—deep, cross-bolts
Cyl. block deck height	229	229	229
Cyl. block length	409.6	409.6	409.6
Cyl. bore spacing	113.28	113.28	113.28
No. of main brgs.—ck/s mat'l.	4—forged steel	4—forged steel	4—forged steel
Ck/shaft main journal dia.	64	64	64
Crankpin journal dia.	58	58	58
Connecting rod centers	151	151	151
Valve system	SOHC—24-valve	SOHC—24-valve	SOHC—24-valve
Valve events	242-252-12	242-252-12	230-240-11
Camshaft drive	Timing belt	Timing belt	Timing belt
Valve head dia.	Intake 35; exhaust 29	Intake 35; exhaust 29	Intake 36.5; exhaust 29
Valve lash adjustment	Hyd. adjuster in rocker arm	Hyd. adjuster in rocker arm	Hyd. adjuster in rocker arm
Oil pump drive	Nose of crankshaft	Nose of crankshaft	Nose of crankshaft
Ignition	Distributorless ignition system	Distributorless ignition system	Distributorless ignition system
Fuel system	SMPI	SMPI	SMPI
Engine drive/orientation	Front-wheel drive/north-south	Rear-wheel drive/north-south	Front-wheel drive/north-south
Engine assembly weight*			179.6 kg (395.9 lb)

* For content, see the Abbreviations and Definitions section near the back of this book.

A Second V-6 from Mitsubishi Motors Corporation—The 2.5-L Single Overhead Camshaft 24-Valve V-6

During the early 1990s, Mitsubishi responded to the growing demand for engines with multiple valves by the design and development of a new small V-6 having four valves per chamber operated by a single camshaft in each cylinder head. Slated for use in Mitsubishi Motors Corporation-built vehicles, the engine was adaptable to the Chrysler E-W transverse engine, front-wheel-drive compartments that were home to the long-running Mitsubishi 3.0-L single overhead camshaft 12-valve V-6 engine, and the Chrysler 3.3-L overhead valve V-6 engine.

The 2497-cc (2.5-L) V-6, with a bore and stroke of 83.5 × 76 mm, was designed with a cast iron, shallow-skirt cylinder block having a 108-mm cylinder spacing and a deck height of 210.5 mm. Its 60-deg included angle between the cylinder banks and the six-throw crankshaft provided equal 120-deg firing intervals. All four main bearing caps were integrated into a single large iron casting that extended the length of the block.

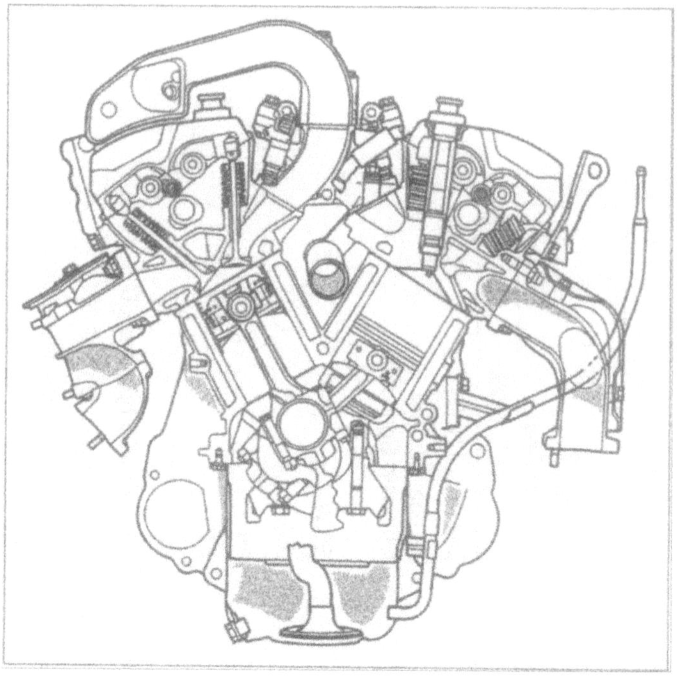

Fig. 13.18 *A cross section of the 1995 2.5-L Mitsubishi Motors Corporation single overhead camshaft 24-valve V-6 engine. The cylinder head arrangement appears somewhat similar to that of the Chrysler 3.5-L single overhead camshaft 24-valve V-6 engine shown on Fig. 13.13.*

The general arrangement of the aluminum cylinder head—with its four valves per chamber, single overhead camshaft, dual rocker arm shafts, and inboard spark plug enclosed in an access tube—was somewhat similar to the arrangement used on the 1993 3.5-L single overhead camshaft 24-valve V-6 engine, as can be seen by comparing the cross section of the 3.5-L V-6 engine shown on Fig. 13.13 with the cross section of the 2.5-L V-6 engine shown on Fig. 13.18. Each of the twelve intake rocker arms of the 2.5-L V-6 engine was equipped with a roller follower for contact with the camshaft lobe and a mini-hydraulic adjuster in the valve tip end. The six exhaust rocker arms were forked, with each arm having a single roller for contacting the camshaft lobe and two mini-hydraulic adjusters at the other end of the arm for contacting the tips of two exhaust valves. Head diameters were 33 mm for the intake valves and 29 mm for the exhaust valves.

Fig. 13.19 *The 1995 Dodge Stratus ES four-door sedan. The 122-kW (164-hp) 2.5-L Mitsubishi Motors Corporation single overhead camshaft 24-valve V-6 engine was a power option for the car.*

A flexible timing belt drove the two camshafts of the engine. The use of a distributorless ignition system obviated the need for a distributor, and the oil pump was driven directly by the nose of the crankshaft. The engine control computer sequentially operated the six fuel injectors, mounted on the lower intake manifold.

A diameter of 64 mm was provided for the four main journals of the nodular iron crankshaft, while the crankpins had a diameter of 54 mm. The forged steel connecting rods were designed with a center distance of 141 mm.

The aluminum upper and lower intake manifold sections together provided long torque-enhancing branches for the intake ports. A single plenum, part of the upper manifold, fed air to the six branches. Installed as standard power in the Chrysler Cirrus LXi four-door sedan and as an option in the smooth-looking Dodge Stratus ES that is shown on Fig. 13.19, the 1995 engine, with a compression ratio of 9.4:1, had an output of 122 kW (164 hp) and 221 N-m (163 lb-ft). Following its introduction, the 1995 Cirrus garnered three prestigious awards: the *Motor Trend* Car of the Year award, the *Popular Science*

Best of What's New award, and the *Popular Mechanics* Design and Engineering award.

The following year, 1996, in its installations that included the new Chrysler Sebring convertible, the power and torque of the 2.5-L V-6 engine went up to 125 kW (168 bhp) and 230 N-m (170 lb-ft). No further changes in output were made through 1998.

Table 13.4 displays the specifications for this Mitsubishi Motors Corporation-built 2.5-L single overhead camshaft 24-valve V-6 engine.

The Chrysler V-6 Engine with Two More Camshafts—The 2.7-L Double Overhead Camshaft 24-Valve V-6

During the mid-1990s, Engineering Management, especially Francois J. Castaing, vice president of Engineering, felt strongly that a smaller-displacement (i.e., smaller than the 3-L or larger engines already in production), compact, lightweight, high-output, double overhead camshaft, 24-valve V-6 engine should be added to the corporate lineup to put Chrysler in the front ranks of the world automotive technology leaders.[7] Such

TABLE 13.4
SPECIFICATIONS FOR THE 2.5-L MMC SOHC 24-VALVE V-6 ENGINE
Note: All dimensions are in metric units (millimeters) rather than customary units (inches).

Model Year	1995	1998
Car line	Dodge, Chrysler	Dodge, Chrysler
Model	Dodge: Stratus Chrysler: Cirrus, Sebring	Dodge: Stratus Chrysler: Cirrus, Sebring
Engine code	EEB	EEB
Displacement (liters)	2.5	2.5
Bore	83.5	83.5
Stroke	76	76
Compression ratio	9.4:1	9.4:1
Power @ rpm	122 kW (164 hp) @ 5900	125 kW (168 hp) @ 5800
Torque @ rpm	221 N-m (163 lb-ft) @ 4350	230 N-m (170 lb-ft) @ 4350
Cyl. block mat'l.—skirt type	Cast iron—shallow "Ladder" main bearing cap	Cast iron—shallow "Ladder" main bearing cap
Cyl. block deck height	210.5	210.5
Cyl. block length	384	384
Cyl. bore spacing	108	108
No. of main brgs.—ck/s mat'l.	4—nodular cast iron	4—nodular cast iron
Ck/shaft main journal dia.	60	60
Crankpin journal dia.	50	50
Connecting rod centers	151	151
Valve system	SOHC—24-valve	SOHC—24-valve
Valve events	252-244-29	252-244-29
Camshaft drive	Timing belt	Timing belt
Valve head dia.	Intake 33; exhaust 28.9	Intake 33; exhaust 28.9
Valve lash adjustment	Hyd. adjuster in rocker arm	Hyd. adjuster in rocker arm
Oil pump drive	Nose of crankshaft	Nose of crankshaft
Fuel system	SMPI	SMPI
Engine assembly weight*	174 kg (383.6 lb) ATX	174 kg (383.6 lb) ATX

* For content, see the Abbreviations and Definitions section near the back of this book.

an engine would be aimed at offering excellent performance, outstanding fuel economy, and low emissions. The result was the design, development, and release for 1998 of an all-new V-6 engine having a displacement of 2736 cc (2.7 L) using a bore and stroke of 86 × 78.5 mm, with both dimensions smaller than the 87.5 × 83 mm bore and stroke of the 2.0-L four-cylinder engine in production since 1995. Basic cylinder block dimensions—cylinder spacing and deck height—were set at 104 and 210 mm, respectively, considerably less than the 113.28-mm cylinder spacing and 229-mm deck height of the 3.2-L and 3.5-L V-6 engine family. With these differences in basic dimensions, especially the different cylinder spacing, the 2.7-L engine is considered to be in an engine family of its own, not related to the 3.2-L and 3.5-L engine family.

Aluminum was specified as the material for the deep-skirt cylinder block, with piston wear surfaces provided by cast-in-place iron bore liners. Six attaching bolts—four vertical and two horizontal tie bolts—were used for each of the four main bearing caps. The two outer vertical bolts of each main bearing cap had a secondary function of attaching an aluminum structural windage tray that also stabilized all four caps for noise reduction. The front of the cylinder block was extended to provide a cavity for a double-row roller timing chain that was used instead of the prior flexible timing belt. The cylinder heads likewise were extended to provide an oil-tight cavity for the timing chain, as can be seen on the cutaway view of the engine shown on Fig. 13.20.

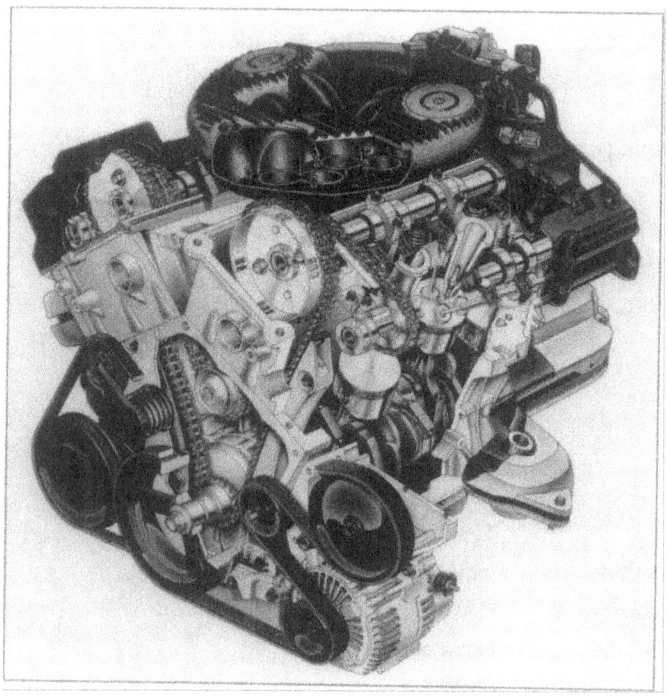

Fig. 13.20 *A cutaway illustration of the 1998 149-kW (200-hp) 2.7-L double overhead camshaft 24-valve all-aluminum V-6 engine. Timing chains were used instead of flexible cog-toothed belts as the means for driving the camshafts.*

Each of the aluminum cylinder heads was equipped with an inboard camshaft for operating its six intake valves and an outboard camshaft for operating its six exhaust valves. The fabricated camshafts, made of steel tubing and hardened steel lobes, provided a weight savings versus conventional solid cast camshafts. The timing chain from the crankshaft operated the two intake camshafts. A short single-row roller chain connected each intake camshaft with its nearby exhaust camshaft. All three chains had slack-side hydraulic tension adjusters. The row of exhaust valves on each cylinder head was positioned 32 deg from the intake valves. With the 33.8-mm-diameter intake valve heads and the 27.8-mm-diameter exhaust valve heads occupying almost all of the combustion chamber area, the chamber inevitably assumed a pent-roof configuration with the spark plug in the center. The stamped-steel valve rocker arms were similar to those used on the 2.0-L and 2.4-L double overhead camshaft four-cylinder engines with their roller followers for contact with the camshaft lobes. The end of the rocker arm opposite the valve end rested on a stationary hydraulic adjuster that provided valve lash adjustment.

Inverted teeth in the back of the main double-row timing chain drove the water pump, which was located in the vee of the cylinder block. In common with other front-wheel-drive engines, the oil pump was driven directly by the nose of the crankshaft.

The four main journals of the forged steel crankshaft had a diameter of 64 mm, while the diameter of the six crankpins, spaced 60 deg apart for equal 120-deg firing, was 54 mm. The connecting rods, made of powdered-metal forgings with cracked caps, had a center distance of 139.8 mm.

The smooth interior surfaces of the glass-fiber-reinforced nylon intake manifold system lowered the resistance to airflow. The manifold was designed in somewhat of a toroidal shape, providing two interconnected plenum chambers and individual runners for the intake valves of each combustion chamber. All were fed inlet air through a single throttle body.

Coil-on-plug ignition was provided for the spark plugs, eliminating the previously used spark plug cables. Engine control, including the control of the sequential multipoint fuel injection, was accomplished by the powertrain control computer.

With a compression ratio of 9.7:1, the output of the 1998 2.7-L double overhead camshaft 24-valve V-6 engine was 149 kW (200 hp) and 258 N-m (190 lb-ft) torque. Its specific power of 74 hp/liter was the highest of any naturally aspirated vee-type production engine built to date by Chrysler. Top honors for specific power were still retained by the 2.0-L double overhead camshaft four-cylinder engine with its 75 hp/liter.

Starting production in time for the 1998 model year, this world-class engine was installed in the Chrysler Concorde pictured on Fig. 13.21 (replacing the 3.5-L single overhead camshaft 24-valve iron block V-6) and in the Dodge Intrepid (replacing

Fig. 13.21 *The 1998 Chrysler Concorde LX four-door sedan. Under the sloping hood was a new 149-kW (200-hp) 2.7-L double overhead camshaft 24-valve all-aluminum V-6 engine.*

both the 3.3-L overhead valve V-6 and the 3.5-L single overhead camshaft iron block V-6 engines).

Table 13.5 shows the specifications of the 2.7-L double overhead camshaft all-aluminum 24-valve V-6 engine.

Chart 13.2 displays time lines of all the overhead camshaft V-6 engines described in this chapter.

TABLE 13.5
SPECIFICATIONS FOR THE 2.7-L DOHC 24-VALVE V-6 ENGINE
Note: All dimensions are in metric units (millimeters) rather than customary units (inches).

Model Year	1998
Car line	Dodge, Chrysler
Model	Dodge: Intrepid Chrysler: Concorde
Engine code	EER
Displacement (liters)	2.7
Bore	86
Stroke	78.5
Compression ratio	9.7:1
Power @ rpm	149 kW (200 hp) @ 5800
Torque @ rpm	258 N-m (190 lb-ft) @ 4850
Cyl. block mat'l.—skirt type	Aluminum—deep Main brg. cap windage tray
Cyl. block deck height	210
Cyl. block length	441
Cyl. bore spacing	104
No. of main brgs.—ck/s mat'l.	4—forged steel
Ck/shaft main journal dia.	64
Crankpin journal dia.	54
Connecting rod centers	139.8
Valve system	DOHC—24-valve
Valve events	252-244-29
Camshaft drive	Chain
Valve head dia.	Intake 33.8; exhaust 27.8
Valve lash adjustment	Hyd. rocker arm pivot
Oil pump drive	Nose of crankshaft
Fuel system	SMPI
Intake manifold	Tri-plenum, active
Engine assembly weight*	152.9 kg (337.1 lb)

* For content, see the Abbreviations and Definitions section near the back of this book.

Chart 13.2
Time Lines—OHV V-6 Engines

Engines, Car Lines, and Displacements (Liters)

Engine	Car Line	1987	1988	1989	1990	1991	1992	1993	1994
3.0-L MMC SOHC 8-valve	Plymouth Model	3.0 Voyager	3.0 Voyager	3.0 Acclaim, Voyager	3.0 Acclaim, Voyager	3.0 Acclaim, Voyager	3.0 Acclaim, Voyager	3.0 Sundance, Acclm, Voyager	3.0 Sundance, Acclm, Voyager
	Dodge Model	3.0 Caravan	3.0 Dynasty, Caravan	3.0 Spirit, Dynasty Caravan	3.0 Daytona, Spirit Dynasty, Caravan	3.0 Daytona, Spirit Dynasty, Caravan	3.0 Daytona, Spirit Dynasty, Caravan	3.0 Shadow, Day, Sprt Dynasty, Caravan	3.0 Shadow, Spirit Caravan
	Chrysler Model		3.0 New Yorker	3.0 New Yorker	3.0 LeB, TC by Mas	3.0 LeBaron	3.0 LeBaron	3.0 LeBaron	3.0 LeBaron
3.2/3.5-L SOHC 24-valve	Dodge Model							3.5 Intrepid	3.5 Intrepid
	Chrysler Model							3.5 Concorde	3.5 Cncrde, NY, LHS
	Eagle Model							3.5 Vision	3.5 Vision

Chart 13.2
(Continued)

Engines, Car Lines, and Displacements (Liters)

Engine	Car Line	1995	1996	1997	1998
3.0-L MMC SOHC 8-valve	Plymouth Model	3.0 Acclaim, Voyager	3.0 Voyager	3.0 Voyager	3.0 Voyager
	Dodge Model	3.0 Spirit, Caravan	3.0 Caravan	3.0 Caravan	3.0 Caravan
	Chrysler Model	3.0 LeBaron			
3.2/3.5-L SOHC 24-valve	Plymouth Model			3.5 Prowler	3.5 Prowler
	Dodge Model	3.5 Intrepid	3.5 Intrepid	3.5 Intrepid	3.2 Intrepid
	Chrysler Model	3.5 Concorde, NY, LHS	3.5 Concorde, NY, LHS	3.5 Concorde, LHS	3.2 Concorde
	Eagle Model	3.5 Vision	3.5	3.5	
2.5-L MMC SOHC 24-valve	Dodge Model	2.5 Stratus	2.5 Stratus	2.5 Stratus	2.5 Stratus
	Chrysler Model	2.5 Cirrus, Sebring	2.5 Cirrus, Sebring	2.5 Cirrus, Sebring	2.5 Cirrus, Sebring
2.7-L DOHC 24-valve	Dodge Model				2.7 Intrepid
	Chrysler Model				2.7 Concorde

CHAPTER 14

A Big, Powerful, All-Aluminum V-10—The Viper Engine

The North American International Auto Show held in January 1989 at Cobo Center in Detroit was the occasion, date, and place for the debut showing of a bright-red two-seat sport-roadster concept car named Viper. The car's emphasis on power was unmistakable—under the hood was an operating, experimental V-10 engine. Curved cooling slits at the rear of the engine compartment allowed a glimpse of tubular headers, and below the door sills, in race car fashion, were dual side-outlet exhaust pipes. The concept car made the auto show debut only eleven months after the idea of creating a super-sports car such as this emerged from the collective thinking of the dynamic Chrysler president Bob Lutz, internationally famous race car driver and performance expert Carroll Shelby, former technical director of the Renault racing program and now Vice President of Chrysler Vehicle Engineering Francois Castaing, and creative Chrysler Vice President of Design Tom Gale.[1]

When work on the concept car started, not only the appearance of the car but also the basic mechanicals had to be put in place so the concept car would be a solid base for the production car, in case the concept was so irresistible that it would be rewarded by the always-hoped-for nod-of-approval for production. At the time, development of a large-displacement V-10 truck engine, described later in Chapter 15, was underway. Also, a running experimental V-10 truck demonstrator engine, fabricated from 5.9-L V-8 cast iron cylinder blocks and heads, was available and was installed in the Viper concept car.

The enthusiastically favorable response to the Viper at the Detroit, Chicago, and New York auto shows in early 1989 was so great that a decision was made to go the next steps—the preparation of a production design and the building of prototype cars and engines for testing. To tackle this challenge, a special Viper team, including people from Engineering, Manufacturing, and Purchasing, was assembled and was assigned space within the former American Motors facilities on Plymouth Road in Detroit, now used by Chrysler Jeep/Truck Engineering. With capable executive engineer Roy Sjoberg in charge of the entire Team Viper, the Viper engine group headed by manager Jim Royer included Charlie Brown, design engineer; Dick Winkles, development engineer; me in the role of consulting engineer (still active two years after officially retiring); and designers Pete Kinsler and Bob Zeimis.

As mentioned, the V-10 truck engine in the concept car was built with a cylinder block and cylinder heads made of cast iron, a material much too heavy to be considered for the production Viper. Only aluminum would be acceptable to meet the extreme performance targets of the Viper. The engineering experience of Lamborghini, an Italian maker of ultra-high-performance sports cars, was available, and a set of V-10 truck engine drawings was sent to Lamborghini, with a request to design an aluminum conversion. The quick completion of the requested conversion was followed by Lamborghini furnishing a small number of machined parts and a number of additional castings.

After this initial brief participation by Lamborghini, the activity and entire responsibility of the Viper engine program was assumed by the Viper engine group. For the production design of the engine, computers running CATIA software were used, the Viper engine group being entirely devoid of traditional drafting tables, pencils, and drafting machines.

Identical to the truck V-10, the ten cylinders of the engine were arranged in two five-cylinder banks with a 90-deg included angle between the banks. The bore and stroke were 101.6 × 98.6 mm (4.00 × 3.88 in.), giving a displacement of 7994 cc (8.0 L, 488 cu. in.). Details of the engine design can be seen on the cross section shown on Fig. 14.1 and on the longitudinal section shown on Fig. 14.2.

The aluminum cylinder block—with a 3.50-in.-deep skirt, a cylinder bore spacing of 4.55 in., a machined length of 28.00 in., and a deck height of 9.48 in.—was fitted with replaceable bore

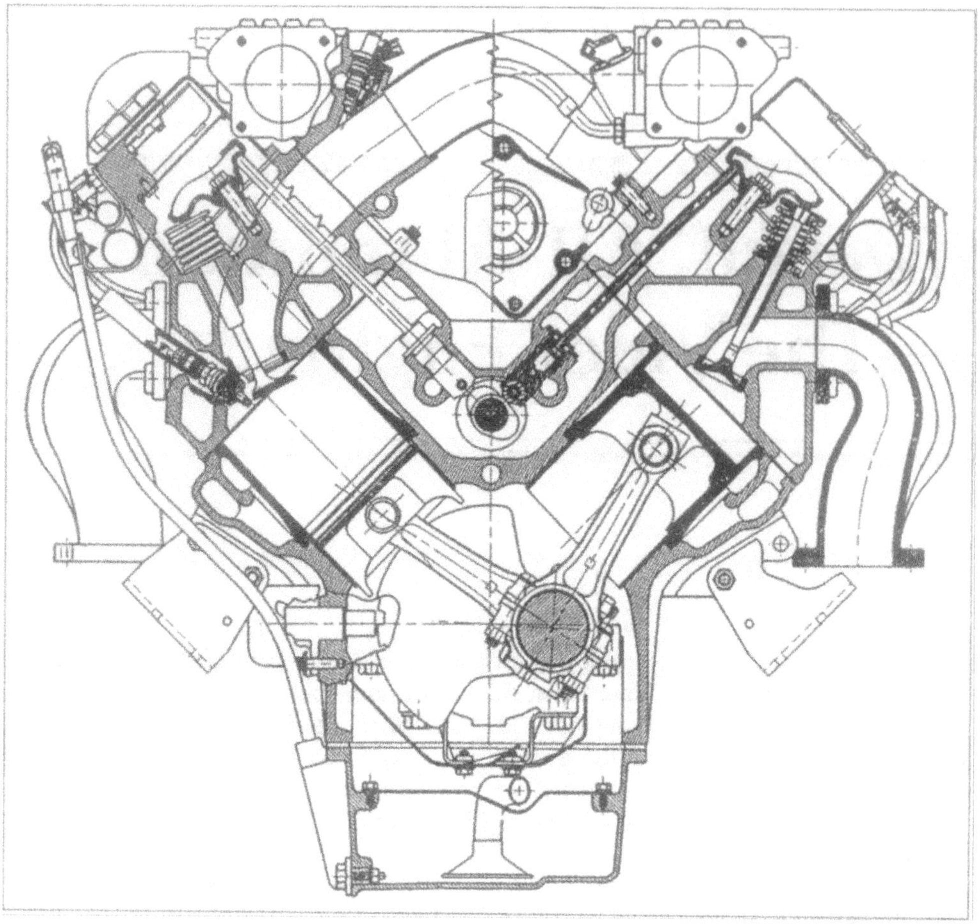

Fig. 14.1 *A cross section of the 1992 Viper 8.0-L overhead valve V-10 engine, with its classic overhead valve arrangement of block-mounted camshaft, tappets, pushrods, and rocker arms. The cast iron bore liners were seated in the cylinder block just below the water jacket.*

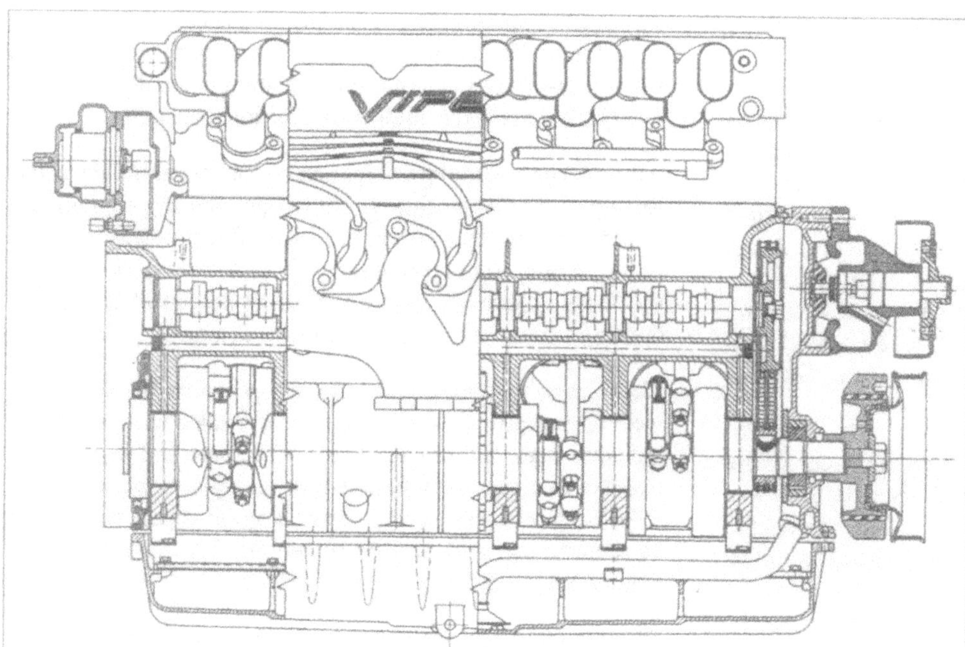

Fig. 14.2 *A longitudinal section of the 1992 Viper 8.0-L overhead valve V-10 engine. The cylinder block of the engine was made of aluminum; the six main bearing caps were made of nodular cast iron.*

liners made of centrifugally cast iron. Dual O-rings at the bottom of each liner provided sealing between the coolant jacket and the crankcase cavity. A longitudinal cast passage outboard of each cylinder bank distributed coolant to the bore liners via a metering hole at each cylinder, ensuring adequate and equal cooling for every cylinder.

Although the six main bearing caps originally were designed as a single aluminum casting with the caps connected by integral panels that were intended to act as a windage tray, dynamometer testing showed problems with pneumatic pulsations in the crankcase bays. This was corrected by changing the design to six separate nodular iron bearing caps with a steel standoff bracket attached to each cap, and by providing a sheet-steel windage tray that bolted to the six standoff brackets.

With sand-cast aluminum used as the cylinder head material, alloy iron inserts were provided for valve stem guides and valve seats. The alloy iron valve guides were press fitted, and the alloy iron valve seat inserts were shrink-fitted into the head. The five intake and five exhaust valves of each cylinder head were in-line, the valves alternating the length of the head. Head diameters were 1.92 and 1.58 in. for the intake and exhaust valves, respectively. In the cross-section view, the valve centerlines were positioned at an 18-deg angle from the cylinder centerlines, a design legacy dating back to the first LA V-8 engine used on the 1964 Plymouth Valiant and Dodge Dart. The cylinder head bolts were arranged in a four-bolt rectangular pattern around each cylinder. All the 9/16-in. cylinder head attaching bolts were 6 in. long, not only for commonality but fundamentally to provide uniform head gasket loading between the inboard and outboard bolts, even though this meant extra-deep threads in the cylinder block for the outer bolts.

A 3/8-pitch double-row roller timing chain drove the camshaft, mounted in the vee of the cylinder block, which provided an aggressive intake valve opening duration, exhaust valve opening duration, and overlap duration of 292.5-288.5-64.5 deg with valve lifts of 0.525 and 0.495 in. for intake and exhaust valves, respectively. For high-speed valve bounce control, dual valve springs were used, giving valve open and closed loads of 300 and 105 lb, respectively. Each hydraulic roller tappet had a valve directing oil into the hollow pushrod for transport to the overhead valve rocker arms.

The forged steel 6.123-in.-center-distance connecting rods of the 5.9-L V-8 engine were found to be acceptable for use in the Viper engine, provided they were shot peened and Magnagloed—appropriate steps for ensuring resistance to high-speed stresses.

The forged aluminum pistons, with a height of only 1.38 in. from the piston pin bore to the top of the piston, were compact, accounting for the cylinder block of the Viper engine having a lower deck height than the related 5.9-L V-8 engine (9.48 versus 9.6 in.), despite the Viper engine having a longer stroke (3.88 versus 3.58 in.).

Forged steel was the material of choice for the five-throw crankshaft, with its robust 3.00-in.-diameter main journals and 2.125-in.-diameter crankpins. The crankpin diameter also showed the small-block V-8 heritage of the V-10 because its 2.125-in. dimension had been used on every A and LA V-8 engine built from the time the Mound Road engine plant assembled its first production engine for 1956 model cars. Ten notches machined into an integral circular disk at the center of the crankshaft were used by a block-mounted sensor to furnish information to the distributorless ignition system ignition control and engine control computers. Faced with twenty-four possible arrangements of the five throws of a V-10 crankshaft, the Viper engine followed the lead of the Dodge Truck V-10 engine by selecting the one arrangement that had complete primary balance and only a small unbalanced secondary couple in the horizontal plane. Seventy-two degrees separated the five crankpins, which, with the 90-deg bank angle, resulted in alternating unequal firing intervals of 90 and 54 deg. There were concerns that with the lightweight aluminum Viper engine, these variations from perfect balance and equal firing would become noticeable. However, early in-car evaluations showed the variations to be virtually indiscernible.

The high-capacity engine oil pump, integrated into the chain case cover and driven directly by the nose of the crankshaft, was fitted with a thermostatically controlled valve that routed engine oil to an air-to-oil cooler mounted ahead of the radiator for prevention of excessive engine oil temperatures during high-speed operation.

The low-profile dual-plenum cross-ram aluminum intake manifold was made up of two base castings, ten thin-walled formed ram tubes, and two cast plenums all brazed together. Each of the base castings had a longitudinal gallery that delivered fuel to the five bottom-feed fuel injectors mounted near the cylinder head joint face. Dual single-barrel, 60-mm throttle bodies were bolted to the front ends of the plenums. At the rear of the manifold, a balance tube connected the two plenums.

The original design intent for exhaust headers for the Viper V-10 was to use fabricated stainless-steel tubing. However, space constraints in the engine compartment resulted in tube formations with weld locations that were difficult to access; therefore, the design of a two-section thin-walled vacuum-cast alloy-steel manifold was prepared. The upper section of the manifold with five separate branches was used on both the left and right sides of the engine. The right and left lower sections, both mounting oxygen sensors, were attached to the upper sections with eight stainless steel bolts and directed the exhaust gas to the exhaust system through a 2-1/2-in. ID spherical outlet connection. These sections and many other details of the engine can be seen in the cutaway drawing, made by superb illustrator David Kimball, that is displayed on Fig. 14.3.

A public showing was made of the Viper almost a year before any production Vipers rolled off the assembly line. The legendary Carroll Shelby took the wheel of a prototype bright-red

Fig. 14.3 *A cutaway illustration of the 1992 Viper 8.0-L overhead valve V-10 engine. The multipiece, low-profile intake manifold and a cast steel two-piece smooth-flowing exhaust manifold are pictured.*

Fig. 14.4 *The 1992 Viper RT/10 roadster photographed at the Chrysler Technology Center in Auburn Hills, Michigan. Under the front-hinged, resin-transfer-mold composite hood was a 298-kW (400-hp) 8.0-L fuel-injected tuned-intake all-aluminum V-10 engine.*

Viper RT/10 and, with an Engineering-built prototype all-aluminum V-10 engine under the hood, led the field of race cars at the start of the May 1991 Indianapolis 500 Race.

On March 2, 1992, the first production Viper engine was shipped from the Mound Road engine plant in Detroit (the engine manufacturing location) to the new Mack assembly plant also in Detroit (the Viper vehicle assembly location), culminating the efforts of the Viper engine team and the Mound Road manufacturing team assigned to the project.[2] With a compression ratio of 9.1:1, the 1992 Viper V-10 engine was rated at 298 kW (400 hp) and 610 N-m (450 lb-ft) torque.

In its meeting on April 7, 1992, the Detroit Section of SAE was host to a presentation on the 1992 Viper RT/10 by Team Viper. Included in the group of speakers were Roy Sjoberg, who presented the car as a whole, and Jim Royer, who presented the design and development of the engine. The presentations, however, were not published.

Sales of the 1992 Viper RT/10 roadster, shown on Fig. 14.4, were a stunning success, giving great satisfaction to all who participated in the program.

Subsequent Changes

A minor change to the engine made in 1995 was a new one-piece cast aluminum intake manifold that replaced the multi-piece, brazed manifold previously described. The output of the engine was unchanged.

In the following year, 1996, the exhaust system of the RT/10 roadster was changed from side outlets at the door sills to outlets that exited at the rear of the car. Although longer, the system had lower backpressure, boosting the output of the engine to 310 kW (415 hp) and 662 N-m (488 lb-ft) torque.

In mid-1996, a second model was added to the Viper roster—a bold-striped GTS coupe shown on Fig. 14.5, powered by a substantially lighter but more powerful version of the Viper V-10 engine. Inside the cylinder block of this second-generation

Fig. 14.5 *The 1996 Viper GTS coupe. The V-10 engine of the closed car was lighter and more powerful than the original Viper V-10 engine.*

engine, the ten original cast iron thick-walled replaceable bore liners were replaced with ten thin-walled non-replaceable cast iron liners shrink-fitted in the block. The original nodular iron main bearing caps that had been attached with four vertical bolts were replaced with cross-bolted caps, still nodular iron, with two vertical bolts and two horizontal bolts clamping the cylinder block skirt walls to the bearing caps for added strength and rigidity. Aluminum volume was reduced throughout the cylinder head with the added benefit gained of a reduction of coolant volume, further lowering the weight of the car and obtaining a faster warmup time.

Bolstered with a longer-duration camshaft, higher valve lifts, and a higher compression ratio (9.6:1 versus 9.1:1), the 80-lb-lighter GTS engine was rated at 336 kW (450 hp) and 664 N-m (490 lb-ft) torque.[3] In 1997, both the Viper RT/10 roadster and the GTS coupe models shared this higher-power engine.

The accomplishment of Team Viper in bringing out the new GTS coupe with its high-performance engine was rewarded with a request from the Indianapolis 500 race organizers that a GTS be made the pace car for the May 1996 event. This time, Chrysler president Bob Lutz, shown with the pace car on Fig. 14.6, led the 33-car field into the start of the race.

Fig. 14.6 *Chrysler president and Viper co-creator Bob Lutz poses proudly with the 1996 Indianapolis 500 pace car, a 1996 Viper GTS coupe he drove on May 26 to give the race its flying start.*

To inform its members about the technical side of the Viper GTS car and engine, the Detroit Section of SAE hosted another Viper evening on November 6, 1996. There, Roy Sjoberg described the engineering of the overall GTS car, and Charlie Brown described the much changed and improved 1996 Viper V-10 engine.

For 1998, the Viper engine received new exhaust manifolds made of stainless-steel tubing—a project that had been set aside during initial development of the V-10 engine but now, with more effort, had become feasible. Power ratings did not change, but the 22 lb of weight reduction was a welcome improvement.

Table 14.1 shows the specifications of the Viper V-10 engine.

Chart 14.1 displays a time line for the Viper V-10 engine.

TABLE 14.1
SPECIFICATIONS FOR THE 8.0-L VIPER V-10 ENGINE

Model Year	1992	Mid-1996	1998
Car line	Dodge	Dodge	Dodge
Model	Viper RT/10	Viper GTS	Viper RT/10, GTS
Engine code	EWB		
Displacement (liters)	8.0	8.0	8.0
Bore	101.6 (4.00)	101.6 (4.00)	101.6 (4.00)
Stroke	98.5 (3.88)	98.5 (3.88)	98.5 (3.88)
Compression ratio	9.1:1	9.6:1	9.6:1
Power @ rpm	298 kW (400 hp) @ 4600	298 kW (450 hp) @ 5200	298 kW (450 hp) @ 5200
Torque @ rpm	610 N-m (450 lb-ft) @ 3600	610 N-m (490 lb-ft) @ 3700	610 N-m (490 lb-ft) @ 3700
Cyl. block mat'l.—skirt type	Aluminum—deep Slip-fit iron bore liners Nod. c.i. caps 4 vertical bolts	Aluminum—deep Shrink-fit iron bore liners Nod. c.i. caps 2 vertical and 2 cross bolts	Aluminum—deep Shrink-fit iron bore liners Nod. c.i. caps 2 vertical and 2 cross bolts
Cyl. block deck height	240.8 (9.48)	240.8 (9.48)	240.8 (9.48)
Cyl. block length	711.2 (28.0)	711.2 (28.0)	711.2 (28.0)
Cyl. bore spacing	115.6 (4.55)	115.6 (4.55)	115.6 (4.55)
No. of main brgs.—ck/s mat'l.	6—forged steel	6—forged steel	6—forged steel
Ck/shaft main journal dia.	76.2 (3.00)	76.2 (3.00)	76.2 (3.00)
Crankpin journal dia.	54 (2.125)	54 (2.125)	54 (2.125)
Connecting rod centers	155.5 (6.123)	155.5 (6.123)	155.5 (6.123)
Valve system	OHV—20-valve	OHV—20-valve	OHV—20-valve
Valve events	292.5-288.5-64.5	290-305-66	290-305-66
Camshaft drive	Chain	Chain	Chain
Valve head dia.	Intake 48.8 (1.92) Exhaust 40.1 (1.58)	Intake 48.8 (1.92) Exhaust 40.1 (1.58)	Intake 48.8 (1.92) Exhaust 40.1 (1.58)
Valve lash adjustment	Roller hyd. tappet	Roller hyd. tappet	Roller hyd. tappet
Oil pump drive	Nose of crankshaft	Nose of crankshaft	Nose of crankshaft
Ignition	Distributorless ignition system Firing intervals: 90 and 54 deg	Distributorless ignition system Firing intervals: 90 and 54 deg	Distributorless ignition system Firing intervals: 90 and 54 deg
Fuel system	SMPI	SMPI	SMPI

Chart 14.1
Time Line—Viper V-10 Engine

Engines, Car Lines, and Displacements (Liters)

Engine	Car Line	1992	1993	1994	1995	1996	1997	1998
		- - Model Years - -						
8.0-L OHV 20-valve V-10	Dodge Model	8.0 Viper	8.0 Viper	8.0 Viper	8.0 Viper	8.0 Viper	8.0 Viper	8.0 Viper

CHAPTER 15

Truck Engines—Hauling the Load

Early Trucks

Maxwell, the forerunner to Chrysler Corporation, produced some trucks powered with the four-cylinder Maxwell engine, using a Maxwell passenger car chassis, prior to the company being absorbed into Chrysler on June 25, 1925. The practice of converting passenger car chassis to truck use appears to have continued after the formation of Chrysler and the demise of the Maxwell name, with the vehicles produced, badged Chrysler, being sold only in the export market.[1] This changed in 1928 when Chrysler introduced a new truck line of its own named Fargo and in the process founded a new subsidiary called the Fargo Motor Corporation. Then, on July 31, 1928, with the sudden purchase of Dodge Brothers by Chrysler, Chrysler Corporation found itself the owner of a well-respected line of trucks badged Graham, engineered by Graham Brothers, by now a subsidiary of Dodge Brothers, which used Dodge engines and was marketed exclusively by Dodge.[2]

The Chrysler Fargo trucks, assembled in its Highland Park complex near the Plymouth and DeSoto assembly lines,[3] used the same engines that powered Plymouth, DeSoto, and Chrysler passenger cars, as will be described later. After two model years, 1929 and 1930, this Fargo line of light- and medium-duty trucks, an obvious duplication of the Dodge and Graham trucks, was cancelled. In 1931, a line of large buses was introduced, bearing the name Fargo and powered by an engine that will be described here later. The Fargo name later was applied to fleet sales, Chrysler Canada-built trucks, and export sales.

Types of Truck Engines

These early trucks used the same engines that were used in passenger cars, with few, if any, changes. During the following years as truck models expanded to heavier service, the basic passenger car engines often were modified with upgraded parts for better endurance and longer service life. In cases with even greater duty requirements or with unique truck installation conditions, truck-only engines were designed and developed.

The next section of this chapter describes truck engines that were derived from passenger cars; these engines were used as is or were equipped with upgraded parts for truck service. Following that section, the next section describes engines that were designed specifically for truck service.

Although certainly of interest but not included here are non-Chrysler-built engines, both gasoline and diesel, which were purchased for factory installation in Dodge trucks, including Caterpillar, Cummins, Detroit Diesel, Hercules, International, Mitsubishi, and Perkins.

Truck Engines Derived from Passenger Car Engines

Plymouth Four-Cylinder Engine

The initial 1929 Fargo trucks, as noted previously, were offered in two sizes: a 1/2-ton Packet and a 3/4-ton Clipper. Power for the Packet, as shown on Fig. 15.1, came from the same 170-cu.-in. four-cylinder engine used in the 1928 Plymouth Q. The 45-hp engine, described in Chapter 4 and still with thermosiphon cooling, was a descendant of the Maxwell 25 four-cylinder engine.

The 1929 Dodge trucks, including the 1/2-ton pickup shown on Fig. 15.2, received a newer and slightly larger displacement version of the same engine, whose primary application was the 1929 Plymouth U passenger car. A longer stroke gave it 175 cu. in., but peak power did not change. Thermosiphon cooling sufficed for the passenger car application of the engine, but for Dodge truck operation, a belt-driven water pump was added for positive coolant flow. This is an early example of a

Fig. 15.1 *A 1929 1/2-ton Fargo Packet. The Maxwell-derived 170-cu.-in. L-head four-cylinder engine used by the Chrysler I-52 and the Plymouth Q powered this businesslike panel delivery truck.*

Fig. 15.3 *The left side of the engine compartment of a 1929 Dodge 1/2-ton pickup truck with its 175-cu.-in. Plymouth U four-cylinder engine. Note the truck-only water pump at the lower left, with its outlet connected to the cylinder block by a rubber hose. Photo by W.L. Weertman.*

passenger car engine upgrade to meet truck needs. The left and right sides of the engine installed in a 1929 Dodge pickup truck are shown in Figs 15.3 and 15.4, respectively. Dodge truck use of four-cylinder engines, similar to passenger car use, ended in the 1932 calendar year when the new high-volume Plymouth and Dodge six-cylinder L-head engines started production at the beginning of the 1933 model year.

Fig. 15.2 *A 1929 Dodge 1/2-ton pickup truck. The Dodge 124 170-cu.-in. four-cylinder engine had been discontinued, and the Plymouth U 175-cu.-in. four-cylinder engine with a water pump added was installed as its powerplant. This vehicle is on display at the Walter P. Chrysler Museum. Photo by W.L. Weertman.*

Fig. 15.4 *The right side of the engine compartment of the 1929 Dodge 1/2-ton pickup truck. Crisp corners identify the engine as the new-for-1929 Plymouth U four-cylinder engine. Photo by W.L. Weertman.*

Small Zeder, Skelton, and Breer L-Head Seven-Main-Bearing Six-Cylinder Engine

Power for the 1929 Fargo 3/4-ton Clipper, as noted previously and as shown on Fig. 15.5, came from one engine, described in Chapter 3—the small Zeder, Skelton, and Breer-designed 196-cu.-in. seven-main-bearing six-cylinder

Fig. 15.5 *A 1929 Fargo 3/4-ton Clipper. Power came from the same 196-cu.-in. Zeder, Skelton, and Breer-designed small seven-main-bearing six-cylinder engine as was used in the 1929 Chrysler Model 65 passenger car.*

Fig. 15.6 *The 1930 Dodge three-ton truck in tractor-trailer service. Its powerplant was the big Zeder, Skelton, and Breer-designed 310-cu.-in. Imperial six-cylinder engine with seven main bearings.*

L-head 65-hp Silver Dome engine, whose primary use was in the 1929 Chrysler Model 65 passenger car. Both the Clipper and the engine were gone at the end of 1930; Dodge trucks had taken over the market slot, and the engine was no longer in production.

Large Zeder, Skelton, and Breer L-Head Seven-Main-Bearing Six-Cylinder Imperial Engine

Starting in 1926, the top-of-the-line Chrysler Imperial passenger car was powered with the largest of the three seven-main-bearing Zeder, Skelton, and Breer six-cylinder engines described in Chapter 3. Initially displacing 289 cu. in., this engine was being built by 1928 with a displacement of 310 cu. in. Its robust structure, together with its output of 100 hp @ 3000 rpm and 203 lb-ft torque @ 1000 rpm, made it exactly the kind of engine Dodge wanted under the hoods of its heaviest two- and three-ton trucks such as the three-ton tractor-trailer rig shown on Fig. 15.6. One durability upgrade to the engine from the passenger car version was the use of helical-tooth timing gears in place of a timing chain. Another upgrade, incorporated in 1932 after exhaust valve seat sinkage problems occurred while in tractor-trailer service hauling iron castings, was the addition of shrink-fitted alloy steel exhaust valve seat inserts in the cylinder block—a change that spread quickly to all Chrysler-built L-head engines. The engine, often called the Big-Six, did so well that it continued to be built for trucks even after 1931 when the Imperial dropped the engine in favor of in-line eight-cylinder engine power. Finally, at the end of 1936, the engine was discontinued, essentially replaced by an even bigger Big-Six engine that is described in the "Truck-Only Engines" section presented later in this chapter.

Dodge Brothers Four-Cylinder Engine

Although the Dodge Brothers Corporation had quit building four-cylinder engines at the time it was purchased by Chrysler in 1928, its long and rich history of four-cylinder-powered trucks deserves mention here because these trucks formed the foundation of its unbroken legacy of excellence in truck engineering, manufacturing, and sales. The original Dodge passenger car four-cylinder 35-hp engine described in Chapter 5 provided power for the early Dodge trucks, such as the 1919 pickup truck shown on Fig. 15.7, with few changes to the engine through 1926. From the beginning, it had a water pump that was lacking on Maxwell and Model T Ford engines of the same era, which certainly helped give it the reputation of being a "Dependable Dodge." The 1928 Dodge pickup truck shown on Fig. 15.8 would have been powered with the totally redesigned 40-hp 212-cu.-in. Dodge 124 four-cylinder engine that had a vertical oil pump and distributor drive. As noted in Chapter 3, this engine lasted only one year, falling victim to the Dodge Brothers plan to become a six-cylinder-only car company in 1929.

Dodge Senior Six-Cylinder Engine

In the early 1920s, Dodge Brothers furnished chassis-cab units to a number of truck conversion builders. With one builder, Graham Brothers, a special bond developed, with Dodge becoming the sole supplier of chassis-cabs to Graham Brothers for conversion and Dodge dealers becoming the exclusive sellers of the finished trucks, as noted at the beginning of this chapter. With the switch by Dodge to six-cylinder engines

Fig. 15.7 *A 1919 Dodge Brothers pickup truck. Power came from the same 35-hp 212-cu.-in. L-head four-cylinder engine used in the Dodge Brothers passenger cars.*

Fig. 15.9 *A 1928 Graham two-ton truck with trailer. The largest engine available, the Dodge Senior 224-cu.-in. six-cylinder, probably would have been a good investment for the owner.*

Fig. 15.8 *A 1928 Dodge delivery truck with roll-down side curtains. The 40-hp 212-cu.-in. Dodge 124 four-cylinder engine provided its motive power.*

Fig. 15.10 *A 1929 Dodge BE 1/2-ton pickup. With the optional 208-cu.-in. Dodge Victory six-cylinder engine under the hood, this nifty pickup would have been quite a hot performer.*

starting with the Dodge Senior model in 1928, described in Chapter 5, Graham trucks immediately benefited from the extra power. Fig. 15.9 shows a 1928 Graham two-ton truck powered by a Dodge Senior 224-cu.-in. 60-hp six-cylinder engine. The engine was discontinued at the end of 1930.

Dodge Victory Six-Cylinder Engine

The smaller and lighter Dodge Victory 208-cu.-in. six-cylinder engine, described in Chapter 5, was placed in truck service as soon as it appeared in 1929, powering trucks ranging from the Dodge BE 1/2-ton pickup shown on Fig. 15.10 to the Graham DE 1-1/2-ton truck. This engine also was discontinued at the end of 1930 when production of all Dodge-engineered six-cylinder engines ended.

Chrysler Jefferson-Built Four-Main-Bearing Deep-Skirt Six-Cylinder Engine

As described in Chapter 6, the first assignment of this four-main-bearing deep-skirt six-cylinder L-head engine was to move the newly created 1929 DeSoto at a brisk rate with its 55-hp 175-cu.-in. displacement. Starting in 1930, the displacement of the engine was increased to 190 cu. in., a size used by both Dodge and DeSoto passenger cars. However, 1930 Fargo one-ton trucks, including the Freighter shown on Fig. 5.11, evidently used the 175-cu.-in. displacement engine with its 55-hp and 110 lb-ft of torque.[4] But Dodge Truck must have liked the larger 190-cu.-in. version because that was the engine specified for the Dodge 1930 two-ton truck, similar to the dual-rear-wheeled stake truck shown on Fig. 15.12, where its 60 hp and 120 lb-ft torque, described in sales literature as providing "flashing performance and maximum fuel economy," could be put to good commercial use.[5] With its deep-skirt block, it was a sturdy engine and continued in truck service through 1936, with displacement topping off at 241.5 cu. in.

Fig. 15.11 *A 1930 Fargo one-ton Freighter. Power under the hood was the Chrysler-built 175-cu.-in. deep-skirt four-main-bearing six-cylinder engine first used on the 1929 DeSoto. The payload on the trailer appears to be four 1929 Pontiacs.*

Fig. 15.12 *A 1930 Dodge Brothers two-ton stake truck. A Chrysler-built 60-hp 190-cu.-in. deep-skirt four-main-bearing six-cylinder engine provided pulling power.*

Chrysler Imperial Nine-Main-Bearing Straight-Eight Engine

When the huge 385-cu.-in. 125-hp 135 lb-ft torque Chrysler Imperial nine-main-bearing straight-eight L-head engine described in Chapter 7 appeared in 1931, it was quickly adapted for a line of large buses named Fargo, built by (or possibly for) the Chrysler Fargo Motor Corporation. In 1932, its use in the buses, such as the inter-city Fargo bus shown on Fig. 15.13, continued, with three- and four-ton G80-series Dodge trucks added to its pulling chores. The year 1936 was the last year of use of this big eight-cylinder engine in Chrysler Imperials. Although the G80 truck is listed through 1937 in annual statistical issues of *Automotive Industries*, serial number records indicate that 1934 was the last year in which any were actually built.

Fig. 15.13 *A 1932 Fargo inter-city motorcoach. Chrysler's largest engine, the 385-cu.-in. Imperial nine-main-bearing straight-eight, provided power for this hefty highway cruiser.*

Plymouth and Dodge Four-Main-Bearing Shallow-Skirt Six-Cylinder Engine

Both Plymouth and Dodge Divisions tooled up the same lightweight six-cylinder L-head engine, described in Chapter 6, in time for 1933 volume production. The Dodge version had a displacement of 201 cu. in. The Plymouth version, with a shorter stroke, had a displacement of 190 cu. in. The Dodge Truck model offering for 1933 included both carryover trucks and new 1/2-ton, 3/4-ton, and one-ton models powered by this four-main-bearing shallow-skirt L-head six-cylinder engine. The 1/2-ton trucks received the 190-cu.-in. engine, and the heavier trucks received the 201-cu.-in. version. Figure 15.14 shows a 1935 medium-duty Dodge truck using the 201-cu.-in.

Fig. 15.14 *A 1935 Dodge medium-duty truck powered by a Dodge-built 201-cu.-in. four-main-bearing L-head six-cylinder engine.*

engine. Built by the hundreds of thousands every year, this compact, easy-to-service, low-cost engine reigned supreme as the base light- and medium-duty Dodge truck engine for 28 years until the close of 1960, when its production and use in light-duty Dodge trucks ceased. The largest 230-cu.-in. displacement size of the engine had been started 19 years earlier, with the 1942 one-ton WD20 and WD21 trucks.[6]

Chrysler Four-Main-Bearing Shallow-Skirt Six-Cylinder L-Head Engine

When the Chrysler four-main-bearing shallow-skirt six-cylinder L-head engine was launched for 1937 Chrysler and DeSoto passenger cars, installations in 1937 Dodge trucks also were started. As described in Chapter 6, the structure of the engine was larger than either the Plymouth and Dodge four-main-bearing six-cylinder L-head engine or the older deep-skirt-block four-main-bearing Chrysler six-cylinder L-head engine that it replaced. This new Chrysler engine proved to be an excellent medium-duty truck engine, with most applications in the 1-1/2- to 2-1/2-ton sizes.

When the engine was first released in 1937, the Chrysler Jefferson Avenue engine machining facility evidently was willing to manufacture a special destroked 218.06-cu.-in. version specifically for Dodge light-duty trucks. Dodge Truck seized the opportunity to drop the small-block Plymouth and Dodge engine and power all its trucks—from the 1/2-ton through the 1-1/2-ton—with this new engine in two displacements of 218.06 and 228 cu. in. That worked for two years; however, in 1939, the smaller, lighter, and less expensive Plymouth and Dodge engine was reinstated to power the 1/2-ton through one-ton light-duty trucks.

During its incredible 32-year lifetime as a truck engine, this Chrysler Jefferson-built four-main-bearing shallow-skirt L-head six-cylinder engine was produced in six displacements: 218.06, 228.1, 241.5, 236.7, 250.6, and 264.5 cu. in. Figure 15.15 shows a 1937 1-1/2-ton cab-over-engine (COE) haulaway truck. This early cab-over-engine model would have had a large "doghouse" next to the driver to accommodate the engine. Later from 1941 through 1955, a 1920s-style updraft carburetor version of the engine was built for use in cab-over-engine trucks, with the low profile of the engine permitting the cab to have a flat floor. The output of the updraft carburetor engine was about 5 hp less than that of the standard downdraft version.

Fig. 15.15 *A 1937 Dodge cab-over-engine truck powered by the new Chrysler-built 228-cu.-in. four-main-bearing shallow-skirt 25-in.-block six-cylinder engine. The car on top of the trailer is a 1937 Dodge four-door sedan.*

Truck versions of this engine were equipped with premium features, depending on the type of duty and the application. The 265.4-cu.-in. engines, introduced in 1953, were fitted with sodium-filled stellite-faced exhaust valves and crankshafts with hardened journals. From 1953 through 1959, an optional version of this 265.4-cu.-in. truck engine, shown on Fig. 15.16, was available with twin carburetors and twin exhaust systems. With twin carburetors, the 1953 265.4-cu.-in. engine had a gross output of 135 hp @ 3600 rpm and 226 lb-ft @ 1600 rpm.

Although passenger car use of this durable L-head six-cylinder engine ceased at the end of 1954 when six-cylinder Chrysler models were no longer offered, truck use continued for another 14 years until the end of 1968, with the four-wheel-drive Dodge Power Wagon being the last vehicle using its power.

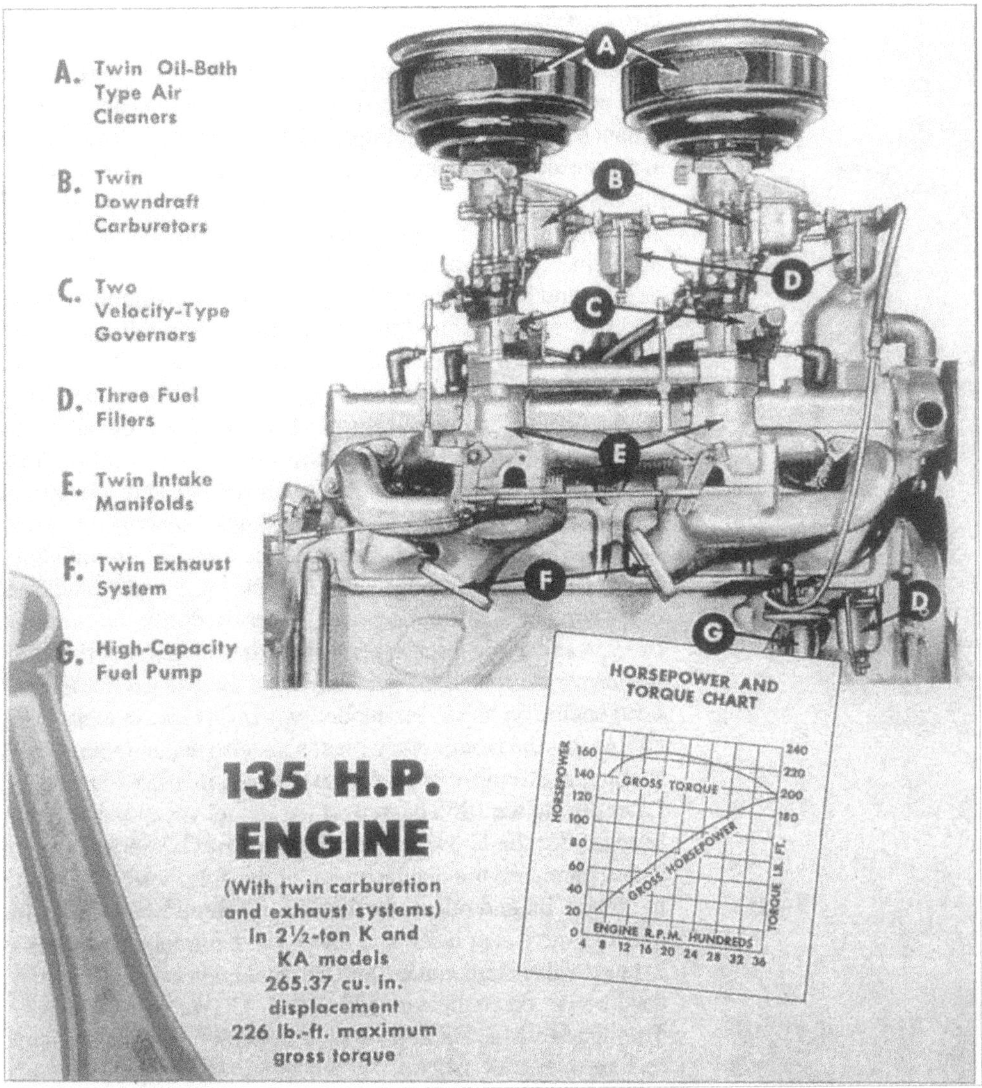

Fig. 15.16 *Sales information for the 1953 twin-carbureted truck version of the Chrysler 265-cu.-in. four-main-bearing shallow-skirt 25-in.-block L-head six-cylinder engine.*

Chrysler Double-Rocker-Shaft Hemi® Head V-8 Engine

When the Chrysler double-rocker-shaft, FirePower Hemi® head V-8 engine, described in Chapter 8, was being prepared for 1951 passenger car production, a separate development program was started for heavy-duty truck use of the engine. After thousands of hours of dynamometer testing and hundreds of thousands of miles of truck testing, the engine was released for the 1954 C-1 series of heavy-duty trucks. During its lifetime of 1954 to 1959, two sizes of the Chrysler Hemi® V-8 truck engine were built: 331 cu. in. from 1954 through 1957, and 354 cu. in. from 1956 through 1959.

Depending on the model year and the tonnage rating, premium features included sodium-filled stellite-faced exhaust valves, exhaust valve rotators, alloy steel intake valves, alloy steel seat inserts for both the intake and exhaust valves, heavy-duty bearings, and gear drive for the camshaft.

With its compact size, high power, good fuel economy, and excellent durability, the engine did an outstanding job as a truck engine. Twin carburetion, using two one-barrel carburetors with governors, was offered on the 331-cu.-in. engine in 1956 and 1957 and on the 354-cu.-in. engine, pictured on Fig. 15.17, from 1956 through 1959. A 354-cu.-in. twin-carburetor Hemi® V-8, rated at 220 hp and 354 lb-ft torque, was put to good use in the 1956 YX model tandem-axle cement truck shown on Fig. 15.18. The 1958 354-cu.-in. engine with twin carburetors was rated at 234 hp and 360 lb-ft torque.

Fig. 15.17 *A 1958 354-cu.-in. Chrysler Power Giant Hemi® V-8 truck engine with twin one-barrel carburetors.*

Fig. 15.18 *A 354-cu.-in. Chrysler Hemi® twin-carburetor V-8 engine provided power for this hard-working 1956 Dodge four-ton YX model tandem-axle truck. Evidently, this truck was early-built for advertising photos. The "Job-Rated" emblem on the side of the fender would be replaced with a forward look and a "Y" emblem by the start of 1956 production.*

Installation of this tough and powerful Chrysler Hemi® engine in Dodge trucks ceased at the end of 1959. It was replaced the following year by a truck-only heavy-duty version of the new B and RB V-8 wedge-head engine built at the Trenton engine plant, which is described later in this chapter.

Dodge-Built V-8 Truck Engines

When the Dodge Main plant started manufacturing Hemi® head and poly-head V-8 engines for passenger cars, as described in Chapter 8, Dodge Truck Engineering was quick to tap into this new source of engines. As related, Dodge Division released its first V-8 engine, a 241-cu.-in. Hemi® head, for its 1953 passenger cars. Truck installations started the following year, 1954, in 1-1/2-, two-, and 2-1/2-ton models. With an output of 133 hp and 220 lb-ft torque, all the V-8s had hydraulic tappets and two-barrel carburetors. For upgraded durability, engines powering the 2-1/2 ton models received sodium-filled exhaust valves, exhaust valve rotators, and trimetal main bearings. Dodge Truck Engineering used its own chronological series designators to coordinate production and service records rather than the passenger-car-based annual model year numerical designators. The truck C-1 series, approximately the same as the 1954 passenger car model year, was the introductory series for the first Dodge-built 241-cu.-in. Hemi® head V-8 truck engine. Later during the C-1 series, during the summer of 1954 and a few months prior to the introduction of the 1955 passenger cars, Dodge Truck replaced the 241-cu.-in. Hemi® head engine on its 1/2-ton models with a 241-cu.-in. poly-head engine that the Dodge Main plant soon also was shipping to the Plymouth assembly plant for installation in 1955 Plymouths. Later during the 1955 passenger car model year, Dodge Truck brought out the C-3 truck series. For this C-3 series, Dodge Truck increased the displacement of the poly-head V-8 engine to 259 cu. in. and released a 270-cu.-in. Hemi® head V-8 with exhaust valve seat inserts, exhaust valve rotators, replaceable exhaust valve stem guides, and trimetal main bearings as standard power on its heavier (35,000-lb. GCW) two-ton trucks. This Dodge Main-built 270-cu.-in. Hemi® head V-8 engine had an output of 142 hp. When the 1956 Dodge passenger cars moved the displacement of the Dodge-built poly-head V-8 engine up to 270 cu. in., Dodge Truck modified its C-3 series engine lineup by dropping the Dodge-built Hemi® V-8 engine and installing the newly available 270-cu.-in. poly-head 172-hp V-8 engine as standard in heavier medium-duty service, including cab-over-engine trucks.

The small 259-cu.-in. Dodge poly-head V-8 engine continued in 1956 light-duty trucks as an option to the standard-equipment 23-in.-block L-head six-cylinder engine. Figure 15.19 shows a 1/2-ton pickup truck equipped with the 259-cu.-in. poly-head V-8 engine.

For its last two years, 1957 and 1958, the Dodge poly-head V-8 engine was built at a displacement of 315 cu. in. In 1958, the Dodge Hemi® head version, at 315-cu.-in. displacement, was reinstated as standard on some medium-duty trucks and optional on others, an assignment that lasted through 1959 when use of the Dodge-built V-8 engines ended. The exit 1959 Dodge Hemi® V-8 engine had a rating of 210 hp and 300 lb-ft torque.

Fig. 15.19 *A 1956 Dodge 1/2-ton pickup truck powered by a Dodge 259-cu.-in. poly-head V-8 engine.*

A-Engine V-8

After its launch as a passenger car engine in 1956, the A-engine V-8, built at both the Mound Road and Windsor engine plants and described in Chapter 8, became a powerplant for Dodge light- and medium-duty trucks, starting with the 1958 Canadian-built and 1959 U.S.-built trucks, replacing the Dodge V-8 engine previously described.

The A-engine V-8 was built in two versions—a light-duty version that was the same as the passenger car engine, and a medium-duty version (shown on Fig. 15.20) that had a number of premium features, including valves made of high-alloy steels, exhaust valve rotators, a roller timing chain, and hydraulic tappets in place of mechanical tappets. Built from 1959 through 1966, the Mound Road engine plant V-8 used in trucks had a single-displacement of 318 cu. in. Figure 15.21 shows a 1959 Dodge pickup truck with the engine available as an option. The Windsor engine plant A-engine, with a slightly smaller bore of 3.875 in. versus 3.91 in. for the Mound Road engine plant engine, displaced 313 cu. in., a size that continued from 1958 through 1964. Its heavy-duty version of the engine powered Canadian-built trucks up to 3 tons.

Fig. 15.20 *A 1959 medium-duty truck version of the Mound Road engine plant-built 318-cu.-in. A-engine V-8.*

Fig. 15.21 *A 1959 Dodge 1/2-ton Sweptline pickup truck. A prospective buyer had a selection of two engines: a 230-cu.-in. Plymouth and Dodge L-head six-cylinder-engine, or a 318-cu.-in. poly-head A-engine V-8.*

LA V-8 Wedge-Head Engine

As described in Chapter 9, the LA V-8 engine, with its in-line valves, wedge combustion chamber, and lightweight castings, was initiated to provide the 1964 Plymouth Valiant and Dodge Dart cars with 273 cu. in. of V-8 power. The 1965 Dodge A-100, a mid-engined compact truck that had been launched in 1964 with Slant Six engines, was judged to be an excellent application of the same V-8 power. When, for 1967, the Mound Road and Windsor engine plants shifted 318-cu.-in. engine production to the LA design, Dodge Truck cancelled the 318-cu.-in. A-engine V-8 and replaced it with the 318-cu.-in. LA V-8 engine. This allowed the A-100 compact truck, such as the truck shown on Fig. 15.22, to boost its optional V-8 engine size from 273 to 318 cu. in. In 1972, Dodge Truck started using the larger 360-cu.-in. LA V-8 engine described in Chapter 9. Also a long-lived engine, it continued in production for Dodge trucks beyond the 1998 final year covered by this book.

Both the 318- and 360-cu.-in. LA engines (identified more recently by their metric sizes of 5.2-L and 5.9-L, respectively) were built at two levels of truck-duty service: a 318-1 and 360-1

Fig. 15.22 *A 1967 Dodge A-100 compact truck. The vehicle came standard with a 170-cu.-in. LG Slant Six engine; however, for more power, the truck could be ordered with a 225-cu.-in. Slant Six or a 318-cu.-in. LA V-8 engine.*

Fig. 15.23 *A 1993 5.9-L (360-cu.-in.) Magnum LA V-8 engine.*

level that had a few upgrades from passenger car engines such as chrome-faced top piston rings, and a 318-3 and 360-3 level that had a host of upgrades that included 18-mm spark plugs in place of the standard 14-mm spark plugs, exhaust valve rotators, upgraded valve steels, heavier-duty bearings, and, in the case of the 318-3 engine, a water-heated crossover in the intake manifold in place of an exhaust gas heated crossover.

In 1987, a 3.9-L 90-deg V-6 spin-off of the LA V-8 engine went into production at the Mound Road engine plant as power for the new Dakota truck. The V-6 engine is described later in this chapter. Lower-friction roller hydraulic tappets replaced the conventional tappets of the 318-cu.-in. engine in 1987. By 1989, carburetors in general were obsolete, and both the 5.2-L and 5.9-L engines were using throttle body fuel injection. Although the Dodge Dakota originally had been designed with barely enough space for four-cylinder and V-6 engines, the 5.2-L V-8 in 1991 somehow was squeezed into the engine compartment, becoming the big power option for the Dakota.

A dramatic increase in outputs of the 5.2-L and 5.9-L engines occurred when the engines were converted to the Magnum design in 1992 and 1993, respectively. By combining sequential multipoint fuel injection with a tuned plenum-ram intake manifold having 18.5-in. branch lengths, improved cylinder head ports and combustion chambers, lower restriction exhaust manifolds, a new progressive-bore throttle body, and pedestal-mounted rocker arms, the engine output increased from 170 to an astounding 230 hp for the 5.2-L V-8 engine, and from 190 to 230 hp for the 5.9-L V-8 engine, which is shown on Fig. 15.23. Torques moved upward from 260 to 280 lb-ft and from 292 to an impressive 325 lb-ft, respectively.

In addition to Dodge trucks, the 1993 5.2-L LA V-8 Magnum engine received an assignment to power the Jeep Grand Cherokee and Grand Wagoneer, replacing the aged AMC 5.9-L V-8 engine that is described in Chapter 19. A camshaft change in 1994 boosted the torque of the 5.2-L Magnum engine but at a slight penalty in power, resulting in a rating of 220 hp and 300 lb-ft torque. A 5.2-L LA V-8 Magnum engine provided the power for the 1998 Dodge Durango shown on Fig. 15.24.

In 1998, which is the ending date covered by this book and is more than three decades after the start of the engine in the 1965 A-100 compact truck, the LA V-8 engine was still pulling loads for Dodge trucks.

Fig. 15.24 *This 1998 Dodge Durango was powered by the optional 5.2-L LA V-8 Magnum engine.*

Slant Six Engines

The Slant Six engine, described in Chapter 11, went to work for 1961 R Series Dodge trucks a year after its 1960 debut in passenger cars. The 170-cu.-in. LG Slant Six was assigned as power for an extra-light-duty D-100 truck and for the forward-control stand-up-driver P-200 delivery van shown on Fig. 15.25. The larger 225-cu.-in. RG Slant Six was assigned to power light-duty trucks up to the 10,000-lb. GVW D-300, including the Dodge Dart D-100 shown on Fig. 15.26. When the 1964 compact 1/2-ton mid-engined A-100, shown on Fig. 15.27, originally was introduced, Slant Six 170- and 225-cu.-in. engines were the only engines available.

Fig. 15.27 *A 1964 Dodge A-100 compact pickup truck. A 225-cu.-in. RG Slant Six engine was available as an option over the standard 170-cu.-in. LG Slant Six engine.*

Fig. 15.25 *A 1961 P-200 Dodge forward-control delivery van. Only Slant Six engines were used—the 170-cu.-in. LG as standard and the 225-cu.-in. RG as optional.*

Initially, both engines were almost identical to the passenger car versions. In 1965, a heavier-duty version of the 225-cu.-in. RG Slant Six engine, designated 225-2, was released for truck use. Equipped with a number of premium features such as stellite-faced exhaust valves with positive rotators, a double-row roller timing chain, and a crankshaft with shot-peened fillets, applications of the engine moved up to include 20,000-lb (GVW) D-500 medium-duty trucks. The release of a new Dodge medium-duty tilt-cab L-600 truck in late 1966 put the engine into truck-tractor service with loads up to a potential 36,000-lb (GCW). The 225-2 heavy-duty engine with its premium features was discontinued at the end of 1973.

The stalwart Slant Six engine continued as the Dodge light-duty pickup truck base engine through 1987, even though the Trenton engine plant, which was the U.S. manufacturer of the engine, stopped building the engine at the end of 1983. The engine source for its final four years was the Toluca engine plant of Chrysler de Mexico.

Fig. 15.26 *A 1961 Dodge Dart D-100 pickup truck. The 140-hp 225-cu.-in. RG Slant Six was its standard engine.*

B and RB V-8 Engines

In the mid-1960s, a trend emerged for pickup trucks whose owners desired performance equal to that of passenger cars. An early example was a 1964 Dodge Custom Sports Special package, available on a D-100 or D-200 pickup truck, which boasted an optional 426-cu.-in. RB V-8 engine. This was the same engine that powered high-performance Plymouth and Dodge passenger cars. In addition, many trucks were being used as recreational vehicles rather than as commercial load haulers, which increased the need for engines that were bigger, had higher horsepower, and were more similar to those of passenger cars.

To meet this demand, the 383-cu.-in. LB V-8 passenger car engine, described in Chapter 9, was released for the 1967 D-100;

the D-200 (shown on Fig. 15.28) and the D-300 truck, with its output in the trucks being 258 hp and 375 lb-ft torque. In 1972, when the passenger car engine increased to 400 cu. in., the truck engine followed suit, and the net ratings of the engine became 200 hp and 320 lb-ft torque. Figure 15.29 shows a 1973 D-100 Dodge truck that offered the 400-cu.-in. LB V-8 engine as optional.

Fig. 15.28 *A 1967 Dodge D-200 3/4-ton pickup truck. For power above either its standard 225-cu.-in. 140-hp Slant Six engine or its 318-cu.-in. 210-hp LA V-8 first option, a 383-cu.-in. 258-hp LB V-8 passenger car engine was available.*

Fig. 15.29 *A 1973 Dodge D-100 Club Cab pickup truck. Four engines were offered: a 225-cu.-in. Slant Six, a 318-cu.-in. LA V-8, a 360-cu.-in. LA V-8, and a 400-cu.-in. LB V-8.*

The RB V-8 passenger car engine, now at 440 cu. in., reappeared in Dodge trucks in 1974 for the start of an eight-year run. In addition to a passenger-car-level version designated 440-1, a heavier-duty version had been developed specifically for powering large Class A motor home chassis, of which Dodge Truck was a prime supplier to the industry. Designated 440-3, this motor home version included in its premium features extra coolant holes straddling the spark plugs for reducing spark plug temperatures, Nimonic exhaust valves, exhaust valve rotators, upgraded intake valves, double-row roller timing chain, stainless steel cylinder head gaskets, heavy-duty connecting rod bearings, and molybdenum-filled top piston rings.

The year 1978 was the last year in which the 400-cu.-in. LB V-8 was used by Dodge Trucks, paralleling its termination by the passenger car divisions. The 440-cu.-in. RB V-8 engine also saw 1978 as its last year in passenger cars and Dodge Truck applications, including motor home chassis.

2.2-L Single Overhead Camshaft Front-Wheel-Drive Four-Cylinder Engine

In 1982 and 1983, Dodge Truck offered a mini-pickup truck version of the subcompact Dodge Omni Charger front-wheel-drive passenger car. Named Rampage, the truck, shown on Fig. 15.30, was powered by the identical transversely mounted 2.2-L single overhead camshaft 84-hp four-cylinder engine, described in Chapter 12 that was used in passenger cars.

Fig. 15.30 *A 1983 Dodge Rampage 2.2. A variation of the Dodge Charger passenger car, the mini-pickup Rampage offered sporty looks, front-wheel drive, and some load-hauling capability.*

Truck-Only Engines

As briefly outlined in the "Types of Truck Engines" section near the beginning of this chapter, a number of Chrysler engines were designed, developed, and produced specifically for truck use. Included in these are the engines that follow.

Big-Six L-Head Six-Cylinder Engine

Although the 310-cu.-in. L-head six-cylinder engine as described earlier in this chapter made an excellent truck engine, it lacked the extra displacement and brute strength needed for high-tonnage heavy-duty truck use. The truck need was resolved with the design, development, and release for 1937 of a new heavy-duty large-displacement L-head six-cylinder engine that was aimed at truck, marine, and industrial applications.

This new L-head engine was the largest six-cylinder engine ever produced by Chrysler, both in displacement and in physical size. For its launch year, its bore and stroke of 3.75 × 5.00 in. gave it a displacement of 331 cu. in., immediately besting the 310 cu. in. of the older engine. Cylinder spacing of the alloy cast iron cylinder block measured 4.75 in. except at the center of the engine, where cylinders #3 and #4 were 5.5625 in. apart. The height of the machined cylinder block from the oil pan rail to the top deck was a towering 16.8175 in., including 3.00 in. devoted to its deep skirt. Its length of 33.75 in. was slightly greater than the 33.1875-in. block length of the Chrysler five-main-bearing straight-eight engine having two additional cylinders. The diameters of the seven main journals and six crankpins of the forged steel crankshaft were 3.00 and 2.3125 in., respectively. This new engine was a really large Big-Six engine. With a compression ratio of 5.2:1, the 1937 engine developed 100 hp and a torque of 230 lb-ft @ 800 rpm, a good match for moving a heavy truck such as the 1937 Airflow tanker shown on Fig. 15.31 and the 1938 three-ton tractor shown on Fig. 15.32.

Fig. 15.32 *This 1938 three-ton Dodge tractor was a good application for the 331-cu.-in. heavy-duty Big-Six engine.*

Fig. 15.31 *A 1937 Dodge Airflow tanker. Behind the waterfall grille was a new 331-cu.-in. Big-Six heavy-duty truck engine.*

Internal details of the engine can be seen on the longitudinal and cross-section views of the engine shown on Figs. 15.33 and 15.34, respectively. The drives for the oil pump and distributor, a vertical drive for the oil pump, and a diagonal drive for the distributor appear to combine the vertical oil pump drive of the three Zeder, Skelton, and Breer-designed early Chrysler engines described in Chapter 3 with the diagonal distributor drive of the later four-main-bearing L-head six-cylinder engines described in Chapter 6. This excellent arrangement gave the advantages of being able to remove the cylinder head without disturbing spark timing, while placing the oil pump in the sump for the best pump efficiency.

Curiously, although downdraft carburetors were in 100% use on passenger cars as of the introductory date of this engine, the engine initially was equipped with only the older updraft carburetor. I can only speculate that Dodge Truck might have anticipated substantial use of the engine in cab-over-engine models that preferred engines with updraft carburetors for minimizing cab intrusion and did not want the burden of developing two versions before release. At any rate, in 1941, during the truck W Series, a downdraft carburetor version was released, which, with a compression ratio increase to 5.8:1, brought the power of the engine up to 106 hp @ 2800 rpm and the torque up to 240 lb-ft @ 800 rpm. In 1947, following a postwar investigation program exploring ways to reduce weight, improve durability, and increase power, the engine underwent an extensive redesign that was coded T135. With no change to the 331-cu.-in. displacement, the cylinder block height was reduced 1.685 in., with connecting rods being shortened from 10.4375 to 8.75 in. The intake and exhaust ports in the cylinder block were rearranged to have six individual intake ports and only two of the six exhaust ports adjacent to each other.

The new port arrangement was accompanied by a total redesign of the intake and exhaust manifolds. The straight manifolds, visible on the longitudinal section shown on Fig. 15.33, were replaced with curved smoother-flowing manifolds with branch separation, a design that can be seen on the engine on Fig. 15.35.

The cylinder head stud pattern also was changed. Originally, each cylinder had six studs in a rectangular arrangement, for a total of 21 studs per engine. The T135 redesign changed this to eight studs per cylinder in somewhat of an elongated hexagon pattern, for a total of 33 studs per engine. In addition, exhaust valve durability received a boost with the release of sodium-filled stellite-faced exhaust valves, the same type of valves that had been developed and used extensively on World

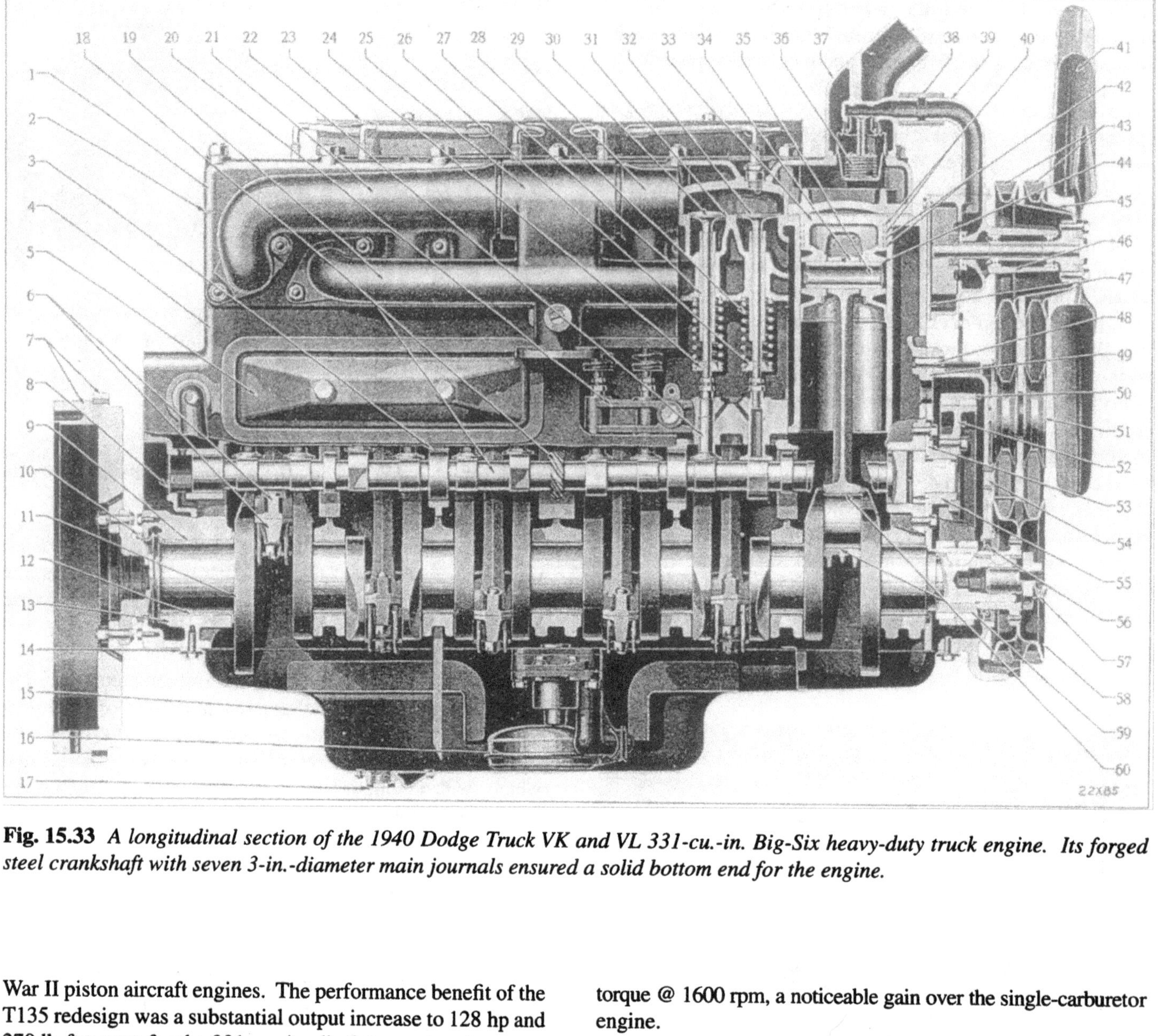

Fig. 15.33 *A longitudinal section of the 1940 Dodge Truck VK and VL 331-cu.-in. Big-Six heavy-duty truck engine. Its forged steel crankshaft with seven 3-in.-diameter main journals ensured a solid bottom end for the engine.*

War II piston aircraft engines. The performance benefit of the T135 redesign was a substantial output increase to 128 hp and 270 lb-ft torque for the 331-cu.-in. displacement engine.

In 1947, a second destroked (4.25-in.) version displacing 281.6 cu. in. was added to the truck engine lineup. For the Dodge Truck B-2 Series in 1950, this 281.6-cu.-in. engine was dropped, and two other sizes were added: a 306-cu.-in. displacement with a 4.625-in. stroke, and a larger 377-cu.-in. displacement with a bore and stroke of 4 × 5 in. The three sizes—306, 331, and 377 cu. in.—carried over through the Dodge Truck 1951–1952 B-3 Series.

In 1950, after a thorough engine dynamometer program headed by engine development manager Harold Welch,[7] a twin-carburetor, twin-exhaust system was released as standard on the largest 377-cu.-in. engine and as optional on the smaller 306- and 331-cu.-in. engines then in production. The gross output rating of the 331-cu.-in. engine with the twin-carburetor twin-exhaust system was 145 hp @ 3000 rpm and 280 lb-ft torque @ 1600 rpm, a noticeable gain over the single-carburetor engine.

The enthusiastic truck sales that followed the introduction of the twin carburetor, twin exhaust system triggered a development program to provide the smaller Chrysler 25-in.-block 265-cu.-in. L-head truck engine with the same type of performance package for 1953.

For its 1953 B-4 Series, the 377-cu.-in. size was dropped, replaced by an even larger size of 413 cu. in. with a bore and stroke of 4.0625 × 5.3125 in., by far the largest six-cylinder engine ever produced by Chrysler. The output of the engine with its standard twin-carburetors twin-exhaust shown on Fig. 15.36 was 171 hp and an impressive 343 lb-ft torque. When the 331-cu.-in. Hemi® V-8 arrived as a truck engine in 1954, its torque level of 294 lb-ft with a two-barrel carburetor was not a match for the torque of the twin-carbureted Big-Six. However, the days of the Big-Six were numbered; it lasted only one more year and was terminated at the end of 1955 because

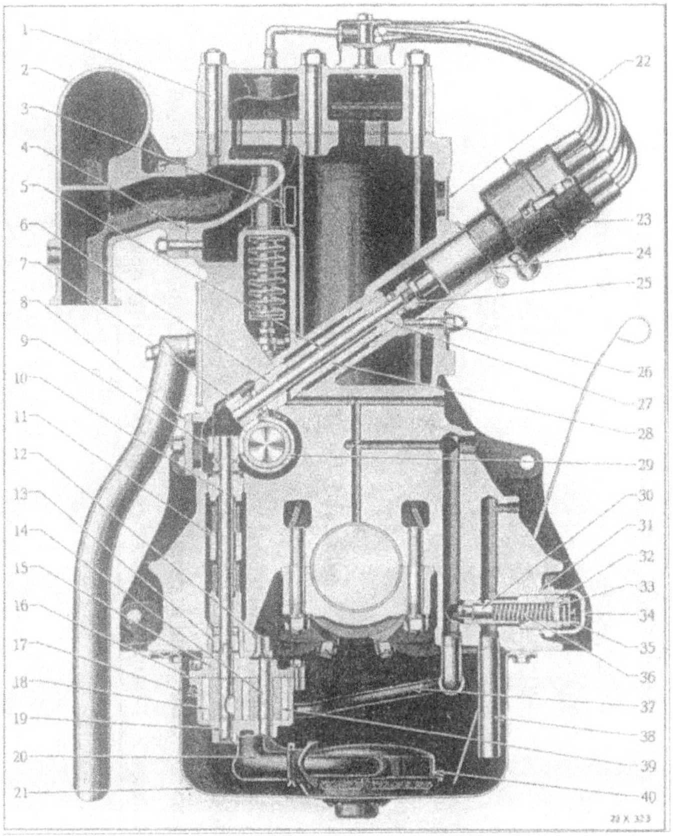

Fig. 15.34 *A cross section through the center of a 1940 331-cu.-in. Big-Six heavy-duty engine, showing the oil pump and the distributor drive.*

Fig. 15.35 *The right side of a circa-1948 331-cu.-in. Big-Six engine installed in a 1938 Dodge Airflow tanker. Note the multipiece curved intake and exhaust manifolds that were part of the T135 engine redesign of 1947. This vehicle is on the roster of the Chrysler Historical Collection. Photo by W.L. Weertman.*

Fig. 15.36 *A right-front view of the 1953 413-cu.-in. Big-Six heavy-duty truck engine with twin one-barrel carburetors and governors.*

the truck engine development engineers had completed work on the twin-carbureted version of the larger 354-cu.-in. Chrysler FirePower Hemi® V-8, mentioned previously, that produced a torque of 354 lb-ft, edging out the Big-Six. The Big-Six engine, machined and assembled at the Dodge Main plant, was replaced by a worthy successor.

Table 15.1 shows specifications for the Big Six L-head six-cylinder engine.

Dodge Diesel Overhead Valve Six-Cylinder Engine

After the start of the initial engineering work on the Big-Six L-head truck gasoline engine, a program was begun to create a diesel version of the engine, with 1939 targeted as the introduction date. The position of Dodge Truck in the heavy-duty truck market certainly would be bolstered by the addition of a Chrysler-produced diesel engine.

Although the bore, stroke, cylinder spacing, and lower end of the engine, including the crankshaft, were common to both the gasoline and diesel versions, the diesel engine required many totally new parts because an overhead valve arrangement was needed to achieve the high compression ratio of the diesel and because diesel combustion meant new combustion chambers, manifolding, fuel pump, and controls.

A key decision preceding the design of the diesel engine was the selection of the type of diesel combustion to be used. The

TABLE 15.1
SPECIFICATIONS FOR THE BIG-SIX L-HEAD DODGE TRUCK ENGINE

Model Year	1937	1947	1947	1950
Truck series	R	W	W	B-2
Model	RK, R, RP	WK, WR	WJ	R, RA
Engine code	T43	T135	T136	T320
Displacement	331	331	281.6	306
Bore	3.75	3.75	3.75	3.75
Stroke	5.0	5.0	4.25	4.625
Compression ratio	5.2:1	6.5:1	6.5:1	6.5:1
BHP @ rpm	100	128 @ 3000	115 @ 3200	122 @ 3200
Torque (lb-ft) @ rpm	230 @ 800	270 @ 1200	225 @ 1200	245 @ 1200
Cyl. block mat'l.—skirt type	Cast iron—deep	Cast iron—deep	Cast iron—deep	Cast iron—deep
Cyl. block deck height	15.505	13.8175	13.8175	13.8175
Cyl. block length	33.75	33.75	33.75	33.75
Cyl. bore spacing	All except 3-4 = 4.75 3-4 = 5.5625	All except 3-4 = 4.75 3-4 = 5.5625	All except 3-4 = 4.75 3-4 = 5.5625	All except 3-4 = 4.75 3-4 = 5.5625
No. of main brgs.—ck/s mat'l.	7—forged steel	7—forged steel	7—forged steel	7—forged steel
Ck/shaft main journal dia.	3.00	3.00	3.00	3.00
Crankpin journal dia.	2.3125	2.3125	2.3125	2.3125
Connecting rod centers	10.4375	8.75	8.75	8.9375
Valve system	L-head	L-head	L-head	L-head
Valve events	220-230-2	248-236-30	248-236-30	248-236-30
Camshaft drive	Chain	Gear	Gear	Gear
Valve head dia.	Intake 1.93; exhaust 1.75	Intake 1.94; exhaust 1.75	Intake 1.94; exhaust 1.75	Intake 1.94; exhaust 1.75
Exhaust valve type	Solid stem	Sodium filled	Sodium filled	Sodium filled
Valve lash adjustment	Mechanical tappet	Mechanical tappet	Mechanical tappet	Mechanical tappet
Oil pump drive	Vertical at center of block	Vertical at center of block	Vertical at center of block	Vertical at center of block
Distributor drive	Diagonal at center of block	Diagonal at center of block	Diagonal at center of block	Diagonal at center of block
Fuel system	Updraft carburetor	1-barrel downdraft carburetor	1-barrel downdraft carburetor	1-barrel downdraft carburetor

(Continued on next page)

TABLE 15.1
(Continued)

Model Year	1950	1951	1951	1953–1954
Truck series	B-2	B-3	B-3	B-4 – C-1
Model	Y, YA	T, V With optional twin carburetion	Y	Y
Engine code	T190, T199	T322, T323	T324, T325	T352
Displacement	377	331	377	413
Bore	4.00	3.75	4.00	4.00625
Stroke	5.00	5.0	5.00	5.3125
Compression ratio	6.5:1	6.5:1	6.5:1	6.5:1
BHP @ rpm	154 @ 3000	145 @ 3000	154 @ 3000	171 @ 3200
Torque (lb-ft) @ rpm	330 @ 1200	280 @ 1300	330 @ 1200	343 @ 1500
Cyl. block mat'l.—skirt type	Cast iron—deep	Cast iron—deep	Cast iron—deep	Cast iron—deep
Cyl. block deck height	13.8175	13.8175	13.8175	13.8175
Cyl. block length	33.75	33.75	33.75	33.75
Cyl. bore spacing	All except 3-4 = 4.75 3-4 = 5.5625	All except 3-4 = 4.75 3-4 = 5.5625	All except 3-4 = 4.75 3-4 = 5.5625	All except 3-4 = 4.75 3-4 = 5.5625
No. of main brgs.—ck/s mat'l.	7—forged steel	7—forged steel	7—forged steel	7—forged steel
Ck/shaft main journal dia.	3.00	3.00	3.00	3.00
Crankpin journal dia.	2.3125	2.3125	2.3125	2.5
Connecting rod centers	8.75	8.75	8.75	8.594
Valve system	L-head	L-head	L-head	L-head
Valve events	244-236-27	248-236-30	Up to: 244-236-27 After: 252-244-28	252-244-28
Camshaft drive	Gear	Gear	Gear	Gear
Valve head dia.	Intake 1.94; exhaust 1.75	Intake 1.94; exhaust 1.75	Intake 1.94; exhaust 1.75	Intake 1.94; exhaust 1.75
Exhaust valve type	Sodium filled	Sodium filled	Sodium filled	Sodium filled
Valve lash adjustment	Mechanical tappet	Mechanical tappet	Hydraulic tappet	Hydraulic tappet
Oil pump drive	Vertical at center of block	Vertical at center of block	Vertical at center of block	Vertical at center of block
Distributor drive	Diagonal at center of block	Diagonal at center of block	Diagonal at center of block	Diagonal at center of block
Fuel system	1-barrel downdraft carburetor	Twin 1-barrel downdraft	Twin 1-barrel downdraft	Twin 1-barrel downdraft

decision was made in favor of the Lanova system that sprayed high-pressure fuel across the combustion chamber into an energy cell where the fuel ignited. The ignited fuel then would exit the energy cell into the main chamber, where its combustion with the air charge would be completed. This indirect ignition system had the advantage of a less rapid pressure rise, providing longer-life durability and smoother operation of the engine than would a direct injection system, while still giving the fuel economy benefits of a diesel cycle.[8]

To incorporate the Lanova chamber, a new cylinder head with overhead valves and cross-flow intake and exhaust manifolds was designed. To withstand the much higher diesel combustion pressures, the head studs were rearranged in a circular pattern, with the studs moved closer to the cylinder bores. With six studs surrounding each combustion chamber, the total number of head studs increased from 21 for the 1939 Big-Six gasoline engine to 31 for the diesel engine.

The new diesel cylinder block casting and machining kept the same camshaft location but provided pushrod clearance holes in place of the valve guide machining needed for the gasoline engine.

A special diesel piston was used that had the crown required for the Lanova chamber and five rings in place of the four rings used on the gasoline engine. A gear drive using helical toothed gears was designed for both the camshaft and the high-pressure diesel fuel injection pump that was mounted solidly to the right side of the cylinder block. Figures 15.37 and 15.38 show a right-side view and a three-quarter front view of the engine, respectively, while the longitudinal section on Fig. 15.39 and the cross sections on Fig. 15.40 show the internal details.

Early 331-cu.-in. diesel engines, such as the one powering the demonstrator truck shown on Fig. 15.41, had an output of 95 hp and 226 lb-ft torque; later engines produced 100 hp and 240 lb-ft torque. The engine displacement of 331 cu. in. remained unchanged throughout its seven-year life span.

After starting production in 1939, engine manufacturing lasted through World War II until 1945. Wartime duty included marine and industrial versions of the engine. Along with the Big-Six gasoline L-head engine, the diesel engine was machined and assembled at the Dodge Main plant.

Table 15.2 shows the specifications for the Dodge Diesel six-cylinder truck engine.

LB and RB Heavy-Duty V-8 Truck Engines

With the knowledge that the Chrysler 354-cu.-in. Hemi® head V-8 engine would be out of production and no longer available for truck use after the end of 1959, an engineering program was initiated to create versions of the new LB and RB wedge-head engines, described in Chapter 9, that would have the requisite strength and durability for powering Dodge medium- and heavy-duty trucks. The engineering design and development programs for a 361-cu.-in. LB and a 413-cu.-in. RB V-8 engine—coded A695 and A696, respectively—culminated in their release for use in 1960 medium- and heavy-duty Dodge trucks. The introduction of these two tough and powerful engines coincided squarely with the Dodge Truck introduction of a new innovative medium- and heavy-duty truck model, the low-cab-forward (LCF) C-model.

The bores and strokes of 4.125 × 3.375 in. for the 361-cu.-in. LB heavy-duty V-8 engine and 4.1875 × 3.75 in. for the 413-cu.-in. RB engine were identical to their passenger car counterparts. Both engines shared many parts with passenger car engines and shared unique, heavy-duty parts with each other. Figure 15.42 shows a heavy-duty 413-cu.-in. RB V-8 display engine, while a longitudinal section and a cross section of the engine appear on Figs. 15.43 and 15.44, respectively.

The major heavy-duty part shared by the LB and RB truck V-8s was the cylinder head, which was designed to provide the extra valve cooling essential to prolonged full-load operation. Exhaust valves were sodium-filled stellite-faced, and equipped with rotators. The combustion chamber faces of the exhaust valves were spray-coated with a nickel-chromium alloy to prevent formation of magnetic scale that could lead to pre-ignition.

Both the intake and exhaust valves were designed with longer stems for better cooling and stability than passenger car valves and had replaceable guides for high-mileage servicing. Spark plugs were located high in the combustion chamber for better cooling and better accessibility. A stainless steel cylinder head gasket added mileage between engine overhauls. Special dished-top pistons were used to lower the 10.0:1 compression ratio of the passenger car to 7.5:1 for the hard-working truck engines. For long life, helical gears drove the camshaft, and the water pump was fitted with heavy-duty bearings. For the most severe service applications, an engine oil cooler and dual oil filters kept the lubrication system in top shape.

The intake manifold floor below the carburetor location was heated with engine coolant rather than exhaust gas to avoid excessive temperatures and thermal stress. To prevent overspeeding, both the 361- and 413-cu.-in. engines were equipped with either a mechanical or a velocity-type governor. Born to pull heavy loads, the heavy-duty 413-cu.-in. RB V-8 engine was under the hood of the 1960 steel-hauling C900 low-cab-forward truck pictured on Fig. 15.45; the heavy-duty 361-cu.-in. LB V-8 engine was assigned to move the load of the 1961 C700 low-cab-forward Dodge truck shown on Fig. 15.46. Both engines were available as power for the compact tilt-cab L-series trucks, such as the 1966 L700 model shown on Fig. 15.47.

Robert S. Rarey and Harold L. Welch co-authored an SAE paper titled "Dodge Truck Heavy-Duty V-8 Engines" to inform the automotive world about the design features and development program of the two engines.[9]

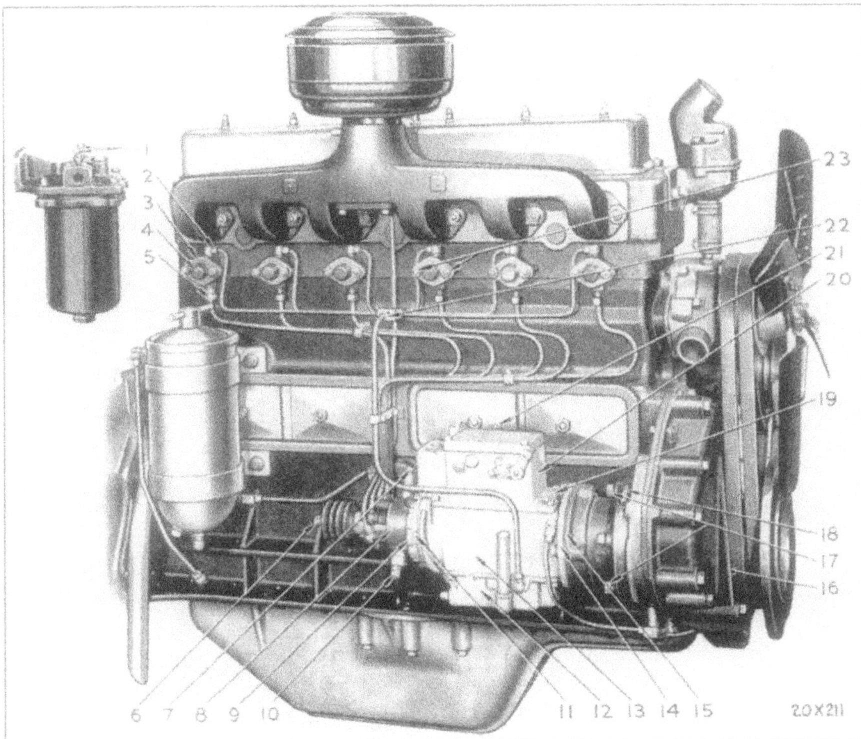

Fig. 15.37 *The right side of the 1939 Dodge Diesel 331-cu.-in. six-cylinder truck engine. The gear-driven injection pump and the high-pressure fuel lines to the six injectors can be seen.*

Fig. 15.38 *A left-front view of the 1939 Dodge Diesel 331-cu.-in. six-cylinder truck engine. Note the compressor for air brakes, belt-driven from the rear of the generator.*

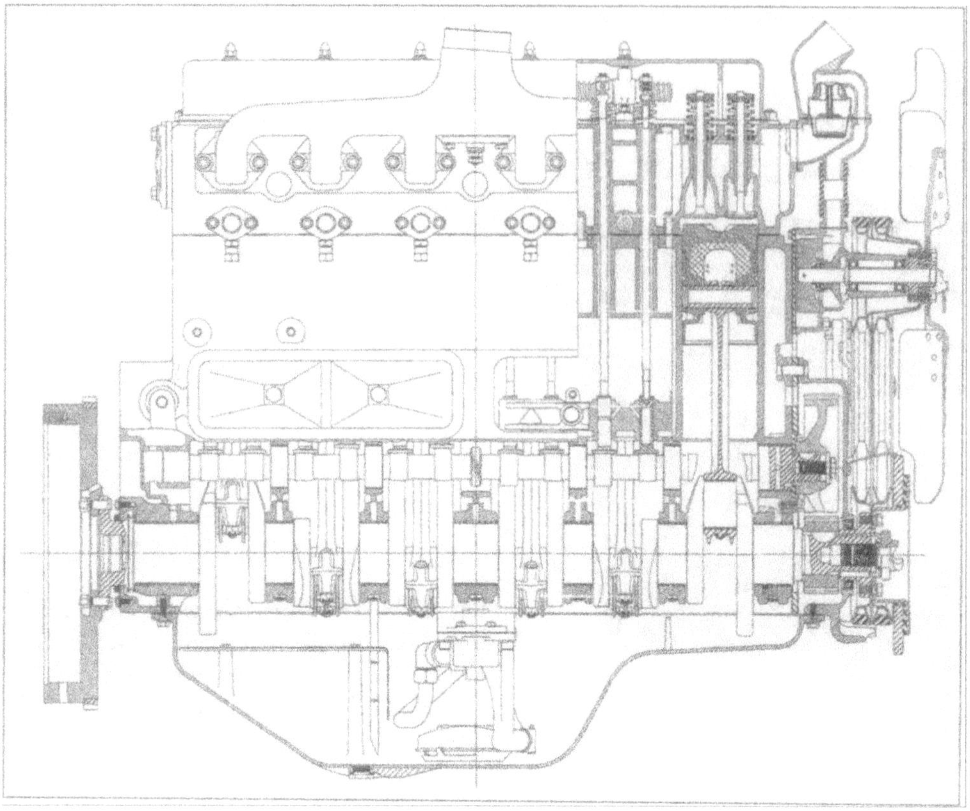

Fig. 15.39 *A longitudinal section of the 1939 Dodge Diesel 331-cu.-in. six-cylinder truck engine. The front portion of this section shows the details of the overhead valvetrain.*

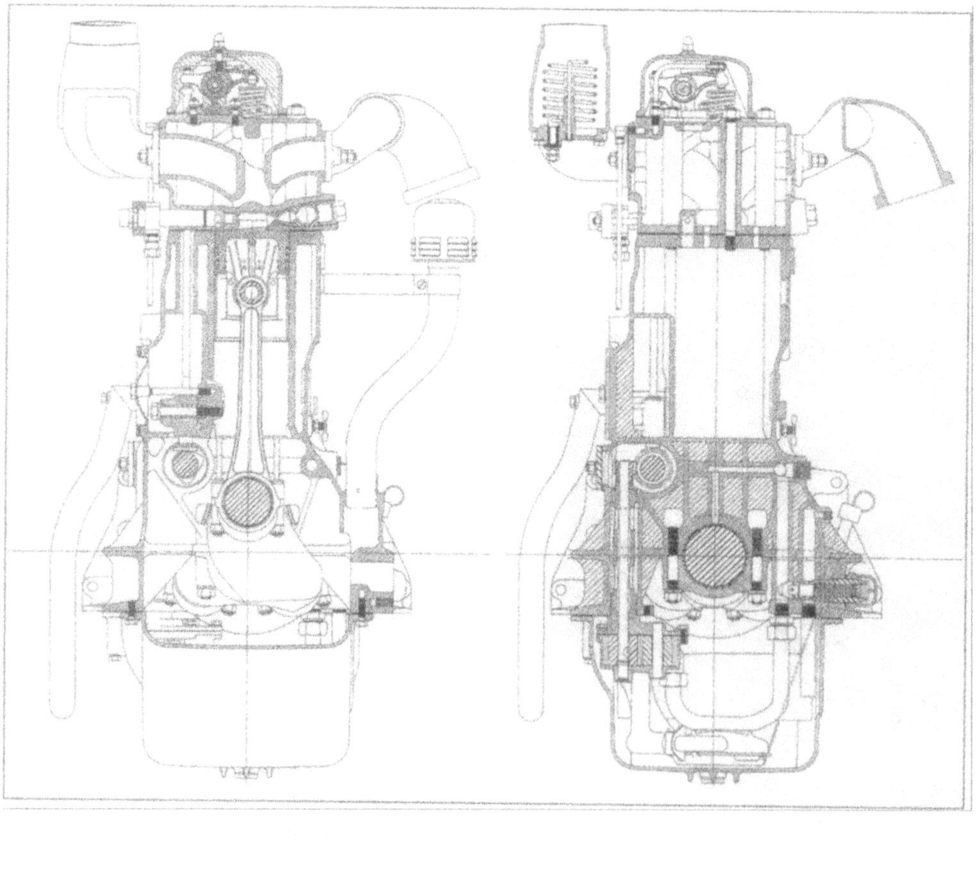

Fig. 15.40 *Cross sections of the 1939 Dodge Diesel 331-cu.-in. six-cylinder truck engine. The view on the left, through the center of the #1 cylinder, shows the diesel injector and the Lanova chamber. The view on the right, through the center of the engine, shows the oil pump drive and the cooling jet for the injector seat in the cylinder head.*

Fig. 15.41 *A 1939 Dodge demonstrator truck powered by a Dodge Diesel 331-cu.-in. six-cylinder engine.*

TABLE 15.2
SPECIFICATIONS FOR THE DODGE DIESEL SIX-CYLINDER TRUCK ENGINE

Model Year	1939	1946
Truck series	Dodge Diesel	Dodge Diesel
Model	TKD, TLD	WKD, WKDA, WLD, WLDA
Engine code	T84	T126
Displacement (cu. in.)	331	331
Bore	3.75	3.75
Stroke	5.00	5.00
Compression ratio	14.5:1	14.75:1
BHP @ rpm	95 @ 2600	100 @ 2600
Torque (lb-ft) @ rpm	226 @ 1000	240 @ 1200-1300
Diesel system	Lanova chamber	Lanova chamber
Cyl. block mat'l.—skirt type	Cast iron—deep	Cast iron—deep
Cyl. block deck height	15.505	15.505
Cyl. block length	33.75	33.75
Cyl. bore spacing	All except 3-4 = 4.75 3-4 = 5.5625	All except 3-4 = 4.75 3-4 = 5.5625
No. of main brgs.—ck/s mat'l.	7—forged steel	7— steel
Ck/shaft main journal dia.	3.00	3.00
Crankpin journal dia.	2.3125	2.3125
Piston—no. of rings	Aluminum—5	Aluminum—5
Connecting rod centers	10.31	10.31
Valve system	OHV—cross-flow	OHV—cross-flow
Valve events	260-244-48	260-244-48
Camshaft drive	Gear	Gear
Valve head dia.	Intake 1.68; exhaust 1.44	Intake 1.68; exhaust 1.44
Valve lash adjustment	Mech.—rocker arm screw	Mech.—rocker arm screw
Oil pump drive	Vertical at center of block Geared to camshaft	Vertical at center of block Geared to camshaft
Fuel injection pump	2000 psi Excello gear-driven	2000 psi Excello gear-driven

Fig. 15.42 *A 1960 heavy-duty 413-cu.-in. RB V-8 truck cutaway display engine.*

All the 361- and 413-cu.-in. LB and RB V-8 truck engines were manufactured by the Trenton engine plant without changes in displacement and few changes in parts until the end of 1977, which was the termination point of the 361-cu.-in. engine and domestic sales of medium-duty trucks. The 413-cu.-in. V-8 engine was kept on the export truck charts for a medium-duty D800 truck until the end of 1982.

Table 15.3 shows specifications for the 361- and 413-cu.-in. LB and RB V-8 heavy-duty engines.

2.2-L and 2.5-L Single Overhead Camshaft North-South Four-Cylinder Engines

When the medium-size Dodge Dakota was created with a planned 1987 introduction, two engines were readied as powerplants: a truck-only version of the 2.2-L single overhead camshaft transversely mounted passenger car engine as was described in Chapter 12, and a new V-6 engine described in the next section.

The major changes for the 2.2-L single overhead camshaft engine for its installation in the Dakota truck chassis centered about it being placed in a longitudinal or north-south orientation rather than the east-west transverse position used on front-wheel-drive passenger cars. Unique new parts included the intake manifold, exhaust manifold, and oil pan. A two-barrel carburetor was specified, the same as for passenger cars, and an

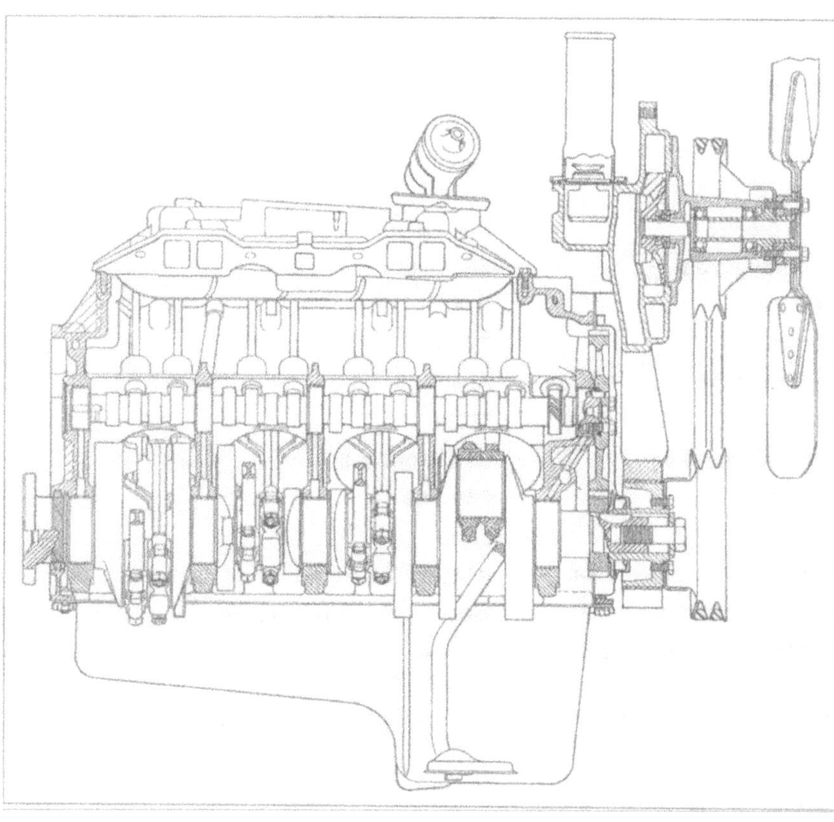

Fig. 15.43 *A longitudinal section of the 1960 heavy-duty 413-cu.-in. RB V-8 engine. The heavy-duty package included the raised water pump with large-size bearings and dual fan belts.*

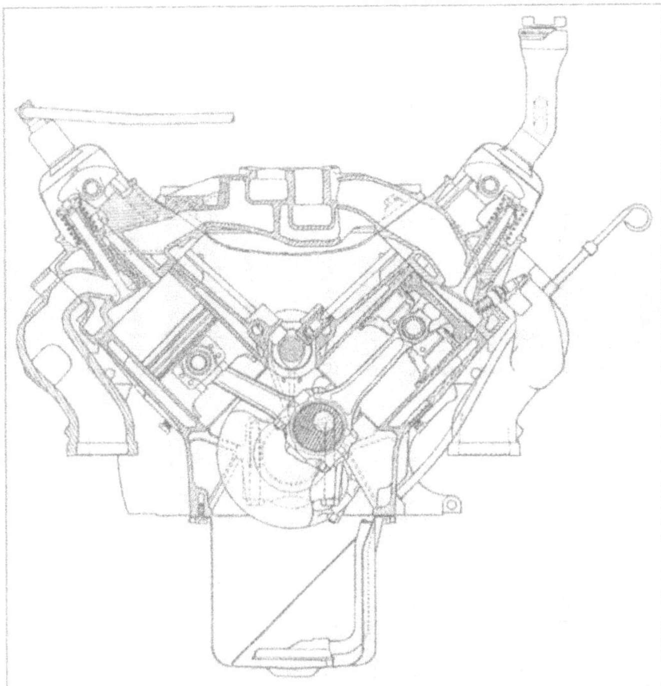

Fig. 15.44 *A cross section of the 1960 heavy-duty 413-cu.-in. RB V-8 engine. Compared to the passenger car 413-cu.-in. RB V-8 cross section shown on Fig. 9.10 in Chapter 9, the spark plugs have been located above the exhaust manifold. Also, the valve guides are replaceable, are longer, and have more cooling.*

Fig. 15.45 *A steel-hauling 1960 C900 low-cab-forward Dodge truck. Two versions of the 413-cu.-in. heavy-duty RB V-8 engine were available: a 217-hp 413-2 version released as the standard engine, and a 228-hp 413-3 version released as optional.*

Fig. 15.46 *A 1961 C700 LCF Dodge truck. Its standard and only engine was the 361-3 heavy-duty LB V-8.*

Fig. 15.47 *A 1966 medium-duty L700 tilt-cab. This bulldog-looking, short-wheelbase truck with excellent city street maneuverability was powered by either a heavy-duty 361-3 LB V-8 or a heavy-duty 413-2 RB V-8 engine.*

electric cooling fan was installed for the radiator, also similar to that of passenger cars. This 1987 engine produced 96 hp @ 5200 rpm and 121 lb-ft torque @ 3200 rpm. Figure 15.48 portrays a 1988 Dakota pickup truck that was marketed with the 2.2-L single overhead camshaft north-south four-cylinder engine as its standard power.

The engine received a displacement increase to 2.5-L in 1989 simply by using the parts designed and developed for the 1989 2.5-L passenger car engine. Although the passenger car 2.5-L engine used balance shafts to eliminate second-order vibrations, Dodge Truck opted for the lighter, less complicated, and lower-cost approach of not using balance shafts on the Dakota

TABLE 15.3
SPECIFICATIONS FOR THE DODGE TRUCK LB AND RB HEAVY-DUTY V-8 ENGINES

Model Year	1960	1960	1977	1977
Engine	LB V8	RB V8	LB V8	RB V8
Engine designator	361-2, 361-3, 361-4	413-1, 413-2	361-3	413-2
Displacement (cu. in.)	361	413	361	413
Bore	4.125	4.1875	4.125	4.1875
Stroke	3.375	3.75	3.375	3.75
Compression ratio	7.5:1	7.5:1	7.5:1	7.5:1
BHP @ rpm	361-2: 178 @ 3500 361-3: 194 @ 3500 361-4: 204 @ 3500	413-1: 217 @ 3500 413-2: 228 @ 3500	186 @ 4000 net	202 @ 3600 net
Torque (lb-ft) @ rpm	361-2: 291 @ 2400 361-3: 310 @ 2400 361-4: 335 @ 2400	413-1: 373 @ 2000 413-2: 400 @ 2000	309 @ 2200 net	374 @ 2200 net
Combustion chamber	Wedge	Wedge	Wedge	Wedge
Cyl. block mat'l.—skirt type	Cast iron—deep	Cast iron—deep	Cast iron—deep	Cast iron—deep
Cyl block deck height	9.98	10.725	9.98	10.725
Cyl. block length	23.46	23.46	23.46	23.46
Cyl. bore spacing	4.8	4.8	4.8	4.8
No. of main brgs.—ck/s mat'l.	5—forged steel	5—forged steel	5—forged steel	5—forged steel
Ck/shaft main journal dia.	2.625	2.75	2.625	2.75
Crankpin journal dia.	2.375	2.375	2.375	2.375
Connecting rod centers	6.358	6.768	6.358	6.768
Valve system	OHV—single rocker shaft	OHV—single rocker shaft	OHV—single rocker shaft	OHV—single rocker shaft
Valve events			260-260-44	260-260-44
Camshaft drive	Gear	Gear	Roller chain	Roller chain
Valve head dia.	Intake 1.88; exhaust 1.5	Intake 1.88; exhaust 1.5	Intake 1.88; exhaust 1.5	Intake 1.88; exhaust 1.5
Exhaust valve type	Sodium filled	Sodium filled	Sodium filled	Sodium filled
Valve lash adjustment	Hydraulic tappet	Hydraulic tappet	Hydraulic tappet	Hydraulic tappet
Oil pump/distributor drive	Diagonal at front of block	Diagonal at front of block	Diagonal at front of block	Diagonal at front of block
Fuel system	2-barrel carburetor	413-1: 2-barrel carburetor 413-2: 4-barrel carburetor	2-barrel carburetor	2-barrel carburetor

Fig. 15.48 *A 1988 Dodge Dakota S pickup truck. Positioned north-south, a 96-hp 2.2-L single overhead camshaft four-cylinder engine was standard power.*

2.5-L north-south version. Also for 1989, the fuel system was changed from a two-barrel carburetor to throttle-body fuel injection. Figures 15.49 and 15.50 show a cross section and a longitudinal section, respectively, of this 1989 2.5-L single overhead camshaft north-south Dakota truck engine.

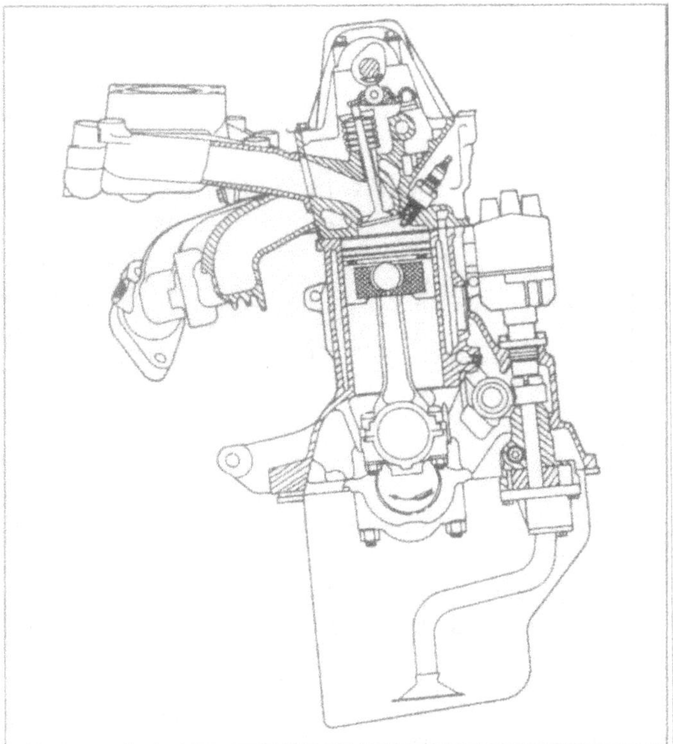

Fig. 15.49 *A cross section of the 1989 2.5-L single overhead camshaft four-cylinder engine adapted for north-south orientation in the Dodge Dakota truck.*

The engine continued in production until the end of 1995, when it was replaced with the excellent AMC 2.5-L overhead valve four-cylinder engine that is described in Chapter 19.

Table 15.4 shows the specifications for the 2.2-L and 2.5-L single overhead camshaft north-south four-cylinder engines.

3.9-L 90-Degree Overhead Valve V-6 Engine

As noted in the preceding section, the new-for-1987 Dodge Dakota truck was powered by two engines: a 2.2-L single overhead camshaft four-cylinder engine as previously described, and a V-6 engine. Engine compartment constraints during the initial design studies of the vehicle ruled out installing the existing LA V-8 engine. But to give the truck broad marketing appeal and extra carrying capacity, an optional engine larger than the four-cylinder engine was considered essential. An all-new engine of any sort was ruled out for both financial and program timing considerations.

The solution that received program approval was a 3.9-L V-6 engine based on the existing Mound Road engine plant-built 5.2-L LA V-8, including the same bore and stroke of 3.91 × 3.31 in. and, importantly, the same 90-deg angle between the cylinder banks. Although many major parts would be unique to the V-6, many other parts would be common with the existing 5.2-L LA V-8 engine, reducing the time and effort needed for the engineering development program while simplifying manufacturing and service complexity.

A basic part of the production engine, the cylinder block, actually used a considerable portion of the V-8 block machining line. This was accomplished by providing the V-6 block casting with a false front that gave it the same length as the V-8 block for initial processing on the block machining line at the Mound Road engine plant. Later down the line, the false front was cut off, and the V-6 block machining was completed on unique tooling. The V-6 machined cylinder block's length of 16.94 in. was exactly one cylinder spacing, 4.46 in., less than the 5.2-L V-8 block's length of 21.40 in. Because the 2.2-L engine's four-cylinder block length of 16.46 in. was almost the same, the 2.2-L four-cylinder and the 3.9-L V-6 were very compatible in the short engine compartment of the Dakota. Figures 15.51 and 15.52 show outside views of the right and left sides, respectively, of the 3.9-L V-6 engine.

Crankshafts of V-8 engines typically are characterized by pairs of connecting rods sharing the same crankpin and giving equal 90-deg firing intervals. However, sharing crankpins on a 90-deg V-6 engine would result in noticeable unequal firing intervals of 90 and 150 deg. To improve this condition on the 3.9-L V-6, the crankpins were split, with the two halves displaced 22 deg apart with a thin connector between them. Narrower connecting rod big-ends and bearings provided space for the crankpin connectors. The resultant alternating firing intervals of 112 and 128 deg were acceptable. To achieve ideal 120-deg equal

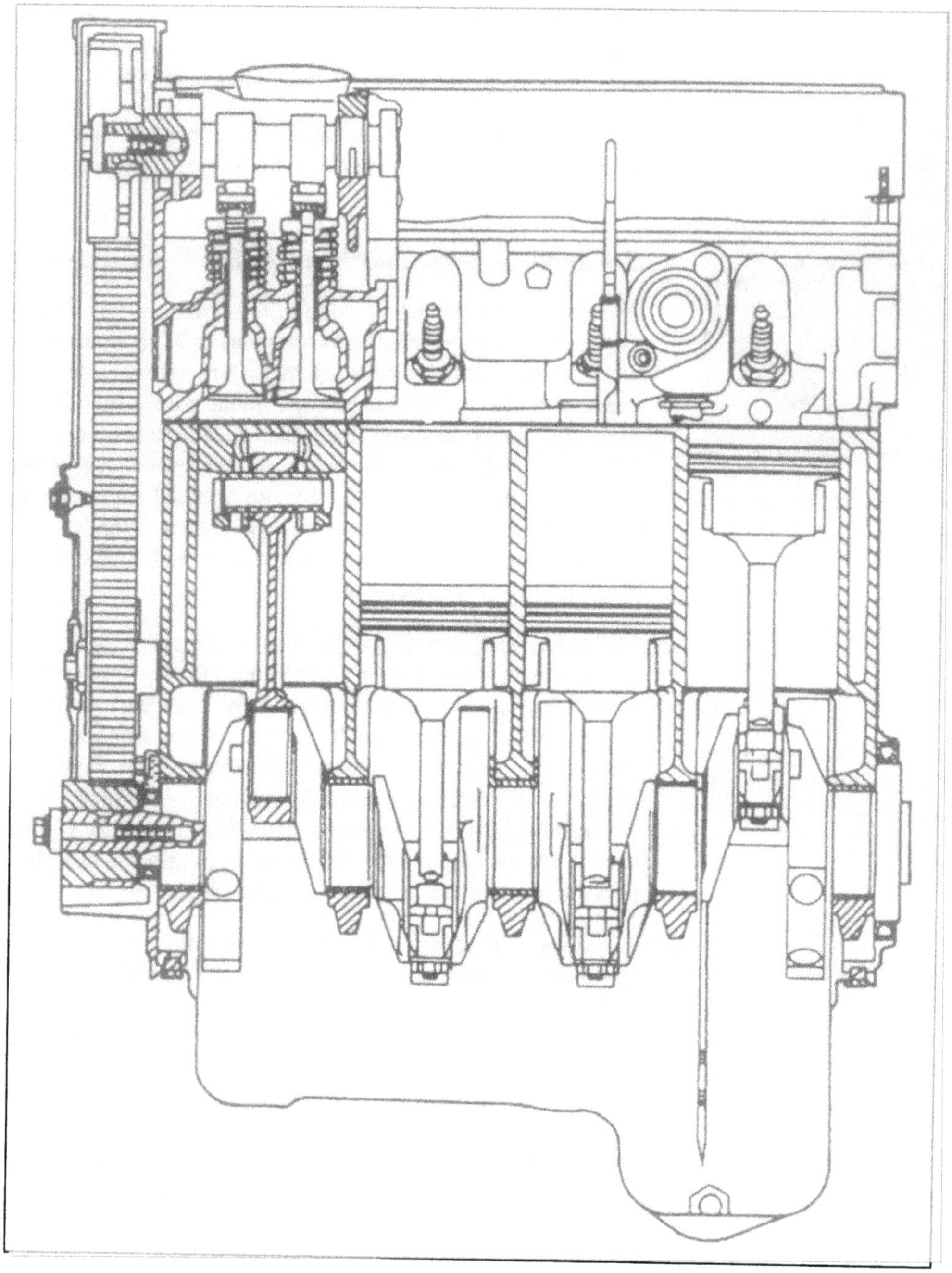

Fig. 15.50 *A longitudinal section of the 1989 2.5-L single overhead camshaft four-cylinder engine adapted for north-south orientation in the Dodge Dakota truck. Note the absence of balance shafts.*

firing intervals would have required 30-deg offset crankpins that would have unacceptably weakened the crankshaft.

The aluminum intake manifold of the 1987 V-6 was designed to have an exhaust gas crossover to provide quick warmup for the fuel/air mixture and to provide a source of exhaust gas, if needed, for emission control exhaust gas recirculation (EGR). The crossover was jacketed with coolant passages to reduce thermal expansion of the manifold for sealing durability. This jacketing can be seen on the longitudinal section of the engine shown on Fig. 15.53.

Equipped with a two-barrel carburetor, a compression ratio of 9.2:1, and valve events of 240-240-10, the 1987 engine had an output of 125 hp @ 4000 rpm and 195 lb-ft torque @ 2000 rpm. That was plenty of pulling power for the Dakota.

The next year, 1988, brought a plethora of changes. The trusty and dependable 225-cu.-in. Slant Six had gone out of production and was no longer available as the base engine for Dodge pickup trucks and B-model vans, so that assignment was handed to the 3.9-L V-6. Ever-tightening emission requirements caused the two-barrel carburetor of the 3.9-L V-6 engine

TABLE 15.4
SPECIFICATIONS FOR THE DODGE TRUCK 2.2-L AND 2.5-L SOHC FOUR-CYLINDER ENGINES

Model Year	1987	1989	1995
Model	Dodge Dakota	Dodge Dakota, Ram Van, Ram Pickup	Dodge Dakota, Ram Van, Ram Pickup
Engine code	EDE	EDM	
Displacement (liters)	2.2	2.5	2.5
Bore	87.5	87.5	87.5
Stroke	92	104	104
Compression ratio	9.5:1	8.9:1	8.9:1
Power @ rpm	69 kW (93 hp) @ 5200	74.6 kW (100 hp) @ 4800	73.9 kW (99 hp) @ 4500
Torque @ rpm	157 N-m (116 lb-ft) @ 2800	183 N-m (135 lb-ft) @ 2800	179 N-m (132 lb-ft) @ 2800
Cyl. block mat'l.—skirt type	Cast iron—shallow	Cast iron—shallow	Cast iron—shallow
Cyl. block deck height	237.8	237.8	237.8
Cyl. block length	418	418	418
Cyl. bore spacing	96	96	96
No. of main brgs.—ck/s mat'l.	5—nodular cast iron	5—nodular cast iron	5—nodular cast iron
Ck/shaft main journal dia.	60	60	60
Crankpin journal dia.	50	50	50
Connecting rod centers	151	151	151
Valve system	SOHC—8-valve	SOHC—8-valve	SOHC—8-valve
Valve events	244-244-28	236-232-8	
Camshaft drive	Timing belt	Timing belt	Timing belt
Valve head dia.	Intake 40.6; exhaust 35.4	Intake 40.6; exhaust 35.4	Intake 40.6; exhaust 35.4
Valve lash adjustment	Hyd. rocker arm pivot	Hyd. rocker arm pivot	Hyd. rocker arm pivot
Oil pump/distributor drive	Geared to auxiliary shaft	Geared to auxiliary shaft	Geared to auxiliary shaft
Fuel system	2-barrel carburetor	TBI	TBI
Balance system	Not provided	Not provided	Not provided
Engine drive/orientation	Rear-wheel drive/north-south	Rear-wheel drive/north-south	Rear-wheel drive/north-south

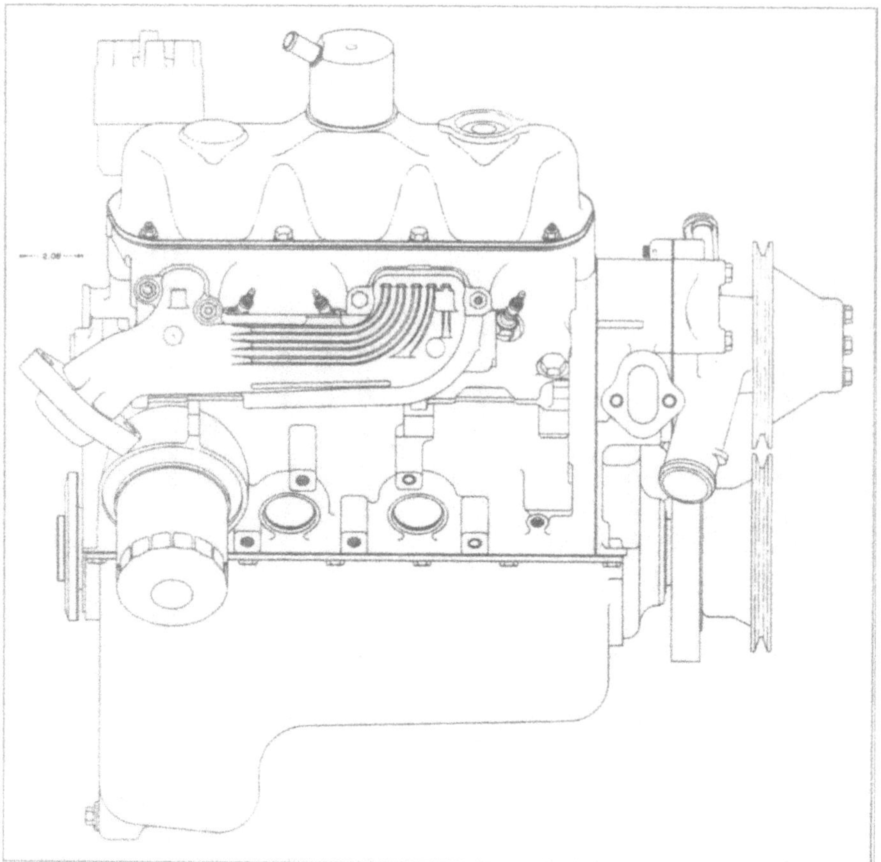

Fig. 15.51 *The right side of the 1987 3.9-L overhead valve V-6 engine.*

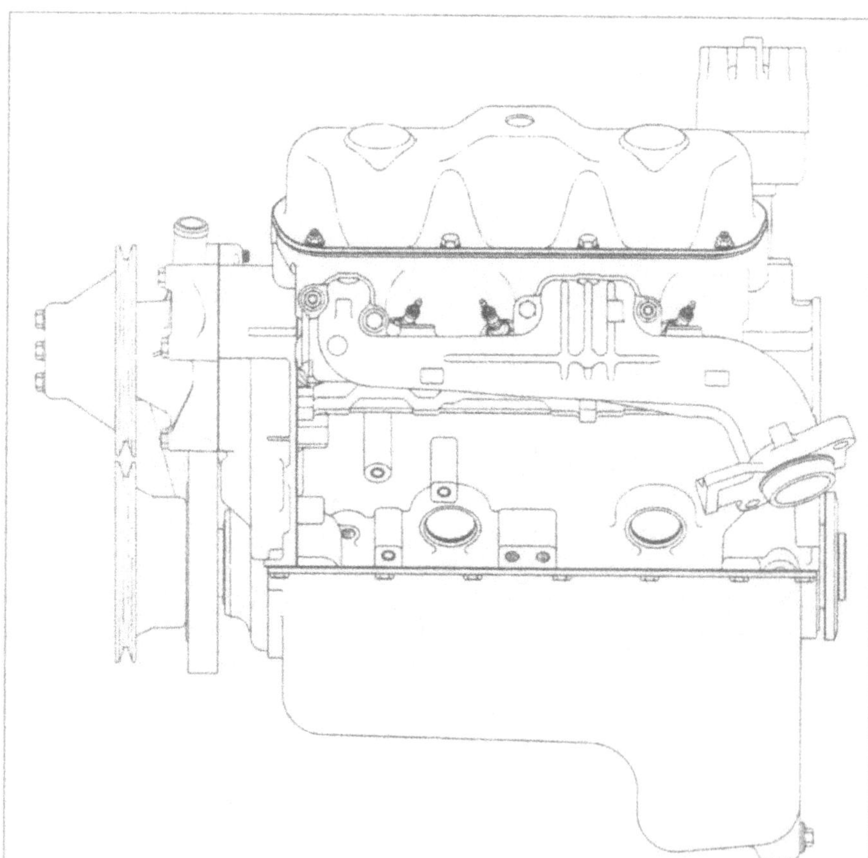

Fig. 15.52 *The left side of the 1987 3.9-L overhead valve V-6 engine.*

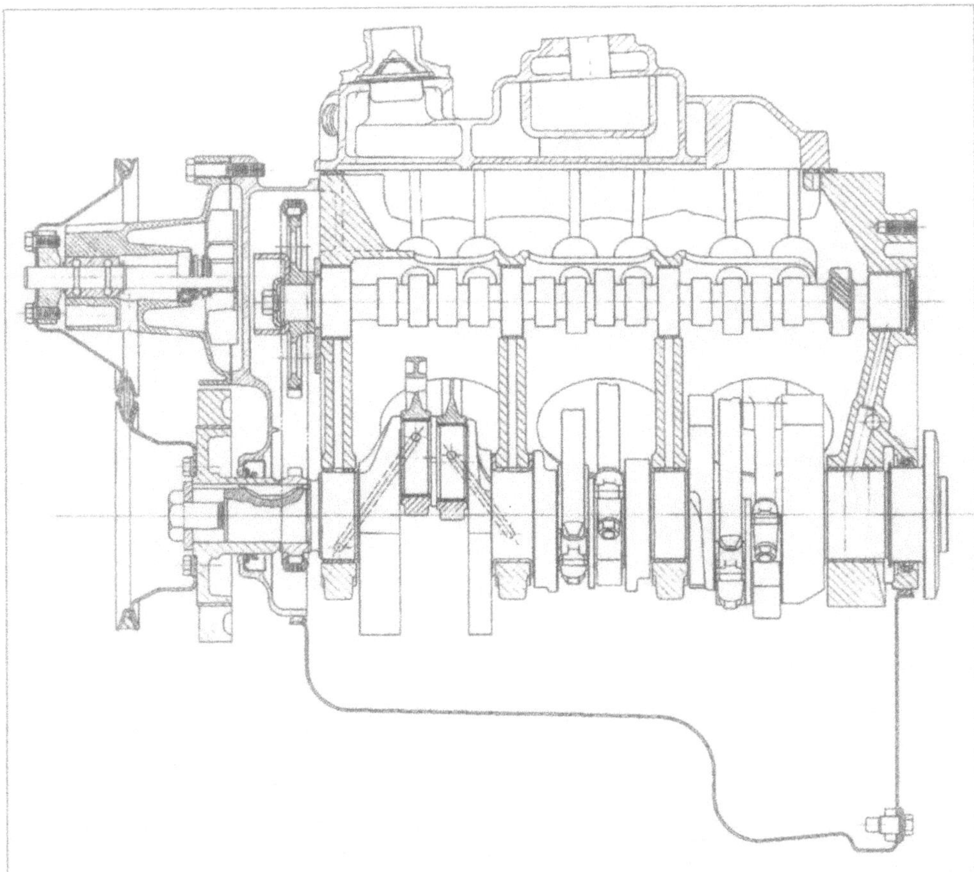

Fig. 15.53 *A longitudinal section of the 1987 3.9-L overhead valve V-6 engine. The exhaust crossover in the intake manifold was water jacketed to reduce thermal expansion. Thin webs between adjacent connecting rod big-ends were required so crank-pin journals could be offset 22 deg for more equal firing.*

to be abandoned in favor of throttle-body fuel injection that gave more precise engine control but did not change output ratings. Finally, the long-used hydraulically adjusted tappets with slightly crowned faces contacting the camshaft lobes of the engine were changed to roller hydraulic tappets for a boost in fuel economy, especially while idling. On Fig. 15.54, which shows a cross section of the 1988 engine, the roller tappets and other details of the engine are visible.

A few years later, in 1992, these changes were overshadowed by a dramatic increase in engine performance with the incorporation of the Magnum package that included sequential multiport fuel injection, tuned intake manifold, reshaped cylinder head ports, new rocker arm pivots, higher valve lifts, and other modifications. The new levels of 180 hp @ 4800 rpm and 225 lb-ft torque @ 3200 rpm put the 3.9-L V-6 engine at the front of its truck engine class.[10]

The engine continued powering Dodge pickup trucks, Dakotas, Durangos, and B-model vans beyond the 1998 ending year covered by this book. Figure 15.55 shows a 1997 Dodge Dakota Sport pickup that had the Magnum 3.9-L V-6 engine specified as standard.

Table 15.5 shows the specifications for the 3.9-L overhead valve V-6 engine.

A Big Gas Engine for Dodge Truck—The 8.0-L Overhead Valve Truck V-10

September 1986 was the recorded start date of the V-10 truck engine, the occasion being a telephone call that I received from Chrysler president Bob Lutz, inquiring if it would be possible to create a "Big Gas" truck engine by adding two cylinders to the existing 5.9-L (360-cu.-in.) LA V-8 engine, the largest engine Chrysler had in production at the time.[11] Keeping the bore and stroke of the 5.9-L engine would result in a V-10 engine with a displacement of 7.4-L.

Competitors were still offering big-block V-8 gasoline truck engines, but Chrysler's own big-block RB V-8 engine had been out of production for ten years. Introduction of a large truck engine would allow Dodge Truck to compete again in this market segment.

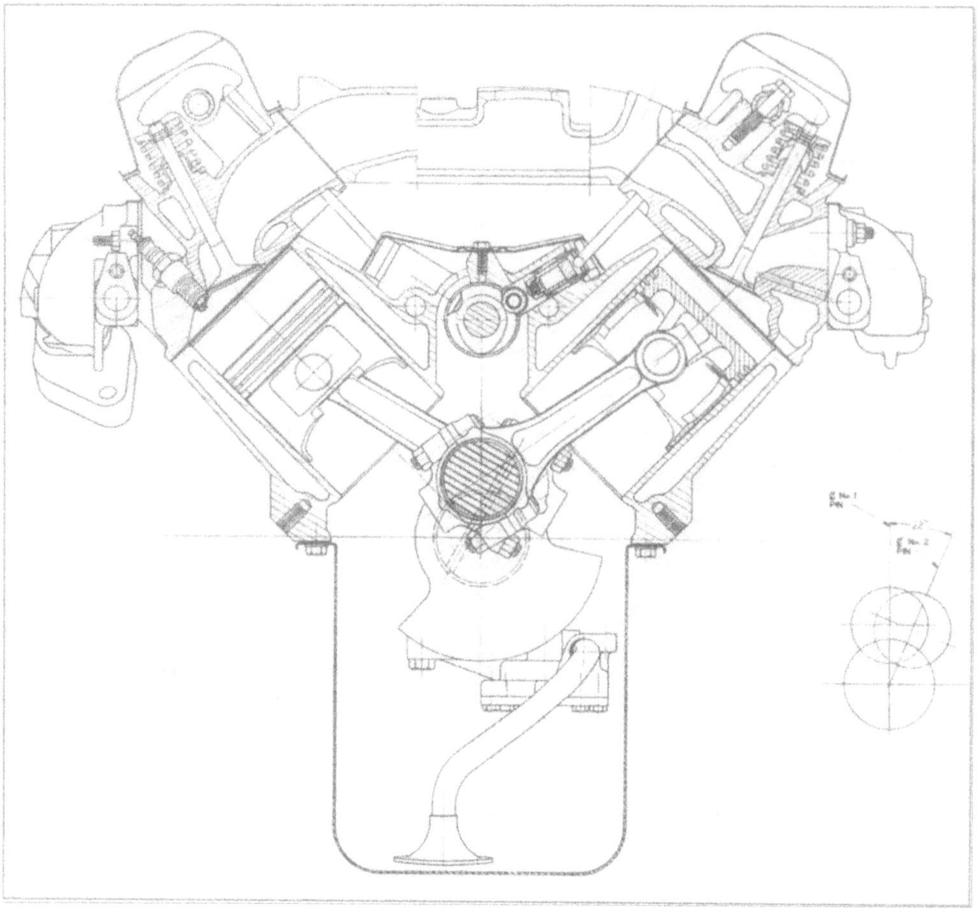

Fig. 15.54 *A cross section of the 1988 3.9-L overhead valve V-6 engine. Roller tappets have been incorporated to reduce engine friction. The small view on the right shows crankpin offset.*

Fig. 15.55 *A 1997 Dodge Dakota 4×2 Sport truck. Standard power was the 3.9-L overhead valve V-6 Magnum engine.*

An analysis of the critical areas of the crankshaft showed that a V-10 crankshaft with five unsplit crankpins would have acceptable stress levels. This conclusion, together with a quick overall design layout, permitted a favorable response to the inquiry. Later, a computer program determined which of the twenty-four possible arrangements of the five crankpins would result in the lowest level of unbalanced forces.

In 1987, Troy Simonsen of Dodge Truck Product Planning had an outside contract engineering company with excellent engine capabilities build a demonstrator V-10 engine using brazed-together 5.9-L LA V-8 cylinder blocks and cylinder heads. All six main bearing caps were integrated into a thick steel bedplate that was bolted to the bottom of the block, giving the block plenty of structural strength. An oversize ten-post distributor was custom-made for the ignition system. A plenum-ram intake manifold and multipoint fuel injection ensured plenty of engine performance. The demonstrator engine, pictured on Fig. 15.56, was installed in the Dodge pickup truck shown on Fig. 15.57 to enable early vehicle evaluations to be made. The response was universally and enthusiastically favorable.

TABLE 15.5
SPECIFICATIONS FOR THE 3.9-L OHV DODGE TRUCK V-6 ENGINE

Model Year	1987	1992	1998
Model	Dodge Dakota	Dodge Dakota, Ram Van, Ram Pickup	Dodge Dakota, Durango, Ram Van, Ram Pickup
Engine code	EHA		
Displacement (liters)	3.9	3.9	3.9
Bore	3.91	3.91	3.91
Stroke	3.31	3.31	3.31
Compression ratio	9.2:1	9.2:1	9.2:1
Power @ rpm	93 kW (125 hp) @ 4000	134 kW (180 hp) @ 4800	131 kW (175 hp) @ 4800
Torque @ rpm	264 N-m (195 lb-ft) @ 2000	305 N-m (225 lb-ft) @ 3200	305 N-m (225 lb-ft) @ 3200
Cyl. block mat'l.—skirt type	Cast iron—shallow	Cast iron—shallow	Cast iron—shallow
Cyl. block deck height	9.6	9.6	9.6
Cyl. block length	16.94	16.94	16.94
Cyl. bore spacing	4.46	4.46	4.46
No. of main brgs.—ck/s mat'l.	4—nodular cast iron	4—nodular cast iron	4—nodular cast iron
Ck/shaft main journal dia.	2.5	2.5	2.5
Crankpin journal dia.	2.125	2.125	2.125
Connecting rod centers	6.123	6.123	6.123
Valve system	OHV—single rocker shaft	OHV—single rocker shaft	OHV—single rocker shaft
Valve events	240-248-26	240-248-26	240-248-26
Camshaft drive	Chain	Chain	Chain
Valve head dia.	Intake 1.78; exhaust 1.52	Intake 1.88; exhaust 1.60	Intake 1.88; exhaust 1.60
Valve lash adjustment	Hydraulic tappet	Roller hydraulic tappet	Roller hydraulic tappet
Oil pump/distributor drive	Vertical at rear of block	Vertical at rear of block	Vertical at rear of block
Fuel system	2-barrel carburetor	SMPI	SMPI
Other		Magnum design	Magnum design

Fig. 15.56 *A 1987 prototype brazed-together 7.4-L overhead valve V-10 truck engine. At the bottom of the cylinder block is a thick steel bedplate, giving the engine plenty of structural strength.*

Fig. 15.57 *A 1987 Dodge Ram pickup truck used as a demonstrator for the V-10 engine. Looking very stock on the outside, the big engine under the hood gave the truck exceptional performance.*

However, other higher-priority programs caused the V-10 to be set aside for a year or so until it was restarted under the direction of the newly formed Jeep Truck engineering group that immediately began a production redesign of the engine, increasing its cylinder bore spacing from 4.46 to 4.55 in. and bumping up its displacement from the 7.4-L of the demonstrator engine to an even 8.0-L by a stroke change from 3.58 to 3.88 in.

This production design included a deep-skirt cast iron cylinder block, cast iron cylinder heads with exhaust valve seat inserts, nodular iron crankshaft with 3-in.-diameter main journals and carryover 2.125-in.-diameter crankpins, roller hydraulic tappets, a dual plenum-ram intake system with 25-in. primary runners, a crankshaft nose-driven oil pump, pedestal-mounted high-lift rocker arms, sequential multipoint fuel injection, and a distributorless ignition. Similar to the all-aluminum Viper V-10 engine described in Chapter 14, the truck V-10 engine had alternating firing intervals of 90 and 54 deg.

The 1994 engine, shown on Fig. 15.58, had a compression ratio of 9.0:1, a power level of 224 kW (300 hp) @ 4000 rpm, and torque of 610 N-m (450 lb-ft) @ 2400 rpm. This engine was under the hood of the dual-rear-wheeled Dodge Ram 3500 pickup truck shown on Fig. 15.59.

Rod Kurowski, Gene Yee, N. Rodenkirch, J. Waterworth, W. McNeil, and Wayne Hill were co-authors of an SAE paper that described the design and development of the V-10 truck engine.[12]

The Mound Road engine plant was the sole manufacturing source of the engine from its 1994 introduction through the 1998 ending date of this book.

Table 15.6 shows the specifications for the V-10 truck engine.

Chart 15.1 shows time lines for all the engines described in this chapter.

Fig. 15.58 *A cutaway illustration of the 1994 8.0-L overhead valve V-10 Dodge truck production engine. The tuned intake system presents a complex of passages, as shown on the illustration.*

Fig. 15.59 *A dual-rear-wheeled 1994 Dodge Ram 3500 pickup truck powered by the 8.0-L overhead valve V-10 engine rated at 224 kW (300 hp) and 610 N-m (450 lb-ft) torque.*

TABLE 15.6
SPECIFICATIONS FOR THE 8.0-L OHV DODGE TRUCK V-10 ENGINE

Model Year	1994	1998
Model	Ram Pickup	Ram Pickup
Engine code		
Displacement (liters)	8.0	8.0
Bore	4.0	4.0
Stroke	3.88	3.88
Compression ratio	9.1:1	9.1:1
Power @ rpm	224 kW (300 hp) @ 4000	224 kW (300 hp) @ 4400
Torque @ rpm	610 N-m (450 lb-ft) @ 2400	610 N-m (450 lb-ft) @ 2800
Cyl. block mat'l.—skirt type	Cast iron—deep	Cast iron—deep
Cyl. block deck height	9.6	9.6
Cyl. block length	28.0	28.0
Cyl. bore spacing	4.55	4.55
No. of main brgs.—ck/s mat'l.	6—nodular cast iron	6—nodular cast iron
Ck/shaft main journal dia.	3.0	3.0
Crankpin journal dia.	2.125	2.125
Connecting rod centers	6.123	6.123
Valve system	OHV—rocker arm pedestals	OHV—rocker arm pedestals
Valve events	246-265-31	246-265-31
Camshaft drive	Chain	Chain
Valve head dia.	Intake 1.93; exhaust 1.6	Intake 1.93; exhaust 1.6
Valve lash adjustment	Roller hydraulic tappet	Roller hydraulic tappet
Oil pump drive	Nose of crankshaft	Nose of crankshaft
Ignition system	Distributorless ignition system	Distributorless ignition system
Fuel system	SMPI	SMPI

Chart 15.1
Time Lines—Truck Engines

- - Model Years and Displacements (cu. in.) - -

Engine Truck series	1918–1927	1928 D	1929	1930 F	1931 F	1932 F	1933 G	1934 H	1935–1936 K - L
Dodge Bros 4-cyl. 3-main bearing, L-head	212								
Dodge Bros 4-cyl. 5-brg, 124, L-head		212							
Dodge Bros 6-cyl. 7-brg, Senior, L-head		224	241.5	241.5					
Dodge Bros 6-cyl. 7-brg, Victory, L-head			208	208					
Plymouth 4-cyl. 3-main bearing, L-head			175	196	196	196			
ZSB 6-cyl. Small block, 7-brg, L-hd.			196						
ZSB 6-cyl. Large block, 7-brg, L-hd.				310	310	310	310	310	310
Chrysler 6-cyl. 4-brg, deep-skirt, L-head				175	212	212 218	218	241.5	241.5
Imperial Straight-8 9-brg, L-head					385 Fargo 90, 91	385 G80	385 G80	385 G80	
Ply/Dodge 6-cyl. 4-brg, L-head							190 201	201 218	201 218

Chart 15.1
(Continued)

- - Model Years and Displacements (cu. in.) - -

Engine Truck series	1937–1938 M - R	1939 T	1940–1941 V - W	1942 W	1943–1945 W	1946 W	1947 W	1948–1949 B-1
Ply/Dodge 6-cyl. 4-brg, L-head		201 218	201 218	218 230	230	218 230	218 230	218 230
Chrysler 6-cyl. L-head 4-brg, shallow skirt	218 228 241.5	228 241.5	228 241.5	237 251	251	237 251	237 251	237 251
Dodge Big-Six Truck 7-brg, L-head	331	331	331	331	331	282 331	282 331	282 331
Dodge Diesel Truck 7-brg, OHV, 6-cyl.		331	331	331	331			

Chart 15.1
(Continued)

- - Model Years and Displacements (cu. in.) - -

Engine / Truck series	1950 B-2	1951–1952 B-3	1953 B-4	1954 C-1	1955 C-3	1956 C-4	1957 K	1958	1959
Ply/Dodge 6-cyl. L-head 4-brg, shallow-skirt	218 230	218 230	218 230	218 230	230	230	230	230	230
Chrysler 6-cyl. L-head 4-brg, shallow-skirt	237 251	237 251	251 265	251 265	251 265	251 265	251 265	251 265	251 265
Dodge Big-Six Truck 7-brg, L-head	307 331 377	307 331 377	307 331 413	413	413				
Chrysler Hemi® Head V-8				331	331	331 354	331 354	354	354
Dodge Hemi® Head V-8				241				315	315
Dodge Poly-Head V-8				241	259 270	259 270	315	315	
A-engine Poly-Head V-8									318

Chart 15.1
(Continued)

- - Model Years and Displacements (cu. in.) - -

Engine / Truck series	1960	1961–1963	1964	1965–1966	1967–1968	1969	1970–1971 F - G	1972 H	1973 J
Ply/Dodge 6-cyl. L-head 4-brg, shallow-skirt	230								
Chrysler 6-cyl. L-head 4-brg, shallow-skirt	251 265	251	251	251	251 Power Wagon				
A-engine V-8	318	318	318	318					
LA V-8				273	318	318	318 360	318 360	318 360
B/RB Heavy-Duty Truck OHV V-8	361 413	361 413	361 413	361 413	361 413	361 413	361 413	361 413	361 413
LG Slant Six		170	170	170	170	170			
RG Slant Six		225	225	225	225	225	198 225	198 225	225
LB V-8					383	383	383	400	400
RB V-8			426 Cust Sprts Pkg	426 Cust Sprts Pkg					440 Motor home

Chart 15.1
(Continued)

- - Model Years and Displacements (cu. in./liters) - -

Engine	1974-1976	1977	1978	1979	1980-1981	1982-1984	1985-1986	1987	1988
LA V-8	318 360	318 360	318 360	318 360	318 360	318 360	318 360	318 360	318 360
B/RB Heavy-Duty Truck OHV V-8	361 413	361 413							
RG Slant Six	225	225	225	225	225	225	225	225	
LB V-8	400	400	400						
RB V-8	440	440	440	440					
2.2-L SOHC 4-cyl. Front-wheel drive						2.2 Rampage			
2.2/2.5-L SOHC 4-cyl. N-S, Rear-wheel drive								2.2 Dakota	2.2 Dakota
3.9-L OHV V-6								3.9 Dakota	3.9 Dakota, B, D, W

Chart 15.1
(Continued)

- - Model Years and Displacements (liters) - -

Engine	1989–1993	1994–1995	1996–1998
LA V-8	5.2 5.9	5.2 5.9	5.2 5.9
2.2/2.5-L SOHC 4-cyl. N-S, Rear-wheel drive	2.5 Dakota	2.5 Dakota	
3.9-L OHV V-6	3.9	3.9	3.9
8.0-L OHV V-10 Truck		8.0	8.0
AMC 2.5-L OHV 4-cyl.			2.5 Dakota

CHAPTER 16

Marine and Industrial Engines—Chrysler Power Afloat and Ashore

Marine Engines

Walter P. Chrysler had been brought up in Kansas, far from any sizable body of water. When he joined Willys Corporation in 1920, he moved to New York City, the location of the company headquarters. First he stayed in hotels, but in 1923, he bought an estate on Long Island Sound. The estate included a boat house and water frontage[1] that evidently paved the way for Walter Chrysler's abiding interest in power boating. Only a year after Chrysler Corporation was founded in 1925, the newly released big seven-main-bearing Chrysler Imperial 80 six-cylinder engine was on the dynamometer, being developed as a marine engine, as can be seen in the rare photo shown on Fig. 16.1. Boat testing of the engine on Long Island Sound started with the completion of two racing-hulled boats. One of them, the twin-propeller *Miss Frolic*, caught at-speed on Fig. 16.2, was powered by the side-by-side Imperial engines shown on Fig. 16.3. *Miss Frolic* performed so well that she was entered in the Gold Cup race of August 22, 1926. Unfortunately, during the race, a propeller broke when she ran over a sunken pile and did not finish. The second and smaller boat, named *Baby Frolic*, was powered with a single Imperial marine engine and took an impressive second place in the Junior Gold Cup race in Detroit.[2]

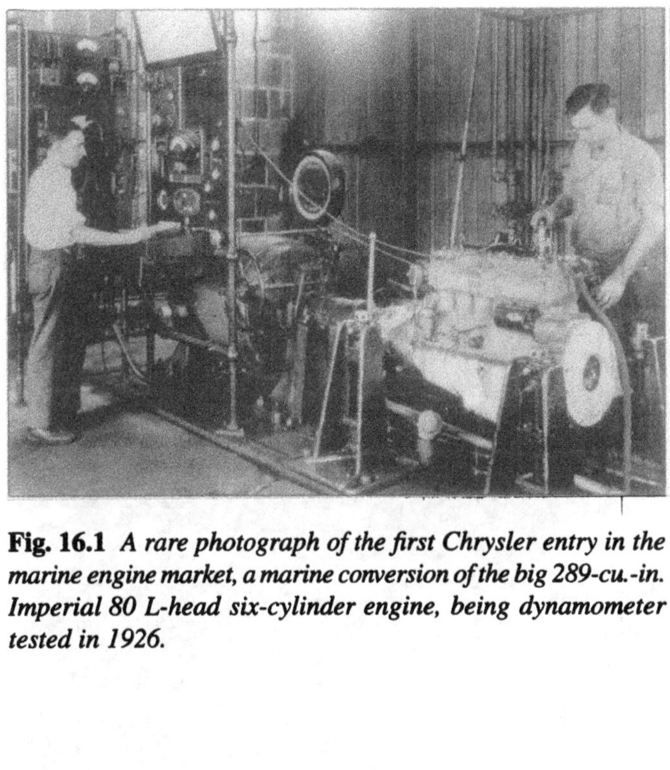

Fig. 16.1 *A rare photograph of the first Chrysler entry in the marine engine market, a marine conversion of the big 289-cu.-in. Imperial 80 L-head six-cylinder engine, being dynamometer tested in 1926.*

With plenty of interest generated from the intended primary customers—boat builders—and with durability and performance tests successfully completed, the Chrysler Marine Engine Division was organized in early 1927 and set up shop inside the sprawling Chrysler Jefferson Avenue plant. By March 1927, sales of the first marine engine of the Chrysler Corporation began—the 288.7-cu.-in. 100-hp Chrysler Imperial six-cylinder marine engine.[3] The next year, 1928, a second engine was added to the lineup, the new engine being a marine version of the original Chrysler seven-main-bearing B-model engine. Named the Chrysler Royal, this engine, shown on Fig. 16.4 and now at 248.9 cu. in., produced 82 hp @ 2800 rpm. The carryover Imperial Marine engine, shown on Fig. 16.5, with a displacement increase to 309.6 cu. in., had a boost in rating to 106 hp @ 2800 rpm. Both engines were available in low-speed, low-horsepower versions, too.[4]

Another significant development that occurred in 1928 was the release of the Chrysler designed, developed, and manufactured reduction gears that made the high-revving Chrysler engines adaptable not only for speedboats with small-diameter propellers but also for large boats that needed slower shaft speeds for their large-diameter propellers. With a spread of available reduction gear ratios, the powerplant assembly now could be

Fig. 16.2 *Chrysler ace test driver Tobe Couture is aboard the sharp-hulled* Miss Frolic, *Gold Cup racer G-15, during boat testing on Long Island Sound.* Miss Frolic *was powered by two Chrysler Imperial marine engines.*

Fig. 16.4 *A 1928 Chrysler Royal Marine six-cylinder engine. A conversion of the original Chrysler B-model engine, now at 249 cu. in., the Royal Marine engine was available at two power levels: 82 hp high duty, and 30 to 50 hp medium duty.*

Fig. 16.3 *The cockpit and engine compartment of* Miss Frolic, *showing her twin Chrysler Imperial 289-cu.-in. six-cylinder marine engines in 1926.*

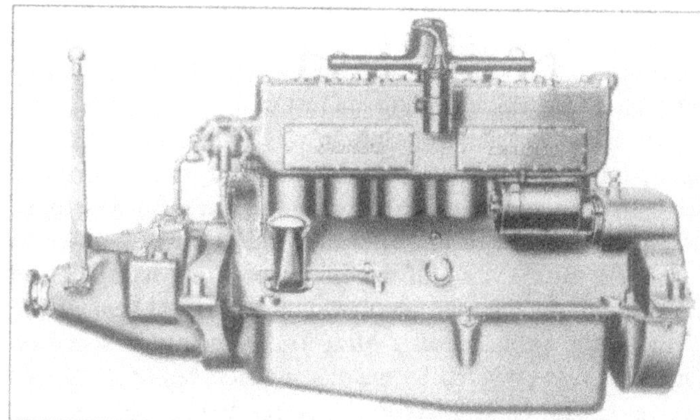

Fig. 16.5 *A 1928 Chrysler Imperial Marine six-cylinder engine. A marine conversion of the 1928 Imperial L80 310-cu.-in. engine, the engine was distinguishable from the smaller Chrysler Royal Marine engine (see Fig. 16.4) by the wider space between the two cylinder block water jacket covers with the cast letters spelling "Chrysler."*

custom built, allowing the engine to run at best power and efficiency for any given boat installation.

In 1931, the two-engine lineup became a three-engine lineup with the addition of a marine version of the huge 385-cu.-in. nine-main-bearing straight-eight Chrysler Imperial car engine. Appropriately named Majestic, the marine version produced 152 bhp. By using high-volume automotive engines as the bases for its marine engines, the Chrysler offerings in the marine engine field combined high quality and high performance with relatively low cost.

As new displacements and engines were added to and deleted from the passenger car and truck rosters, the Marine Division selectively revised its offerings. Two of the many boats that were powered with Chrysler marine engines as original equipment in the 1940s are the speedy-looking 30-ft Gar Wood shown on Fig. 16.6, powered by a pair of Chrysler Royal 323.5-cu.-in. straight-eight L-head engines, and the elegant 45-ft Wheeler shown on Fig. 16.7, powered by twin Chrysler Majestic 331-cu.-in. Big-Six heavy-duty engines.

Figures 16.8, 16.9, 16.10, and 16.11, respectively, show marine engines of 1950: a Chrysler Ace Special M-46 217.7-cu.-in. 105-hp marine engine based on the Dodge and Plymouth 23-in.-block six-cylinder L-head engine, a Chrysler Crown

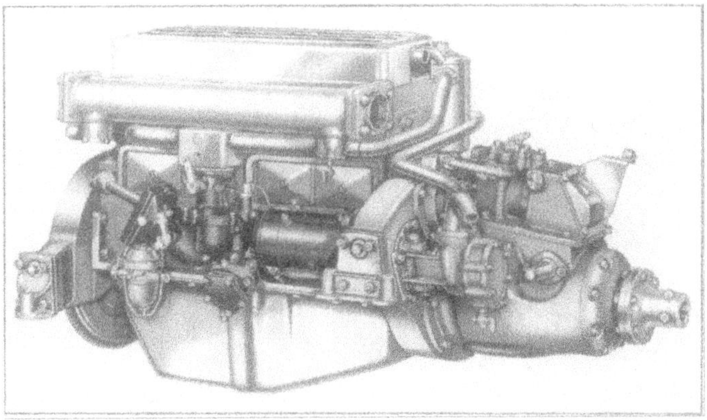

Fig. 16.8 *The 1950 218-cu.-in. Chrysler Ace Special M-46 six-cylinder marine engine, based on the Plymouth and Dodge 23-in.-block L-head six-cylinder engine.*

Fig. 16.6 *A 1940s 30-ft Gar Wood speedboat powered by twin 323.5-cu.-in. Chrysler Royal Marine straight-eight engines.*

Fig. 16.9 *The 1950 251-cu.-in. Chrysler Crown Special M-47 six-cylinder marine engine with twin updraft carburetors, based on the Chrysler 25-in.-block L-head six-cylinder engine.*

Fig. 16.7 *A 1940s 45-ft Wheeler cabin cruiser. Propulsion was provided by twin Chrysler Majestic Marine six-cylinder engines that were marine conversions of the 331-cu.-in. heavy-duty Big-Six truck engine.*

Fig. 16.10 *The 1950 323.5-cu.-in. Chrysler Royal Special M-48 marine engine, based on the Chrysler five-main-bearing L-head straight-eight engine.*

Fig. 16.11 *The 1950 331-cu.-in. Chrysler Majestic M-49 twin-carburetor marine engine, based on the Dodge truck heavy-duty L-head Big-Six engine.*

Special M-47 250.6-cu.-in. twin-carburetor 125-hp marine engine based on the Chrysler 25-in.-block six-cylinder L-head engine, a Chrysler Royal Special M-48 323.5-cu.-in. 165-hp marine engine based on the Chrysler straight-eight five-main-bearing L-head engine, and a Chrysler Majestic M-49 331-cu.-in. 160-hp marine engine based on the Big Six L-head heavy-duty truck engine.

A typical conversion of a passenger car engine to a marine engine involved a number of changes for the different installation and operating conditions of a hull-encased waterborne powerplant. The most apparent among these was a need to change the passenger car normally single-wall exhaust manifold to a water-jacketed manifold to eliminate any hot surfaces that might ignite gasoline fumes that could be present in the usually confined space of the engine compartment. Updraft carburetors were specified for L-head marine engines even after passenger car engines had switched to downdraft, to keep the height of the engine as low as possible for compact, decked-over installations. Instead of a large passenger-car air cleaner, an air filter equipped with a spark arrestor was used.

Other changes to the engine for marine use included an oil pan made of cast iron rather than sheet steel for corrosion resistance in the usually damp environment of the engine space at the bottom of the boat. Also for corrosion resistance, valve springs were plated or made of stainless steel, and the passenger car steel cylinder block core hole plugs were replaced with brass core hole plugs. For the extended durability expected of a marine engine, early engines used timing gears in place of the automotive timing chain. The function of engine coolant circulation usually was taken over by a low-mounted marine water pump. Wedge-shaped adapters were provided to keep the carburetors level, despite any downward angle of the engine for the boat installation. Special engine mounts were provided that conformed to the standardized stringer width for engine placement in the boat. A clutch and clutch housing were attached to the output end of the engine, followed by a reverse gear and a reduction gear, unless the engine was specified with straight 1:1 drive. Reverse rotation engines, built with special camshafts, starter motors, and generators and alternators were produced for use in boats where the extra power and extra maneuverability of twin engines and twin propellers were desirable.

The L-head marine engines usually were positioned in a reverse manner from the customary flywheel-to-rear orientation of passenger car and truck engines. With the propeller shaft needing the engine to be placed at a considerable downward angle as noted, by having the large-diameter flywheel toward the bow of the boat, the entire engine and driveline could be located closer to the bottom of the boat. Because of this reversal, the L-head engine manifolds appear on the as-installed left side of the Chrysler marine engines rather than on the usual right side.

When Chrysler passenger car engines began their fundamental change from L-head to overhead valve systems, starting with the 1951 Chrysler 331-cu.-in. Hemi® head V-8, marine conversions of the new overhead valve engines soon followed. Figure 16.12 shows a mid-1950s Chrysler M-45 331-cu.-in. Hemi® head V-8 marine engine equipped with tandem four-barrel carburetors. More recent V-8 engines included the 1966 273-cu.-in. 195-hp LA V-8 marine engine shown on Fig. 16.13, the 413-cu.-in. RB heavy-duty V-8, the 440-cu.-in. RB, and the 1967 426-cu.-in. 525-hp RB Hemi® V-8 shown on Fig. 16.14, which used the same dual-plenum short cross-ram intake manifold and dual four-barrel carburetors as the 426 RB Hemi® V-8 drag race engine.

Fig. 16.12 *The 1950s 331-cu.-in. Chrysler M-45 Hemi® V-8 marine engine with tandem four-barrel carburetors.*

Fig. 16.13 *The 1966 273-cu.-in. Chrysler Commando 195 marine engine, based on the 273-cu.-in. LA V-8 engine.*

Fig. 16.14 *The 1967 525-hp 426-cu.-in. RB Hemi® V-8 marine engine with dual four-barrel carburetors.*

In a scenario that is somewhat reminiscent of the beginning of the Chrysler Marine Engine Division but this time with better race results, an unlimited hydroplane Gold Cup racing boat—*Miss Chrysler Crew*, U-77 shown on Fig. 16.15, powered by two of the 426 Hemi® RB V-8 marine engines, modified with a longer 4.125-in. stroke to displace 473-cu.-in., supercharged, fuel injected, and made race-ready by performance engine specialist Keith Black[5]—entered the July 2, 1967 World Championship race on the Detroit River. Using the power estimated at 1000 hp per engine, owner and pilot Bill Sterett

Fig. 16.15 *The unlimited hydroplane Gold Cup racing boat,* Miss Chrysler Crew, *U-77, with owner-driver Bill Sterett at the wheel and two supercharged RB Hemi® V-8 engines furnishing power in 1967.*

went airborne in the first turn of the final race but recovered and led the field to victory at the finish line with an average speed of 100.671 mph.[6]

The Slant Six engine, released for passenger cars in 1959, also was readied for marine use. Its standard marine conversion appears on Fig. 16.16. The broad aspects of converting engines for marine application were presented to SAE by Marine Division engineer Neil Newman in a paper titled "The Metamorphosis of a Powerplant—From Passenger Car to Pleasure

Fig. 16.16 *The Chrysler Marine 225-cu.-in. Supercrown 150 Slant Six engine. This engine was designed to be installed in a passenger car orientation.*

Boat."[7] The paper also contained a detailed account of how the low-silhouette Space Saver 225-cu.-in. RG Slant Six marine engine was designed and developed. Figures 16.17 and 16.18 show an outside view and a cross section, respectively, of this unique marine engine.

In the mid-1960s, the marine product line was broadened with the Chrysler purchase of the West Bend Company, a maker of outboard engines, and with the purchase of the Lone Star Boat Company that made various sizes and types of fiberglass-hulled boats.

Marine engines and industrial engines were built at the Chrysler Jefferson Avenue plant since the beginning of marine engine operations in the 1920s. In 1953, the Marine and Industrial Division and its engine assembly moved to a newly built plant in Trenton, Michigan, that later was to become the Trenton engine plant. After two more moves in the Detroit area—first to an office building on Russell Street and then back to the Chrysler Jefferson Avenue plant—the Marine and Industrial Division relocated its offices, engineering, and engine building operations in March 1963 to a facility at Marysville, Michigan, and in 1978, to Beaver Dam, Wisconsin.[8] The year 1980 was the last year for the Chrysler Marine and Industrial Division. Under the terms of the U.S. government-approved Chrysler Corporation Loan Guarantee Act that provided Chrysler with federal loan guarantees to assist its financial recovery, the division was sold.[9]

Fig. 16.17 *The 225-cu.-in. Space-Saver Slant Six marine engine. The box-like casting nearest the viewer forms the oil sump of the engine.*

Industrial Engines

In 1931, four years after Chrysler started building and selling engines for marine use, engines converted to industrial power requirements were produced, with the first customer being the Ready-Power Company of Detroit, which installed the engine in a factory floor truck.[10] With a broad range of passenger and premium-featured truck engines available for conversion, a wide spectrum of industrial applications could be covered. The myriad of uses for the Chrysler industrial engines over the years included power for tractors, motor-generator sets, road machinery, fire pumps, cement mixers, street sweepers, agricultural machinery, forklift trucks, welders, irrigation pumps, air compressors, portable sawmills, cranes, shovels, and motor homes. Figure 16.19 shows a factory scene with a forklift truck powered by a Chrysler industrial engine. Figure 16.20 shows a 331-cu.-in. Industrial Diesel six-cylinder engine at an irrigation well pumping station.

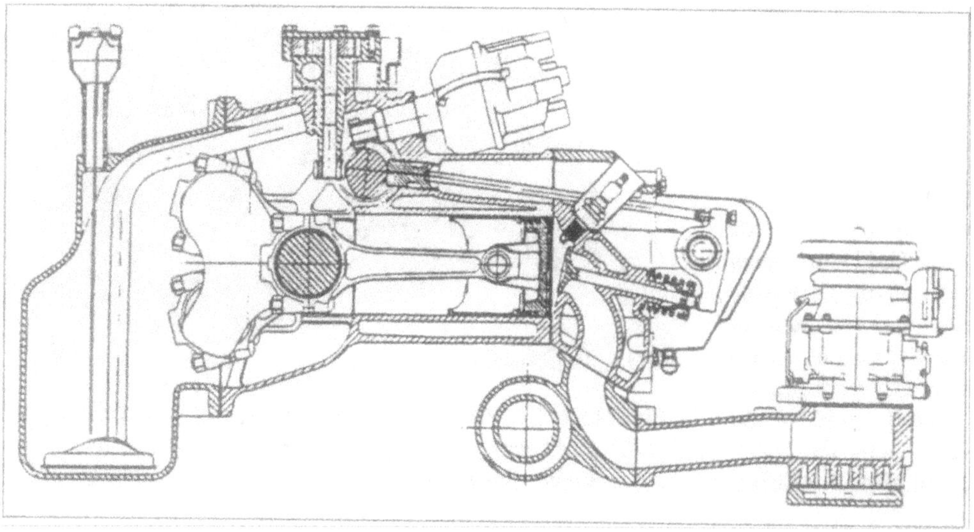

Fig. 16.18 *A cross section of the Space Saver Slant Six marine engine, showing the location of the horizontal cylinder bores, the oil pump and distributor on top, and the manifolds and carburetor at the bottom of the engine assembly. Viewed from the front, the Slant Six engine has been rotated 120-deg clockwise from its passenger car position.*

Fig. 16.19 *Power for forklift trucks, such as the Clark truck shown here, was a popular application of Chrysler Industrial six-cylinder engines during the 1940s.*

Fig. 16.20 *A 331-cu.-in. Chrysler Industrial Diesel six-cylinder engine at a wellhead, pumping irrigation water.*

The Marine and Industrial Division designed and built a series of standardized power package units and, when needed, made custom installations. A Chrysler industrial engine catalog of 1948 shows engines of seven different displacements, including an Industrial Model 5, which was a 218-cu.-in. Plymouth L-head six-cylinder shown on Fig. 16.21; an Industrial Model 12, which was a 323.5-cu.-in. Chrysler five-main-bearing L-head straight-eight shown on Fig. 16.22; and the biggest engine offered, an Industrial Model 14, the 331-cu.-in. heavy-duty Big-Six shown on Fig. 16.23. By 1972, all these engines were gone, and the Industrial Engine catalog of that year listed the H225, an industrial version of the 225-cu.-in. Slant Six engine, and six V-8 engines ranging from the LH 318, an industrial version of the 318-cu.-in. LA V-8, to the big H440, an industrial version of the 440-cu.-in. RB V-8 engine.

Fig. 16.21 *The Chrysler Industrial Model 5 power unit in 1948. The engine was an industrial version of the 218-cu.-in. Plymouth and Dodge L-head six-cylinder engine.*

Fig. 16.22 *The Chrysler Industrial Model 12 power unit. The Chrysler 323.5-cu.-in. L-head straight-eight, with durability upgrades, was the engine used in 1948.*

Fig. 16.23 *The Chrysler Industrial Model 14 six-cylinder engine. This was an industrial version of the 331-cu.-in. heavy-duty Dodge Truck Big-Six engine, circa 1948.*

As mentioned in the preceding Chrysler Marine Engine section, Chrysler Industrial Engine operations ceased after 1980 in compliance with the terms of the Chrysler Corporation Loan Guarantee Act.

CHAPTER 17

Chrysler Power Goes to War—Chrysler Engines During World War II

Wartime Truck, Marine, and Industrial Engines

As a prelude to vastly greater production of military trucks and vehicles for the U.S. Army during World War II, Dodge Truck contracted to build army trucks starting in 1940. These included two types of army trucks: a four-wheel-drive 1-1/2 ton look-alike to a civilian truck, powered with a Chrysler 25-in.-block 228-cu.-in. L-head engine; and a 1/2-ton pickup truck with a modified military-style body powered by a 201-cu.-in. Dodge 23-in.-block L-head six-cylinder engine. For 1941 army vehicles, still produced in relatively low quantities, the Dodge L-head engine moved up to 218 cu. in. and later to 230 cu. in., all the trucks having an unmistakable military appearance and four-wheel drive, as illustrated in Fig. 17.1, which shows a 1941 Dodge-built 1/2-ton Army Command Car powered by a 230-cu.-in. L-head engine. Halfway through the 1942 model year when the United States made its full-scale entry into World War II immediately following the bombing of Pearl Harbor on December 7, 1941, the Dodge Truck plant on Mound Road in Warren, Michigan devoted its entire capacity to building 1/2-ton four-wheel-drive military trucks; starting on April 3, 1944, a handful of carefully rationed civilian trucks were built alongside the olive-drab-painted vehicles.[1] The predominant power for the approximately half-million Dodge army trucks produced during World War II was the 23-in.-block Dodge L-head six-cylinder engine at displacements of 218 and 230 cu. in. An exception to this was an order of 15,000 three-ton, two-wheel-drive Dodge military trucks built for the Chinese government, such as the one shown on Fig. 17.2, which were desperately needed to transport supplies over a tortuous mountain road from India to China, known as the Burma Road. The biggest Dodge truck engine available, the 331-cu.-in. seven-main-bearing heavy-duty L-head Big-Six as described in Chapter 15, was the engine selected for this tough assignment.

Substantial contributions to the war effort were made by the Chrysler Marine and Industrial Division. The Chrysler line of marine engines became available to the U.S. Navy, including the 331-cu.-in. heavy-duty six-cylinder diesel engine. Working closely with the Navy, the Marine Division came up with two innovative floating powerplants. One of the powerplants, called

Fig. 17.1 *A 1941 Dodge 1/2-ton Command Car. One of thousands built for the U.S. Army during World War II, it was powered by a Dodge 218-cu.-in. L-head six-cylinder engine. This vehicle is on display at the Walter P. Chrysler Museum. Photo by W.L. Weertman.*

Fig. 17.2 *Chrysler Chief Engineer of Truck Testing, O.M. "Frenchy" Raes (left), and Captain Tung-sung Kwei (right) of the Chinese Army confer about the 331-cu.-in. Big-Six powered three-ton Dodge military truck, one of 15,000 built for the Chinese Army during World War II.*

a Marine Tractor, had a single Chrysler Royal 323.5-cu.-in. straight-eight L-head marine engine inside a steel pontoon equipped with a rudder, steering controls, and engine controls for the single propeller. Similar to an ordinary outboard motor but much larger in size, it could not operate alone but was designed to always be attached to one or more cargo-carrying pontoons to form a complete craft. The other floating powerplant, shown on Fig. 17.3 and called a Sea Mule, was designed as a complete craft. Looking like a short tugboat, the Sea Mule was built in several sizes, all powered with Chrysler Royal 323.5-cu.-in. straight-eight marine engines in either single- or twin-engine installations, as pictured on Fig. 17.4. It could both pull and push and was built by the thousands at the Chrysler Jefferson Avenue plant for the Army as well as the Navy.

Fig. 17.3 *The Chrysler Marine World War II Sea Mule.*

Chrysler Industrial engines also went to war, being adaptable to new war-related duties such as powering air-raid sirens and firefighting pumps, with about 30,000 engines built in 1943 alone.[2]

Fig. 17.4 *The engine compartment of the Model 4-A Sea Mule, showing the installation of twin Chrysler 323.5-cu.-in. straight-eight marine engines.*

Five-Bank Tank Engine

The building of the Chrysler Detroit Tank Arsenal was being completed when the United States entered World War II in December 1941. The only medium tank ready for production, characterized by its main 75-mm cannon being mounted on the right side of the hull with limited horizontal traverse, was the U.S. Army M3 tank, shown on Fig. 17.5, powered with a Continental Motors-built 975-cu.-in. radial air-cooled nine-cylinder aircraft-type engine. Lack of a design of a turret with the strength and capacity for handling a large gun at the time when the tank originally was designed led to the unusual placement of the main cannon. With escalating wartime demands for aircraft engines, the Continental radial engine immediately became in short supply.

However, Chrysler did have its 251-cu.-in. 25-in.-block six-cylinder L-head engine available in large numbers. Under the overall direction of Executive Engineer Harry T. Woolson, the Engine Design department, headed by Mel Carpentier, came

Fig. 17.5 *The U.S. Army M3 tank with a side-mounted 75-mm cannon. The innovative Chrysler five-bank engine replaced the original aircraft-type radial air-cooled engine.*

up with an innovative powerplant that was coded A57, which combined five of these engines into a single assembly. One of these five-bank engine assemblies, serial no. 10647, was located in Argentina only a few years ago. It was restored and is on display at the Walter P. Chrysler Museum, as can be seen on Figs. 17.6 and 17.7, which show left-rear and right-rear views of the engine, respectively. With relatively minor changes to the

Fig. 17.6 *A left-rear view of the World War II 1253-cu.-in. five-bank tank engine. The engine, on display at the Walter P. Chrysler Museum, carries the serial no. 10647. A single water pump, at the center of the engine assembly, sends coolant to the five engines through individual hoses. Photo by W.L. Weertman.*

Fig. 17.7 *A right-rear view of the World War II five-bank tank engine that is on display at the Walter P. Chrysler Museum. The black-painted vertical pipe shown is the downpipe for the high-mounted carburetor that feeds the intake manifold of the lowest engine on this side. Photo by W.L. Weertman.*

tank, which did include lengthening the engine compartment, the five-bank Chrysler engine replaced the radial engine, with so-equipped tanks having the special designation of M3A4.

The following are highlights of the details of the design of the engine assembly, which were obtained from a report in the World War II files of the Chrysler Historical Collection[3] and from archival microfilm files there.

The cornerstone of the design of the five-bank engine assembly was a large cast iron crankcase that formed the base structure of the powerplant assembly. Two of the Chrysler 251-cu.-in. L-head six-cylinder engines were bolted to the lower portion of the crankcase, with their cylinders angled 7-1/2 deg above horizontal. Two more engines were bolted to the crankcase above the first two, with their cylinders 27 deg above horizontal. The fifth engine was bolted vertically at the top of the crankcase, giving the entire assembly somewhat of a pentagon appearance. All five engines were mounted with their crankshaft output flanges to the front and their noses to the rear of the tank. Two oil pumps were located in the sump at the bottom of the powerplant assembly: a scavenge pump to transfer spent lubricating oil to a remote reservoir, and a pressure pump, fed from the reservoir, which delivered high-pressure oil to all five engines.

Each engine had its own ignition coil and distributor that was driven from the nose of its camshaft. The tank engines used the same heavy-duty gear drive for the camshaft as was used on the marine and industrial 251-cu.-in. L-head engines. Originally equipped with five belt-driven water pumps, the design later was changed to a single centrally mounted auxiliary-shaft-driven water pump with five outlets—one for each engine. The single-barrel carburetors of the first engines to be produced were mounted directly on the intake manifolds. This arrangement later was changed to locate all five carburetors at the top of the engine assembly, each connected to its respective engine with a downpipe, a design reflected in the engine on display.

A gear case, bolted to the front of the engine assembly, held five drive gears that meshed with a single driven gear, which in turn drove the output shaft of the powerplant, all gears having herringbone teeth. A coupling connected each crankshaft output flange with one of the drive gears.

The combined 30 cylinders displacing 1253 cu. in. produced 445 hp and a tank-moving 1060 lb-ft of torque. After completing 109 M3A4 tanks with the five-bank engine, with the first one built on June 3, 1942,[4] the M3 tank was replaced with a new and far superior M4 Sherman tank that had its 75-mm cannon mounted in a top turret, with 360-deg traverse capability. Figure 17.8 shows a five-bank engine being lowered into a partially finished M4 tank at the Chrysler Tank Arsenal. Two completed tanks, designated M4A4 with their five-bank engines, can be seen inside the Tank Arsenal as shown on Fig. 17.9. The first production tank of this type rolled off the end of the line on June 30, 1942. From April 9, 1942 to September 1943, 9,965 five-bank engines were built, with 7,500 installed in production tanks and the balance built as spare engines.[4] Many of the tanks were sent to the British Army, where the engine was well regarded.[5]

Fig. 17.8 *Installing the Chrysler five-bank tank engine into an M4 tank at the Chrysler Tank Arsenal in Warren, Michigan, on August 11, 1942.*

Fig. 17.9 *Two M4A4 tanks powered by Chrysler five-bank engines at the end of the assembly line at the Chrysler Tank Arsenal on September 8, 1942.*

Table 17.1 shows the specifications of this unique five-bank engine.

XIV 2220 Inverted V-16 Aircraft Engine—Sixteen Cylinders of Airborne Power

As early as May 23, 1940, the U.S. Army Air Forces (AAF) expressed interest in having Chrysler design and develop a 2000-hp liquid-cooled aircraft engine with a small frontal area.[6] Design studies were prepared for a sixteen-cylinder, inverted V-type engine with a reduction gear train located at the center of the engine. Engineering coded A45, the engine achieved the desired small frontal area, fitting within a 33.5-in.-diameter circle, but was lengthy at 10 ft 5 in. Following design completion, prototype components were procured and brought to the Engineering Research laboratories in the Chrysler Engineering complex at Highland Park, Michigan. They were individually tested as a prelude to multicylinder engine procurement and testing.

Initial dynamometer development tests were run on single-cylinder and twin-cylinder test engines that replicated a single bay of the complete engine. The first multicylinder engine was started and run under its own power on July 30, 1942.[7] Figure 17.10 shows the fifth multicylinder engine built during the program, now on display at the Walter P. Chrysler Museum, carrying serial no. D-000005.

The engine had a bore and stroke of 5.8 × 5.25 in., giving a displacement of 2220 cu. in. The Army Air Forces designator of the engine, XIV-2220, stood for "Experimental, Inverted, V-type, 2220 cu. in."[8] The designator usually was shortened to XI-2220, which later became official.[9] The overall design

TABLE 17.1
SPECIFICATIONS FOR THE WORLD WAR II FIVE-BANK TANK ENGINE

Years of manufacture	1942–1943
Used in	M3A4 and M4A4 medium Army tanks
Engineering program code	A57
Displacement (cu. in.)	1253
Bore	3.4375
Stroke	4.50
Compression ratio	6.6:1
BHP	445
Torque (lb-ft)	1060
Engine assembly	Five Chrysler 250.6-cu.-in. L-head engines (30 cylinders) mounted on a common cast iron crankcase
Engine internal specifications	Same as 1942 Chrysler 250.6-cu.-in., four-main-bearing, shallow-skirt, six-cylinder engines except as follows:
Camshaft drive	Gear
Oil pumps	Central pressure and scavenge pumps driven by accessory shaft
Water pump	Early engines—individual pumps; later engines—central water pump
Fuel system	Five individual one-barrel carburetors
Output drive	The five engines were coupled to five herringbone gears that meshed with a single large herringbone gear mounted on the output shaft.

Fig. 17.10 *A front-left view of the XI-2220 inverted V-16 aircraft engine, serial no. D-000005, on display at the Walter P. Chrysler Museum. The large round objects near the front of the engine are magnetos—one for each bank. Flex lines and tubing enclose the high-voltage wires to the dual spark plugs in each cylinder. Photo by W.L. Weertman.*

arrangement of the engine can be seen in Figures 17.11 and 17.12, which show a cross section and a longitudinal section, respectively. A single aluminum casting formed the entire crankcase with its two banks of cylinders located 60 deg apart. Two four-throw steel crankshafts were used, bolted together end-to-end at the center of the engine with a spur gear sandwiched between them. The crankshaft throws of the combined eight-throw crankshaft, in the same way as found on a V-8 engine, were 90 deg apart. The gear between the crankshafts meshed with a larger spur gear that drove both the propeller shaft, located in the front half of the engine, and an auxiliary shaft, located in the rear half of the engine. The reduction ratio of the gear set was 2.457:1.

The spacing between the cylinder centers was 7.62 in., except for cylinder pairs #4 and #5 at the center of the engine, which were 14.62 in. apart. The cylinders for the two banks were located exactly opposite each other, so cylinder pairs were numbered 1 through 8, with an R or L added to specify the right- or left-bank location. Ten main bearing caps were used for the 3.25-in.-diameter main bearings, and each cap was

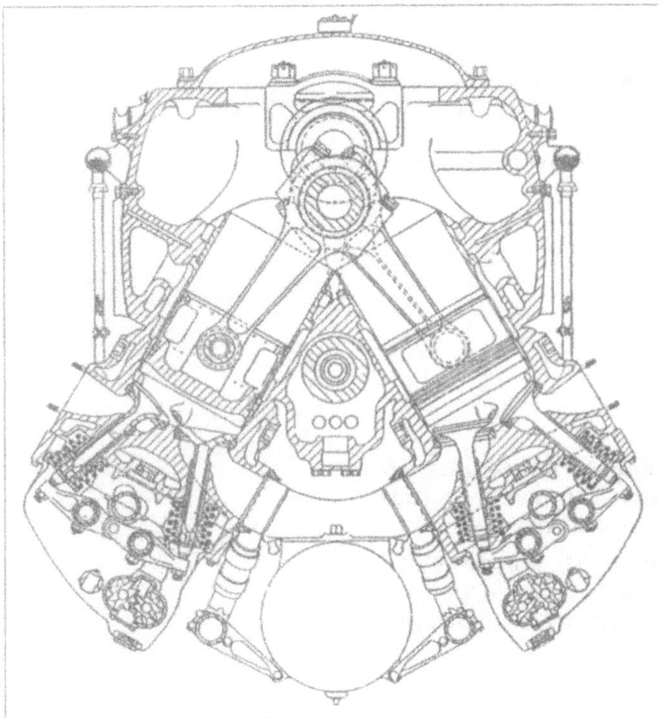

Fig. 17.11 *A cross section of the XI-2220 inverted V-16 aircraft engine. Each cylinder had its own cylinder head with a screwed-in steel cylinder sleeve. Oil scavenge pumps can be seen at the bottom of the cylinder head covers (also called cam box covers).*

attached with two vertical 3/4-in.-diameter studs and nuts. In addition to the attaching studs and nuts, two long horizontal 3/8-in.-diameter tie bolts extended across the crankcase through clearance holes in each cap. When tightened, the tie bolts clamped the crankcase to the sides of the main bearing caps for a strong and rigid assembly.

Instead of the usual automotive practice of attaching the cylinder heads with a number of bolts around each combustion chamber, a different design was used that did not require any head bolts. Each cylinder had its own individual aluminum cylinder head and forged steel barrel. As illustrated in the section view on Fig. 17.13, the barrel screwed into the cylinder head using an interference-fit thread. The sixteen head and barrel assemblies were attached individually to the crankcase by a large ring nut that threaded onto the lower end of the barrel from inside the crankcase. Figure 17.14 shows an outside view of an early cylinder head and barrel.

In common with many aircraft engines, the cylinder head design incorporated a hemispherical combustion chamber with two valves per chamber. Sodium cooling was applied to both the intake and exhaust valves, and, for ignition reliability, two spark plugs were used in each combustion chamber, a customary aircraft engine practice. The valves were operated by rocker arms with a lash adjusting screw and lock nut on the valve tip end and a roller follower on the camshaft end. Each overhead camshaft (using upright engine terminology) consisted of two four-cylinders-long camshafts bolted together at the center of the engine, with a gear clamped between the two. A long camshaft housing, running the length of the engine, was attached to the tops of the eight cylinders in each bank. A camshaft cover, often called a cam box cover and of similar length, completed the oil-tight enclosure of the camshaft and valve operating system.

As mentioned, the two banks of the engine were exactly opposite each other without the bank offset used in V-8 engines. This was possible by the use of "blade and fork" connecting rod lower ends, which had two bearing segments of the rod from one bank straddling a single bearing segment of the rod from the opposite bank.

A large double-row Timken roller bearing located near the front end of the propeller shaft absorbed the radial and thrust loading of the propeller. A sleeve bearing, called a transfer

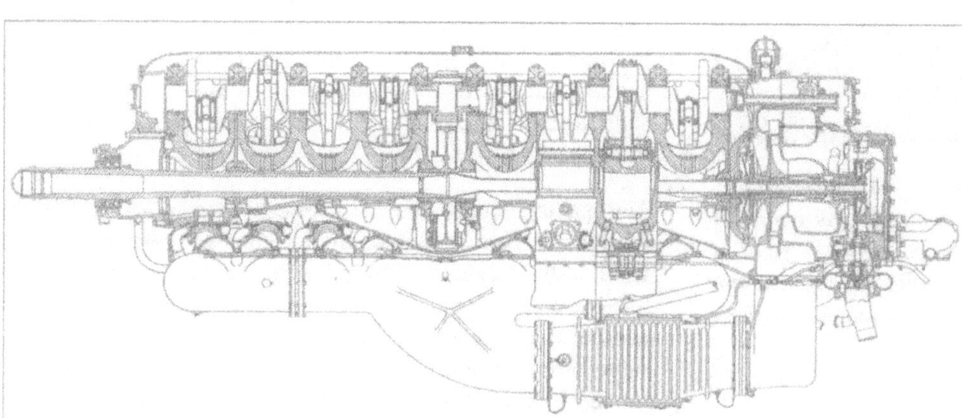

Fig. 17.12 *A longitudinal section of the XI-2220 inverted V-16 aircraft engine. A gear train located at the center of the engine transmitted power to the long and hollow propeller shaft on the left. The thinner shaft on the right transmitted power to the auxiliaries, including the centrifugal supercharger shown in the section. The ribbed box below the engine was an aftercooler for the compressed fuel/air intake charge.*

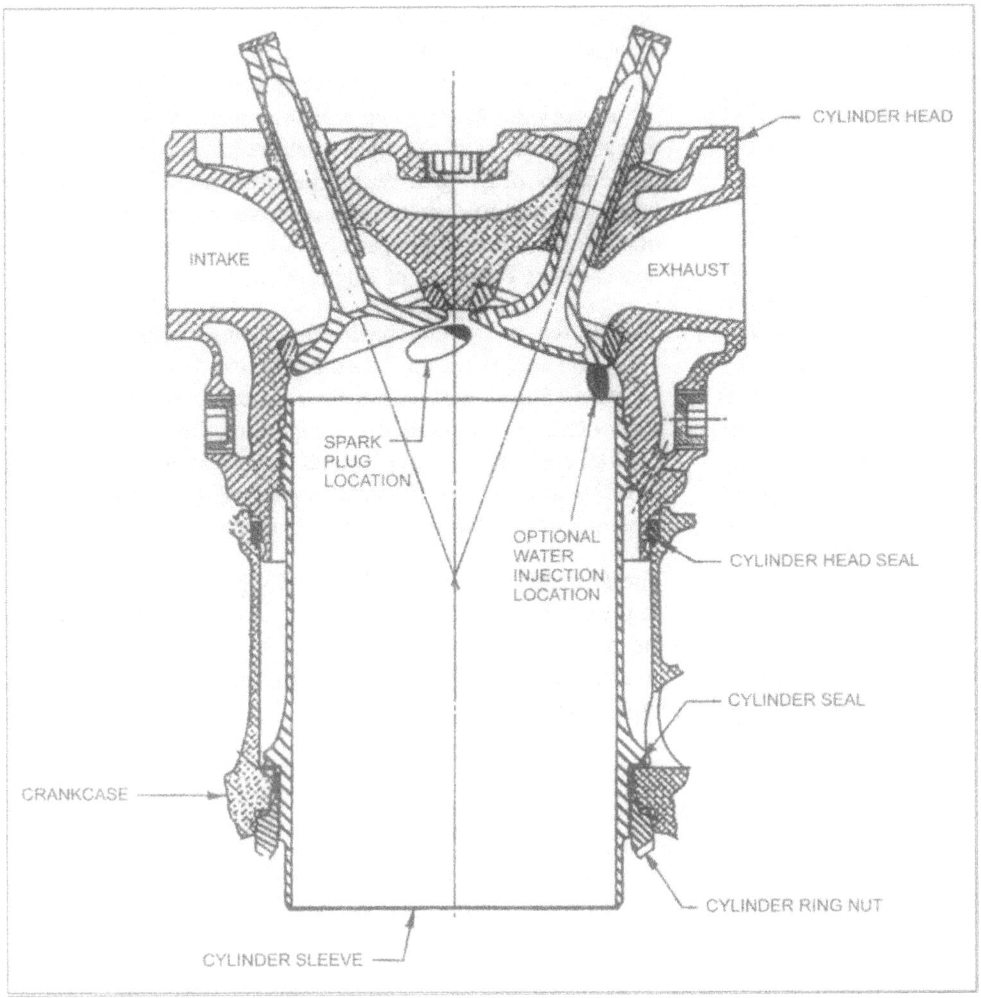

Fig. 17.13 *The cylinder assembly of an XI-2220 aircraft engine, shown in an upright position rather than an inverted position. With this design, cylinder head bolts were not required.*

Fig. 17.14 *An XI-2220 early cylinder head assembly on test. A sensor occupies one of the spark plug holes.*

bearing, was located farther aft on the propeller shaft. The rear end of the propeller shaft was splined into the hub of the reduction gear. Of interest, at full load, the 54-in.-long, hollow, and alloy steel propeller shaft deflected 4.9 deg in twist that, multiplied by the reduction gear ratio, gave an automatic 12 deg of spark retard.[10]

The auxiliary driveshaft, already mentioned, drove the centrifugal supercharger and all the accessories mounted to the accessory housing at the rear of the engine. The gearing for the entire engine was a complex affair, requiring some 40 gears, as can be seen on Fig. 17.15. The early decision to locate the main reduction gear set at the center rather than at one end of the engine (the usual design practice) was based on a desire to avoid potential gear and propeller problems originating from crankshaft torsional vibrations.[11] In addition to dynamometer testing, the engines were tested at rated power with propellers to simulate airplane service, as shown on Fig. 17.16.

The main reduction gear set at the center of the engine had the formidable task of transmitting 3000 hp at 3400 rpm—2500 hp for the propeller (raised from the original 2000 hp at the request of the Army Air Forces), and 500 hp for the supercharger, oil pumps, water pump, and other accessories. Syd Terry, in charge of gear design, with the help of Earle Buckingham, a consultant from Massachusetts Institute of Technology, designed the gear teeth profiles so that two teeth were always in contact, resulting in the gears being about half the width of conventionally designed gears, with savings in both weight and space.[12] Other refinements in the gear design included provision for inertia loads and, where necessary, contour adjustments for deflections under load. Later in his career at Chrysler, Syd Terry rose to become Vice President of Engineering.

The lubrication system consisted of a single large gear-driven 65-psi oil supply pump; a number of small scavenge pumps, including one at each end of the cam box covers; dual oil-to-air coolers; and a single oil storage tank located high on the firewall behind the engine, which had a downpipe connecting it to the suction side of the oil supply pump. This dry sump system was appropriate for an aircraft engine.

The gear-driven centrifugal supercharger, located at the rear of the engine, had a single 11-in.-diameter impeller. During the program, two-stage and axial-flow superchargers were investigated; however, the final selection was the single-stage supercharger.

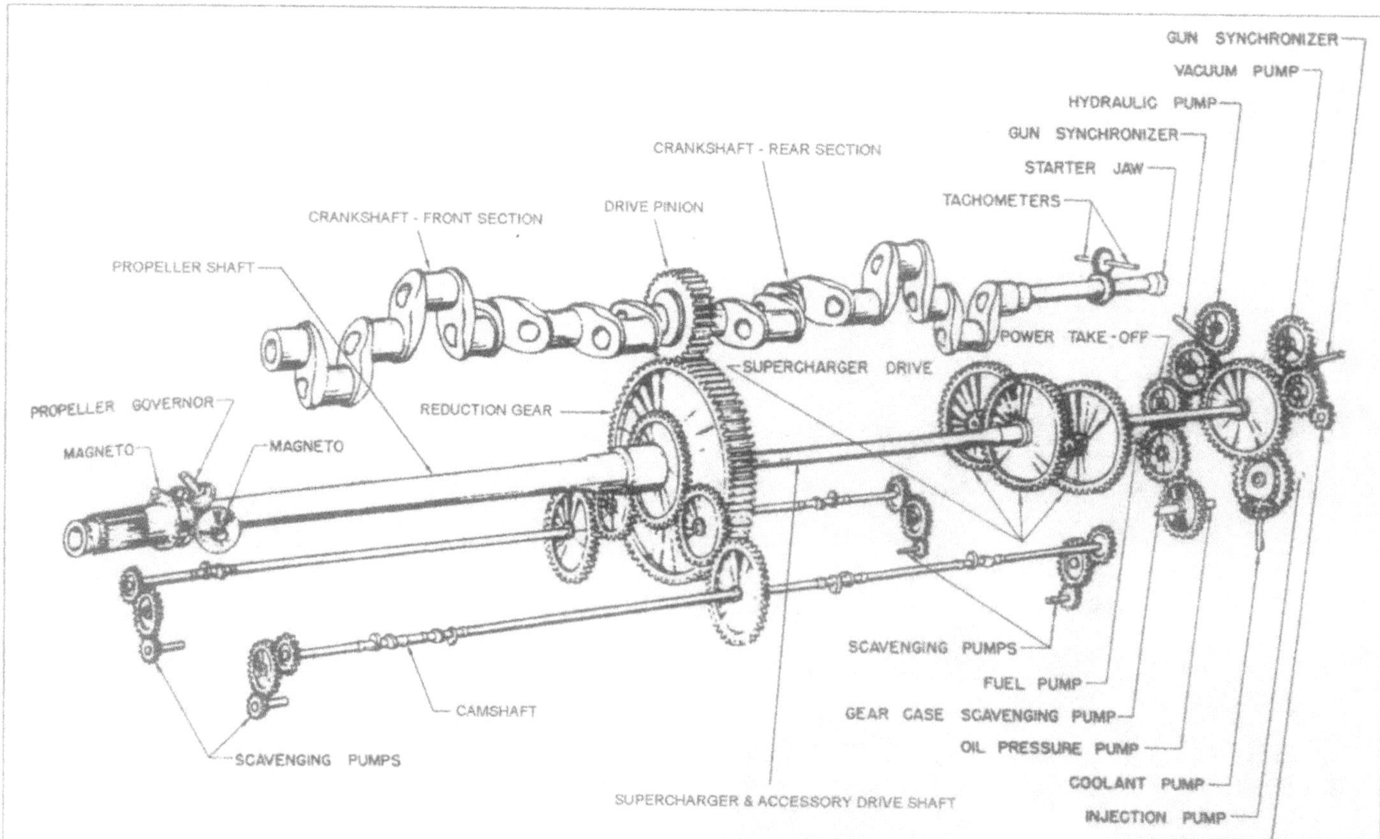

Fig. 17.15 *A diagram of the XI-2220 V-16 engine gear system. Gear design was a major engineering task associated with the engine.*

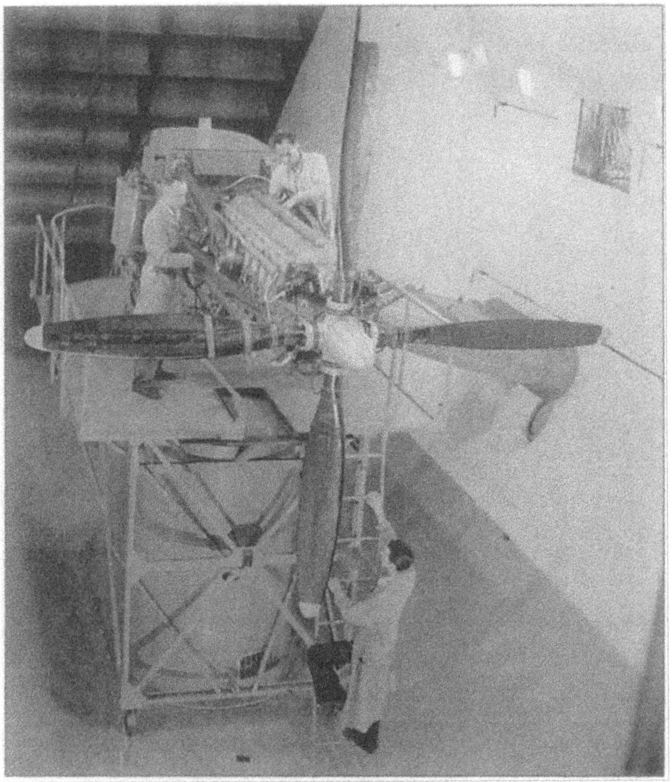

Fig. 17.16 *A wind-tunnel stand used for testing the XI-2220 engine, complete with the intended 13-ft-diameter Curtiss-Electric four-blade propeller.*

The ignition system consisted of two large, circular, and gear-driven Wico magnetos located at the front of the engine; high-tension cables; and two simultaneously firing spark plugs in each cylinder head, as already mentioned. From each magneto, the high-tension cables were enclosed in a long metal conduit that ran the length of the cylinder block, just below the crankcase cover rail. At each cylinder, dual flexible metallic hoses led the cables from the conduit to the spark plugs. Because of the 60-deg bank angle, the sixteen cylinders, and the 90-deg crankshaft throws, the cylinders fired at alternating 30- and 60-deg intervals.

Not long after the start of the program, the Army Air Forces informed Chrysler that, after considering a number of different aircraft, it had decided that a Republic Aviation P-47 Thunderbolt fighter airplane would be the most appropriate aircraft to use for flight testing of the engine.[7] With this decision, contacts with Republic Aviation were established, and installation design work by Chrysler, including scale models, was begun. Figure 17.17 shows a diagram of the flight test airplane.

The intake air for the engine entered the airplane through two oval-shaped openings in the fuselage behind the pilot. The air flowed downward through twin ducts to the exhaust gas turbosupercharger (years later, the word generally was shortened to "turbocharger" or simply "turbo") located below the openings. The compressed air then was routed forward through dual air-to-air intercoolers for temperature reduction and then through dual ducts to the Bendix pressure carburetor at the rear of the engine.

The fuel/air mixture from the carburetor entered the single-stage centrifugal supercharger of the engine for a second pressure boost, followed by a pass through the liquid-cooled aftercooler mounted to the bottom of the engine for a second temperature reduction. A large elbow connected the intercooler outlet to the intake manifold of the engine for distribution to the sixteen cylinders. A charge pressure of 72 in. Hg (35 psi) was needed to achieve the 2500-hp output target of the engine. The primary function of the turbocharger was to counteract the debilitating effect of lower air density at higher altitudes and to maintain the 72-in.-Hg intake manifold pressure capability up to an altitude of least 30,000 ft.

A considerable amount of development was done with direct cylinder fuel injection that looked promising, to the extent that the build level designation XI-2220-5 was pre-assigned to any multicylinder fuel-injected engines that might be built.[8] However, fuel injection work was terminated in a major program reduction that took place in April 1945[7] before any fuel-injected multicylinder engines were built.

Two separate liquid cooling systems were used for the engine. To cool the engine, a gear-driven engine-mounted pump circulated the ethylene glycol coolant mixture through the engine and then through a ram-air-cooled radiator that was part of a large radiator package located behind the pilot, from which the coolant returned to the pump inlet. Another system cooled the intake charge air twice—the first time after it left the turbocharger outlet, and the second time after it exited the supercharger outlet, as already described. A smaller, gear-driven, and engine-mounted pump did the task of circulating the ethlyene glycol of this system through the coolers and through its radiator, also in the package behind the pilot where ram air carried away its heat. The expansion tanks for both cooling systems were mounted above the engine, and both systems used movable air doors on the sides of the fuselage behind the pilot to regulate coolant temperatures. A major engineering program all by itself, the combined engine and intake charge cooling systems required a staggering amount of complex ductwork, heat exchangers, piping, and controls that were hidden under the aluminum skin of the aircraft.

Although the selection of the P-47 as the flight test airplane had been made early in the program, the decision on where the installation was to be done and which company would be in charge was not made until August 1944 when, with agreement by all parties, the Army Air Forces decided that the installation would be made at the Republic Aviation Modification Center at Evansville, Indiana and that Chrysler would be the prime contractor, with Republic a subcontractor to Chrysler for any work that Chrysler would want Republic to perform.[7]

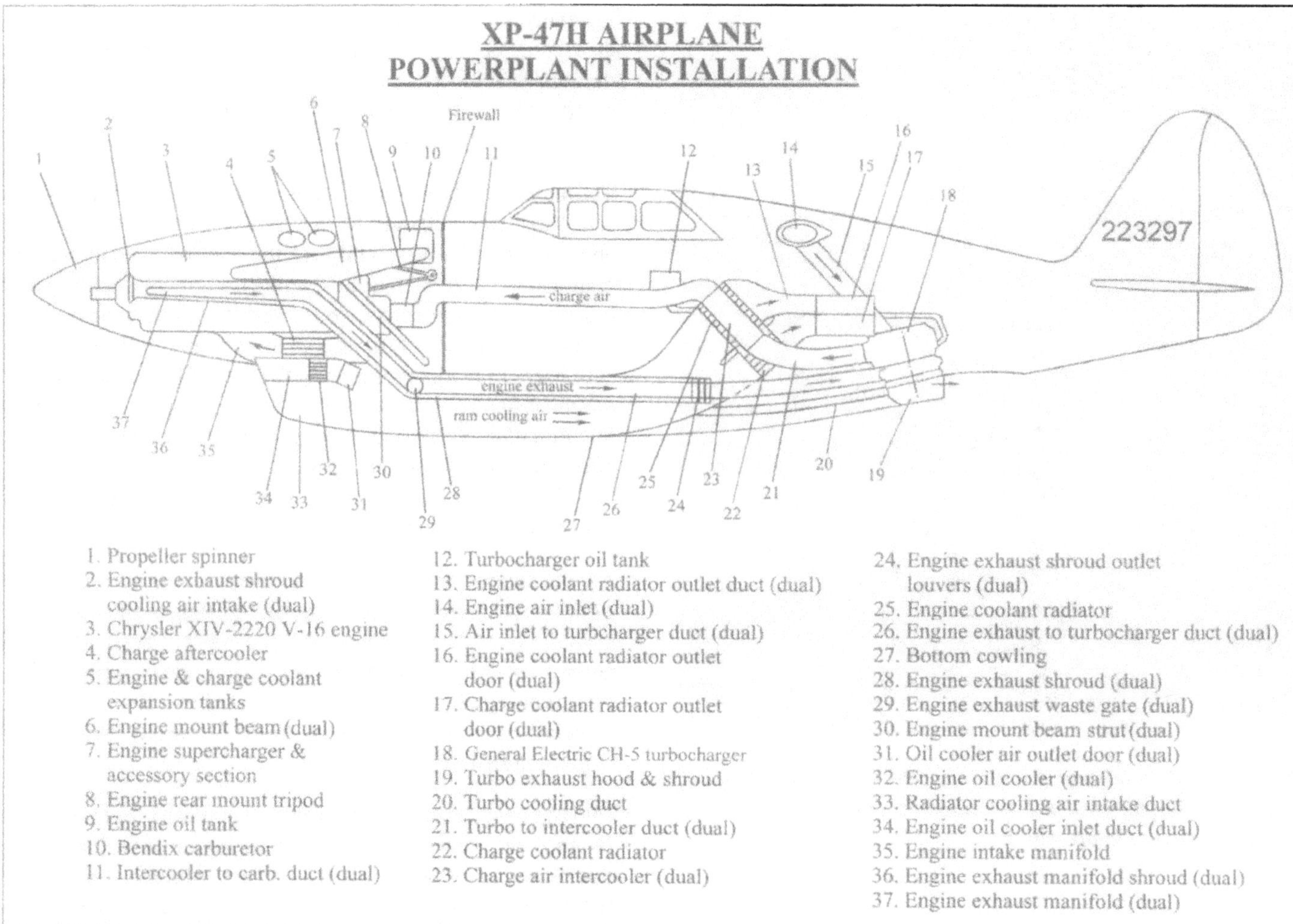

Fig. 17.17 *An installation diagram of a flight test airplane with an XI-2220 engine. Modifying the airplane to accept a liquid-cooled engine in place of its original air-cooled engine required lowering the underside of the fuselage to accept the large ductwork and heat exchangers that were needed.*

Chrysler engineers arrived at the Modification Center on September 4, 1944 to start work. The first of two airplanes to be modified, furnished by the Army Air Forces, arrived on September 17, 1944.[7] The second arrived in October. The Chrysler engineer in overall charge of the installation was Carl Stamm, with Chrysler engineer Charles Williams in charge of the airplane engine test program, assisted by Paul Wise and others in the Chrysler group.[13] Jack Riley was the Republic engineer in charge of the flight test program.

As delivered, the P-47D fighter airplanes were powered by a Pratt & Whitney R-2800 air-cooled radial twin-row 2000-hp engine, with a General Electric type C-1 turbocharger mounted in the fuselage behind the pilot, and dual air-to-air intercoolers for temperature reduction of the charge air exiting the turbocharger.

The redesign and modification needed for the installation of the Chrysler XI-2220 engine required removal and replacement of the entire structure forward of the firewall, lowering and enlarging the ductwork and cowling below the fuselage, replacing the original intercoolers with the radiator package previously mentioned, and replacing the original turbocharger with a larger-capacity General Electric type CH-5 turbocharger, already in use on the more recent Model P-47N aircraft.

Figure 17.18 shows the first Thunderbolt P-47 airplane to be converted, serial no. 42-23297, with its first Chrysler V-16 engine, serial no. D-000002, in place. This engine, shipped from Chrysler Engineering to the Republic Aviation Modification Center on January 29, 1945, was to be used for buildup and initial ground tests. The converted airplanes were given a special designator of XP-47H.

However, while this activity was taking place at Evansville and the engine was undergoing continued development at Chrysler Engineering Research in Highland Park, the advances of the Allied armies in Europe were going so well that the importance and priority of the Chrysler efforts on the XI-2220 engine program were greatly diminished. On April 25, 1945, with

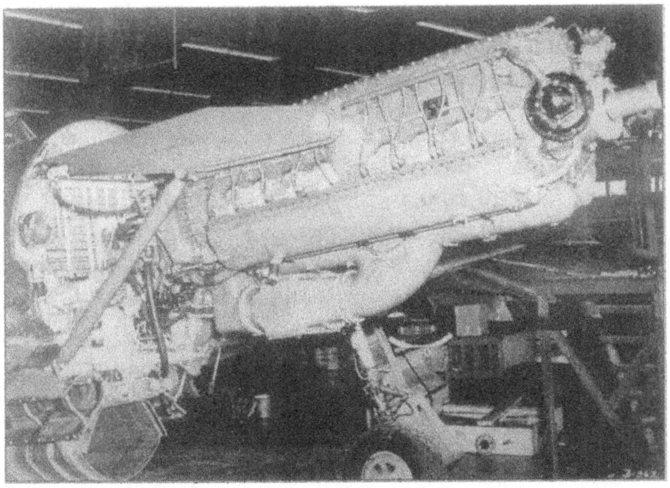

Fig. 17.18 *Conversion of the Thunderbolt P-47 airplane, serial no. 42-23297, into the XP-47H, with the initial XI-2220 engine, serial no. D-000002, in place. The large duct beneath the fuselage transported ram air to a heat exchanger behind the pilot. Republic Aviation Modification Center, Evansville, Indiana.*

the conversion of the first airplane not yet completed, the Air Technical Service Command advised Chrysler that the entire engine program would be reduced to only the activity needed to support the flight tests. This program limitation effectively terminated any further development on the engine, including work on fuel injection and superchargers. It also removed the engine from any consideration for production. One of the reasons given by the Air Technical Service Command was that gas turbine engines were being developed rapidly for future aircraft.[7]

The airplane conversion work continued at the Republic Aviation Modification Center. On May 31, 1945, with the process nearing completion, the group picture shown on Fig. 17.19 was taken. About one month later, on June 28, 1945, the conversion was completed, and aircraft XP-47H, serial no. 42-23297, powered with the Chrysler XI-2220 engine, serial no. D-000002, started ground tests.

On July 6, 1945, flight test engine D-000006 was accepted by the Army Air Forces at Chrysler Engineering and was shipped to the Republic Aviation Modification Center, followed by the acceptance and shipping of a second flight test engine, D-000001, on July 13, 1945. When ground testing was completed, engine D-000002 was removed from the aircraft as planned, and engine D-000006 was installed. The airplane was instrumented with sensors and onboard recorders to record conditions as the pilot took the airplane through the prescribed Army Air Forces speed, climbing, and maneuver tests. Note that the primary objective of the flight tests was to find and resolve any engine operation problems in the airborne environment. Although the converted XP-47H with its sharply pointed nose looked faster than the blunt-nosed standard P-47 airplane with its radial air-cooled engine, no wind-tunnel work had been done to assess actual drag levels.

On July 27, 1945, all was in readiness for the first flight of the XP-47H airplane, which was scheduled as a short test to ensure

Fig. 17.19 *The joint Chrysler and Republic Aviation task group poses with Carl Breer, who is wearing a suit and is standing with his hand in his pocket, on May 31, 1945. Breer was the Director of Research for Chrysler at the time.*

that everything was working properly. A few minutes after a successful takeoff, as shown on Fig. 17.20, the perhaps overly exuberant pilot put the airplane in a horizontal roll, a maneuver he was specifically warned not to do because development for inverted flight had not been completed. The oil pressure dropped, and the pilot immediately landed the airplane with a total of six minutes of airtime. Later inspection showed that a single main-bearing-cap stud was loose and that the main bearing shells were in need of replacement, but no major damage had been done. With repairs accomplished quickly, flights resumed on July 31 without incident. On the following day, two flights were made, completing the engine slow time schedule with completely satisfactory operation. However, with four engine hours now accumulated, inspection revealed five loose main-bearing-cap studs.[14] This was a serious condition because a loose main bearing cap can lead to a broken crankshaft.

Fig. 17.20 *The first flight of the XP-47H airplane powered by the Chrysler XI-2220 inverted V-16 engine on July 27, 1945, at the Republic Aviation Modification Center in Evansville, Indiana.*

A decision was made to return flight test engine D-000006 to Chrysler Engineering for repair and to install the second flight test engine D-000001 in the airplane. As a precaution, locks were installed on all main-bearing-cap studs to ensure that the loosening would not recur. Testing resumed, and much data were accumulated, including engine cooling conditions under various rates of climb. A number of problem areas that needed improvement were revealed.

Local over-temperature conditions, corrected with better heat shields, were among the minor problems. However, a persistent and serious problem of pressure surging in the turbocharger outlet ducts was encountered.[14] With limited development resources available, the surging, which was transmitted through the entire intake system to cause sudden engine power fluctuations, was considerably reduced but was not eliminated.

The engine charge air inlet openings originally were flush with the outer skin of the airplane. However, after the first series of tests, forward-facing scoops were added to the openings to improve airplane performance at low engine revolutions per minute. The in-flight photograph of the XP-47H airplane 42-23297 on Fig. 17.21 shows a forward-facing scoop.

Fig. 17.21 *The XP-47H airplane in the process of being flight tested in Evansville, Indiana, on September 28, 1945.*

November 5, 1945 was an eventful day for the engine program. The first flight of the day started routinely, but the pilot returned after 35 minutes of flight time, complaining of turbocharger surge at 28,500 ft as he attempted to climb to 30,000 ft. A decision was made to have the test pilot, Tom Cushman, make another flight, this time going to only 20,000 ft for level flight data. Then disaster struck. According to Chrysler engineer Paul Wise, who witnessed the flight, the aircraft took off, made a 180-deg turn, and flew parallel to the runway at 500 ft. Then the engine suddenly became silent.[15] The propeller shaft had broken,[14] and with the magnetos totally out of sync, the engine had stopped running. Fortunately, the break was to the rear of the transfer bearing, so the freewheeling propeller remained in place. Thus, the pilot was able to make a non-injury dead-stick landing, hitting the ground short of the runway and then bouncing onto the runway.[15] The engine, serial no. D-000001, was removed and returned to Chrysler Engineering; the airplane later was scrapped.[16] A report of the propeller shaft failure, prepared by the Air Technical Service Command after inspection of the failed shaft at Chrysler Research,[17] concluded that boost pressure surges caused engine misfiring that imposed such high torques on the propeller shaft that it yielded in twisting. A seam in the steel became the locus of crack formation that finally propagated to complete failure. An analysis by Chrysler, contained in a letter to the Air Technical Service Command,[18] indicates general agreement with the Air Technical Service Command report, with the addition that a turbocharger size mismatch might be contributing to the surges. Had the war been in process, vigorous corrective action would have been taken. However, with Japan accepting surrender on August 14 and with President Truman proclaiming September 2 as VJ Day, there was no interest by anybody in doing any more engineering work on the engine. After being rebuilt, engine D-000001 was placed in storage at Chrysler Engineering.

During the successful flight tests that were completed, the airplane reached a true speed of 382 mph, using 2440 hp at 20,000 ft altitude. Level flights above 20,000 ft were not performed, but calculations presented in the flight test report[19] showed the airplane might achieve 414 mph at 30,000 ft, with the engine producing its full 2500 hp. As noted, the airplane was not optimized for drag.

While the flight tests of the first airplane, serial no. 42-23297, were underway, the second airplane, serial no. 42-23298, was modified to XP-47H configuration. Engine D-000006, which had been repaired at Chrysler Engineering after its four hours of use in the first airplane, was installed. The airplane, shown on Fig. 17.22, was completed and ground tested on December 12, 1945, with its much-less-heralded first flight being a 30-minute checkout, taking place on December 17, 1945.[7] This marked the end of the XI-2220 engine program.

Fig. 17.22 *The second XP-47H airplane, serial no. 42-23298, with XI-2220 engine D-000006 installed, is ready for its first flight at the Republic Aviation Modification Center on December 12, 1945. Collection of Paul Wise.*

Sometime later, the second airplane, serial no. 42-23298, was flown by the Army Air Forces from the Republic Aviation Modification Center to Freeman Field in Seymour, Indiana, and then to Orchard Place Airport in Chicago, Illinois (now O'Hare International Airport), where it was dismantled.[20] Its engine, serial no. D-000006, and the front framework of the airplane were sent separately to the Smithsonian Institution in Washington, D.C. Figure 17.23 shows a photo of the engine, reunited with the frames and longitudinals of the front of the airplane, at the Smithsonian Graber Preservation, Restoration, and Storage Facility in Silver Hill, Maryland.

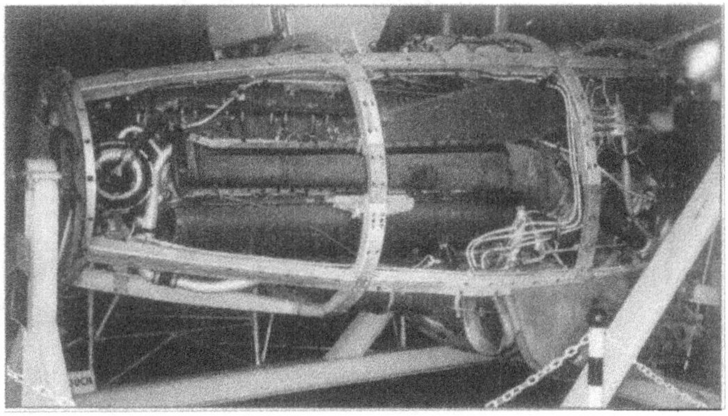

Fig. 17.23 *The XI-2220 engine D-000006 mounted inside the framework from the XP-47H airplane serial no. 42-23298 at the Smithsonian Graber Preservation, Restoration, and Storage Facility at Silver Hill, Maryland. Photo by W.L. Weertman.*

Engine D-000001, which had suffered the broken propeller shaft, was rebuilt as noted. In 1969, the engine was donated by Chrysler to the New England Air Museum at Bradley International Airport in Windsor Locks, Connecticut, where it is on display, mounted in a rollover stand furnished by Chrysler Engineering Research.

A third engine, serial no. D-000005, had been sent from Chrysler Engineering to the Army Air Forces Wright Field in Dayton, Ohio during the program for display purposes. When the program ended, it was transferred to the aeronautical collection of the U.S. Air Force Museum. Later, this engine was sold to an outside collector and had two subsequent owners. The engine then was located and purchased by Chrysler in 1996 and was totally reconditioned. It is now on display at the Walter P. Chrysler Museum, as shown in Fig. 17.10.

During the program, each engine carried a specification level that was changed as major engine parts and components were revised. All three existing engines are specification 11 engines, indicating that they were carbureted, had a compression ratio of 6.3:1, were equipped with a radial geared supercharger running at a ratio of 6.46:1, and were built with a single-wall crankcase,[21] a design that overcame earlier cracking problems. According to Chrysler records, the other three XI-2220 engines of the total of six engines built were scrapped.

Although the development of the engine was not completed and the engine never achieved production status, it was the largest and most powerful engine ever developed by Chrysler, and the knowledge and experience gained during the program was invaluable. Its use of a hemispherical combustion chamber was a key precursor to the use of the hemispherical chamber in the Chrysler V-8 engines of the early 1950s.

Many of the engineers who worked on the program went on to make significant contributions to future Chrysler engines. Among these were Syd Terry, already mentioned; Robert S. Rarey, who became Chief Engineer of Engine Design and Development; R. Dean Engle, who became Chief Engineer of Dodge Division, Chief Engineer of Power Train Manufacturing, and Chief Engineer of Engine Design and Development, succeeding Bob Rarey; and Ev Moeller, who became Assistant Chief Engineer of Engine Development. George Huebner, assistant to Carl Breer, later became Director of Research, with the responsibility for the development of the Chrysler automotive gas turbine program that is described in Chapter 18. Herb Bevans, then a dynamometer test engineer, later became Chief Engineer of Chrysler Proving Grounds, followed by the position of Executive Engineer of Powertrain Engineering.

Table 17.2 shows the specifications for this inverted V-16 fighter aircraft engine.

TABLE 17.2
SPECIFICATIONS FOR THE WORLD WAR II XI-2220 INVERTED V-16 AIRCRAFT ENGINE

Program years	1940–1945
Used in	Experimental XP-47H fighter aircraft
Engineering program code	A45
Displacement (cu. in.)	2220
Bore	5.8
Stroke	5.25
Compression ratio	6.25:1
BHP @ rpm	2500 @ 3400
Combustion chamber	Hemispherical
Crankcase mat'l.—skirt type	Aluminum—deep
Cyl. bore spacing	All 7.62, except at center of engine—14.62
Cyl. head type/mat'l.	16 separate—aluminum
No. of main brgs.—ck/s mat'l.	10—forged steel
Ck/shaft main journal dia.	3.25
Crankpin journal dia.	3.00
Connecting rod centers	9.25
Piston mat'l.—no. of rings	Forged aluminum—5
Valve system	SOHC—32-valve, double rocker shaft
Valve events	296-296-92
Camshaft drive	Gear train at center of engine
Valve head dia.	Intake 2.984; exhaust 2.687
Valve lash adjustment	Mech.—rocker arm screw
Fuel system	Bendix dual venturi pressure carburetor
Ignition system	Dual Wico magnetos 2 spark plugs per cylinder
Power augment	G.E. CH-5 turbocharger 11-in. centrifugal supercharger Dual intercoolers + aftercooler
Engine weight*	2460 lb

* Includes base engine, carburetor, magnetos, and ignition distribution system.

Aircraft Engines by the Thousands—Producing the Wright R-3350 Radial Air-Cooled Engine

Only a few days after the December 7, 1941 attack on Pearl Harbor, the U.S. War Production Board told Chrysler Corporation that it was interested in having Chrysler become a production source of Wright Aeronautical aircraft engines. This was followed, on February 27, 1942, with a contract from the Army Air Corps for building complete engine assemblies of the Wright model R-3350-23 engine[22] that was in the final stages of entering production at Wright Aeronautical plants. The engine, which was quite different from any automotive engine with which the Chrysler manufacturing engineers were familiar, was a Wright Cyclone two-row radial 18-cylinder air-cooled engine, with many parts made of aluminum and magnesium. The "R" in its Army Air Forces designator, R-3350, stood for "Radial" and the "3350" its displacement in cubic inches, obtained from a bore and stroke of 6.125 × 6.3 in. The major components of the engine are shown on Fig. 17.24 in the exploded view of the engine.

Chrysler accepted the contract, assigning it to Dodge Division, which immediately launched the biggest manufacturing program in Chrysler history. Because none of the existing Chrysler plants had sufficient space for the project, plans were laid for an all-new plant. A suitable site was found near Chicago, Illinois, and with enough information to get started from the Dodge design staff and Albert Kahn, the architect, ground was broken on June 4, 1942 for the first of many buildings that would occupy the area.

When completed, the complex, which was named Dodge Chicago, had nineteen major buildings totaling six million square feet of floor space. Included were an aluminum foundry, a magnesium foundry, a plant for steel forgings, vast machining and assembly areas, and forty-eight multi-story test cells for proof-testing the finished engines before shipment.[23]

However, the engineering development of the engine was still in process when the production program started, resulting in literally thousands of major and minor changes to part designs during the tooling. It was a major headache to the manufacturing engineers but had to be accommodated. By October 1943, finished parts were ready in some of the Dodge Chicago departments, and these were shipped to Wright Aeronautical for use in its engine production plants that were building the same engine.[22]

January 17, 1944 was the triumphant day when Dodge Chicago shipped its first R-3350 Model BA engine following assembly, testing, and acceptance by the Air Corps.[22] By the end of the year, 6,053 engines had been shipped. Figure 17.25 shows the 10,000th engine, completed in March 1945. Until

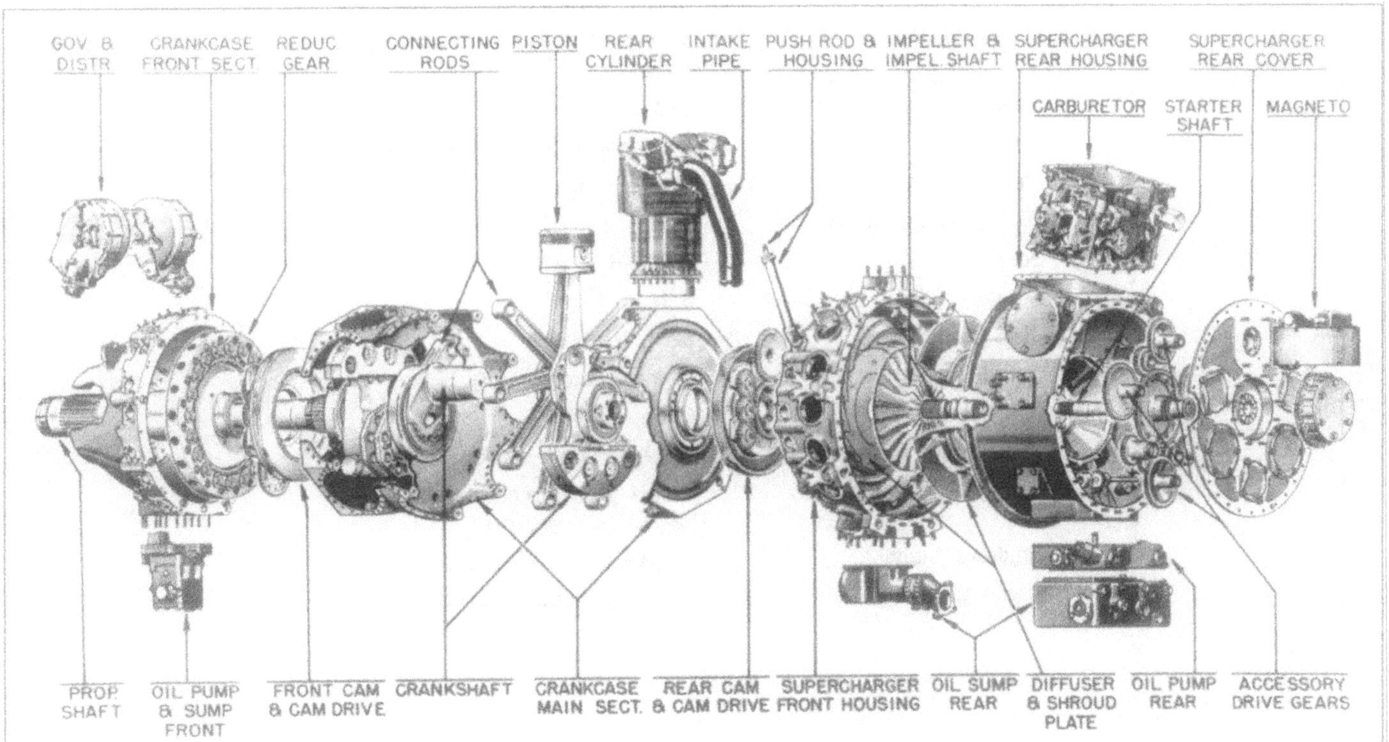

Fig. 17.24 *An exploded view of the Wright Cyclone R-3350 Model BA radial air-cooled 18-cylinder aircraft engine.*

Fig. 17.25 *The 10,000th Wright Cyclone R-3350 engine produced by the Dodge Chicago plant. L.L. (Tex) Colbert, left, was general manager. With him is W.H. Eddy, production manager, on March 23, 1945.*

October 1944, the engines were built with carburetors (Model R-3350-23). After that date, the engines were fuel injected, and the model designator was changed to R-3350-57.[8]

The R-3350 engine originally had been created by Wright to power the B-29 Superfortress long-range heavy bomber; by far, that was its principal application. With each B-29 using four engines, the high volume of engines built by Dodge Chicago was vital to the war effort, especially the war in the Pacific, where most of the B-29s were based. When World War II ended on August 14, 1945, Dodge Chicago almost immediately ceased production, having built 18,413 engines.[23]

Table 17.3 shows the specifications of this wartime engine.

TABLE 17.3
SPECIFICATIONS FOR THE WORLD WAR II WRIGHT R-3350 AIRCRAFT ENGINE

Engine name	Wright Aeronautical Cyclone aircraft engine
Engine type	Aircooled, two-row, 18-cylinder, radial
Years of manufacture	1942–1945
Used in	Primarily B-29 heavy bomber
Manufacturing location	Dodge Chicago
Displacement (cu. in.)	3350
Bore	6.125
Stroke	6.3
BHP @ rpm	2200 @ 2800
Combustion chamber	Hemispherical
Valve system	OHV—36-valve
Valve lash adjustment	Mech.—rocker arm screw
Fuel system	Initial: pressure carburetor Later: fuel injection
Ignition system	2 spark plugs per cylinder
Power augment	Centrifugal supercharger

CHAPTER 18

A Different Type of Engine—The Chrysler Gas Turbine

The near-monopoly position of the internal combustion piston engine as the primary power source for ground-based vehicles has not been without challengers over the many decades of its use. A serious challenge to dethrone the piston engine came from Chrysler Engineering in a series of designs and developments spanning 35 years. Deserving a more comprehensive presentation than the brief summary contained in this chapter, what follows is intended only to give recognition to the gas turbine engine as a strong effort involving a corps of excellent engineers and to place it alongside the other engines described in this book.

The Chrysler Engineering automotive gas turbine investigation started in 1944, during World War II, as an Engineering Research program to determine the fundamental requirements for a small gas turbine that might have use for Chrysler. The program included the design of a compact automotive-size regenerative gas turbine, investigation of combustion techniques, and testing of turbine materials.[1] Late in the same year, the Power Plant Division of the Bureau of Aeronautics, hearing about the Chrysler program, inquired about the possibility of having Engineering Research adapt the design to the much larger size requirements of Navy aircraft. After agreement was reached on the specifics, including that the gas turbine powerplant would be a regenerator type and would drive an aircraft propeller, a contract was signed, and the initial design work by Engineering Research started in July 1945. However, during a review of this design in October 1945, representatives of the Bureau of Aeronautics requested higher goals for part-load fuel economy, which caused a completely new design to be prepared. This second design had a seventeen-stage axial flow compressor, a two-stage power turbine, ten burners, and a stationary regenerator.[1] Designed to produce 1000 hp at the propeller shaft, the engine, coded A86 by Engineering and XT-36-D2 by the Bureau of Aeronautics, completed an official preliminary 50-hour flight rating test on March 3, 1949.[2] Figure 18.1 shows a photo of the engine that completed the

Fig. 18.1 *The A86 Navy turboprop gas turbine after completion of a preliminary 50-hour flight rating test.*

test. However, by the end of the year, this Navy gas turboprop engine program, as it came to be known and which was a direct predecessor to the Chrysler automotive gas turbine program, was cancelled.[3]

Using experience gained from this work, Engineering Research, headed by George Huebner, launched in the early 1950s another automotive-sized gas turbine program, coded A249. This program placed heavy emphasis on basic research of materials that retained strength at high temperatures and on designs that conserved the heat of combustion. Development progressed on engine components and on complete engine assemblies, using bench tests, dynamometer tests, and vehicle tests to the extent that on March 25, 1954, Chrysler announced that a successful road test had been made of a 1954 Plymouth Belvedere two-door hardtop powered with an experimental gas turbine engine.[3]

A significant advantage of the Chrysler gas turbine over most existing gas turbines was the use of a regenerator, already mentioned, for capturing otherwise discarded energy. This regenerator transferred heat from the exhaust gas to the intake air on its way to the burner for a substantial boost in fuel economy and a welcome reduction in tailpipe exhaust gas temperatures.

Two years later, in March 1956, a cross-country trip from New York to Los Angeles was made in a 1956 Plymouth four-door sedan powered by a similar engine that had upgraded components. The turbine engine at this stage, designated first generation, developed 100 hp. Further improvements in compressor efficiency, regenerator efficiency, burner efficiency, and metallurgy led to a second-generation 200-hp turbine engine.

Development continued, and three vehicles—a Turboflite concept car, a 1960 Plymouth four-door hardtop, and a 1960 2-1/2 ton Dodge truck, all powered with a third-generation CR2A turbine—were shown to the news media on February 28, 1961. This third-generation turbine engine, with an output of 140 hp and 375 lb-ft torque, was equipped with variable nozzle vanes in the second-stage turbine section, an improvement over the fixed vanes used in prior engines.[3]

Following the third-generation turbine, a major design revision was made to replace the single circular rotating regenerator—located on top of the turbine housing and used on all automotive turbines to date—with dual smaller regenerators, also circular and rotating, that were located on the right and left sides of the turbine housing. A cutaway view and an outside view of this new turbine, designated fourth generation, are shown on Figs. 18.2 and 18.3, respectively.

On May 14, 1963, a special Chrysler-designed and Ghia-built eye-catching metallic-bronze turbine car, shown on Fig. 18.4, was unveiled to the public. Powered with a fourth-generation engine, 50 of these cars were built and loaned to individuals outside the corporation, in all parts of the country, as a way of obtaining widespread evaluation and certainly good public

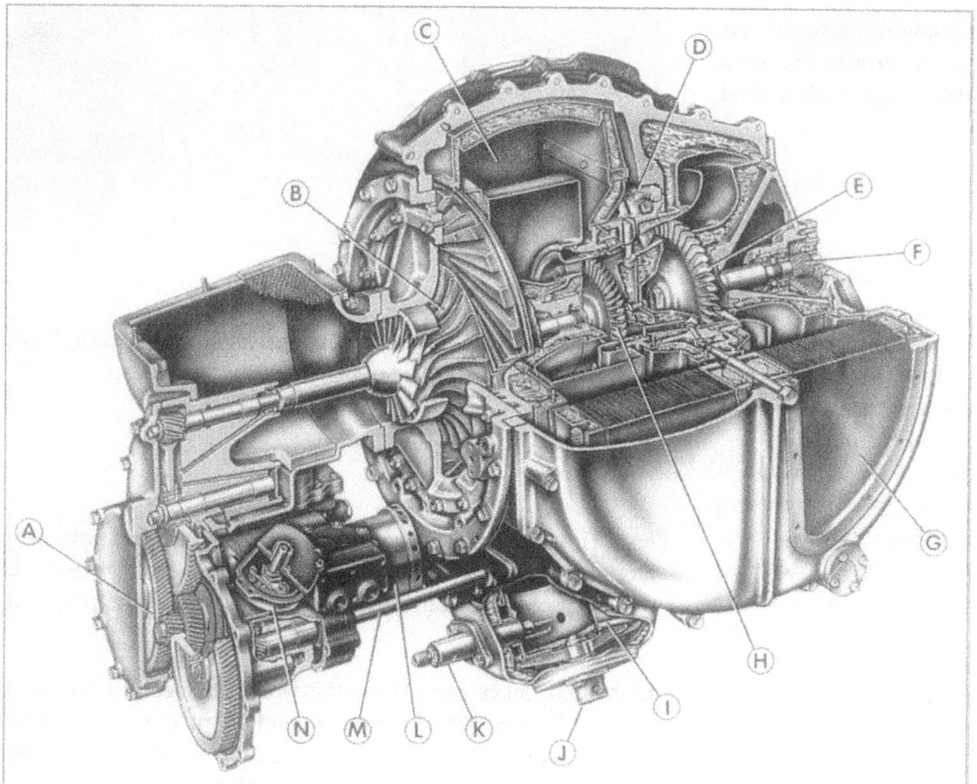

MAIN COMPONENTS OF THE FOURTH GENERATION GAS TURBINE ENGINE

(A) accessory drive; (B) compressor; (C) right regenerator; (D) variable nozzle unit; (E) power turbine; (F) reduction gear; (G) left regenerator; (H) compressor turbine; (I) burner; (J) fuel nozzle; (K) igniter; (L) starter-generator; (M) regenerator drive shaft; (N) ignition unit.

Fig. 18.2 *A cutaway illustration of the Chrysler fourth-generation automotive gas turbine engine.*

Fig. 18.3 *The fourth-generation gas turbine engine on display at the Walter P. Chrysler Museum. Photo by W.L. Weertman.*

Fig. 18.4 *The Ghia-built Chrysler turbine car was powered by a 130-hp fourth-generation turbine engine. Fifty of these cars were used for evaluation throughout the nation.*

relations for the potential production turbine-engined car. After their return to Engineering, many were assigned to Engineering personnel (including me) for periods of several days to continue the evaluation. Smooth in operation, on acceleration these cars had a unique windup sound similar to that of a jet aircraft engine.[4] In January 1964, Research Engineering presented four SAE papers describing progress to date and the latest designs and development, including the Ghia-built cars.[5–8] Authors of these papers included George J. Huebner, Jr., William Chapman, Amedee Roy, Frederick Hagen, Claude Belleau, Gerald DeClaire, and Al Bell.

More development followed, especially in the area of replacing the expensive high-alloy metals with temperature-resistant ceramics. Over the next fifteen years, three additional generations of turbine engines were identified. The seventh and final generation had a single ceramic regenerator, water injection at wide-open throttle, and electronic controls. However, despite major efforts, the turbine engine could not overcome the part-throttle fuel economy disadvantage it had versus the entrenched piston engines. When the worldwide oil shortages hit in the early 1970s and again in the late 1970s, this was a significant factor in the termination of the Chrysler gas turbine engine program in 1979.

Table 18.1 shows the specifications for the Chrysler automotive gas turbine engine.

TABLE 18.1
SPECIFICATIONS FOR CHRYSLER GAS TURBINE ENGINES

Year	1954–1956	1959	1960–1962	1963–1965
Car line	Turbine car	Turbine car	Turbine car Dodge truck	Ghia-built 50 turbine cars
Turbine version	1st generation	2nd generation	3rd generation CR2A	4th generation
Pressure ratio			4:1	4:1
BHP @ rpm	100	200	140 @ 4570	130 @ 3600
Torque (lb-ft) @ rpm			375 @ 0	425 @ 0
Compressor	Single stage, centrifugal	Single stage, centrifugal	Single stage, centrifugal	Single stage, centrifugal
Compressor turbine	Single stage, axial	Single stage, axial	Single stage, centrifugal	Single stage, centrifugal
Power turbine	Single stage, axial Fixed nozzle vanes	Single stage, axial Fixed nozzle vanes	Single stage, axial Variable nozzle vanes	Single stage, axial Variable nozzle vanes
Regenerator	Single, circular, mounted on top of engine	Single, circular, mounted on top of engine	Single, circular, mounted on top of engine	Dual, circular, mounted on right and left sides of engine
Engine weight			204 kg (450 lb)	186 kg (410 lb)

(Continued on next page)

TABLE 18.1 (Continued)

Year	1966	1966–1973	1974–1978
Car line	Dodge Charger	Plymouth and Dodge	Dodge Aspen turbine concept car
Turbine version	5th generation	6th generation	7th generation
Pressure ratio			
BHP @ rpm		150	104
Torque (lb-ft) @ rpm			375 @ 0
Compressor	Single stage, centrifugal		
Compressor turbine	Single stage, centrifugal		Free rotor
Power turbine	Single stage, axial Variable nozzle vanes		
Regenerator	Dual, circular, mounted on right and left sides of engine	Dual, circular, mounted on right and left sides of engine	Single, circular, ceramic
Other	Similar to 4th generation but with larger regenerators and higher cycle temperatures	Accessories driven from both compressor turbine and power turbine	Water injection at wide-open throttle; all accessories driven by power turbine; electronic control

Chapter 19

American Motors Engines Join the Chrysler Power Lineup

Introduction

When American Motors Corporation (AMC) merged with Chrysler Corporation on August 5, 1987, it brought with it a long and rich history of cars and engines dating back to the beginning of the automobile industry in the early 1900s. A direct predecessor company of AMC, the Thomas B. Jeffery Company, formed in 1902, quickly became an industry leader with its 1902 Rambler model powered by a single cylinder engine, followed by the addition of a two-cylinder engine in 1904 and a four-cylinder engine in 1906 when the single-cylinder engine was dropped.[1] The creation of AMC was the result of the merger of Nash-Kelvinator and Hudson Motor Car Company in 1954. In 1970, Jeep was added to AMC by purchase of the Kaiser Jeep Corporation.[2]

Two years after Charles W. Nash purchased the Thomas B. Jeffery Company in 1916 and founded Nash Motors, a new car that was named Nash was introduced and was powered by an overhead valve six-cylinder engine. Later, in 1930, Nash Motors introduced an overhead valve straight-eight engine.[1]

Fig. 19.1 *A cutaway view of the World War II Jeep 60-hp 134-cu.-in. Go-Devil L-head four-cylinder engine on display at the Walter P. Chrysler Museum. Photo by W.L. Weertman.*

The L-head four-cylinder Go-Devil engine, pictured on Fig. 19.1, of another predecessor company, Willys-Overland, powered the renowned military Jeep of World War II[3] and continued powering it for many years after the end of the war. AMC predecessors also contributed a solid record of auto racing success using high-performance versions of production engines, especially during the early 1950s when stock car tracks were dominated by the Hudson Hornet Twin H dual-carburetor six-cylinder L-head engine, shown on Fig. 19.2. This engine had a bore and stroke of 308 cu. in., the largest six-cylinder engine in passenger car use in the industry at the time.[4]

An innovative engine in the history of AMC was the 1961 Rambler Classic Six overhead valve engine with its cylinder block made of die-cast aluminum.[5] In production for four years, this contemporary of the Chrysler Slant Six aluminum engine was cancelled at the end of 1964,[1] outlasting the Chrysler aluminum block engine by two years.

Certainly the many different and interesting engines produced by AMC and predecessor companies deserve their own individual presentations. However, this overview of Chrysler engines is, of necessity, limited to the five AMC engines that were in production at the time of the merger with Chrysler. These five engines are described next.

Fig. 19.2 *This 1953 Hudson Hornet 308-cu.-in. L-head six-cylinder engine is on display at the Walter P. Chrysler Museum. With twin carburetors, this engine gave the Hudson Hornet the power it needed to dominate the stock car tracks of the early 1950s. Photo by W.L. Weertman.*

AMC 4.2-L (258-Cubic-Inch) Overhead Valve Six-Cylinder Engine

The AMC 4.2-L (258-cu.-in.) overhead valve in-line six-cylinder engine displayed on Fig. 19.3, with a bore smaller than its stroke at 95.3 × 98.3 mm (3.75 × 3.895 in.), was the most senior of the AMC engines at the time of the merger with Chrysler; the basic engine design had changed little since its introduction on the 1964-1/2 AMC Rambler Typhoon two-door hardtop. An SAE paper describing the engine was presented in June 1964.[6]

The design of this engine was classic, with the cylinders of the shallow-skirt cast iron cylinder block positioned straight upright and evenly spaced at 4.38 in., the camshaft located on the right side of the block, loop-flow combustion chambers, in-line valves, and the intake and exhaust manifolds both located on the left side of the cast iron cylinder head to provide plenty of exhaust heat to the intake manifold for quick warmup. Similar to many other six-cylinder overhead valve engines, the camshaft of this engine was driven by a timing chain. The distributor, rotated by a gear on the camshaft, was mounted on the right side of the cylinder block. The design of the seven-main-bearing crankshaft included diameters of 2.5 in. for the main journals and 2.094 in. for the crankpins.

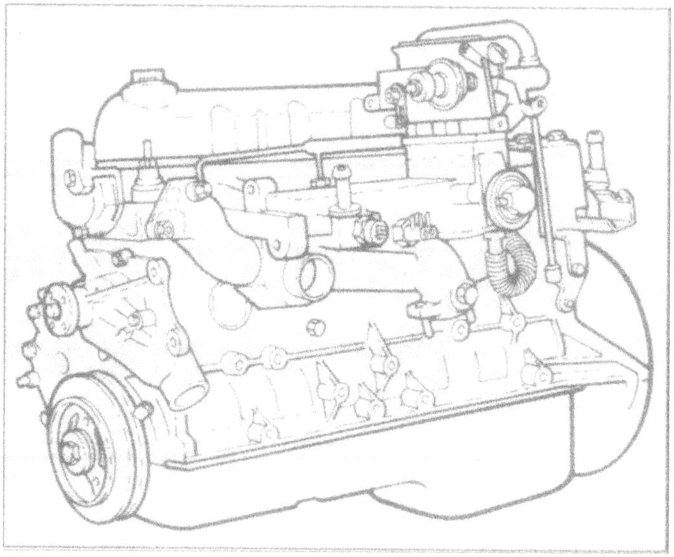

Fig. 19.3 *The 1985 AMC 4.2-L (258-cu.-in.) overhead valve six-cylinder engine.*

Valve head diameters for the intake and exhaust valves were 1.787 and 1.406 in., respectively.

The engine had two assignments in 1988: an optional engine for the Jeep Wrangler, and the standard and only engine for the AMC Eagle, a four-wheel-drive four-door station wagon, shown on Fig. 19.4, during the last year of production of the car in 1988. The long run of production for this 4.2-L overhead

Fig. 19.4 *The 1987 AMC Eagle station wagon. The AMC 4.2-L overhead valve six-cylinder engine was the standard and only power for this four-wheel-drive vehicle. The design of the engine dated back to 1964.*

valve six-cylinder engine—built in displacements of 2.26-L (199 cu. in.), 3.8-L (232 cu. in.), and 4.2-L (258 cu. in.)—came to a close at the end of 1990 when it was pulled from Jeep. It was replaced by an excellent and newer related-but-different AMC engine, the 4.0-L overhead valve six, which is described later in this chapter.

Table 19.1 shows the specifications of the AMC 4.2-L overhead valve six-cylinder engine.

AMC 5.9-L (360-Cubic-Inch) Overhead Valve V-8 Engine

The 5.9-L (360-cu.-in.) overhead valve V-8 engine illustrated on Fig. 19.5 was the second most senior AMC engine at the time of the merger with Chrysler, with its general design dating back to mid-1966 when, displacing 290 cu. in. and named Typhoon V-8, it debuted as an optional engine on the AMC Rambler American. It was designed to be considerably lighter and more compact than the overhead valve V-8 engine that it replaced, which was a 1956 vintage V-8 with a deep-skirt cylinder block. It retained the cylinder spacing of 4.75 in. of the earlier engine, which gave it the authority of a big-block V-8. Following its 1966 release, the engine was built at displacements of 5.0-L (304 cu. in.), 5.6-L (343 cu. in.), 5.9-L (360 cu. in.), 6.4-L (390 cu. in.), and 6.6-L (401 cu. in.), the largest size it attained during its many years of production. A 1968 SAE paper co-authored by G. Fred Leydorf, Jr. and G.E. Onishi provides a description of the high-performance 390-cu.-in. version of the engine.[7]

The 5.9-L (360-cu.-in.) engine had a bore and stroke of 4.08 × 3.44 in. A short-skirt cast iron cylinder block, cast iron cylinder heads with in-line overhead valves, and a crankshaft with 2.75-in. main journals and 2.094-in. crankpin journals

TABLE 19.1
SPECIFICATIONS FOR THE 4.2-L AMC OHV SIX-CYLINDER ENGINE

Model Year	1988	1990
Car line	Jeep, Eagle	Jeep
Model	Jeep: Wrangler, J-10 truck Eagle	Wrangler
Engine code		ERF
Displacement (liters)	4.2	4.2
Bore	3.75	3.75
Stroke	3.895	3.895
Compression ratio	9.2:1	9.2:1
Power @ rpm	84 kW (112 hp) @ 3000*	84 kW (112 hp) @ 3000
Torque @ rpm	285 N-m (210 lb-ft) @ 2000*	285 N-m (210 lb-ft) @ 2000
Cyl. block mat'l.—skirt type	Cast iron—shallow	Cast iron—shallow
Cyl. block deck height	9.49	9.49
Cyl. bore spacing	4.38	4.38
No. of main brgs.—ck/s mat'l.	7—nodular cast iron	7—nodular cast iron
Ck/shaft main journal dia.	2.5	2.5
Crankpin journal dia.	2.094	2.094
Connecting rod centers	5.875	5.875
Valve system	OHV—rocker pedestals	OHV—rocker pedestals
Valve events	262-262-34	262-262-34
Camshaft drive	Chain	Chain
Valve head dia.	Intake 1.79; exhaust 1.41	Intake 1.79; exhaust 1.41
Valve lash adjustment	Hydraulic tappet	Hydraulic tappet
Oil pump/distributor drive	Geared to camshaft at center of block	Geared to camshaft at center of block
Fuel system	2-barrel carburetor	2-barrel carburetor

* Wrangler, J-10 truck.

Fig. 19.5 *The AMC 5.9-L (360-cu.-in.) overhead valve V-8 engine. The engine originally was launched by AMC during 1966.*

put it in the mainstream of V-8 engine design of the era. The 4.08-in. bore permitted intake and exhaust valve head diameters of 2.025 and 1.68 in., respectively. At the time of the merger, use of the engine was confined to three vehicles in the Jeep line: the Jeep Grand Wagoneer shown on Fig. 19.6, the J-10 pickup truck, and the J-20 pickup truck. With its two-barrel carburetor and a compression ratio of 8.25:1, the engine produced 144 hp and 280 lb-ft for these vehicles.

Fig. 19.6 *The 1988 Jeep Grand Wagoneer. Power for this four-wheel-drive vehicle that was at the top of the Jeep line came from a Kenosha-built 144-hp 5.9-L overhead valve V-8 engine.*

However, the obvious redundancy of this engine with the Chrysler small-block 5.9-L LA V-8 engine caused the inevitable to occur. The AMC 5.9-L V-8 was cancelled at the end of 1991, and LA V-8 engines were assigned to fulfill the V-8 power tasks of the Jeep line.

Table 19.2 shows the specifications of the AMC 5.9-L overhead valve V-8 engine.

AMC 2.5-L (150-Cubic-Inch) Overhead Valve Four-Cylinder Engine

The 2.5-L (150-cu.-in.) overhead valve four-cylinder and 4.0-L (242-cu.in.) overhead valve six-cylinder engines (described next) were designed as a two-engine set to obtain the benefits of maximum part interchangeability and minimum tooling costs.[8] By shifting the design in the direction of larger bores and shorter strokes than the older six-cylinder engine, the engines were launched on a course to become truly outstanding performers.

The first of the two engines to appear, the 2.5-L four-cylinder engine, started production in mid-1983, going under the hoods of American Eagle Series 50 sedans and station wagons and replacing a Pontiac-built 2.5-L four-cylinder engine that had been used since 1980. In 1984, use of the four-cylinder AMC-built 2.5-L engine spread to the Jeep line, replacing the Pontiac 2.5-L four-cylinder engine there also. Of interest, this was the first AMC-built four-cylinder engine to power the Jeep.

This AMC 2.5-L overhead valve, four-cylinder engine (not related to the Pontiac 2.5-L engine in any way) had a straightforward arrangement, using an upright position for the cylinder block, overhead in-line valves, loop-flow cylinder head combustion chambers, and intake and exhaust manifolds located on the left side of the cylinder head. The large bore relative to the stroke, 98.4 × 81 mm (3.876 × 3.188 in.), gave room for a 48.5-mm (1.91-in.) intake and 38.3-mm (1.50-in.) exhaust valve. Both the shallow-skirt cylinder block with its 4.38-in. cylinder spacing and the cylinder head were made of cast iron. The five main journals of the nodular iron crankshaft had a diameter of 2.5 in.; the four crankpins were 2.094 in. in diameter. As installed in the 1984 Jeep line, with a compression ratio of 9.2:1 and a single-barrel electronic feedback carburetor, the engine was rated at a modest 86 hp @ 3650 rpm and 132 lb-ft torque @ 2000 rpm. However, in 1986, just before the Chrysler/AMC merger, the engine went through a remarkable performance transformation that included the replacement of the carburetor with throttle-body fuel injection, bringing its power up to an outstanding 117 hp @ 5000 rpm, with a small gain in torque to 135 lb-ft @ 3500 rpm. In 1991, when its fuel system was changed to sequential multipoint fuel injection, output again increased to 91.8 kW (123 hp) and 188 N-m (139 lb-ft) torque in the Jeep Wrangler, and even higher, 97 kW (130 hp) and 202 N-m (149 lb-ft torque), in the Cherokee and Comanche. Figures 19.7 and 19.8 show the external and

TABLE 19.2
SPECIFICATIONS FOR THE 5.9-L AMC OHV V-8 ENGINE

Model Year	1988–1991
Car line	Jeep
Model	Grand Wagoneer
Engine code	
Displacement (liters)	5.9
Bore	4.08
Stroke	3.44
Compression ratio	8.25:1
Power @ rpm	107 kW (144 hp) @ 3200
Torque @ rpm	380 N-m (280 lb-ft) @ 1500
Cyl. block mat'l.—skirt type	Cast iron—shallow
Cyl. block deck height	9.21
Cyl. bore spacing	4.75
No. of main brgs.—ck/s mat'l.	5—nodular cast iron
Ck/shaft main journal dia.	2.75
Crankpin journal dia.	2.094
Connecting rod centers	5.875
Valve system	OHV—rocker pedestals
Valve events	263.5-263.5-41.5
Camshaft drive	Chain
Valve head dia.	Intake 2.025; exhaust 1.68
Valve lash adjustment	Hydraulic tappet
Oil pump/distributor drive	Diagonal at front of block
Fuel system	2-barrel carburetor

Fig. 19.7 *The AMC 2.5-L overhead valve four-cylinder engine. With sequential multipoint fuel injection, the engine produced 97 kW (130 hp) as installed in the 1991 Jeep Cherokee.*

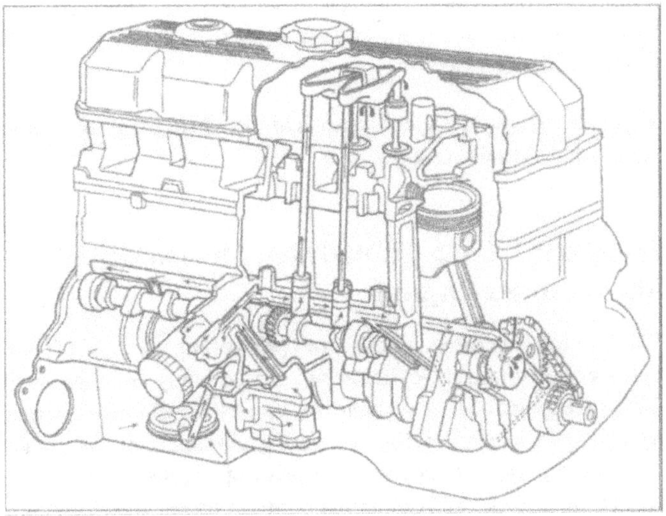

Fig. 19.8 *A drawing of the AMC 2.5-L overhead valve four-cylinder engine, showing lubrication and valve systems.*

internal features of the engine, respectively. In 1988, the engine provided the base power for three models: the Jeep Wrangler, pictured on Fig. 19.9; the Jeep Cherokee; and the Comanche 1/2-ton pickup truck.

Fig. 19.9 *The 1993 Jeep Wrangler S (for Standard). A 123-hp 2.5-L AMC overhead valve multipoint fuel injected four-cylinder engine was its standard and only power.*

Tough and powerful, with output exceeding that of the Chrysler 2.5-L single overhead camshaft four-cylinder engine, this former AMC 2.5-L overhead valve engine was given the assignment in 1996 to become the base power for the Dodge Dakota truck, replacing the Trenton engine plant-built 2.5-L SOHC four-cylinder engine.

Table 19.3 shows the specifications of the AMC 2.5-L overhead valve four-cylinder engine.

AMC 4.0-L (242-Cubic-Inch) Overhead Valve Six-Cylinder Engine

This engine, the other member of the two-engine 2.5-L four-cylinder and 4.0-L six-cylinder set, shared many features with the preceding 2.5-L overhead valve, four-cylinder engine, including bore size, cylinder spacing, connecting rods, and valvetrain parts. Its seven-main-bearing crankshaft had the same main journal and crankpin diameters as the 2.5-L engine. Similar to the 2.5-L engine, its shallow-skirt block and cylinder head were made of cast iron. The AMC Kenosha engine plant had two cylinder block machining lines—one for each block—and a single shared cylinder-head machining line. The sharing was made possible by providing the four-cylinder head casting with a dummy extension, bringing it up to the same length as the six-cylinder head for transport through the machining stations. After going through almost all the machining operations, the extension was sawed off, leaving the necessary four cylinders.

A criteria for the design of the two-engine set was that advantage should be taken of the machine tools already in use for machining parts for the 4.2-L six-cylinder engine to minimize investment in all-new tooling.[8] Use of the same 4.38-in. cylinder spacing was a must to accomplish this, and, attesting to the skill of the AMC engine designers, the new engines showed no performance compromise because of this requirement.

Released for the 1987 Jeep Cherokee (shown on Fig. 19.10), Wagoneer, and Comanche truck, the 4.0-L engine appearing on Fig. 19.11, with its common bore of 3.876 in., had a longer stroke of 3.41 in. versus the 3.19-in. bore of the 2.5-L engine. To accommodate the longer stroke, the cylinder block deck height was slightly higher at 9.432 in. versus 9.32 in. for the 2.5-L engine. Because it arrived in the late 1980s, the engine was equipped from the beginning with state-of-the-art sequential multiport fuel injection, bypassing the carburetor and throttle-body fuel injection phases that the 2.5-L engine went through, as previously related. The output of this engine was an impressive 132 kW (177 hp) @ 4500 rpm and 304 N-m (224 lb-ft) torque @ 2500 rpm. By 1991, its output had increased further to 142 kW (190 hp), with torque remaining virtually the same at 305 N-m (225 lb-ft). In 1991, the engine first was installed in the Jeep Wrangler, marking the end of the line for the venerable 4.2-L six-cylinder engine. Many changes were made between 1996 and 1998 to improve noise, vibration, and harshness qualities of the engine, including a revised camshaft lobe profile to reduce valve seating velocity,

Fig. 19.10 *The 1987 Jeep Cherokee. The new 173-hp 4.0-L AMC overhead valve six-cylinder engine with sequential fuel injection was an optional powerplant for this vehicle.*

TABLE 19.3
SPECIFICATIONS FOR THE 2.5-L AMC OHV FOUR-CYLINDER ENGINE

Model Year	1988	1991	1998
Car line	Jeep, Eagle	Jeep	Jeep, Dodge
Model	Jeep: Wrangler, Cherokee, Comanche Eagle: Premier	Wrangler, Cherokee	Jeep: Wrangler, Cherokee Dodge: Dakota
Engine code	EPC		
Displacement (liters)	2.5	2.5	2.5
Bore	3.876	3.876	3.876
Stroke	3.188	3.188	3.188
Compression ratio	9.2:1	9.2:1	9.2:1
Power @ rpm	90 kW (121 hp) @ 5000*	97 kW (130 hp) @ 5250**	90 kW (125 hp) @ 5400**
Torque@ rpm	191 N-m (141 lb-ft) @ 3500*	202 N-m (149 lb-ft) @ 3250**	203 N-m (150 lb-ft) @ 3250**
Cyl. block mat'l.—skirt type	Cast iron—shallow	Cast iron—shallow	Cast iron—shallow
Cyl. block deck height	9.32	9.32	9.32
Cyl. block length			
Cyl. bore spacing	4.38	4.38	4.38
No. of main brgs.—ck/s mat'l.	5—nodular cast iron	5—nodular cast iron	5—nodular cast iron
Ck/shaft main journal dia.	2.5	2.5	2.5
Crankpin journal dia.	2.094	2.094	2.094
Connecting rod centers	6.125	6.125	6.125
Valve system	OHV—rocker pedestals	OHV—rocker pedestals	OHV—rocker pedestals
Valve events	270-270-46	270-270-46	270-270-46
Camshaft drive	Chain	Chain	Chain
Valve head dia.	Intake 1.91; exhaust 1.5	Intake 1.91; exhaust 1.5	Intake 1.91; exhaust 1.5
Valve lash adjustment	Hydraulic tappet	Hydraulic tappet	Hydraulic tappet
Oil pump/distributor drive	Vertical at rear of block	Vertical at rear of block	Vertical at rear of block
Fuel system	TBI	SMPI	SMPI

* Cherokee and Comanche.

** Cherokee.

Fig. 19.11 *The AMC 4.0-L overhead valve six-cylinder engine. Released in 1987, the engine had a close design relationship with the 2.5-L overhead valve four-cylinder engine.*

isolated valve cover fasteners, stiffer rocker arm pedestals, the addition of a main bearing brace connecting all seven main bearing caps to reduce main bearing cap motion, a lightweight piston with a barrel-shaped skirt for quieter idle, stiffer rocker arms, and a modified valve spring for better dynamics.

As of the 1998 ending date covered by this book, this engine continued to power Jeep vehicles.

Table 19.4 shows the specifications of the AMC 4.0-L overhead valve six-cylinder engine.

An All-Aluminum Engine from Europe—The 3.0-L PRV/AMC Single Overhead Camshaft V-6

Prior to the merger with Chrysler on August 5, 1987, AMC had a partnership with Renault of France. Renault was involved with two other European companies—Peugeot, S.A. of France, and Volvo of Sweden—in a joint venture to build engines. The joint-venture engine was an all-aluminum, single overhead camshaft engine with V-6 and V-8 configurations. For tooling commonality, both versions were designed with the same 90-deg bank angle. However, demand for the V-8 version did not materialize, and it was never produced.

The location of the engine production factory was the town of Douvrin in northern France. Volvo was the partner selected to manage the facility. Using the initials of the parent companies' names, the engine was titled the PRV. The engine had many build variations, each customized for the vehicles of the individual parent company.

An AMC and Renault legacy, the 1988 Eagle Premier ES that appears on Fig. 19.12 was built in Brampton, Ontario, and was powered with a 3.0-L version of the PRV engine having a bore and stroke of 93 × 73 mm. Figure 19.13 displays a cutaway view of the engine. The deep-skirt aluminum cylinder block with cylinder centers 108 mm (4.24 in.) apart had open water

TABLE 19.4
SPECIFICATIONS FOR THE 4.0-L AMC OHV SIX-CYLINDER ENGINE

Model Year	1988	1991	1998
Car line	Jeep	Jeep	Jeep
Model	Cherokee, Wagoneer, Comanche	Wrangler, Cherokee	Wrangler, Cherokee
Engine code	ERB		
Displacement (liters)	4.0	4.0	4.0
Bore	3.876	3.876	3.876
Stroke	3.41	3.41	3.41
Compression ratio	8.8:1	8.8:1	8.8:1
Power @ rpm	132 kW (177 hp) @ 4500*	132 kW (190 hp) @ 4750*	132 kW (190 hp) @ 4600*
Torque @ rpm	304 N-m (224 lb-ft) @ 2500*	304 N-m (225 lb-ft) @ 4000*	304 N-m (225 lb-ft) @ 3000*
Cyl. block mat'l.—skirt type	Cast iron—shallow	Cast iron—shallow	Cast iron—shallow
Cyl. block deck height	9.43	9.43	9.43
Cyl. bore spacing	4.38	4.38	4.38
No. of main brgs.—ck/s mat'l.	7—nodular cast iron	7—nodular cast iron	7—nodular cast iron
Ck/shaft main journal dia.	2.5	2.5	2.5
Crankpin journal dia.	2.094	2.094	2.094
Connecting rod centers	6.125	6.125	6.125
Valve system	OHV—rocker pedestals	OHV—rocker pedestals	OHV—rocker pedestals
Valve events	270-270-46		
Camshaft drive	Chain	Chain	Chain
Valve head dia.	Intake 1.91; exhaust 1.5	Intake 1.91; exhaust 1.5	Intake 1.91; exhaust 1.5
Valve lash adjustment	Hydraulic tappet	Hydraulic tappet	Hydraulic tappet
Oil pump/distributor drive	Geared to camshaft at center of block	Geared to camshaft at center of block	Geared to camshaft at center of block
Fuel system	MPI	SMPI	SMPI

* Cherokee.

Fig. 19.12 *The 1988 Eagle Premier ES. Power came from a 112-kW (150-hp) 3.0-L all-aluminum single overhead camshaft V-6 engine built in Douvrin, France.*

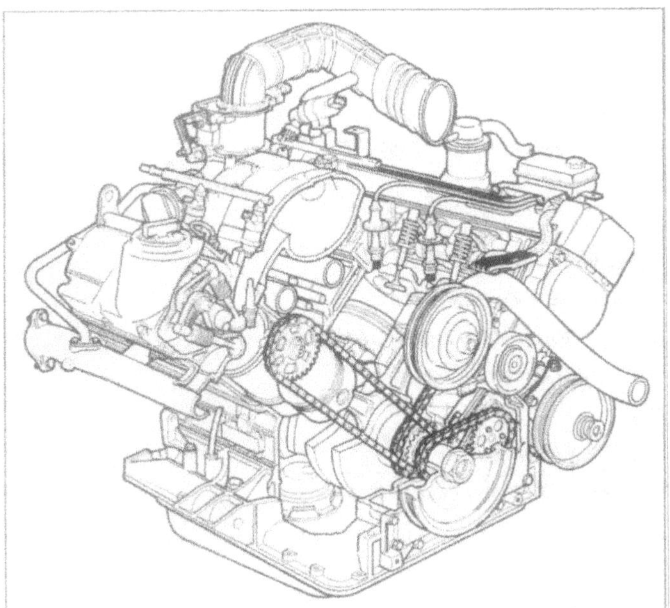

Fig. 19.13 *A cutaway illustration of the 3.0-L PRV single overhead camshaft V-6. AMC use of this French-built engine with its hemispherical combustion chambers and multipoint fuel injection started with the 1988 model.*

jackets fitted with replaceable cast iron bore liners. The oil pump, driven by a short chain from the crankshaft, was integrated into the front lower-left section of the block.

Two timing chains driven by the crankshaft were used to drive the single overhead camshaft installed in each aluminum cylinder head. The hemispherical combustion chambers used two valves per cylinder operated by rocker arms that pivoted about a single rocker shaft in each cylinder head located directly above the camshaft. Valve lash adjustment was done automatically by a mini-hydraulic lash adjuster in the valve tip end of each rocker arm.

In 1990, a second model powered by the engine was released—an Eagle Premier ES look-alike carrying the name of Dodge Monaco. Three years later, at the end of 1992, the Eagle Premier and the Dodge Monaco, and with them the PRV engine, all were canceled by Chrysler.

Table 19.5 shows the specifications for the PRV and AMC V-6 engine.

Chart 19.1 displays time lines for the five AMC engines discussed in this chapter.

TABLE 19.5
SPECIFICATIONS FOR THE 3.0-L PRV AND AMC SOHC V-6 ENGINE

Model Year	1988–1992
Car line	Eagle, Dodge
Model	Eagle: Premier 1988–1992 Dodge: Monaco 1990–1992
Engine code	EFP
Displacement (liters)	3.0
Bore	93
Stroke	73
Compression ratio	9.3:1
Power @ rpm	112 kW (150 hp) @ 5000
Torque @ rpm	232 N-m (171 lb-ft) @ 3750
Combustion chamber	Hemispherical
Cyl. block mat'l.—skirt type	Aluminum—deep Replaceable iron bore liners
Cyl. block deck height	220.8
Cyl. bore spacing	108
No. of main brgs.—ck/s mat'l.	4—forged steel
Ck/shaft main journal dia.	70
Crankpin journal dia.	60
Connecting rod centers	146.15
Valve system	SOHC—single rocker shaft
Valve events	252-248-26
Camshaft drive	Chain
Valve head dia.	Intake 45.3; exhaust 38.5
Valve lash adjustment	Hyd. adjuster in rocker arm
Oil pump drive	Chain driven from nose of crankshaft
Distributor drive	From nose of left camshaft
Fuel system	MPI
Engine assembly weight*	166.82 kg (367 lb)

*For content, see the Abbreviations and Definitions section near the back of this book.

Chart 19.1
Time Lines—AMC Engines

- - Model Years and Displacements (Liters) - -

Engine	1988	1989	1990	1991	1992	1993 – 1995	1996 – 1998
2.5-L OHV 4-cyl.	2.5 Jeep, Eagle Prem	2.5 Jeep, Eagle Prem	2.5 Jeep	2.5 Jeep	2.5 Jeep	2.5 Jeep	2.5 Jeep, Dakota
4.0-L OHV 6-cyl.	4.0 Jeep (except Wrangler)	4.0 Jeep (except Wrangler)	4.0 Jeep (except Wrangler)	4.0 Jeep	4.0 Jeep	4.0 Jeep	4.0 Jeep
4.2-L OHV 6-cyl.	4.2 Jeep YJ, 4wd Eagl	4.2 Jeep Wrangler	4.2 Jeep Wrangler				
5.9-L OHV V-8	5.9 Grand Wagoneer	5.9 Grand Wagoneer	5.9 Grand Wagoneer	5.9 Grand Wagoneer			
3.0-L PRV/AMC SOHC V-6	3.0 Eagle Premier	3.0 Eagle Premier	3.0 Eagle Premier Dodge Monaco	3.0 Eagle Premier Dodge Monaco	3.0 Eagle Premier Dodge Monaco		

Epilogue

Designing, developing, and putting into production new engines and making improvements to existing engines is a continuum that is vital to every automotive company. Although the time span covered by this book ended in 1998, the engine engineering process of Chrysler has continued unabated with excellent single overhead camshaft V-6 and V-8 engines for Dodge Truck, and a truly impressive V-8 Hemi® head engine for both trucks and passenger cars being more recent arrivals.

With engines such as these now on the roster, the Chrysler legacy of engine technological excellence, which started with the Zeder, Skelton, and Breer engines of the 1920s, remains as a source of pride for all those involved.

As a young graduate engineer, when I entered the open door that Chrysler, and with it the automotive industry, offered to me in 1947, I traveled down the road—fast at times, really fast at times, mostly smooth, sometimes bumpy, but never dull—until I finally and figuratively turned off the key in 1993 to enjoy a less frenetic pace. It has been quite a ride.

Addendum
Chrysler Emission Control Systems
1961 to 1976

C - California
A - All Models
S - Some Models

SYSTEMS AND COMPONENTS INCLUDED	MODEL YEAR															
	61	62	63	64	65	66	67	68	69	70	71	72	73	74	75	76
BASIC ENGINE																
Camshaft Overlap											S	S	S	S	S	S
Camshaft Timing							S	S	S	S	S	S	S	S	S	S
Combustion Chamber Shape							S	S	S	S	S	S	S	S	S	S
Cylinder Head Gasket Shape							S	S	S	S	S	S	S	S	S	S
Lower Compression Pistons										S	S	S	S	S	S	S
Intake Manifold Design									S	S	S	S	S	S	S	
Intake Port Shape										S	S	S	S	S	S	S
Exhaust Manifold Heat Control Valve								S	S	A	A	A	A	A	A	A
Coolant Thermostat Control Temperature									S	S	S	S	S	A	A	A
Single Plane Intake Manifold (318 V-8 Engine)										S	S	S				
Exhaust Gas Recirculation - Floor Jets												C	S			
Proportional Exhaust Gas Recirculation													S	A	A	A

(Continued on next page)

SYSTEMS AND COMPONENTS INCLUDED	MODEL YEAR															
	61	62	63	64	65	66	67	68	69	70	71	72	73	74	75	76
CARBURETION																
Idle Mixture Limiter						C	C	A	A	A	A	A	A	A	A	A
Carburetor and Choke Calibration						S	S	A	A	A	A	A	A	A	A	A
PCV Carburetor Inlet Location						S	S	S	S	S	S	S	S	S	S	S
Off-Idle Mixture Adjustment										S	S	S				
Fixed Bypass Air Idle System										S	S	S	S	S	S	S
Choke Modulation									S	S	S	S	S	S	S	S
Electric Assist Choke													A	A	A	A
Hot Idle Compensator								S	S	S	S	S	S	S		
Single Idle Adjustment Screw (V-8 Engine)								S	S							
Plastic Carburetor Main Body											S	S	S	S	S	S
Smaller Primary, Larger Secondary Venturi											S	S	S	S	S	S
Triple Boost Venturi											S	S	S	S	S	S
Solenoid Throttle Stop										S	S	S	S	S	S	S
Solid Fuel Metering System														S	S	S
Idle Enrichment System															S	S
Altitude Compensation															C	S
DISTRIBUTOR																
Vacuum Advance Control Valve						C	C	S	C							
Spark Advance Curves						C	C	S	S	A	A	A	A	A	A	A
Spark Timing Accuracy						A	A	A	A							
Distributor Designed for Improved Mfg Accuracy										A	A	A	A	A	A	A
Solenoid Idle Retard										S	S					
Solenoid Spark Advance - Start Only												S	S			
Solenoid Controlled Vacuum Advance											C	C				
Orifice Spark Advance Control													A	A	S	S
ELECTRONIC LEAN BURN SYSTEM																
Throttle Position Advance																S
Air Temperature Advance																S
Rate of Throttle Opening Advance																S
Vacuum Advance Time Control																S
Throttle Closed Retard																S
Coolant Control of Timing Advance																S
EXTERNAL CONTROLS																
Heated Inlet Air										S	A	A	A	A	A	A
Air Injection System												C	C	C	S	S
Catalytic Converter															S	S
POSITIVE CRANKCASE VENTILATION																
Standard System	C	C	A	S	S	S	S									
Closed System				C	C	C	C	A	A	A	A	A	A	A	A	A
EVAPORATIVE EMISSION CONTROLS																
Crankcase Storage										C	A				S	S
Charcoal Canister												A	A	A	A	S
Carburetor Bowl Vent										C	A	A	A	S	A	A
Thermal Expansion Tank										C	S					
Domed Fuel Tank														S	S	S
Vapor Separator															S	S
Pressure-Vent Filler Cap										C	A	A	A	A	A	A
Overfill Limiting Valve												A	A	A	A	

Reproduced from Engineering Technical Data booklet: "Chrysler's Cleaner Air System for 1976 Models".

References

Chapter 1 Chrysler-Willys—The First "Chrysler" Cars and Engines

1. "Chrysler's Name Is Given to New Plant," *Automobile Topics*, October 30, 1920.
2. Curcio, Vincent, *Chrysler—The Life and Times of an Automotive Genius*, Oxford University Press, New York, NY, 2000.
3. "J.N. Willys Acquires Chrysler's Services," *Automobile Topics*, January 10, 1920.
4. "Willys Corp. Names Six 'Chrysler,' " *Automobile Topics*, April 24, 1920.
5. "Plan to Market Chrysler Six," *Automobile Topics*, August 21, 1920.
6. Breer, Carl, *The Birth of Chrysler Corporation and Its Engineering Legacy*, Society of Automotive Engineers, Warrendale, PA, 1995.
7. Couture, A.B. (Tobe), unpublished memoirs on file at the Chrysler Corporate Collection, Detroit, MI.
8. Skelton, Owen R., unpublished memoirs on file at the Chrysler Historical Collection, Detroit, MI.
9. "Chrysler Production to Start in Spring," *Automotive Industries*, December 9, 1920.
10. "Maxwell Plant Ordered Sold," *Automotive Industries*, April 21, 1921.
11. "Off to New Start, Maxwell Reduces," *Automotive Industries*, May 19, 1921.
12. "Chrysler Out of Overland," *Automobile Topics*, February 18, 1922.

Chapter 2 Zeder, Skelton, and Breer's Engines A, B, C, and D

1. Breer, Carl, *The Birth of Chrysler Corporation and Its Engineering Legacy*, Society of Automotive Engineers, Warrendale, PA, 1995.
2. Skelton, Owen R., unpublished memoirs on file at the Chrysler Historical Collection, Detroit, MI.
3. "Fred M. Zeder," Advance Press Service, March 2, 1925, on file at the Bentley Historical Library, University of Michigan, Ann Arbor, MI.
4. "Franklin Not to Buy Plant at Elizabeth," *Motor World*, March 1, 1922.
5. Curcio, Vincent, *Chrysler—The Life and Times of an Automotive Genius*, Oxford University Press, New York, NY, 2000.
6. "Clement Studebaker to Head New Line," *Automobile Topics*, March 4, 1922.
7. "Zeder Lends Name to Car and Company," *Automobile Topics*, March 11, 1922.
8. "May Buy Chrysler Rights," *Motor World*, March 15, 1922.
9. "Plan Chrysler Six," *Motor Age*, March 16, 1922.
10. "Chrysler Refuses His Name to Car Developed by Him," *Motor Age*, March 23, 1922.
11. "Durant High Bidder for Elizabeth Plant," *Automotive Industries*, June 15, 1922.
12. Chrysler Engineering Technical Report Index Card for Report 4406.7, Chrysler Historical Collection, Detroit, MI.

13. Chrysler, Walter P., *Life of an American Workman*, Curtis Publishing Company, New York, NY, 1937, and Dodd, Mead & Co., New York, NY, 1950.

14. *Lester-Steele Handbook of Automotive Specifications 1915—1942*, Lester-Steele Publishing Co., Deerfield Beach, FL, 1984.

15. "Tubular Backbone Frame Reinforcement Used on New Flint. This construction and seven-bearing crankshaft are outstanding features," *Automotive Industries*, July 19, 1923.

16. "Flint," sales literature, National Automotive History Collection, Detroit Public Library, Detroit, MI.

17. "Combustion Session" [of the SAE Summer Meeting], *Automotive Industries*, June 2, 1921.

18. " 'Model 90' Embodies Many Features New to Locomobile Practice. Engine is high speed type with L-head cylinders, detachable head and pressure lubrication," *Automotive Industries*, November 26, 1925.

Chapter 3 Zeder, Skelton, and Breer's Three Engineering Masterpieces—Engines B, E, and H

1. "Combustion Chamber Design," *Automotive Industries*, June 28, 1923.

2. "Combustion Session" [of the SAE Summer Meeting], *Automotive Industries*, June 2, 1921.

3. Breer, Carl, *The Birth of Chrysler Corporation and Its Engineering Legacy*, Society of Automotive Engineers, Warrendale, PA, 1995.

4. *Lester-Steele Handbook of Automotive Specifications 1915–1942*, Lester-Steele Publishing Co., Deerfield Beach, FL, 1984.

5. "Chrysler Six Appears with Seven Bearing Crankshaft," *Automotive Industries*, December 27, 1923.

6. Skelton, Owen R., unpublished memoirs on file at the Chrysler Historical Collection, Detroit, MI.

7. Curcio, Vincent, *Chrysler—The Life and Times of an Automotive Genius*, Oxford University Press, New York, NY, 2000.

8. Historical pace car information posted on the Web site of the Indianapolis Motor Speedway.

Chapter 4 Maxwell, Chrysler, and Plymouth Four-Cylinder Engines

1. Curcio, Vincent, *Chrysler—The Life and Times of an Automotive Genius*, Oxford University Press, New York, NY, 2000.

2. Breer, Carl, *The Birth of Chrysler Corporation and Its Engineering Legacy*, Society of Automotive Engineers, Warrendale, PA, 1995.

3. Lentzke, James T., *Standard Catalog of Chrysler 1914–2000*, Kraus Publications, Iola, WI, 2000.

4. "History of Various Buildings at Chrysler Highland Park," report in "Highland Park" file at the Chrysler Historical Collection, Detroit, MI.

Chapter 5 Dodge Power Joins Chrysler

1. Fourth Annual Report of Chrysler Corporation, year ended December 31, 1928.

2. Butler, Don, "Warmed-Over Dodges," Part V of the series on Dodge Brothers cars, *Cars & Parts*, September 1979.

3. Butler, Don, "The Dodge Brothers—Good Enough Is Not Acceptable," Part I of the series on Dodge Brothers cars, *Cars & Parts*, April 1979.

4. McPherson, Thomas A., *The Dodge Story*, Crestline Publishing Co., Glen Ellyn, IL, 1975.

5. Butler, Don, "Dodge Style Takes Backseat to Mechanical Excellence," Part IV of the series on Dodge Brothers cars, *Cars & Parts*, August 1979.

6. *A Pictorial History of Chrysler Corporation Cars*, Chrysler Corporation, 1966, revised 1968, 1973, and 1975.

Chapter 6 A New L-Head Six-Cylinder Design Takes Over

1. Harry T. Woolson's position as head of engine design is evidenced by his initials appearing in the "App Eng" (i.e., Approving Engineer) box in the title blocks of the engine drawings, the space that in earlier years showed the initials of Owen R. Skelton.

2. "L. Irving Woolson," BBDO [advertising agency] news release for DeSoto Division, January 11, 1956, in "Biographical" files at the Chrysler Historical Collection, Detroit, MI.

3. Curcio, Vincent, *Chrysler—The Life and Times of an Automotive Genius*, Oxford University Press, New York, NY, 2000.

4. Breer, Carl, *The Birth of Chrysler Corporation and Its Engineering Legacy*, Society of Automotive Engineers, Warrendale, PA, 1995.

Chapter 7 Engines of Luxury—The Straight-Eights

1. Langworth, Richard M. and Norbye, Jan P., *The Complete History of Chrysler Corporation 1924–1985*, Beekman House, New York, NY, 1985.

2. "De Luxe 8 by Chrysler," *Automobile Trade Journal*, June 1931.

3. Lentzke, James T., *Standard Catalog of Chrysler 1914–2000*, Kraus Publications, Iola, WI, 2000.

Chapter 8 Hemispheres and Polyspheres—The First Generation of V-8s

1. Breer, Carl, *The Birth of Chrysler Corporation and Its Engineering Legacy*, Society of Automotive Engineers, Warrendale, PA, 1995.

2. "Single Cylinder 3-3/8" × 4-1/4" Engine," Engineering Division Technical Report 4406.295, October 9, 1947, on file at the Chrysler Historical Collection, Detroit, MI.

3. Zeder, James C., "New Horizons in Engine Development," SAE Paper No. 520262, *SAE Quarterly Transactions*, Vol. 60, No. 4, October 1952, Society of Automotive Engineers, Warrendale, PA, 1952.

4. Young, Anthony, *HEMI—History of the Chrysler Hemi V-8 Engine and Hemi-Powered Cars*, Motorbooks International, Publishers & Wholesalers, Osceola, WI, 1991.

5. My conversation with Fred Shrimpton, Chrysler engine designer and design supervisor.

6. Drinkard, William E. and Carpentier, Mel L., "Development Highlights and Unique Features of New Chrysler V-8 Engine," SAE Paper No. 510196, *SAE Quarterly Transactions*, Vol. 59, No. 3, July 1951, Society of Automotive Engineers, Warrendale, PA, 1951.

7. Zeder, James C., "Why the Efficiency of Firepower," *Motor Trend*, December 1951.

8. 1951 race results posted on the Formula 2 Web site.

9. 1953 race results posted on the Formula 2 Web site.

10. Cushing, Harry, "Potent Threat for Indianapolis," *Motor Trend*, December 1952.

11. My conversations and information exchanges with Joe Freeman, race car owner, August–September 2004.

12. My conversation with Pete Hagenbuch, retired Chrysler Engine Development engineer.

13. Gunnell, John, *Standard Catalog of American Cars 1946–1975*, Kraus Publications, Iola, WI, 2002.

14. Lentzke, James T., *Standard Catalog of Chrysler 1914–2000*, Kraus Publications, Iola, WI, 2000.

15. Dodge Division Confidential Price Bulletin J10, December 22, 1955, on file at the Chrysler Historical Collection, Detroit, MI.

16. Dodge Division Confidential Price Bulletin J11 Supplement, December 22, 1955, on file at the Chrysler Historical Collection, Detroit, MI.

17. 1955 Plymouth sales literature, on file at the Chrysler Historical Collection, Detroit, MI.

18. Welch, Harold L., "The New Plymouth V-8 Engine," SAE Paper No. 560019, *SAE Transactions*, Vol. 64, 1956, Society of Automotive Engineers, Warrendale, PA, 1956.

19. My observations while touring the Mound Road engine plant site in November 1954 after being assigned the position of Resident Engine Engineer.

20. Boyce, Terry, "Fury Drive Report," *Special Interest Autos*, April–May 1972.

21. Couture, A.B. (Tobe), unpublished memoirs, on file at the Chrysler Historical Collection, Detroit, MI.

Chapter 9 Wedge-Head V-8s—B, RB, and LA Engines

1. Bob Rarey, former Assistant Chief Engineer–Engine Design, and me, Manager–Engine Design, collective memories from the days when the B and RB engine was being designed.

2. "The Inside Story on Chrysler's New V-8," *Popular Science*, October 1957.

3. Rarey, R.S. and Moeller, E.G., "Chrysler Corporation's New V-8 Engine," SAE Paper No. 590013 (formerly SAE Paper No. 32B), *SAE Transactions*, Vol. 67, 1959, Society of Automotive Engineers, Warrendale, PA, 1959.

4. My memories of meetings with Trenton engine plant and Engine Division representatives at the time regarding how to produce a 383-cu.-in. engine for Chrysler cars within the block machining limitations at the Trenton engine plant.

5. My memories of discussions with Product Planning when the 400-cu.-in. LB engine was proposed in 1968.

6. "1962 Engineering Data Book," internal Chrysler Engineering Technical Information publication, January 1, 1962, on file at Chrysler Historical Collection, Detroit, MI.

7. My memories of numerous unsuccessful design studies that tried to fit the poly-head V-8 A-engine into the Valiant engine compartment.

8. Weertman, W.L. and Beckman, E.W., "Chrysler Corp.'s New 273 cu. in. V-8 Engine," SAE Paper No. 640132 (formerly SAE Paper No. 826A), Society of Automotive Engineers, Warrendale, PA, 1964.

9. Dean, J.W. and Casebeer, H.M., "Chrysler Corporation's 340 cu. in. V-8 Engine," SAE Paper No. 680019, Society of Automotive Engineers, Warrendale, PA, 1968.

10. Chrysler-Plymouth Division news release regarding the AAR 'Cuda, February 19, 1970.

11. My memories of foundry problems encountered when first trying to cast the W2 cylinder heads.

12. Internal letter from me to M.E. Walden of the Technical Computer Center (TCC) regarding race use of W2 cylinder heads machined by the Technical Computer Center, February 10, 1971.

13. Moriarty, Frank, *Supercars—The Story of the Dodge Charger Daytona and Plymouth SuperBird*, Howell Press, Charlottesville, VA, 1995.

14. My conversations with John Wehrly, retired manager of Race Engine Development, April 27, 2005.

Chapter 10 Power Personified—The 426 Hemi® Engine

1. My memories of the initiation of the 426 Hemi® engine design.

2. My notes on the startup of the 426 Hemi® engine.

3. Weertman, W.L. and Lechner, R.J., "Chrysler Corporation's New Hemi Head High Performance Engines," presented at the Detroit Section SAE meeting, April 11, 1966, SAE Paper No. 660342, Society of Automotive Engineers, Warrendale, PA, 1966.

4. Shepard, Larry S. and Gingerella, Michael J., *HEMI® ENGINES*, Chrysler Motors Corporation, Mopar Performance Parts, 1988.

5. My conversations with Bob Cahill, Product Programming Manager, and Tom Hoover, Engineering Racing Programs Coordinator, prior to the cancellation of the Street Hemi® engine.

6. E-mail from Bob Cahill to me, June 11, 2006.

7. E-mail from Bob Cahill to me, June 10, 2006.

8. My conversation with Tom Hoover on June 9, 2006.

Chapter 11 Overhead Valve In-Line Sixes

1. My conversations with John Hurst, retired Chrysler engineer, who at the time was a design engineer in the Advance Engine Design group, Engine Design department, regarding the origin of slanting the engine.

2. Bob Rarey's and my collective sharing of memories about the decisions that added the 225-cu.-in. Slant Six to the 170-cu.-in. Slant Six Valiant program.

3. My conversation with Ray Latham, Advance Engine Design Supervisor, during the initial design of the Slant Six engine.

4. Bob Rarey's and my memories from the days of the initial design of the Slant Six engine.

5. Loofbourrow, A.G., Exner, V.M., and Sinclair, R.M., "The Valiant—A New Motoring Concept," SAE Paper No. 600075 (formerly SAE Paper No. 140B), Society of Automotive Engineers, Warrendale, PA, 1960.

6. My conversations with Tom Hoover, retired Engineering Race Program Coordinator, about dynamometer development of the Hyper-Pak Slant Six engine.

7. Moeller, E.G., Weertman, W.L., and Eriksen, H.E., "Chrysler Corporation's Die-Cast Aluminum Slant Six Engine," SAE Paper No. 610411 (formerly SAE Paper No. 307C), Society of Automotive Engineers, Warrendale, PA, 1961.

8. "Final Detailed Results—Mobil Economy Run," published each year by the United States Auto Club; information used for results cited from 1962 through 1968.

9. Internal review report of the Australian Six-Cylinder Engine Product Engineering Program, October 15, 1968.

10. Davis, Tony, *The Valiant Book*, Marque Publishing Company, Wyoming, New South Wales, Australia.

11. Farmer, Gavin, "1971–81 Australian Chrysler Valiant: Sounding Retreat," *Collectible Automobile*, February 2003.

Chapter 12 A New Era—Driving the Front Wheels with Four-Cylinder Engines

1. My memories of technical meetings with Volkswagen engineers in Wolfsburg, West Germany, in August 1975.
2. "1978 Car of the Year," *Motor Trend*, February 1978.
3. My memories of engineering meetings held to define the 2.2-L engine arrangement.
4. Weertman, W.L. and Dean, J.W., "Chrysler Corporation's New 2.2 Liter 4-Cylinder Engine," SAE Paper No. 810007, Society of Automotive Engineers, Warrendale, PA, 1981.
5. Chrysler News Relations Office press release regarding the Saltillo, Mexico engine plant, October 8, 1979.
6. U.S. Patent #4,703,725, "Mounting of an Engine Balancing Device," inventor Willem L. Weertman, November 3, 1987.
7. My memories as a participant in the meeting described that directed the use of the water-cooled turbocharger housing.
8. Section titled "Chrysler Automotive Firsts" of internal report titled "1990 Engineering Features," dated May 1989, on file at the Chrysler Historical Collection, Detroit, MI.
9. Allen, Floyd E. and Rinschler, Gordon L., "Turbocharging the Chrysler 2.2 Liter Engine," SAE Paper No. 840252, Society of Automotive Engineers, Warrendale, PA, 1984.
10. Allen, Floyd E. and Witte, Thomas W., "Turbocharging the Chrysler 2.5 Liter Engine," SAE Paper No. 900852, Society of Automotive Engineers, Warrendale, PA, 1990.
11. My notes while serving as consulting engineer to the Chrysler Maserati 2.2-L double overhead cam turbocharged engine program.

Chapter 13 V-6 Engines—More Power for Front-Wheel-Drive Cars

1. My memories of negotiations with Mitsubishi Motors Corporation engineers prior to the signing of the purchase agreement for V-6 engines.
2. My memories of design studies and presentations made to senior Chrysler management regarding the Chrysler-built V-6 engine for front-wheel-drive vehicles.
3. Section titled "3.3-L V-6 Engine" of internal report titled "1990 Engineering Features," dated May 1989, on file at the Chrysler Historical Collection, Detroit, MI.
4. Witte, Thomas W. and Wright, Keith G., "Chrysler 3.5 Liter V-6 Engine," SAE Paper No. 930875, Society of Automotive Engineers, Warrendale, PA, 1993.
5. U.S. Patent #5,590,628, "Multi-Mode Intake Manifold Assembly for Internal Combustion Engines," inventors Michael P. Patyi, William E. Hughes, Mary T. Chappell, Frederick B. Hummer, and Donald H. Kerska, January 7, 1997.
6. U.S. Patent #5,632,239, "Method of Distributing Air in an Engine Intake Manifold," inventors Michael P. Patyi, William E. Hughes, Mary T. Chappell, Frederick B. Hummer, and Donald H. Kerska, May 27, 1997.
7. My conversation with Floyd E. Allen, retired Vice President—Power Train Product Engineering, regarding origins of the 2.7-L double overhead cam engine, June 11, 2005.

Chapter 14 A Big, Powerful, All-Aluminum V-10—The Viper Engine

1. "Viper! Dodge Performance Close-Up, Salesperson Preview," Viper sales literature, Chrysler Corporation, 1992.
2. My notes regarding the start of Viper engine production.
3. Unpublished transcript of a presentation about the Viper GTS engine to the SAE Detroit Section on November 6, 1996 by Charlie Brown III, manager of the Viper Engine group.

Chapter 15 Truck Engines—Hauling the Load

1. Mroz, Albert, *The Illustrated Encyclopedia of American Trucks and Commercial Vehicles*, Kraus Publications, Iola, WI, 1996.
2. Curcio, Vincent, *Chrysler—The Life and Times of an Automotive Genius*, Oxford University Press, New York, NY, 2000.
3. "The History of Chrysler Center 1908–2000, Highland Park, Michigan," Chrysler pamphlet on file at the Chrysler Historical Collection, Detroit, MI.
4. 1930 Fargo one-ton freighter sales literature, National Automotive History Collection, Detroit Public Library, Detroit, MI.
5. 1930 Dodge Brothers Second Series Truck sales literature, Chrysler Historical Collection, Detroit, MI.
6. Gunnell, John, *Standard Catalog of Light-Duty Dodge Trucks 1917–2002*, Kraus Publications, Iola, WI, 2002.

7. My conversation with Harold Welch, retired Chrysler engineer, regarding his experiences in engine development, November 12, 2001.

8. My conversations with Ray Latham, supervisor of Advance Engine Design, regarding the Dodge diesel engine, circa 1964.

9. Welch, H.L. and Rarey, R.S., "Dodge Truck Heavy-Duty V-8 Engines," SAE Paper No. 600127 (formerly SAE Paper No. S257), Society of Automotive Engineers, Warrendale, PA, 1960.

10. Bunn, Don, *Dodge Trucks*, Motorbooks International, Osceola, WI, 1996.

11. My notes. The date also was recorded in a letter dated December 23, 1992 from me to Bob Lutz regarding the V-10 engine.

12. Kurowski, R., Yee, G., Rodenkirch, N., Waterworth, J., McNeil, W., and Hill, W., "Chrysler 8.0-Liter V-10 Engine," SAE Paper No. 933033, Society of Automotive Engineers, Warrendale, PA, 1993.

Chapter 16 Marine and Industrial Engines—Chrysler Power Afloat and Ashore

1. Curcio, Vincent, *Chrysler—The Life and Times of an Automotive Genius*, Oxford University Press, New York, NY, 2000.

2. "The Chrysler Marine Engine," fourteen-page review of the Chrysler Marine Engine Division, circa 1948, on file at the Chrysler Historical Collection, Detroit, MI.

3. "Background of Chrysler Marine Division," eight-page review of the Chrysler Marine Division, circa 1965, on file at the Chrysler Historical Collection, Detroit, MI.

4. Chrysler Confidential Bulletin to Distributors and Dealers, No. 383, April 12, 1928.

5. "Sterett, 'Miss Chrysler Crew' Renew Unlimited Boat Racing 'Rebellion,' " Chrysler Marine Products Operations News Release, 1967.

6. "1967 UIM World Championship, Detroit River, Detroit, MI, July 2, 1967, U-77, Miss Chrysler Crew Takes World Cup," information posted on Web site of Union Internationale Motonautique.

7. Newman, Neil A., "The Metamorphosis of a Powerplant—From Passenger Car to Pleasure Boat," SAE Paper No. 648304, Society of Automotive Engineers, Warrendale, PA, 1964.

8. Memories of Fred Kaehler, former Chief Engineer, Marine and Industrial Division, as related to me.

9. "DaimlerChrysler History—The Year 1980," information posted on the Web site of Chrysler LLC.

10. "Chrysler Industrial Engines in Peace and War," eight-page review of the Chrysler Industrial Engine Division formation and operations, circa 1943, on file at the Chrysler Historical Collection, Detroit, MI.

Chapter 17 Chrysler Power Goes to War—Chrysler Engines During World War II

1. Bunn, Don, *Dodge Trucks*, Motorbooks International, Osceola, WI, 1996.

2. "Orders, Shipments, Corporation Contracts and Inventory," report regarding 1943 wartime production of Chrysler Division, on file at the Chrysler Historical Collection, Detroit, MI.

3. "Notes on Materiel for Medium Tank M3 Multiple Engine Chrysler Corporation Model A57A," 1942 report prepared by Chrysler Corporation, on file at the Chrysler Historical Collection, Detroit, MI.

4. "Diary of a War Project—The Multibank Engine," letter written on June 5, 2000 by Gene Weiss, former Chrysler engineer and historian, with attached chronological notes regarding the Chrysler World War II tank-building program and the five-bank tank engine, on file at the Chrysler Historical Collection, Detroit, MI.

5. Placard of a five-bank tank engine on display at the Walter P. Chrysler Museum, Auburn Hills, MI.

6. Letter from K.T. Keller, president of Chrysler Corporation, to Major General Charles E. Branshaw, AAF Materiel Command, regarding the XI-2220 engine program, August 13, 1943, on file at the National Archives and Records, Record Group 342.

7. Executive Chronology File—XI-2220 Aircraft Engine, World War II files of the Chrysler Historical Collection, Detroit, MI.

8. "Model Designations of USAF Aircraft Engines," publication of Air Materiel Command, posted on the Website of the Aircraft Engine Historical Society.

9. "Case History of Chrysler XIV-2220 Engine," Air Technical Service Command, on microfilm at the Department of the Air Force, Air Force Historical Research Agency, Maxwell Air Force Base, AL.

10. "Summary of Development," XI-2220 Aircraft Engine, Chrysler Technical Report #A-60402.19, reported by L.X. Chapin, dated August 21, 1945, on file at the Daimler-Chrysler Corporate Historical Collection, Detroit, MI.

11. "XI-2220 Aircraft Engine," program status report, December 28, 1943, on file at the DaimlerChrysler Corporate Historical Collection, Detroit, MI.

12. A.G. Herreshoff, unpublished memoirs titled "The XI-2220 Aircraft Engine," June 7, 1973, on file at the Chrysler Historical Collection, Detroit, MI.

13. My conversation with E. Paul Wise, former Chrysler engineer and participant in the XI-2220 program, December 21, 2002.

14. "Preliminary Flight Tests—XP-47H Airplane," Chrysler Technical Report #A-60008.6, reported by E. Paul Wise on November 27, 1945, on file at the Chrysler Historical Collection, Detroit, MI.

15. E-mail from E. Paul Wise to W. L. Weertman regarding XI-2220 information, November 12, 2005.

16. AAF Individual Aircraft Record Card for Aircraft Serial No. 42-23297, obtained from Air Force Historical Research Agency, Maxwell Air Force Base, AL.

17. AAF Air Technical Service Command Memorandum Report TSEPP5-503-3236, December 4, 1945; purpose: inspect the failure of the propeller shaft of No. D-000001 Chrysler engine from the XP-47H airplane.

18. Letter from G.J. Huebner, Jr. Chrysler Chief Engineer of Research to Commanding General, Air Technical Service Command, January 10, 1945, regarding Engine D-000001, Propeller Shaft Failure.

19. "Functional Flight Tests on the XP-47H Airplane," Republic Aviation Corporation Flight Test Report No. FR-60, reported by J.F. Riley, January 18, 1946.

20. My conversation with Kenneth D. Wilson, aviation historian and photographer, Evansville, IN, February 9, 2003.

21. "Model Designations of U.S.A.F. Aircraft Engines," publication of Air Material Command, posted on the Website of the Aircraft Engine Historical Society.

22. Executive Chronology File—Dodge Chicago plant, World War II files of the Chrysler Historical Collection, Detroit, MI.

23. Stout, Wesley W., *Great Engines and Great Planes*, Chrysler Corporation, Detroit, MI, 1947.

Chapter 18 A Different Type of Engine—The Chrysler Gas Turbine

1. "Preliminary Design Studies Report. A-86 Turbo-Prop Power Plant," Chrysler Engineering Research Report, September 1946, on file at the Chrysler Historical Collection, Detroit, MI.

2. "A-86A Turbo-Prop Power Plant, Official AN-E-25 50-Hour Preliminary Flight Rating Test," Chrysler Technical Report 60409.18-04, reported by Frank R. Swaney, May 27, 1949, on file at the Chrysler Historical Collection, Detroit, MI.

3. "History of Chrysler Corporation's Gas Turbine Vehicles," report prepared by Chrysler Engineering Office Technical Information, January 1979, on file at the Chrysler Historical Collection, Detroit, MI.

4. My memories of driving the turbine car.

5. Huebner, George J., Jr., "The Chrysler Regenerative Turbine-Powered Passenger Car," presented January 1964 at the SAE Engineering Congress and Exposition, Detroit, MI, SAE Paper No. 640414 (formerly SAE Paper No. 777A), Society of Automotive Engineers, Warrendale, PA.

6. Chapman, William I., "Chrysler's Gas Turbine Car—Powerplant Design Characteristics," presented January 1964 at the SAE Engineering Congress and Exposition, Detroit, MI, SAE Paper No. 640402 (formerly SAE Paper No. 777B), Society of Automotive Engineers, Warrendale, PA.

7. Roy, Amedee, Hagen, Frederick, A., and Belleau, Claude, "Chrysler's Gas Turbine Car—Materials Requirements," presented January 1964 at the SAE Engineering Congress and Exposition, Detroit, MI, SAE Paper No. 640417 (formerly SAE Paper No. 777C), Society of Automotive Engineers, Warrendale, PA.

8. DeClaire, Gerald and Bell, Albert, H., "Chrysler's Gas Turbine Car—Laboratory Procedures and Development Methods," presented January 1964 at the SAE Engineering Congress and Exposition, Detroit, MI, SAE Paper No. 640418 (formerly SAE Paper No. 777D), Society of Automotive Engineers, Warrendale, PA.

Chapter 19 American Motors Engines Join the Chrysler Power Lineup

1. Gunnell, John, *Standard Catalog of American Motors 1902–1987*, Kraus Publications, Iola, WI, 2002.

2. "Teaming Up to Be the Best! The History Behind the Chrysler/American Motors Merger," Chrysler Corporation, Detroit, MI, 1987.

3. Placard of a Willys-Jeep World War II engine on display at the Walter P. Chrysler Museum, Auburn Hills, MI.

4. Placard of the Hudson Hornet Twin H engine on display at the Walter P. Chrysler Museum, Auburn Hills, MI.

5. Adamson, J.F., Burke, C.E., Potter, D.V., and Zechel, W.J., "The American Motors New Die Cast Aluminum Engine," SAE Paper No. 610409 (formerly SAE Paper No. 307A), Society of Automotive Engineers, Warrendale, PA, 1961.

6. Potter, D.V., Leydorf, G.F. Jr., and Lawler, R.L., "The New Rambler Six Engine—Torque Command 232," SAE Paper No. 640276 (formerly SAE Paper No. 884B), Society of Automotive Engineers, Warrendale, PA, 1964.

7. Leydorf, G.F. Jr. and Onishi, G.E., "The American Motors AMX-390 V-8 Engine," SAE Paper No. 680443, Society of Automotive Engineers, Warrendale, PA, 1968.

8. My conversations with Rod Kurowski, former AMC Chief Engineer of Engine Engineering.

Sources

The principal sources of information for this book about Chrysler engines reside in the files, vehicles, and engines of the Chrysler Historical Collection. In the files are owner's manuals dating back to even before the formation of Chrysler Corporation; service manuals from the early 1930s when these manuals were first issued; sales literature year by year and model by model; catalogued photographs of cars, trucks, and engines; and product information sent to car dealers and their salespersons.

The vehicles of the Collection, under protective covers, are arranged in neat rows. In chronology, they span from the earliest days of the automobile industry to the latest Viper models. In size, they range from a diminutive Snow Runner marketed at one time by the Chrysler Marine and Industrial Division to a towering diesel-powered Dodge truck that was retired after years of hauling semi-trailers. The separate engines include many former display engines with placards intact.

Additional resources of the Collection include records from the Engineering Technical Data department, which for many years published annually an "Engineering Features" booklet that described in detail the changes made in the vehicles for the forthcoming model year. Specialized collections at the Chrysler Historical Collection contain information about Chrysler manufacturing plants and a wealth of information about the contributions of Chrysler to the war effort during World War II.

A truly outstanding asset of the Chrysler Historical Collection, especially in the area of technical engineering information, is its vast microfilm file. Recorded over a period of thirty years—from 1944 to 1974—the microfilms contain photo reproductions of the engineering drawings, tracings, parts lists, and layouts that were in the main Engineering blueprint vault at the time. Because the vault contained drawings of all Chrysler products since the formation of the corporation, plus the drawings that Chrysler Corporation inherited from both Dodge Brothers and Maxwell (including Maxwell's predecessor, the E.R. Thomas-Detroit Company of 1906) and some Chrysler/Willys drawings, the microfilms are a great historical treasure for the Chrysler Group, Chrysler LLC, and the automotive industry.

Another significant asset of the Chrysler Historical Collection is its collection of the actual drawings and tracings made by Zeder, Skelton, and Breer and their design staff from the time they first joined Walter P. Chrysler at Willys in New Jersey in 1920 until the time when they rejoined him at Maxwell in Detroit in 1922.

A public source of excellent information, and always a pleasure to visit, is the Walter P. Chrysler Museum located in Auburn Hills, Michigan, on the grounds of the Chrysler Headquarters campus. Engines available for close-up inspection range from four-cylinder L-heads of the 1920s to the huge 10-ft-long, 2220-cu.-in. inverted V-16 fighter aircraft engine developed during World War II.

In gathering data for this book, considerable use was made of records of the former Engine Design department. These records now reside with an engine group at the Chrysler Technology Center.

The National Automotive History Collection of the Detroit Public Library, mentioned in the Acknowledgments section at the beginning of this book, proved invaluable as a source of information about cars other than Chrysler products and as a source of trade publication news of the 1920s, when the automobile industry was in a constant state of agitation during its formative years, roiled by swings in the nation's economic condition.

A final and truly invaluable source of information was in the form of the many memories of people with whom I had worked and with whom I communicated during the writing of this book. Their memories, little dimmed by the passing years, were able to put back into sharp focus the events of long ago, with instant recall of engine program numbers and numerous details. Invariably with interest expressed in my project, they were glad to contribute to the knowledge base needed for this book.

Abbreviations and Definitions

access	Accessory.
AHRA	American Hot Rod Association.
Al, alum.	Aluminum.
AMA	Automobile Manufacturers Association, an early name for the Motor Vehicle Manufacturers Association (MVMA). *See* MVMA.
bbl.	Barrel. Usually used in defining the number of barrels, throats, or venturis of a carburetor.
BDC	Bottom dead center. The lowest point of the piston during its up-and-down travel in the cylinder bore.
brg.	Bearing.
carb.	Carburetor.
ca/shaft or ca/s	Camshaft.
cc	Cubic centimeters.
CHC	Chrysler Historical Collection.
c.i.	Cast iron.
ck/shaft or ck/s	Crankshaft.
COE (c.o.e.)	Cab over engine. A truck design that positions the cab directly above the engine to obtain a short distance from the front bumper to the back of the cab.
conv.	Convertible.
crankpin	The portion of the crankshaft on which the connecting rod bearings ride. Invariably ground to an exacting diametral tolerance, the crankpins usually are lapped and polished to promote long bearing life.
cross flow	A combustion chamber design having the intake charge and exhaust gas enter and leave from opposite sides of the cylinder head, in effect crossing from one side of the cylinder head to the other through the chamber.
ctr.	Center.

cu. in.	Cubic inches.
cwt.	Counterweight.
cyl.	Cylinder.
deck height	A cylinder block dimension, measured along the cylinder bore axis, from the crankshaft centerline to the top of the block.
deg	Degree.
dia.	Diameter.
DIS	Distributorless ignition system.
distr.	Distributor.
DOHC	Double (or dual) overhead cam. A cylinder head design that has two camshafts for operating the valves, invariably one for the intake valves and the other for the exhaust valves.
DRS	Double rocker shaft. An overhead valve system that has two rocker shafts in each cylinder head, one shaft for mounting the intake valve rocker arms, and the other for mounting the exhaust valve rocker arms.
engine assembly weight	Also known as "total engine dressed mass (weight)." The total weight of the base engine plus add-ons including the starter, alternator, carburetor (or fuel injection equivalent), air cleaner, ignition system, fuel pump, engine-mounted emission controls, standard fan and drive belts, oil filter, mounts attached to the engine, and engine-mounted throttle controls. The base engine includes the manifolds and the water pump. Oil and coolant are not included.
E-W	East-west. Used to describe an engine orientation in a front-wheel-drive car having the engine centerline across the car with the nose of the crankshaft pointing to the right as viewed from the driver's seat.
exh.	Exhaust.
freewheeling clearance	An attribute of an engine that is designed so that the valves will not contact the pistons, even if the camshaft and crankshaft are not in proper phase with each other.
GCW	Gross combination weight. The maximum loaded weight of a truck and trailer.
GVW	Gross vehicle weight. The maximum loaded weight of a truck.
hemi	Short for hemispherical. Used to describe a combustion chamber that has the shape of a portion of a sphere. When used as a proper adjective, as in "Hemi engine" or "Hemi head," the word "Hemi" is a Chrysler Holding LLC trademark and is indicated as "Hemi®."
hp	Horsepower. A customary measure of power.
hsg.	Housing.
hyd.	Hydraulic.

int.	Intake.
kW	Kilowatt. A metric measure of power.
L	Liter.
lb-ft	Pound-feet. A customary measure of torque.
LCF	Low cab forward. A truck design featured by Dodge Truck that combined a high degree of shortness (but not as much as a cab-over-engine [COE] design) with excellent serviceability due to its swing-out fenders and high-opening hood.
L-head	An engine that has its valves mounted in the cylinder block above the camshaft. The combustion chamber in the cylinder head together with the cylinder bore have the approximate shape of an inverted letter "L," giving the engine its name.
loop flow	A combustion chamber design in which the intake charge and exhaust gas enter and leave respectively from the same side of the cylinder head, in effect making a loop through the chamber.
mat'l.	Material.
mech.	Mechanical.
mm	Millimeter.
MPI	Multipoint fuel injection. A fuel system that provides each cylinder with an individual fuel injector.
MRE	Mound Road engine plant. The Chrysler Engine Division plant located on Mound Road in Detroit, Michigan.
mtd.	Mounted.
MVMA	Motor Vehicle Manufacturers Association. The vehicle specification sheets of this organization are an excellent source of detailed information.
NASCAR	National Association for Stock Car Auto Racing.
NHRA	National Hot Rod Association.
N-m	Newton-meter. A metric measure of torque.
Nodular cast iron	A type of cast iron having higher strength and greater ductility than ordinary cast iron. During the melting process, a procedure is used to cause the carbon particles in the iron to become nodular in shape rather than in the form of flakes. Also called ductile cast iron.
N-S	North-south. Used to describe an engine orientation having the engine centerline parallel to the car centerline, with the nose of the crankshaft pointing toward the front of the car. Almost universal on rear-wheel-drive cars, it also is used on some front-wheel-drive cars such as the Chrysler Concorde.
OHC	Overhead camshaft. The camshaft and the valves are mounted in the cylinder head.

OHV	Overhead valve. An engine with its valves mounted in the cylinder head. Not stated but understood is that a so-equipped engine uses cylinder block-mounted tappets and pushrods to operate its valves. Prior to the mid-1950s, the phrase "overhead valve" often was used to describe an engine with its valves in the cylinder head (as opposed to its valves being in the cylinder block), regardless of how the valves were operated.
opt.	Optional.
PCV	Positive crankcase ventilation. An arrangement that draws crankcase fumes into the intake system of the engine, rather than allowing them to escape to the atmosphere.
pre-ignition	An uncontrolled igniting of the fuel/air charge caused by a heat source other than the spark across the gap of the spark plug. Occurring before the desired point during the upward travel of the piston, the condition can lead quickly to piston failure and engine destruction.
rpm	Revolutions per minute.
shrink-fit	A method of obtaining an extremely tight fit between two parts by heating the outer part and/or cooling the inner part before the two are joined. In the case of the exhaust valve seat inserts of L-head engines, the circular steel inserts were chilled in dry ice before being hammered into the counterbores at the top of the cylinder block.
skirt	The cylinder block outside walls that extend below the crankshaft centerline. As used in the text, the walls of a shallow-skirt block would extend about 0.5 in. or less; the walls of a deep-skirt block would extend more than this amount.
SMPI	Sequential multipoint fuel injection. The individual injectors are energized in sequence related to the firing order of the engine.
SOHC	Single overhead camshaft. An overhead camshaft design with one camshaft used in each cylinder head.
SRS	Single rocker shaft. An overhead valve system that uses one shaft in each cylinder head for mounting all the valve rocker arms.
std.	Standard.
tandem	An arrangement of carburetors having one positioned behind the other on the intake manifold. The manifold used sometimes is called a runner manifold or a log manifold.
TBI	Throttle-body fuel injection. Also called single-point fuel injection. All the fuel is injected into the intake air stream at a single location upstream of all the cylinders.
TDC	Top dead center. The highest point of the piston during its up-and-down travel in the cylinder bore.
TEP	Trenton engine plant. The Chrysler Engine Division plant located in Trenton, Michigan.
Thermosiphon	An early engine cooling system that did not use a water pump. Instead, it relied on the engine water, heated by the combustion chambers and cylinder walls, rising to flow to the top tank of the radiator, replaced by colder water flowing from the radiator bottom tank into the water jacket of the cylinder block.

valve events — A description of the number of degrees when the intake valve is open, when the exhaust valve is open, and when both valves are open (the "overlap"). Best displayed in a diagram; for brevity, the specification tables show valve events as three hyphenated numbers. The first number is the crankshaft degrees the intake valve is open, the second number is the degrees the exhaust valve is open, and the third number is the degrees of overlap when both valves are open. If the third number is shown as negative, it means the exhaust valve has closed before the intake valve starts to open. The camshaft in this diagram is shown with events of 220-230-2 in the engine specifications listed in Table 3.3 of Chapter 3.

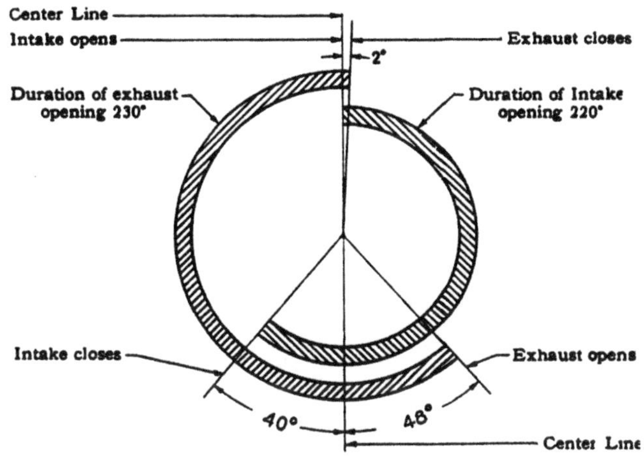

Valve Events Timing Diagram
1927 Chrysler H60 Six-cylinder Engine

vert. — Vertical.

WEP — Windsor engine plant. The Chrysler Canada engine plant located in Windsor, Ontario, Canada.

ZSB — Zeder, Skelton, and Breer. Individually, Fred M. Zeder, Owen R. Skelton, and Carl Breer, the three exceptionally gifted engineers who, working together as a single entity, executed perfectly their vital role as the designers and developers of the successful 1924 Chrysler car that led directly to the formation of the Chrysler Corporation.

Conversion Factors

Customary to Metric

Volume:

cubic inches (cu. in.) × 16.387 = cubic centimeters (cc)

cubic inches (cu. in.) × 0.016387 = liters (L)

Power:

horsepower (hp) × 0.7457 = kilowatts (kW)

Torque:

pound-feet (lb-ft) × 1.3558 = Newton-meters (N-m)

Weight (mass):

pounds (lb) × 0.4536 = kilograms (kg)

Metric to Customary

Volume:

cubic centimeters (cc) × 0.06102 = cubic inches (cu. in.)

liters (L) × 61.02 = cubic inches (cu. in.)

Power:

kilowatts (kW) × 1.341 = horsepower (hp)

Torque:

Newton-meters (N-m) × 0.73756 = pound-feet (lb-ft)

Weight (mass):

kilograms (kg) × 2.2046 = pounds (lb)

Index

About the Author

Willem L. (Bill) Weertman's entire professional career was spent with Chrysler Corporation starting in 1947 when, holding a Bachelor of Engineering degree from Yale University, he was admitted to the Chrysler Institute of Engineering for its two-year Master of Automotive Engineering program. At Yale, Weertman was elected to the honorary engineering society of Tau Beta Pi. Graduation from the Chrysler Institute of Engineering was followed by a position on the staff of the Resident Engineering department at the Plymouth Lynch Road assembly plant. Interrupted by two years of active duty as an officer in the U.S. Navy during the Korean conflict, in 1954 Weertman was appointed Resident Engineer of the Plymouth V-8 engine plant, coincident with the original startup of the plant as a high-volume world-class producer of engines.

In November 1955, Weertman was promoted to Managing Engineer of Engine Design at the Chrysler Central Engineering Division in Highland Park, Michigan, working under Assistant Chief Engineer–Engine Design Robert S. (Bob) Rarey. The next years were filled with fast-paced engine design activity—changes and improvements to engines in production, finishing the designs and launching the new big-block B and RB V-8 engines, designing and launching the new Slant Six G and RG engines, and designing truck, marine, and industrial versions of the new engines.

With the promotion of Rarey to Chief Engineer in 1962, Weertman became Assistant Chief Engineer–Engine Design, leading the design effort for the LA V-8 engine series, the outstanding 426 Hemi® competition and street engine, and an all-new six-cylinder engine for manufacture in Australia. In 1976, Weertman became Chief Engineer of Engine Design and Development, a title that was changed slightly to Chief Engineer–Engine Engineering in 1980. During these years until his retirement in 1987, Weertman was responsible for all the new engines needed to power the corporation's new car lines of front-wheel-drive vehicles and for adapting and testing the engines purchased from outside companies.

Even after retirement from Chrysler in 1987, Weertman continued to be professionally active. As a consultant on several engine programs, he assisted in the final development of a Maserati double overhead camshaft, turbocharged version of the Chrysler 2.2-L four-cylinder engine and in the design and development of the phenomenal Viper V-10 all-aluminum engine.

Weertman has been a member of SAE International since April 1955, becoming a Life Member in 1995. He received certification as a Professional Engineer in Mechanical Engineering from the State of Michigan on November 4, 1960.

www.ingramcontent.com/pod-product-compliance
Lightning Source LLC
LaVergne TN
LVHW062139200926
848484LV00007B/1833